The Complete Guide to High-End Audio

Fourth Edition

Robert Harley

Acapella Publishing
Carlsbad, California

Fourth Edition
First Printing—2010

International Standard Book Number:
0-9786493-1-1

Cover Concept and Design: *Torquil Dewar*

Printed in the United States of America

Trademark Notice

This book comments on and describes various audio and video components. Many such products are identified by their tradenames. In most cases, these designations are claimed as legally protected trademarks by the companies manufacturing the products. Use of a term in this book should not be regarded as affecting the validity of any trademark or service mark. You should investigate a claimed trademark before using it for any purpose other than to refer to the product.

Warning Disclaimer

This book is designed to provide the most current, correct, and clearly expressed information regarding the subject matter covered. It is sold with the understanding that the publisher and author are not engaged in rendering audio/video system design.

It is not the purpose of this manual to reprint all the information otherwise available to the author and/or publisher, but to complement, amplify, and supplement other texts. You are urged to read all available material and learn as much as possible about audio, video, and home theater to tailor the information to your individual needs. For more information, readers are encouraged to consult with professional specialty audio and video retailers or manufacturers for information concerning specific matters before making any decision.

Every effort has been made to make this manual as complete and accurate as possible. However, there may be mistakes both typographical and in content. Therefore, this text should be used as a general guide and not as the ultimate source of audio and video information. Furthermore, this text contains information current only up to the printing date.

The purpose of this manual is to educate and entertain. The author and Acapella Publishing shall have neither liability nor responsibility to any person or entity with respect to any loss or damage caused, or alleged to be caused, directly or indirectly by the information contained in this book.

If you do not wish to be bound by the above, you may return this book to the place of purchase for a full refund.

Contents

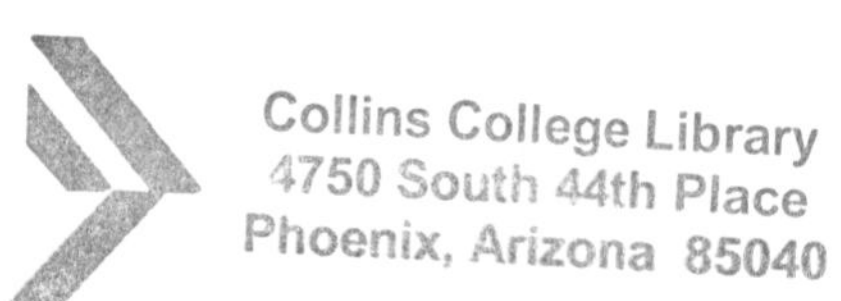

Foreword

By Keith Jarrett

Music is the sonic motion of intention. With words, sound can be divorced from meaning by taking away the physical quality of speech. But music's meaning is *in* its physical quality: its sound. When a musician plays something a certain way and we can't hear the intent (the reason) behind it, we are hearing wasted motion, and register it as such because we haven't been given enough clues about the intent. We can then grow to think that everything is only gesture, and miss the real thing.

The media through which we hear music (our systems, rooms, etc.) cannot be separated from our ability to experience the music. It isn't the same *music* on a different system because we cannot separate music's rhetoric (its words) from its physical reality (its delivery). This makes the "delivery systems" (our stereos) more important than we might think they are. Can they *tell* us what the musicians on the recording are telling us?

As a musician, I often—too often—had the following experience: I would play a concert, hear the tape afterward, and wonder what was missing. I would remember incredible things in the concert that just weren't on the tape. The notes were there, but notes are not music. Where was the music, the *intention*?

We could think of it this way: On the tape, the rhetoric had no meaning. Had I trusted the tape and not my memory of the actual event, I would have never grown to understand that, even though the sound is on tape, it doesn't mean you've recorded the *music*. If you've heard a certain CD on a certain system, it doesn't necessarily mean that you've heard what's on the CD. We must learn to trust the responses of our own system—our ears—to music systems. Of course, this demands that we be in touch with ourselves—no easy thing.

People to whom music is important need to get close to the intention in a recording, and there's only one way to do this in the home: learn about the world of audio equipment. Use your (and others') ears to help remove whatever hinders you from the musical experience on the recording. Of course, it's not only the reproduction side that needs care—but that's the only side the listener has control over.

For instance, it's demonstrable that by merely flipping a two-pronged AC plug on a CD player, or even a turntable, a record you thought you didn't like can become a favorite—just because the polarity was wrong. Since music cannot be divorced from its emotional content, the *sound* of a record can determine whether you think you like the *music*. And vice versa, when you can't listen to music you really think you like because of how it was *recorded*.

Obviously, the musical experience is a delicate, complex thing, and we humans are more sensitive than we sometimes think. But we have the option to tune our music systems to better balance the equation. We can get closer to what we want if we *know* what we want.

There are stereo components that approximate the musical experience at many different price levels. We all know what our financial limitations are; but, given the desire to improve our systems, we *can* do it.

It by no means follows that musicians have to be audiophiles. Though I've been recording since 1965, I didn't seriously think about much of this until the last decade. But audiophiles and music lovers push the envelope, and we all benefit. Also, the more serious audiophiles are determined to keep their minds and ears open, keep learning, and try to remain patient during the process. Doing this thing right can take time.

There are a lot of people out there listening to all of these components for us. I recommend *using* this fact, and *carefully* reading others' evaluations, until you can tell whether a reviewer's preferences in sound match your own priorities. You *can* sort of get to know these guys over a period of time.

But, of course, it's *your* ears that count. I think you should pay attention to their needs. After all, we're talking nutrition in an age of diet soft drinks.

Preface to the Fourth Edition

This new fourth edition of *The Complete Guide to High-End Audio* represents the most extensive overhaul of the book since its introduction in 1994. The reason is simple: today's music listener is faced with an unprecedented array of new technologies for acquiring, storing, and accessing music. As recently as 2004 (the book's last update), CD was the dominant medium, SACD and DVD-Audio were battling to be the CD's high-resolution successor, downloadable music was severely compromised in sound quality and thus of little interest to the audiophile, and only a few hardy souls dared to use a PC as a music server.

Just six years later, everything is different. Downloadable music is no longer synonymous with horribly compressed MP3 sound quality—it's the serious music lover's path to spectacular-sounding high-resolution digital audio. The format war between SACD and DVD-A was rendered moot by the death of physical media. Although SACD is still a great format with many years of life left, the Internet allows us to transcend physical formats and download high-resolution files, independent of the dictates of the mass market or the need to standardize on a disc format. Linn Products, the company that revolutionized high-performance audio in 1972 with its LP12 turntable, ceased production of CD players in early 2010. Its customers had moved on to accessing music via the Internet. Further evidence of this trend comes from the president of Polygram records, who stated that 90% of all CDs sold are played just once—to be ripped to a computer hard drive. We are in an exciting new age.

But it's not just high-res digital, music servers, downloadable music, and hard-drive storage that have dramatically changed the audiophile landscape. Technology advances have greatly improved the performance of the traditional cone-based dynamic loudspeaker. Switching amplifiers, with their cool operation, compact size, and high output power, are beginning to be taken seriously. That venerable old format, the vinyl record, is seeing a remarkable resurgence in popularity (LP sales doubled in 2009 over 2008). Today's turntables, tonearms, and cartridges extract even more musical information from the LP's grooves. And CD playback has improved immeasurably, largely the result of inventive new digital filters. The establishment of Blu-ray Disc as a mass-market format has wonderful implications for the music lover. The format can contain not only high-definition video, but also up to eight channels of high-resolution digital audio with perfect bit-for-bit accuracy to the source. Concert performances on Blu-ray are nothing short of spectacular. Finally, although the laws of physics haven't been repealed, this fourth edition incorporates new techniques and products for optimizing your listening room and tweaking your system for the best possible sound.

As has been the case with previous editions, each chapter progresses from the most basic information to deeper technical discussions. When you've reached your own comfortable depth, simply skip to the beginning of the next chapter where the material becomes easier. This book is meant to be used as a reference rather than read linearly from start to finish.

Keep in mind that you don't need to understand the technical aspects of audio to enjoy music in your home. I've presented the technical content for those readers interested in knowing how audio works, and to make this book truly The Complete Guide to High-End Audio.

Robert Harley
Carlsbad, California

About the Author

Robert Harley is the author of *Home Theater For Everyone, Introductory Guide to High-Performance Audio Systems,* and Editor-in-Chief of *The Absolute Sound* magazine. *The Absolute Sound,* founded in 1972, is the world's most respected journal of high-end audio. Robert Harley's more than 1000 published equipment reviews and articles on music and home-theater sound reproduction have helped thousands of enthusiasts improve their home-entertainment systems.

Robert Harley holds a degree in recording engineering and has taught a college degree program in that field. He has worked as a recording engineer and studio owner, compact disc mastering engineer, technical writer, and audio journalist. Before joining *The Absolute Sound* and *The Perfect Vision* in 1999, he was Technical Editor of *Stereophile* magazine for eight years, and also served in that capacity at *Fi: The Magazine of Music and Sound.*

1 What Is High-End Audio?

High-end audio is about passion—passion for music, and for how well it is reproduced. High-end audio is the quest to re-create in the listener's home the musical message of the composer or performer with the maximum realism, emotion, and intensity. Because music is important, re-creating it with the highest possible fidelity is important.

High-end audio products constitute a unique subset of music-reproduction components that bear little similarity to the "stereo systems" sold in department stores. A music-reproduction system isn't a home appliance like a washing machine or toaster; it is a vehicle for expressing the vast emotional and intellectual potential of the music encoded on our records and CDs. The higher the quality of reproduction, the deeper our connection with the music.

The high-end ethos—that music and the quality of its reproduction matter deeply—is manifested in high-end audio products. They are designed by dedicated enthusiasts who combine technical skill and musical sensitivity in their crafting of components that take us one step closer to the original musical event. High-end products are designed by ear, built by hand, and exist for one reason: to enhance the experience of music listening.

A common misperception among the hi-fi–consuming public is that high-end audio means high-*priced* audio. In the mass-market mind, high-end audio is nothing more than elaborate stereo equipment with fancy features and price tags aimed at millionaires. Sure, the performance may be a little better than the hi-fi you find at your local appliance store, but who can afford it? Moreover, high-end audio is seen as being only for trained, discriminating listeners, snobs, or gadget freaks—but not for the average person on the street.

High-end audio is none of these things.

First, the term "high-end" refers to the products' *performance,* not their price. Many true high-end systems cost no more—and often less—than the all-in-one rack systems sold in department stores. I've heard many inexpensive systems that capture the essence of what high-quality music reproduction is all about—systems easily within the budgets of average consumers. Although many high-end components *are* high-priced, this doesn't mean that you have to take out a second mortgage to have high-quality music reproduction in your home. A great-sounding system can be less expensive than you might think.

Second, high-end audio is about communicating the musical experience, not adding elaborate, difficult-to-operate features. In fact, high-end systems are much easier to use than mass-market mid-fi systems. This is because the high-end ethic eliminates useless features, instead putting the money into sound quality. High-end audio is for music lovers, not electronics whizzes.

Third, *anyone* who likes music can immediately appreciate the value of high-quality sound reproduction. It doesn't take a "golden ear" to know what sounds good. The differences between good and mediocre music reproduction are instantly obvious. The reaction—usually pleasure and surprise—of someone hearing a true high-end audio system for the first time underscores that high-end audio can be appreciated by everyone. If you enjoy music, you'll enjoy it more through a high-end system. It's that simple.

Finally, the goal of high-end audio is to make the equipment "disappear"; when that happens, we know that we have reached the highest state of communication between musician and listener. High-end audio isn't about equipment; it's about music.

The high-end credo holds that the less the musical signal is processed, the better. Any electronic circuit, wire, tone control, or switch degrades the signal—and thus the musical experience. This is why you won't find graphic equalizers, "spatial enhancers," "subharmonic synthesizers," or other such gimmicks in high-end equipment. These devices are not only departures from musical reality, they add unnecessary circuitry to the signal path. By minimizing the amount of electronics between you and the musicians, high-end audio products can maximize the directness of the musical experience. Less is more.

Imagine yourself standing at the edge of the Grand Canyon, feeling overwhelmed by its grandeur. You experience not only the vastness of this massive sculpture carved deep into the earth, but all its smaller features jump out at you as well, vivid and alive. You can discern fine gradations of hue in the rock layers—distinctions between the many shades of red are readily apparent. Fine details of the huge formations are easily resolved simply by your looking at them, thus deepening your appreciation. The contrasts of light and shadow highlight the apparently infinite maze of cracks and crevasses. The longer and closer you look, the more you see. The wealth of sensory input keeps you standing silently at the edge, in awe of nature's unfathomable beauty.

Now imagine yourself looking at the Grand Canyon through a window made of many thicknesses of glass, each one less than perfectly transparent. One pane has a slight grayish opacity that dulls the vivid hues and obliterates the subtle distinctions between similar shades of color. The fine granular structure of the next pane diminishes your ability to resolve features in the rock. Another pane reduces the contrast between light and shadow, turning the Canyon's immense depth and breadth into a flat canvas. Finally, the windowframe itself constricts your view, destroying the Canyon's overall impact. Instead of the direct and immediate reality of standing at the edge of the Grand Canyon, what you see is gray, murky, lifeless, and synthetic. You may as well be watching it on television.

Hearing reproduced music through a mediocre playback system is like looking at the Grand Canyon through those panes of glass. Each component in the playback chain—CD player, turntable, preamplifier, power amplifier, loudspeakers, and the cables that connect them—in some way distorts the signal passing through it. One product may add a coarse, grainy character to instrumental textures. Another may reduce the dynamic contrasts between loud and soft, muting the composer's or performer's expression. Yet another may cast a thick, murky pall over the music, destroying its subtle tonal colors and overlaying all instruments with an undifferentiated timbre. Finally, the windowframe—that is, the electronic and mechanical playback system—diminishes the expanse that is the musicians' artistic intent.

Fig. 1-1 Each component in an audio system can be thought of as a piece of glass through which we experience music. (Courtesy AudioQuest)

High-end audio is about removing as many panes of glass as possible, and making those that remain as transparent as they can be. The fewer the panes, and the less effect each has on the information passing through it, the closer we get to the live experience and the deeper our connection with the musical message.

Why are high-end audio products more transparent windows on the musical event than mass-market "stereo systems"? High-end products are designed to *sound* good—that is, like the real thing. They're not necessarily designed to perform "well" according to some arbitrary technical specification. The true high-end designer *listens* to the product during its development, changing parts and trying different techniques to produce the most realistic sound possible. He combines technical skill with musical sensitivity to create a product that best conveys the musical experience. This dedication often becomes a zealous pursuit, involving many hundreds of listening hours and painstaking attention to every factor that influences the sound. Often, a more expensive part will be included to improve the product's sound, while the retail price remains the same. The higher cost of this musically superior part comes off the company's bottom line. Why? Because the high-end designer cares deeply about music and its reproduction.

Conversely, mass-market audio components are often designed to look good "on paper"—on the specification sheet—sometimes at the expense of sound quality. A good example of this is the "THD wars" of the 1970s and '80s. THD stands for Total Harmonic Distortion, a specification widely used by uneducated consumers as a measure of amplifier quality. (If you've done this, don't worry; before I learned more about audio, I, too, looked at THD figures.) The lower the THD, the better the amplifier was perceived to be. This led the electronics giants to produce products with vanishingly low THD numbers. It became a contest to see which brand had the most zeros after the decimal point in its THD specification (0.001%, for example). Many buyers bought receivers or amplifiers solely on the basis of this specification.

Although low THD is a worthy design goal, the problem arose in *how* those extremely low distortion figures were obtained. A technique to reduce distortion in amplifiers is called "feedback"—taking part of the output signal and feeding it back to the input. Large amounts of feedback reduce THD, but cause all kinds of other problems that degrade the amplifier's musical qualities. Did the electronics giants care that the large amounts of negative feedback induced to reduce their products' THD measurements actually made those products sound *worse*? Not a chance. The only thing that mattered was making a commodity that would sell in greater quantity. They traded musical performance for an insignificant technical specification that was sold to the public as being important. Those buyers choosing components on the basis of a specification sheet rather than listening ended up with poor-sounding systems. Ironically, the amplifiers that had the lowest THDs probably had the lowest quality of sound as well.

This example illustrates the vast difference between mass-market manufacturers' and high-end companies' conceptions of what an audio component should do. High-end manufacturers care more about how the product sounds than about how it performs on the test bench. They know that their audience of musically sensitive listeners will buy on the basis of sound quality, not specifications.

High-end products are not only designed by ear, but are often hand-built by skilled craftspeople who take pride in their work. The assemblers are often audiophiles themselves, building the products with as much care as if the products were to be installed in their own homes. This meticulous attention to detail results in a better quality of construction, or *build quality*. Better build quality can not only improve a product's sound, but increase its long-term reliability as well. Moreover, beautifully hand-crafted components can inspire a pride in ownership that the makers of mass-produced products can't hope to match.

High-end audio products are often backed by better customer service than mid-fi products. Because high-end manufacturers care more about their products and customers, they generally offer longer warranties, more liberal exchange policies, and better service. It is not uncommon for high-end manufacturers to repair products out of warranty at no charge. This isn't to say you should expect such treatment, only that it sometimes happens with high-end and is unthinkable with mass-market products. High-end companies care about their customers.

These attributes also apply to high-end specialty retailers. The high-end dealer shares a passion for quality music reproduction and commitment to customer service. If you're used to buying audio components at a mass-market dealer, you'll be pleasantly surprised by a visit to a high-end store. Rather than trying to get you to buy something that may not be right for you, the responsible high-end dealer will strive to assemble a system that will provide the greatest long-term musical pleasure. Such a dealer will put your musical satisfaction ahead of this month's bottom line.

Finally, most high-end products are designed and built in America by American companies. In fact, American-made audio components are highly regarded throughout the world. More than 40% of all American high-end audio production is exported to foreign countries, particularly the Far East. This is true even though high-end products cost about twice as much abroad as they do in the U.S., owing to shipping, import duties, and importer profit. The enthusiasm for American high-end products abroad is even more remarkable when one remembers the popular American misperception that the best audio equipment is made in Japan.

On a deeper level, high-end products are fundamentally different from mass-market products. From their conception, purpose, design, construction, and marketing. In all these differences, what distinguishes a high-end from a mass-market product is the designer's caring attitude toward music. He isn't creating boxes to be sold like any other commodity; he's making musical instruments whose performance will affect how his customers experience music. The high-end component is a physical manifestation of a deeply felt concern about how well music is reproduced, and, by extension, how much it is enjoyed by the listener.

The high-end designer builds products he would want to listen to himself. Because he cares about music, it matters to him how an unknown listener, perhaps thousands of miles away, experiences the joy of music. The greater the listener's involvement in the music, the better the designer has done his job.

A digital processor designer I know epitomizes this dedication. He had specified a premium-quality resistor at a certain point in his new design. This resistor cost $1 rather than the pennies most resistors cost. Just as the design was about to go into production, he looked even harder for any changes that would improve the product's sound. For fun, he tried an exotic $10 resistor in the circuit in place of the $1 resistor. He was surprised at how much better the product sounded with this change, and couldn't bear to see the product shipped with the $1 resistors. The company made the product with the $10 resistors although the retail price had already been established based on the parts cost using $1 resistors. High-end designers try to add quality to, rather than subtract cost from, their products.

To the high-end designer, electronic or mechanical design isn't merely a technical undertaking—it's an act of love and devotion. Each aspect of a product's design, technical as well as musical, is examined in a way that would surprise those unaccustomed to such commitment. The ethos of music reproduction goes to the very core of the high-end designer's being; it's not a job he merely shows up for every day. The result is a much more powerful and intimate involvement in the music for the listener than is possible with products designed without this dedication.

What *is* high-end audio? What is high-end sound? It is when the playback system is forgotten, seemingly replaced by the performers in your listening room. It is when you feel the composer or performer speaking across time and space *to you*. It is feeling a physical rush during a musical climax. It is the ineffable roller-coaster ride of emotion the composer somehow managed to encode in a combination of sounds. It is when the physical world disappears, leaving only your consciousness and the music.

That is high-end audio.

If you're a newcomer to audio, you may first want to read the introductory sections of Chapters 4–13 before shopping for a system. Each of these chapters is devoted to a specific component, with an introductory section explaining the component's function, terminology, and features. Later sections in each chapter go into specific detail about selecting that component on musical and technical bases.

In this chapter, we'll look at how to choose a high-end system. The major topics include:

- ***Choosing the System Best Suited to Your Needs***
- ***Setting your budget***
- ***Complete vs. incremental purchases***
- ***Allocating your budget to specific components***
- ***Upgrading a single component***
- ***How to read magazine reviews of audio equipment***
- ***System matching***
- ***Do's and don'ts of selecting components***
- ***Your relationship with the retailer***

Choosing the System Best Suited to Your Needs

Your first major decision will be whether your audio system will be used for 2-channel music listening, multichannel music, home theater, or a combination of all three. This fundamental choice will affect the types of components you choose as well as the allocation of your budget to specific components. We'll cover multichannel music and home-theater audio in their own chapters later, but keep in mind that spreading out your budget over five or seven loudspeakers and channels of amplification to drive them will necessitate buying products of lower quality than if you had invested your entire budget in just two loudspeakers and amplifier channels. Nonetheless, the general advice in this chapter on choosing an audio system applies equally well to any system configuration. Specific advice on choosing home theater and multichannel audio equipment appears in Chapters 12 and 13, respectively.

Just as a pickup truck is better suited to the farmer and a compact car to the city dweller, a hi-fi system ideal for a small New York City apartment would be entirely inadequate in a large suburban home. The hi-fi system must not only match your musical taste, as described in the next chapter, but must also suit your room and listening needs. (The following section is only an overview of how to choose the best system. More detailed information on how to select specific components is contained in Chapters 4–13.)

Many of the guidelines are fairly obvious. First, match the loudspeaker size to your listening room. Large, full-range loudspeakers don't work well in small rooms. Not only are large loudspeakers physically dominating, they tend to overload the room with bass energy. A loudspeaker that sounds fine in a 17′ by 25′ room will likely be thick, boomy, and bottom-heavy in a 12′ by 15′ room. The bass performance you paid dearly for (it's expensive to get correct deep-bass reproduction) will work against you if the loudspeaker is put in a small room. For the same money, you could buy a superb minimonitor whose build cost was put into making the upper bass, midrange, and treble superlative. You win both ways with the minimonitor: your room won't be overloaded by bass, and the minimonitor will likely have much better soundstaging and tonal purity. There are other benefits: minimonitors, with their limited low-frequency extension, are less likely to annoy neighbors. You can thus listen to music louder without bothering anyone. Further, placement is much easier in small rooms.

Conversely, a minimonitor just won't fill a large room with sound. The sense of power, dynamic drive, deep-bass extension, and feeling of physical impact so satisfying in some music just doesn't happen with minimonitors. If you've got the room and the budget, a full-range, floorstanding loudspeaker is the best choice.

This is just one example of how the system you choose should be carefully tailored to your specific needs.

Setting Your Budget

How much you should spend on a music playback system depends on two things: your priorities in life and your financial means.

Let's take the priorities first. One person may consider a $5000 stereo system an extravagance, yet not bat an eye at blowing $15,000 on a two-week European vacation. Conversely, another person of similar means would find a $15000 vacation a waste of money when there is so much great hi-fi on the wish list. The first person probably considers music as merely a dispensable diversion while driving to work. To the second person, however, enjoying and appreciating music is a vital aspect of human existence. How much of your disposable income you should spend on a hi-fi system is a matter of how important music is to you, and only you can decide that. Owning a good hi-fi system will probably elevate music listening to a much higher priority in your life.

The second factor—your financial means—can often suggest a hi-fi budget. Surveys of readers of audiophile magazines suggest that about 10% to 15% of one's annual income is an average expenditure. In one recent survey, 25% of respondents owned systems costing less than $10,000; 31% owned systems costing $10,000 to $35,000; 39% owned systems costing $35,000 to $75,000; and 5% of these magazine readers had more than $75,000 invested in their audio systems. These are loose generalizations; high-end systems range in cost from less than $1500 to more than $500,000. And don't let these figures intimidate you: readers of audiophile magazines are dedicated enthusiasts who are not necessarily representative of all music lovers. You can put together a musically satisfying system for much less than the figures cited.

I strongly urge you, however, to establish a significant budget for your high-end system. The expenditure may seem high at the time, but you will be rewarded night after night and year after year with your favorite music wonderfully reproduced. A year or two from now, as you enjoy your system, the money spent will have been forgotten, but the pleasure will continue—a good music-playback system is a lasting and fulfilling investment. Moreover, if you buy a quality system now, you won't want to sell it or trade it in for something better later. It is sometimes false economy to "save" money on a less than adequate system. Do it right the first time.

There's another way of determining how much to spend for a hi-fi system: Find the level of quality you're happy with and let that set your budget. Visit your dealer and have him play systems of various levels of quality. You may find yourself satisfied with a moderately priced system—or you may discover how good reproduced music can sound with the best equipment, and just *have* to have it.

When setting a budget for your music system, you must also consider whether the system will be used only for 2-channel music reproduction, or will be called upon to reproduce movie soundtracks as part of a home-theater system. Maintaining high quality with home theater's six speakers instead of two, and buying a multichannel surround-sound preamplifier rather than a 2-channel preamp, can significantly increase the amount of money you'll need to spend. There are, however, ways to add multichannel music reproduction and film-soundtrack playback without greatly escalating the system cost. We'll briefly touch on this subject in this chapter, and delve into it in more detail in Chapter 12.

The Complete vs. the Incremental Purchase

After you've established a budget, you must decide which of the following three ways of buying a hi-fi is best for you:

1) Buy an entire system made up of the finest components.
2) Buy an entire system made up of components within a limited budget.
3) Buy just a few components now and add to the system as finances permit.

The first option doesn't require much thought—just a large bank account. The purchaser of this system needn't worry about budgets, upgrading, and adding components later. Other elements of buying a high-end system (I'll talk about these later) do apply to the cost-no-object audiophile: allocating the budget to specific components, dealing with the retailer, and home auditioning. But this sort of listener isn't under the same financial constraints that force the tough choices inherent in the other two options. Just choose a top-notch system and start enjoying it.

Most of us, however, don't enjoy such luxury, and must live within the set budgets of options 2 or 3. Option 2 is to spend the entire budget on a complete system now. In option 3, you'll spend your entire budget on just a few, higher-quality components and add the other pieces later as money becomes available. We'll call option 2 the *complete purchase*, and option 3 the *incremental purchase*. There are advantages and disadvantages to each approach.

Buying an entire system at once means that the overall budget must be spread among all the components that make up a hi-fi system. Consequently, you'll have less to spend on each component. You may not get the quality of components you'd hoped for, but the system will be complete and you can start enjoying it right away. This type of purchase is more suited to the music lover who doesn't want to think about equipment, but wants a system he can set up, forget, and use to enjoy music.

The listener more inclined to treat audio as a hobby, or who has her sights set on a more ambitious system, will buy just a few components now and add to the system as finances permit. This listener may spend the same amount of money as our first listener, but on just a pair of loudspeakers instead of on a whole system. She will use her old receiver or integrated amplifier, turntable, disc player, or portable music player until she can afford the electronics and sources she really wants. Her system will be limited by the receiver's performance; she won't immediately get the benefit of her high-end loudspeakers. But when she *does* buy her dream electronics, she'll have a truly first-rate system. This approach requires more patience, commitment, and, in the long term, more money. But it is one way to end up with a superlative system.

The audiophile buying a system piece by piece can benefit from rapidly changing technology. She can start with components that don't change much over the years—power amplifiers, for example—and wait to buy those products likely to get better and cheaper over time: disc players, digital-to-analog converters (DACs), music servers , and so on.

Another benefit of adding components to your system one at a time is the ability to audition components in that system before buying. Rather than putting together a whole system in a store showroom based on your own auditioning or a salesperson's recommendations, the incremental buyer can carefully audition components and choose the best musical match for the rest of the system. This is a big advantage when assembling a hi-fi best suited to your musical tastes. This piece-by-piece approach is more for the hobbyist, and demands a deeper level of commitment to audio. It also requires a great deal more patience.

In short, if you want to get a good system, forget about the hardware, and just enjoy music, buy the entire system now. Given the same initial expenditure, you won't end up with the same-quality system as if you'd bought pieces slowly, but you'll be spending less money in the long run and can have the benefit of high-end music playback immediately. Moreover, you can go about your life without thinking about what piece of audio hardware to buy next. The complete purchase is recommended for the music lover who isn't—and doesn't want to be—an audio hobbyist. Many music lovers taking this approach, however, find themselves upgrading their systems piece by piece after discovering the rewards of owning a high-end audio system.

Conversely, the music lover on a budget who has set her sights on a more ambitious system, and who takes a more active role in audio equipment, will probably build a system gradually. This listener will more likely read product reviews in magazines, visit the dealer often, and take equipment home for evaluation. By doing so she'll become a better listener and a more critical audiophile, as well as develop a broader knowledge of audio equipment. Whichever approach you take, the information in the rest of this chapter—selecting a system suited to your listening, allocating the budget to specific components, and dealing with the retailer—applies equally well.

Value vs. Luxury Components

High-end audio components run the gamut from utilitarian-looking boxes to lavish, gold-plated, cost-no-object shrines. The packaging doesn't always reflect the quality of the electronics inside, but rather the manufacturer's product philosophy. Some companies try to offer the best sound for the least money by putting excellent electronics in inexpensive chassis. These are the so-called "value" products. Manufacturers of "luxury" components may put the same level of electronics in a lavish chassis with a 1"-thick front panel, lacquer-filled engraving, expensive metalwork, and custom machined input jacks or terminal posts.

A designer of *very* expensive electronics once told me that he could sell his products for *half* the price if he used a cheap chassis. He felt, however, that the level of design and execution in his electronics deserved no less than the ultimate in packaging.

Some buyers demand elegant appearance and a luxury feel; others merely want the best sound for the least money. To some music lovers, appearance and elegant packaging are secondary to sound quality; they don't care what it looks like so long as it sounds good. Conversely, some audiophiles are willing to pay for gorgeous cosmetics, battleship build quality, and all the trimmings that make some products exude elegance and luxury. There's an undeniable pride of ownership that accompanies the finest-built audio components.

When choosing high-end components, match your needs with the manufacturer's product philosophy. That way, you won't waste money on thick faceplates you don't care about, or spend money on a product that doesn't do justice to your home decor.

Another type of manufacturer puts mediocre or even poor electronics in a fancy, eye-catching package. Their market is the less musically sophisticated buyer who chooses on the basis of appearance and status (or the company's past reputation) rather than sound quality. These so-called "boutique" brands are not high-end, however, and should be avoided.

A related topic is convenience vs. cost. Some products include convenience features that add to the component's price without improving its sound quality. When choosing components, consider the tradeoffs between sound quality, convenience, and price.

Allocating Your Budget to Specific Components

There are no set rules for how much of your total budget you should spend on each component in your system. Allocating your budget between components depends greatly on which components you choose, and your overall audio philosophy. Mass-market mid-fi magazines have been telling their readers for years to spend most of a hi-fi budget on the loudspeakers because they ultimately produce the sound. This thinking also suggests that all amplifiers and digital sources sound alike; why waste money on expensive amplifiers and disc players?

The high-end listener makes different assumptions about music reproduction. A fundamental tenet of high-end audio holds that if the signal isn't good at the beginning of the reproduction chain, nothing downstream can ever improve it. In fact, the signal will only be degraded by any product it flows through. High-end audio equipment simply minimizes that degradation. If your DAC or music server is bright, hard, and unmusical, the final sound will be bright, hard, and unmusical. Similarly, the total system's performance is limited by the resolution of the worst component in the signal path. You may have superb loudspeakers and an excellent turntable and cartridge, but they'll be wasted with a poor-quality preamp in the signal chain.

Quality matching between components is essential to getting the most sound for your budget. High-quality loudspeakers at the end of a chain containing a bad-sounding component can even make the system sound worse than lower-quality loudspeakers: The high-resolution loudspeakers reveal all the imperfections of the electronics upstream of them. This situation has been likened to having a large picture window in your home. If the view is of the Northern California coastline, you want that window to be as clean and transparent as possible. But if the window overlooks a garbage dump, you'd prefer that it somewhat obscure the view.

I've listened to $400 loudspeakers driven by $30,000 worth of electronics, and $158,000 loudspeakers driven by budget integrated amplifiers. I can state categorically that the electronics and source components are every bit as important as the loudspeakers. Although the loudspeakers significantly influence the overall sound, high-quality source components (turntable and digital source), good electronics (preamplifier and power amplifier), and excellent cables are essential to realizing a musical high-end system.

Since the first three editions of this book, however, I've had experiences with very good-sounding and affordable integrated amplifiers driving moderately expensive loudspeakers, and the results have been excellent. As explained in Chapters 5 and6, loudspeaker sensitivity (how loudly the speaker will play for a given amount of amplifier power) greatly affects how large an amplifier you need to achieve a satisfying volume. High-sensitivity speakers need very little amplifier power. And because amplifier power is costly, it follows that a system with high-sensitivity speakers can be driven by lower-cost amplification—*provided that the amplification is of high quality*. Fortunately, manufacturers have recently responded to the need for relatively inexpensive high-performance amplification by designing integrated amplifiers with outstanding sound quality, but with lower output powers. By putting the preamplifier and power amplifier in one chassis and cutting back on power output, manufacturers can put their high-end circuits in lower-priced products. The trick is to find those bargain integrated amplifiers that deliver truly high-end sound, and mate them with loudspeakers that not only have the appropriate sensitivity, but are also a good musical match. This approach will get you the best sound for the least money. If, however, cost is secondary to sound quality—that is, you're willing to spend more for an improvement in sound—then buy the best separate preamplifier and power amplifier you can find.

For the following exercise, I assembled an imaginary 2-channel system of the components I'd choose if my audio budget totaled $10,000. This hypothetical system follows a traditional audiophile approach. Here are the costs per item:

Preamplifier	$2000
Power amplifier	2000
Digital source	1300
Loudspeakers	4000
Interconnects and cables	700
Total	**$10,000**

As you can see, loudspeakers consumed 40% of the budget, the digital source took up another 13%, and the preamp and power amplifier each received 20%. The remaining 7% was spent on interconnects and cables. These numbers and percentages aren't cast in stone, but they're a good starting point in allocating your budget. If you wanted to include a turntable, tonearm, and cartridge, the budget for the other components would have to be reduced.

Another approach with this budget would be to buy an inexpensive disc player ($500) and put the $800 saved into better loudspeakers or electronics. Then, as finances permitted, you could add an outboard DAC and drive it with the disc player's digital output jack (see Chapter 7). You'll have music in the meantime, and end up with a better-sounding system in the long run.

The 40% figure for loudspeakers is very flexible. Keep in mind that there's an important threshold in loudspeaker performance at about $2500: Loudspeakers costing a little more than $2500 are often disproportionately better than those costing a little less than $2500, so you may want to adjust your budget allocations to cross this threshold. As described in Chapter 6, many moderately priced loudspeakers outperform much more expensive models. Use Chapter 6's guidelines on choosing loudspeakers to get the most performance for your loudspeaker dollar.

Following the earlier discussion of matching a superb but low-powered integrated amplifier with high-sensitivity speakers, here's another example of how I might allocate a $10,000 budget:

Integrated amplifier	$3000
Digital source	$1000
Loudspeakers	$5500
Interconnects and cables	$500
Total	**$10,000**

Again, the key to putting so much of the budget into loudspeakers is extremely careful matching of the amplifier's power output power to the loudspeaker's sensitivity (and impedance curve, explained in Chapter 6), along with finding those few integrated amplifiers that deliver the musicality of expensive separates, but simply have lower output powers. Here's an extreme case: I lived with a system for about a month (during a product review) that included $11,000 loudspeakers driven by a $1500 integrated amplifier and the result was musical magic. It takes a lot of searching to find these synergistic combinations—or a great dealer who has discovered these ideal matches for you. I must stress that this approach only works with certain components, and is useful for getting the best sound for the least money. It is not the ideal strategy when the best possible sound is your goal.

Here's another sample budget, this one based on a maximum expenditure of $2000:

Amplification	$750
Digital source	400
Loudspeakers	750
Interconnects and cables	100
Total	**$2000**

Again, I selected components that experience suggested would be a good match, and tallied the percentages after choosing the components. Interestingly, the breakdown was similar to that in the first example: 37% on loudspeakers, 20% on a digital source, 37% on amplification, and 5% on interconnects and cables.

I've heard systems at this price level that are absolutely stunning musically. When carefully chosen and set up, a $2000 high-end system can achieve the essence of what high-quality music reproduction is all about—communicating the musical message. I've even heard a whole system with a list price of $850 that was musical and enjoyable. The point isn't how much you spend on a hi-fi, but how carefully you can choose components to make a satisfying system within your budget.

These guidelines are for a dedicated 2-channel music system, not a multichannel music and home-theater system. The different requirements of home theater suggest a somewhat different budget allocation. If most of your listening time with a multichannel system is devoted to music, the budget guidelines are very similar to those described earlier, but with a shift of money from electronics to loudspeakers. That is, you should spend a bit more of your total budget on speakers. That's because home theater and multichannel music require five or six loudspeakers, and spreading out a fixed amount of money over five speakers rather than two inevitably results in compromised performance. If music is your primary consideration, put most of your speaker budget into the left and right speakers, with less devoted to the center, surround speakers, and subwoofer. Those readers who spend more time watching movies should spend even less on electronics and source and more on speakers. That's because the center channel loudspeaker plays an important role in the home-theater experience (it reproduces nearly all the dialogue), and choosing a quality model is of paramount importance.

In my experience, the sonic differences among amplifiers, preamplifiers, and source components are much less pronounced when we watch movies than when we listen to music. So much of our attention is taken away from the aural experience by the overwhelming visual sensory input. I've used audio/video receivers in my system (during product reviews) and found most of them lacking when reproducing music, but more than adequate for film-soundtrack playback.

Whether choosing a dedicated 2-channel music system or a combined music and home-theater package, you should save some of your budget for an AC power conditioner and accessories. I advise against buying a power conditioner and accessories when you buy the system. Take the system home, get it set up and optimized, *then* add a power conditioner and start experimenting with accessories. Here's why: AC conditioners don't always make an improvement. In fact, they can degrade the sound. There are many variables with AC power conditioners, including the quality of AC from your wall, the method of AC conditioning, and the number and nature of the components plugged into the conditioner. It is therefore best to try the conditioner at home before buying.

There's another good reason for adding an AC line conditioner later: By getting to know how your system sounds *without* an AC conditioner, you'll be better able to judge if the conditioner is an improvement. Remember that a change in sound isn't always for the better. The same logic holds true for accessories such as cones, feet, and tube dampers: You'll be in a much better position to judge their effectiveness—or lack of it—by knowing your system intimately before installing accessories. Set aside some of your budget—perhaps a few hundred dollars—for accessories. If they don't make a difference, you've now got a few hundred dollars to spend on records or CDs. (A full survey of accessories is included in Chapter 15.)

Upgrading a Single Component

Many audiophiles gradually improve their systems by replacing one component at a time. The trick to getting the most improvement for the money is to replace the least good component in your system. A poor-sounding preamp won't let you hear how good your music server is, for example. Conversely, a very clean and transparent preamp used with a grainy and hard digital source will let you hear only how grainy and hard the digital source is. The system should be of similar quality throughout. If there's a quality mismatch, however, it should be in favor of high-quality source components.

Determining which component to upgrade can be difficult. This is where a good high-end audio retailer's advice is invaluable—he can often pinpoint which component you should consider upgrading first. Another way is to borrow components from a friend and see how they sound in your system. Listen for which component makes the biggest improvement in the sound. Finally, you can get an idea of the relative quality of your components by carefully reading the high-end audio magazines, particularly when they recommend specific components.

In Chapter 1, I likened listening to music through a playback system to looking at the Grand Canyon through a series of panes of glass. Each pane distorts the image in a different way. The fewer and more transparent the panes are, the clearer the view, and the closer the connection to the direct experience.

Think of each component of a high-end audio system as one of those panes of glass. Some of the panes are relatively clear, while others tend to have an ugly coating that distorts the image. The pane closest to you is the loudspeaker; the next closest pane is the power amplifier; next comes the preamp; and the last pane is the signal source (disc player, music server, turntable). Your view on the music—the system's overall transparency—is the sum of the panes. You may have a few very transparent panes, but the view is still clouded by the dirtiest, most colored panes. This idea is shown graphically on page 3 earlier.

The key to upgrading a hi-fi system is getting rid of those panes—those components—that most degrade the music performance, and replacing them with clearer, cleaner ones. This technique gives you the biggest improvement in sound quality for the money spent.

Conversely, putting a very transparent pane closest to you—the loudspeaker—only reveals in greater detail what's wrong with the power amplifier, preamplifier, and source components. A high-resolution loudspeaker at the end of a mediocre electronics chain can actually sound worse than the same system with a lower-quality loudspeaker.

Following this logic, we can see that a hi-fi system can never be any better than its source components. If the first pane of glass—the source component—is ugly, colored, and distorts the image, the result will be an ugly, colored, and distorted view.

As you upgrade your system, you can start to see that other panes you thought were transparent actually have some flaws you couldn't detect before. The next upgrade step is to identify and replace what is *now* the weakest component in the system. This can easily become an ongoing process.

Unfortunately, as the level of quality of your playback system rises, your standard of what constitutes good performance rises with it. You may become ever more critical, upgrading component after component in the search for musical satisfaction. This pursuit can become an addiction and ultimately *diminish* your ability to enjoy music. The next chapter includes an editorial I wrote for *Stereophile* magazine examining this subject.

How to Read Magazine Reviews

I've deliberately avoided recommending specific products in this book. By the time you would have read the recommendations, the products would likely have been updated or discontinued. The best source for advice on components is product reviews in high-end audio magazines. Because most of these magazines are published monthly, they can stay on top of new products and offer up-to-date buying advice. Many magazine reviewers are highly skilled listeners, technically competent, and uncompromising in their willingness to report truthfully about audio products. There's a saying in journalism that epitomizes the ethic of many high-end product reviewers: "without fear or favor." The magazine should neither fear the manufacturer when publishing a negative review, nor expect favor for publishing a positive one. Instead, the competent review provides unbiased and educated opinion about the sound, build quality, and value of individual products. Because high-end reviewers hear lots of products under good conditions, they are in an ideal position to assess the relative merits and drawbacks and report their informed opinions. The best reviewers have a combination of good ears, honesty, and technical competence.

You'll notice a big difference between reviews in high-end magazines and reviews in the so-called "mainstream slicks." The mass-market, mainstream magazines are advertiser-driven; their constituents are their advertisers. Conversely, high-end magazines are usually reader-driven; the magazines' goal is to serve their readers, not their advertisers. Consequently, high-end magazines often publish negative reviews, while mass-market magazines generally do not. Moreover, high-end magazines are much more discriminating about what makes a good component that can be recommended to readers. Mass-market magazines cater to the average person in the street, who, they believe (mistakenly, in my view), doesn't care about aspects of music reproduction important to the audiophile. The high-end product review is thus not only more honest, but much more discriminating in determining what is a worthy product. If you're reading a hi-fi magazine that never criticizes products, beware. Not all audio components are worth buying; therefore not all magazine reviews should conclude with a recommendation. Also beware of so-called "consumer" publications that regard an audio system as an appliance, not as a vehicle for communicating an art form. In their view, the improvements in performance gained by spending more are not worth the money, so they recommend the least expensive products with the most features. They take a similar approach to reviewing cars: why spend lots of money on a depreciating asset when all cars perform an identical function—moving you from point A to point B? It is simply not within their ethos to value spirited acceleration, tight handling, a luxurious interior, or the overall feel of the driving experience.

As someone who makes his living reviewing high-end audio products, I'll let you in on a few secrets that can help you use magazine reviews to your advantage. First, associate the review you're reading with the reviewer. Before reading the review, look at the byline and keep in mind who's writing the review. This way, you'll quickly learn different reviewers' tastes in music and equipment. Seek the guidance of reviewers with musical sensibilities similar to your own.

Second, don't assign equal weight to all reviews or reviewers. Audio reviewing is like any other field of expertise—there are many different levels of competence. Some reviewers have practiced their craft for decades, while others are newcomers who lack the seasoned veteran's commitment to the profession. Consider the reviewer's reputation, experience, and track record when giving a review credence. Also consider the reviewer's standards for what makes a good product. Just as there's a big difference between a movie recommendation given by Joel Siegel (*Good Morning America*) and Andrew Sarris (*The Village Voice*), so too are there differences in product recommendations from high-end reviewers. The rise of the internet has allowed anyone with a computer to publish audio-equipment reviews. A few of the on-line magazines employ skilled listeners and adhere to high journalistic standards; others are merely vehicles for audiophiles seeking free use of review equipment, self-aggrandizement, or both.

Finally, listen to the products yourself. If two components are compared in a review, compare those products to hear if your perceptions match the reviewer's. Even if you aren't in the market for the product under review, listening to the components will sharpen your skills and put the reviewer's value judgments in perspective.

A common mistake among audiophiles looking for guidance is to select a component on the basis of a rave review without fully auditioning the product for themselves. The review should be a starting point, not some final judgment of component quality. It's much easier to buy a product because a particular reviewer liked it than it is to research the product's merits and shortcomings for yourself. Buying products solely on the basis of a review is fraught with danger. Never forget that a review is nothing more than one person's opinion, however informed and educated that opinion might be. Moreover, if the reviewer's tastes differ from yours, you may end up with a component you don't like. We all have different priorities in judging reproduced sound quality; what the reviewer values most—soundstaging, for example—may be lower in your sonic hierarchy. Your priorities are the most important consideration when choosing music-playback components. Trust your own ears.

Sharp-eyed readers can find out which products a reviewer really thinks are special. Most reviewers list the reference system with which they audition new products. When a particular product shows up in the reference system in review after review, you can be sure that the writer thinks that product is especially good.

Some magazine readers view all reviews with suspicion because the reviewer enjoyed the free loan of the product. That is, they believe there's a *quid pro quo* that guarantees the manufacturer a positive review because the reviewer got to use the product for several months. This belief is patently false. Reviewers have access to virtually any products they want, and have no allegiance to the manufacturer who lent it to them. This may be hard to believe from a reader's standpoint, but after receiving, unpacking, setting-up, and learning about hundreds of products, getting new equipment often becomes a chore rather than cause for excitement.

I'd also like to address the issue of long-term equipment loans. Most reviewers have reference systems composed of products they don't own. The manufacturers or distributors of those products choose to allow the reviewer to keep the product for many months or even a year, giving the product exposure in issue after issue. Reviewers need reference-quality equipment with which to judge the product under evaluation—reference-quality equipment they can't afford to buy. The long-term loan is mutually beneficial; the reviewer has access to the best tools available, the manufacturer enjoys the prestige—and presumably increased sales—conferred by the reviewer's repeated mentions, and the reader gets the most accurate descriptions of the products reviewed within the context of the reference system. Moreover, because so many manufacturers see the value in long-term loans, reviewers have access to virtually any products they want. This removes the putative pressure to give positive reviews only to those manufacturers who agree to long-term loans. If a reviewer keeps a product in his system, it's because that product sounds good.

It would be a violation of journalistic ethics, however, if a deal were struck between manufacturer and reviewer that allowed the reviewer to keep the product as his own at the conclusion of the review. In my twenty-one years' experience as a reviewer and magazine editor, I have never been approached by a manufacturer in an untoward way, and have never been aware of such behavior (at least on the magazines I've worked for), and would not tolerate any such ethical lapses by a writer on our staff.

A final word about product reviews: Don't stop enjoying music if a component you've already purchased gets a negative review. You enjoyed music through that product before the review appeared; why should another person's opinion diminish your musical satisfaction? Positive or negative, magazine reviews carry far too much weight in readers' minds and in the marketplace. (Remember, you heard this from a professional reviewer.)

In sum, reviews can be very useful, provided you:

- Get to know individual reviewers' sonic and musical priorities. Find a reviewer on your sonic and musical wavelength, then trust his or her opinions.

- Compare your impressions of products to the reviewer's impressions. This will not only give you a feel for the reviewer's tastes and skill, but the exercise will make you a better listener.

- Don't assign the same weight to all reviews. Consider the reviewer's reputation and experience in the field. How many similar products has the reviewer auditioned? If a reviewer has heard virtually every serious phono cartridge, for example, his or her opinion will be worth more than that of the reviewer who has heard only a few models.

- Don't buy—or summarily reject—products solely on the basis of a review. Use product recommendations as a starting point for your own auditioning. Listen to the product yourself to decide if that product is for you. Let *your* ears decide.

System Matching

It is a truism of high-end audio that an inexpensive system can often outperform a more costly and ambitious rig. I've heard modest systems costing, say, $1500 that are more musically involving than $50,000 behemoths. Why?

Part of the answer is that some well-designed budget components sound better than ill-conceived or poorly executed esoteric products. But the most important factor in a playback system's musicality is *system matching*. System matching is the art of putting together components that complement each other sonically so that the overall result is a musicality beyond what each of the components could achieve if combined with less compatible products. The concept of synergy—that the whole is greater than the sum of the parts—is very important in creating the best-sounding system for the least money.

System matching is the last step in choosing an audio system. You should have first defined the system in terms of your individual needs, set your budget, and established a relationship with a local specialty audio retailer. After you've narrowed down your choices, which products you select will greatly depend on system matching.

Knowing what components work best with other components is best learned by listening to a wide range of equipment. Many of you don't have the time—or access to many diverse components—to find out for yourselves what equipment works best with other equipment. Consequently, you must rely on experts for general guidance, and on your own ears for choosing specific equipment combinations.

The two best sources for this information are magazine reviews and your local dealer. Your dealer will have the greatest knowledge about products he carries, and can make system-matching recommendations based on his experience in assembling systems for his customers. Your dealer will likely have auditioned the products he sells in a variety of configurations; you can benefit from his experience by following his system-matching recommendations.

The other source of system-matching tips is magazine reviews. Product reviews published in reputable magazines will often name the associated equipment used in evaluating the product under review. The reviewer will sometimes describe his or her experiences with other equipment not directly part of the review. For example, a loudspeaker review may include a report on how the loudspeaker sounded when driven by three or four different power amplifiers. The sonic characteristics of each combination will be described, giving the reader an insight into which amplifier was the best match for that loudspeaker. More important, however, the sonic descriptions and judgments expressed can suggest the *type* of amplifier best suited to that loudspeaker. By *type* I mean both technical performance (tubed vs. transistor, power output, output impedance, etc.) and general sonic characteristics (hard treble, forward presentation, well-controlled bass, etc.).

Let's say the reviewer drove the loudspeakers with four amplifiers: a low-powered but sweet-sounding integrated amplifier, a high-output-impedance tubed design, a medium- to high-powered inexpensive solid-state unit, and a massive solid-state amplifier that requires two people to lug it into the listening room. The reviewer reports that the integrated amplifier just didn't have enough power to produce sufficient volume, and that the sound lacked dynamics. The high-output-impedance tubed amplifier was mushy in the bass and had a reduced sense of pace and rhythm. The inexpensive solid-state amplifier had terrific bass control, but its forward presentation and grainy treble made it less than ideal with the loudspeaker under review. (All of these terms—like *forward, grainy,* and *pace*—are defined in the next chapter.) Finally, the reviewer concludes that the solid-state behemoth is the only amplifier suitable for this particular loudspeaker.

This doesn't mean that the most expensive amplifier will always work the best with all loudspeakers. I can think of loudspeakers in which this scenario would be completely different. Another loudspeaker would sound just fine with the integrated amplifier, suggesting that using the huge solid-state unit would be overkill. However, if the loudspeaker was a little tizzy in the treble and lean in the midbass, the tubed amplifier would tend to ameliorate these tendencies. Finally, if the loudspeaker wasn't that sensitive to treble grain, but needed to be driven by an amplifier with control and authority in the bass, the inexpensive solid-state unit would be a good choice—and the most cost-effective.

These reports of system matching can extend beyond the specific products reported on in the review. A fairly good idea of which type of sonic and technical performance is best suited to a particular product can be gained from a careful reading of product reviews. For example, you may conclude that a particular loudspeaker needs to be driven by a large, high-current amplifier. This knowledge can then point you in the right direction for equipment to audition yourself; you can rule out low-powered designs. You can also get a feel for how professional reviewers assemble systems in the annual Recommended Systems feature in *The Absolute Sound.*

By reading magazine reviews, following your dealer's advice, and listening to combinations of products for yourself, you can assemble a well-matched system that squeezes the highest musical performance from your hi-fi budget.

Do's and Don'ts of Selecting Components

Some audiophiles are tempted to buy certain products for the wrong reasons. For example, many high-end products are marketed on the basis of some technical aspect of their design. A power amplifier may, for example, be touted as having "over 200,000 microfarads (μF) of filter capacitance," "32 high-current output devices," and a "discrete JFET input stage." While these may be laudable attributes, they don't guarantee that the amplifier will produce good sound. Don't be swayed by technical claims—listen to the product for yourself.

Just as you shouldn't make a purchasing decision based on specifications, neither should you base your decision solely on brand name. Many high-end manufacturers with solid reputations sometimes produce mediocre-sounding products. A high-end marque doesn't necessarily mean high-end sound. Again, let your ears be your guide. I'm often pleasantly surprised to find moderately priced products that sound as good—or very nearly as good—as products costing two or three times as much.

You should, however, consider the company's longevity, reputation for build quality, customer service record, and product reliability when choosing components. High-end manufacturers run the gamut from one-man garage operations to companies with hundreds of employees and advanced design and manufacturing facilities. The garage operation may produce good-sounding products, but may not be in business next year. This not only makes it hard to get service, but also greatly lowers the product's resale value.

High-end manufacturers also have very different policies regarding service. Some repair their products grudgingly, and/or charge high fees for fixing products out of warranty. Others bend over backward to keep their valued customers happy. In fact, some high-end audio companies go to extraordinary lengths to please their customers. One amplifier manufacturer who received an out-of-warranty product for repair not only fixed the amplifier free of charge, but replaced the customer's scratched faceplate at no cost! It pays in the long run to do business with manufacturers who have reputations for good customer service.

Another factor to consider before laying down your hard-earned cash is how long the product has been on the market. Without warning, manufacturers often discontinue products and replace them with new ones, or update a product to "Mark II" status. When this happens, the value of the older product drops immediately. If you know an excellent product is about to be discontinued, you can often buy the floor sample at a discount. This is a good way of saving money, provided the discount is significant. You end up with a lower price, plus all the service and support inherent in buying from an authorized and reputable dealer rather than a private party.

The best source of advance information on new products and what's about to be discontinued are reports in audio magazines from the annual Consumer Electronics Show held every January.

Your Relationship with the Retailer

If, in the past, you've bought audio equipment only from mass-market retailers, you should expect to have an entirely different sort of relationship with a high-end dealer. The good specialty audio retailer doesn't just "move" boxes of electronics; he provides you with the satisfaction of great-sounding music in your home. More than just an equipment dealer, he's usually a dedicated audio and music enthusiast himself—he *knows* his products, and is often the best person to advise you on selecting equipment and system setup. The great dealers search the entire audio landscape for the very best brands and are extremely selective in choosing which lines to carry.

Consider the very different relationships between seller and buyer in the following scenarios. In the first, a used-car dealer in downtown Los Angeles is trying to sell a car to someone from out of town. The seller has only one shot at the buyer, and he intends to make the most of it. He doesn't care about return business, the customer's long-term satisfaction with the purchase, or what the customer will tell his friends about the dealer. It will be an adversarial relationship from start to finish.

Then consider a new-car dealer in Great Falls, Montana, selling a car to another Great Falls resident. For this dealer, return business is vital to his survival. So is customer satisfaction, quality service, providing expert advice on models and options, finding exactly the right car for the particular buyer, and giving the customer an occasional ride to work when he drops off his car for service. He knows his customers by name, and has developed mutually beneficial, long-term relationships with them.

If buying a mass-market hi-fi system is like negotiating with a used-car dealer in downtown L.A., selecting a high-end music system should be made within a relationship similar to the one enjoyed between the Montana dealer and his customers.

Take the time to establish a relationship with your local dealer. Make friends with him—it'll pay off in the long run. Get to know a particular salesperson and, if possible, the store's owner. Tell them your musical tastes, needs, lifestyle, and budget—then let them offer equipment suggestions. They know their products best, and can offer specific component recommendations. The good stores will regard you as a valued, long-term customer, not someone with whom they have one shot at making a sale. Don't shop just for equipment—shop for the retailer with the greatest honesty and competence.

Keep in mind, however, that dealers will naturally favor the brands they carry. Be suspicious of dealers who badmouth competing brands that have earned good reputations in the high-end audio press. The best starting point in assembling your system is a healthy mix of your dealer's recommendations and unbiased, competent magazine reviews.

The high-end retailing business is very different from the mass-market merchandising of the low-quality "home entertainment" products sold in appliance emporiums. The specialty retailer's annual turnover is vastly lower than that of the mid-fi store down the street. Consequently, the specialty retailer's profit margin must be larger for him to stay in business. Don't expect him to offer huge discounts and price cuts on equipment to take a sale from a mid-fi store. Because the high-end dealer offers so much more than just pushing a box over the counter, his prices just can't be competitive. Instead, you should be prepared to pay full list price—or very close to it.

Here's why. After paying his employees, rent, lights, heat, insurance, advertising, and a host of other expenses, the specialty audio retailer can expect to put in his pocket about five cents out of every dollar spent in his store. Now, if he discounts his price by even as little as 5%, he is essentially working for free, and only keeping his doors open a little longer. If the dealer offers a discount or marks down demonstration or discontinued units, you should take advantage of these opportunities. But don't expect the dealer to discount; he deserves the full margin provided by the product's suggested retail price.

In return for paying full price, however, you should receive a level of service and professionalism second to none. Expect the best from your dealer. Spend as much time as you feel is necessary auditioning components in the showroom before you buy. Listen to components at home in your own system before you buy. Ask the retailer to set up your system for you. Exploit the dealer's wide knowledge of what components are best for the money. Use his knowledge of system matching to get the best sound possible on a given budget. And if one of your components needs repair, don't be afraid to ask for a loaner until yours is fixed. The dealer should bend over backward to accommodate your needs.

If you give the high-end dealer your loyalty, you can expect this red-carpet treatment. This relationship can be undermined, however, if, to save a few dollars, you buy from a competitor or by mail a product that your dealer also sells. If the product purchased elsewhere doesn't sound good in your system, don't expect your local dealer to help you out. Further, don't abuse the home audition privilege. Take home only those products you're seriously considering buying. If the dealer let everyone take equipment home for an audition, he'd have nothing in the store to demonstrate. The home audition should be used to confirm that you've selected the right component through store auditioning, magazine reviews, and the dealer's recommendations. The higher price charged by the dealer may seem hard to justify at first, but in the long run you'll benefit from his expertise and commitment to you as a customer.

If you don't live close to any high-end dealers, there are several very good mail-order companies that offer excellent audio advice over the phone. They provide as much service as possible by phone, including money-back guarantees, product exchanges, and component-matching suggestions. You can't audition components in a store, but you can often listen to them in your system and get a refund if the product isn't what you'd hoped it would be.

In short, if you treat your dealer right, you can expect his full expertise and commitment to getting you the best sound possible. There's absolutely no substitute for a skilled dealer's services and commitment to your satisfaction.

Used Equipment

A used audio component often sells for half its original list price, making used gear a tempting alternative. The lower prices on used high-end components provide an opportunity to get a high-quality system for the same budget as a less ambitious new system. Moreover, buying used gear lets you audition many components at length. If you find a product you like, keep it. If you don't like the sound of your used purchase, you can often sell it for no less than what you paid.

There are two ways of buying used equipment: from a dealer, and from a private party, usually over an internet audio-trading site. A retailer may charge a little more for used products, but often offers a short warranty (60 or 90 days), and sometimes exchange privileges. Buying used gear from a reputable retailer is a lot less risky than dealing with a private party (unless you're buying from a friend).

The audiophile inclined to buy used equipment can often get great deals. Some audiophiles simply *must* own the latest and greatest product, no matter the cost. They'll buy a state-of-the-art component one year, only to sell it the next to acquire the current top of the line. These audiophiles generally take good care of the equipment and sell it at bargain-basement prices. If you can find such a person, have him put your name at the top of his calling list when he's ready to sell. You can end up with a superb system for a fraction of its original selling price.

A few pitfalls await the buyer of used equipment, though the disadvantages listed below apply mostly to buying from a private party rather than from a reputable dealer. First, there are no assurances that the component is working properly; the product could have a defect not apparent from a cursory examination. Second, the used product could be so outdated that its performance falls far short of new gear selling for the same price—or less—than the used component. This is especially true of digital products—DACs, disc players, and music servers. Third, a used product carries no warranty; you'll have to pay for any repair work. Finally, you must ask why the person is selling the used equipment. All too often, a music lover doesn't do his homework and buys a product that doesn't satisfy musically or work well with the rest of his system. If you see lots of people selling the same product, beware—it's a sign that the product has a fundamental musical flaw. Finally, buying used equipment from a private party eliminates everything that makes your local specialty retailer such an asset when selecting equipment: You don't get the dealer's expert opinions, home audition, trade or upgrade policy, dealer setup, warranty, dealer service, loaner units, or any of the other benefits you get from buying new at a dealer.

Approach used components with caution; they can be a windfall—or a nightmare.

Product Upgrades

Many manufacturers improve their products and offer customers the option of upgrading their components to current performance. This is the "Mark II" designation on some components. The dealer can usually handle sending your component back to the factory. Some manufacturers prefer to deal with the customer directly, saving the dealer markup and keeping the upgrade price lower.

I have ambivalent feelings about manufacturer upgrades. Some upgrade programs provide lasting value to a company's customers. Enjoying the benefits of a company's advances without selling a component at a loss and buying a new one can be a wonderful bonus. Conversely, some manufacturers think of an upgrade program as a profit center, charging large amounts for even minor improvements. Consider a company's track record when evaluating potential future upgrades. For products that are upgradeable via software updates, ask how much these revisions will cost. Some manufacturers charge nothing, or a nominal fee, for updated software.

Component Selection Summary

When choosing a high-end system or component, follow these ten guidelines:

1) Establish your budget. Buy a component or system you'll be happy with in the long run, not one that will "do" for now. Do it right the first time.

2) Be an informed consumer—learn all you can about high-end audio. Study magazine reviews, visit your local specialty retailer, and read the rest of this book. Do your homework.

3) Develop a relationship with your dealer. He can be the best source of information in choosing components and assembling a system.

4) Find components that work synergistically. Again, your dealer knows his products and can offer suggestions.

5) Select products based on their musical qualities—not technical performance, favorable reviews, specifications, price, or brand name.

6) Choose carefully; many lower-priced components can outperform higher-priced ones. Take your time and maintain high standards—there are some great bargains out there.

7) Buy products from companies with good reputations for value, customer service, and reliability. Also, match the company's product philosophy (i.e., cost-no-object vs. best value for the money) to your needs.

8) When possible, listen to prospective components in your system at home before buying.

9) Follow the setup guidelines in Chapters 14 and 15 to get the most from your system. Enlist the aid of your dealer in system setup.

10) Add accessories *after* your system is set up.

If you read the rest of this book, subscribe to one or more reputable high-end magazines, and follow these guidelines, you'll be well on your way to making the best purchasing decisions—and having high-quality music reproduction in your home.

One last piece of advice: After you get your system set up, forget about the hardware. It's time to start enjoying music.

3 Becoming a Better Listener

Critical listening—the practice of evaluating the quality of audio equipment by careful analytical listening—is very different from listening for pleasure. The goal isn't to enjoy the musical experience, but to determine if a system or component sounds good or bad, and what *specific characteristics* of the sound make it good or bad. You want to critically examine what you're hearing so that you can form judgments about the reproduced sound. You can then use this information to evaluate and choose components, and to fine-tune a system for greater musical enjoyment.

Evaluating audio equipment by ear is essential—today's technical measurements simply aren't advanced enough to characterize the musical performance of audio products. The human hearing mechanism is vastly more sensitive and complex than the most sophisticated test equipment now available. Though technical performance is a valid consideration when choosing equipment, the ear should always be the final arbiter of good sound.

Moreover, the musical significance of sonic differences between components can only be judged subjectively. This is best expressed by Michael Polanyi in his book, *Personal Knowledge*:

Whenever we find connoisseurship operating within science or technology we may assume it persists only because it has not been possible to replace it by a measurable grading.

Because connoisseurship is used in the evaluation of audio products, the process is more an aesthetic endeavor than a purely technical one. Good technical performance can contribute to high-quality musical performance, but it doesn't tell you what you really want to know: how well the product communicates the musical message. To find that out, you must listen. I have auditioned hundreds of audio products for review and measured their technical performance in a test laboratory. My experience overwhelmingly indicates that much more about the quality of an audio component can be learned in the listening room than in the test lab.

Many newcomers to high-performance music reproduction—and even a small fringe group of experienced audiophiles—question the need for listening to evaluate products. They believe that measurements can tell them everything they need to know about a product's performance. And since these measurements are purely "objective," why interject human subjectivity through critical listening?

The answer is that the common measurements in use today were created decades ago as design tools, not as descriptors of sound quality. The test data generated by a typical mix of audio measurements were never meant to be a representation of musical reality, only a rough guide when designing. For example, an amplifier circuit that had 1% harmonic distortion was probably better than one with 10% harmonic distortion. It doesn't follow that a harmonic distortion specification in any way describes the sound of that amplifier.

A second problem is that audio test-bench measurements attempt to quantify a variety of two-dimensional phenomena: how much distortion the product introduces at a given signal level, its frequency response, noise level, and other factors. But music listening is a *three*-dimensional experience that is much more complex than any set of numbers can hope to quantify. How can you reduce to a series of mathematical symbols the ability of one power amplifier, and not another, to make the hair on your arms stand up? Or the feeling that a vocalist is singing directly *to you*?

I once read that a motorcycle manufacturer attempted to quantify the "feel" of prototype motorcycles by attaching hundreds of sensors at many positions around the motorcycle and putting the motorcycle on a dynamometer in a wind tunnel. The large amount of data collected was then processed to guide the engineers in designing improvements in the motorcycle's handling and overall riding experience. After investing a significant amount of money in the project, the company went back to relying on the comments of experienced test riders who could describe the experience, in "subjective" terms, of how the motorcycle "felt."

Judging music-reproduction equipment is no different. No matter how many measurements are gathered about the product's technical performance, they still don't tell you how well that product communicates the music. If I had to choose between two unknown CD players as my main source of music for the next five years, I'd rather have ten minutes with each player in the listening room than ten hours with each in the test lab. Today's measurements are crude tools that are inferior to the most powerful test instrument ever devised: the human brain.

A type of vacuum-tubed amplifier called "single-ended triode" (described in Chapter 5) illustrates the limitations of measurement to describe the musical attributes of audio components. All single-ended triode amplifiers have technical performance that is, by any standard, laughably bad. They have high distortion, absurdly low output power (often less than ten watts per channel), and can drive very few loudspeakers. Nonetheless, single-ended triode amplifiers can have a directness of musical communication that must be heard to be believed. Clearly, there are qualities of the audio signal beyond the reach of existing measurements.

Richard Heyser, one of the greatest conceptual thinkers in audio, created a stunningly simple device that revealed the fallacy of relying solely on measurements to judge the quality of audio equipment. The device was a small box with an input jack on one end and an output jack on the other end. When connected to test equipment, the box measured perfectly; it had no distortion, no frequency-response aberrations, and no noise. Indeed, the measurements suggested that all the box contained was a piece of wire connecting input to output.

But when you listened to music through the box, the sound was so distorted it was nearly unintelligible. The secret to the Heyser Box is simple: when the box detected the pure, single-frequency sinewaves used in audio testing, a relay inside the box simply connected the input to the output. But when it detected a music signal, with its complex waveform, the relay was repeatedly opened and closed quickly so that the output "chattered."

The Heyser Box brilliantly illustrates the danger of relying purely on measurement rather than listening when evaluating audio equipment performance. Moreover, the Heyser Box illustrates the fallacy of testing products in ways that bear no relationship to how those products will be used in the real world.

I must point out, however, that some sophisticated audio designers have developed tests and measurements that are far more advanced than the decades-old parameters routinely used in the audio industry. These advanced measurements can, in some cases, reliably predict specific aspects of audible performance. As their creators are the first to admit, these measurements don't quantify the overall musical experience, nor are they a replacement for critical listening. The measurements are instead tools that help the designer more quickly or more reliably achieve some specifically sought sonic attribute in a product (less treble hardness, for example). Unfortunately, designers consider these sophisticated techniques trade secrets and don't share them with the audio engineering community at large.

Given a massive research budget, it may one day be possible to more reliably predict sound quality through measured performance. But that day is a long way off. Until then, we must listen.

Knowing what sounds good and what doesn't is easy; virtually anyone can tell the difference between excellent and poor sound. But discovering *why* a product is musically satisfying or not, and the ability to recognize and describe subtle differences in sound quality, are learned skills. Like all skills, that of critical listening improves with practice: The more you listen, the better a listener you'll become. As your ear improves, you'll be able to distinguish smaller and smaller differences in reproduced sound quality—and be able to describe *how* two presentations are different, and why one is better.

This chapter defines the language of critical listening, describes what to listen for, and outlines the procedures for setting up valid listening comparisons. It will either get you started in critical listening, or help you become a more highly skilled listener.

Audiophile Values

A general discussion of audiophile values is important in understanding the next sections of this chapter. Here are some broad statements about what distinguishes good from superlative sound quality, and audiophile value in reproduced music.

Good sound is only a means to the end of musical satisfaction; it is not the end itself. If a neighbor or colleague invites you over to hear his hi-fi system, you can tell immediately whether he's a music lover or a "hi-fi buff" more interested in sound than in music. If he plays the music very loud, then turns it down after 30 seconds to seek your opinion (approval), he's probably not a music lover. If, however, he sits you down, asks what kind of music *you* like, plays it at a reasonable volume, and says or does nothing for the next 20 minutes while you both listen, it's likely that this person holds audiophile values or simply cares a lot about music.

In the first example, the acquaintance tried to impress you with *sound*. In the second case, your friend also wanted to impress you with his system, but by its ability to express the *music*, not shake the walls. This is the fundamental difference between "hi-fi enthusiasts" and music lovers.

(You can use the same test to immediately tell what kind of hi-fi store you're in. If anyone pulls out a CD of trains, sonic booms, Shuttle launches, or jet takeoffs, run for cover.)

I've noticed an unusual trend when playing my system for friends and acquaintances not involved in audio. I sit them down in the "sweet spot" (the seating position that provides the best sound) and put on some music I think they'll like. Rather than sitting there for the whole piece or song, they tend to jump up immediately and tell me how good it sounds. They've apparently been conditioned by the kind of razzle-dazzle demonstrations in some hi-fi stores or friends' houses. When *you* begin listening—at someone's house or a dealer's showroom—don't feel the need to express an opinion about the sound. Sit and listen attentively with eyes closed, letting the *music*—not the sound—tell you how good the system is.

When listening in a group, don't be swayed by others' opinions. If they're skilled listeners, try to understand what they're talking about. Listen to their descriptions and compare their impressions with your own. In fact, this is the best way to recognize the specific sonic characteristics we'll talk about later in this chapter. But don't go along with what everyone else says. If you don't hear a difference between two digital cables, for example, don't be afraid to say so. In addition, you should be completely truthful when asked for an opinion about a system. If the sound is bad, say so.

All audio components affect the signal passing through them. Some products add artifacts (distortion) such as a *grainy treble* or a *lumpy bass*. Others subtract parts of the signal—for example, a loudspeaker that doesn't go very low in the bass. (Listening terms are defined later in this chapter.) A fundamental audiophile value holds that sins of commission (adding something to the music) are far worse than sins of omission (removing something from the music). If parts of the music are missing, the ear/brain system subconsciously fills in what isn't there; you can still enjoy listening. But if the playback system adds an artificial character to the sound, you are constantly reminded that you're hearing a reproduction and not the real thing.

Let's illustrate this sins-of-commission/omission dichotomy with two loudspeakers. The first loudspeaker—a three-way system with a 15" woofer in a very large cabinet—sells for a moderate price in a mass-market appliance store; it plays loudly and develops lots of bass. The second loudspeaker sells for about the same price, but is a small two-way system with a 6" woofer. It doesn't play nearly as loudly, and produces much less bass. While you need a refrigerator dolly to move the first loudspeaker, you can almost hold the second loudspeaker in your outstretched hand.

The behemoth loudspeaker has some problems: The bass is boomy, thick, and overwhelming. All the bass notes seem to have the same pitch. The very prominent treble is coarse and grainy, and the midrange has a big peak of excess energy that makes singers sound as if they have colds.

The small loudspeaker has no such problems. The treble is smooth and clear, and the midrange is pure and open. It has, however, very little bass by comparison, won't play very loudly, and doesn't produce a physical sensation of sound hitting your body.

The first loudspeaker commits sins of commission, *adding* unnatural artifacts to the sound. The bass peaks that make it sound boomy, the grain overlaying the treble, and the midrange colorations are all additive distortions.

The second loudspeaker's faults, however, are of omission. It *removes* certain elements of the music—low bass and the ability to play loudly—but leaves the remainder of the music intact. It doesn't add grain to the treble, thickness to the bass, or colorations to the midrange.

There's no doubt that the second loudspeaker will be more musically satisfying. The first loudspeaker's additive distortions are not only much more musically objectionable, they also constantly remind you that you are listening to artificially reproduced music. The second loudspeaker's flaws are of a nature that allows you to forget that you're listening to loudspeakers. In the reproduction of music, addition is far worse than subtraction.

Another audiophile value holds that even small differences in the quality of the musical presentation are important. Because music matters to us, we get excited by *any* improvement in sound quality. Moreover, there isn't a linear relationship between the magnitude of a sonic difference and its musical significance. A quality difference can be sonically small but musically large.

While reviewing a revelatory new state-of-the-art digital-to-analog converter, I listened to a piece of music I'd heard hundreds of times before. The piece, performed by a five-member group, had vocals and very long instrumental breaks. During the instrumental breaks, the vocalist played percussion instruments. Through lesser-quality DACs, the percussion had always been just another sound fused into the music's tapestry; I'd never heard it as a separate instrument played by the vocalist. The group seemed to become a four-piece ensemble when the vocalist wasn't singing; I never heard the percussion as separate from the rest of the music.

The new DAC was particularly good at resolving individual instruments and presenting them not as just more sounds homogenized into the overall musical fabric, but as distinct entities. Consequently, when the instrumental break came, I heard the percussion as a separate, more prominent instrument. In my mind's eye, and for the first time, the vocalist never left—she remained "on stage," playing the percussion instruments. By just this "small" change in the presentation, the band went from being a quartet to a quintet during the instrumental breaks. The "objective" difference in the electrical signal must have been minuscule; the subjective musical consequences were profound.

This is why small differences in the musical presentation are important—*if* you care deeply enough about music and about how well it is reproduced. "Small" improvements can have large subjective consequences. This example highlights the inability of measurements alone to characterize audio equipment performance. Measurements on the DAC in question indicated no technical attributes that would have contributed to my perceptions. More fundamentally, how can a number representing some aspect of the DAC's technical performance begin to describe the musical significance of the change I heard?

Much of music's expression and meaning can be found in such minutiae of detail, subtlety, and nuance. When such subtlety is conveyed by the playback system, you feel a vastly deeper communication with the musicians. Their intent and expression are more vivid, allowing you to more deeply appreciate their artistry. For example, if you compare two performances of Max Reger's Sonata in D Major for solo violin—one competent, the other superlative—you could say that, on an objective basis, they were virtually identical. Both performers played the same notes at about the same tempo. The difference in expression is in the nuances—the inspired subtleties of rubato, tempo, emphasis, articulation, and dynamics that bring the performance to life and convey the piece's musical meaning and intent. This example is analogous to the difference between mediocre and superb music playback systems, and why small differences in sound quality can matter so much. High-end audio is about reproducing these nuances so that you can come one step closer to the musical expression.

The sad but universal truth about audio equipment is that, any time you put a signal into an audio component, it *never* comes out better at the other end. You therefore want to keep the signal path as simple as possible, to remove any unnecessary electronics from between you and the music. This is why inserting equalizers and other such "enhancers" into the signal chain is usually a bad idea—the less done to the signal, the better. The advent of digital technology, however, has made possible some beneficial signal processing. (An example is digital room correction, described in Chapter 14.)

Pitfalls of Becoming a Critical Listener

There are dangers inherent in developing critical listening skills. The first is an inability to distinguish between critical listening and listening for pleasure. Once started on the path of critiquing sound quality, it's all too easy to forget that the reason you're involved in audio is because you love music, and to start thinking that every time you hear music, you must have an opinion about what's right and what's wrong with the sound. This is the surest path toward a condition humorously known as *Audiophilia nervosa*. Symptoms include constantly changing equipment, playing only one track of a CD or LP at a time instead of the whole record, changing cables for certain music, refusing to listen to great music if it happens to be poorly recorded, and in general "listening to the hardware" instead of to the music.

But high-end audio is about making the hardware disappear. When listening for pleasure—which should be the vast majority of your listening time—forget about the system. Forget about critical listening. Shift into critical-listening mode only when you need to make a diagnostic judgment about the sound quality, or just for practice to become a better listener. Draw the line between critical listening and listening for pleasure—and know when to cross it and when *not* to cross it.

There is also the related danger that your standards of sound quality will rise to such a height that you can't enjoy music unless it's "perfectly" reproduced—in other words, to the point that you can't enjoy music, period. Although it's not very high-quality reproduction, I get a great deal of pleasure from my car stereo—don't let being an audiophile interfere with *your* enjoyment of music, anytime, anywhere. When you can't control the sound quality, lower your expectations.

Sonic Descriptions and their Meanings

The biggest problem in critical listening is finding words to express our perceptions. We hear things in reproduced music that are difficult to identify and put into words. A listening vocabulary is essential not only to conveying to others what we hear, but also to recognizing and understanding our own perceptions. If you can attach a descriptive name to a perception, you can more easily recognize that perception when you experience it again.

A perfect example of the bond between words and perceptions is found in Michael Polanyi's *Personal Knowledge*. His description of the radiology student's experience is identical to that of the audiophile becoming a skilled critical listener:

Think of a medical student attending a course in the X-ray diagnosis of pulmonary diseases. He watches in a darkened room shadowy traces on a fluorescent screen placed against a patient's chest, and hears the radiologist commenting to his assistants in technical language, on the significant features of these shadows. At first the student is completely puzzled. For he can see in the X-ray picture of a chest only the shadows of the heart and the ribs, with a few spidery blotches between them. The experts seem to be romancing about figments of their imagination; he can see nothing that they are talking about. Then as he goes on listening for a few weeks, looking carefully at ever new pictures of different cases, a tentative understanding will dawn on him; he will gradually forget about the ribs and begin to see the lungs. And eventually, if he perseveres intelligently, a rich panorama of significant details will be revealed to him; of physiological variations and pathological changes, of scars, of chronic infections and signs of acute disease. He has entered a new world. He still sees only a fraction of what the experts can see, but the pictures are definitely making sense now and so do most of the comments made on them. He is about to grasp what he is being taught; it has clicked. Thus, at the moment when he has learned the language of pulmonary radiology, the student will also have learned to understand pulmonary radiograms. The two can only happen together. Both halves of the problem set to us by an unintelligible text, referring to an unintelligible subject, jointly guide our efforts to solve them, and they are solved eventually together by discovering a conception which comprises a joint understanding of both the words and the things.

Just as the student learns to understand pulmonary radiograms at the moment he learns the language of pulmonary radiology, the audiophile learns to identify specific sonic characteristics when he learns the language of critical listening. The bond between words and ideas is inextricable. That's why listening with more skilled listeners is the best way to learn to identify differences yourself.

By describing in detail the specific sonic characteristics of how electronic components change the sound of music passing through them, I hope to attune you to recognizing those same characteristics when you listen. After reading this next section, listen to two products for yourself and try to hear what I'm describing. It can be any two products—if you have a portable music player, hook it up to your system and compare it to your home CD player. The important thing is to start listening analytically. If you don't hear the sonic differences immediately, keep listening. The more you listen, the more sensitive you'll become to those differences.

Each listener hears a little differently. What I listen for during critical listening may be different from another listener's sonic checklist. Moreover, the different values placed on different aspects of the sonic presentation are matters of taste. I notice this first-hand when I occasionally spend time listening critically in my listening room with visiting manufacturers and designers of high-end equipment—many of them highly skilled listeners. While we share many commonalities in determining what sounds good, there is a wide range of perception about what aspects of the presentation are most important.

For example, some listeners value correct reproduction of timbre above all else. A product may have outstanding bass, spectacular soundstaging, and clean treble, but if the timbre is synthetic or hard, those other good qualities won't matter to that listener. Another listener may find soundstaging of utmost importance. Finally, a layer of treble grain that is hardly noticed by one listener may be unacceptable to another. We all hear differently, and value different things in music reproduction. Not only that, but a single listener's musical perceptions and acuity vary with the time of day, as well as his mood and state of relaxation.

There's also evidence that people perceive soundstage information (the impression of music occupying a physical space) in two different ways. When you sit between two loudspeakers reproducing the same signal, your ears each send the same signal to your brain. You thus interpret the sound as coming from a point exactly between the loudspeakers. If the sound suddenly becomes slightly louder in the left loudspeaker, your brain concludes that the sound source is slightly left of center. This method of localization is called "intensity stereo": the information your brain uses to localize the sound is the difference in volume, or intensity, between your left and right ears.

A second method of localizing sounds and forming a soundstage in your head uses the phase (timing) information in a recording. Consider our two loudspeakers reproducing the same signal, which you hear as existing between the two speakers. If the left-channel signal is delayed slightly (a few thousandths of a second), you hear the image shift to the right. Although both loudspeakers are producing the same volume of signal, the image appears to come from the loudspeaker that produces the sound first. The two signals are said to have a different phase (timing) relationship to each other.

Some listeners are much more attuned to phase information than others. In recordings made in acoustic spaces (as opposed to recording studios), the phase information is enormously complex. As sounds reflect from the walls of the acoustic space, they create an almost infinitely dense pattern of very slight delays. It is this temporal (timing) information that cues your brain as to the depth of images, the focus of the images, and the overall size and scale of the concert hall. If a hi-fi system doesn't accurately resolve this temporal information, the soundstage is foreshortened and image placement is vague—to some listeners. Other listeners may not hear the difference between a system that resolves these spatial cues and one that does not. That isn't to say that one way of listening is better than another, only that it appears that two separate mechanisms exist.

The recording that, in my experience, is most revealing of a system's ability to correctly preserve phase information is John Rutter's *Requiem* on the Reference Recordings label. When this disc is played back through a first-rate system, one's ability to hear the placement of individual images, along with the sense of the hall's size and characteristics, is stunning. Even a slight degradation of this recording's phase information reduces these qualities. (See Appendix A, "Sound and Hearing," for more on human hearing.)

The following description of what I listen for is a reflection of my own tastes, mixed with what I've learned from other skilled listeners. You may have your own sonic priorities, but the important thing is that we all hear and describe the same characteristics with the same language. The descriptive terms presented here can be found in my and other writers' product reviews for *The Absolute Sound*. There is general agreement about what these terms mean, but shades of meaning vary among individual listeners.

You should also know that recordings made with audiophile techniques are more revealing of some aspects of reproduced sound than recordings made for mass consumption. For example, a recording of classical music made in a concert hall with very few microphones, a simple signal path, and high-quality recording equipment will likely reveal more about a component's soundstaging performance than a pop recording made in a studio. Similarly, most mass-market recordings have little low bass and almost no dynamic range so that they sound "good" on a 4" car-stereo speaker. For these reasons, some of the sonic terms described in this chapter apply much more to audiophile-quality recordings than to mass-market ones.

When I listen critically to an audio component to characterize its sonic and musical performance and form judgments about that product, I first follow the setup procedures described later in this chapter. They are vital to forming accurate impressions of products, but secondary to the following discussion.

The sonic terms and characteristics defined here apply to audio components in general. Advice on what to listen for when choosing specific components is included in Chapters 4–13. Note that we often describe differences in terms of differences in frequency response (i.e., "too much treble"). The component may have a flat (accurate) measured treble response, but the distortions it introduces give the impression of too much treble.

It's also useful to understand the broad terms that describe the audio frequency band. The range of human hearing, which spans ten octaves from about 16Hz (cycles per second) to 20,000Hz, or 20 kilohertz (20kHz), can be divided into the specific regions described below. Note that these divisions are somewhat arbitrary; you can't say specifically that the lower treble begins at 2000Hz and not 2500Hz, for example. The table nonetheless provides a rough guideline for understanding the relationship between frequency ranges and their descriptive names.

Frequency Range		
Lower Limit	**Upper Limit**	**Description**
16	40	Deep Bass
40	100	Midbass
100	250	Upper Bass
250	500	Lower Midrange
500	1000	(Middle) Midrange
1000	2000	Upper Midrange
2000	3500	Lower Treble
3500	6000	Middle Treble
6000	10,000	Upper Treble
10,000	20,000	Top Octave

This rough guide will help you understand the following terms and definitions. A full characterization of how a product "sounds" will include aspects of each of the following sonic qualities.

Tonal Balance

The first aspect of the musical presentation to listen for is the product's overall *tonal balance.* How well balanced are the bass, midrange, and treble? If it sounds as though there is too much treble, we call the presentation *bright*. The impression of too little treble produces a *dull* or *rolled-off* sound. If the bass overwhelms the rest of the music, we say the presentation is *heavy* or *weighty*. If we hear too little bass, we call the presentation *thin, lightweight, uptilted,* or *lean*.

A product's tonal balance is a significant—and often overwhelming—aspect of its sonic signature.

Overall Perspective

The term *perspective* describes the apparent distance between the listener and the music. Perspective is largely a function of the recording (particularly the distance between the performers and the microphones), but is also affected by components in the playback system. Some products push the presentation *forward,* toward the listener; others sound more *distant,* or *laid-back.* The forward product presents the music in front of the loudspeakers; the laid-back product makes the music appear slightly behind the loudspeakers. Put another way, the forward product sounds as though the musicians have taken a few steps toward you; the laid-back product gives the impression that the musicians have taken a few steps back.

Another way of describing perspective is by row number in a concert hall. Some products seem to "seat" the listener at the front of the hall—in Row D, say. Others give you the impression that you're sitting farther back; say, in Row S.

Several other terms describe perspective. *Dry* generally means lacking reverberation and space, but can also apply to a forward perspective. Other watchwords for a forward presentation are *immediate, incisive, vivid, palpable,* and *present.* Terms associated with laid-back include *relaxed, easygoing,* and *gentle.*

Products with a forward presentation produce a greater sense of an instrument's *presence* before you, but can quickly become fatiguing. Conversely, if the presentation is too laid-back, the music is uninvolving and lacking in immediacy. If the product under evaluation distorts the perspective captured during the recording, I prefer that it err in the direction of being laid-back. When a product is overly immediate, I feel as though the music is coming at me, assaulting my ears. The reaction is to close up, to try to keep the music at arm's length. The ideal midrange presentation has a sense of *palpable presence,* that feeling of the musicians existing right before you, but without sounding aggressive.

Conversely, a laid-back presentation invites the listener in, pulling her gently forward *into* the music, allowing her the space to explore its subtleties. It's like the difference between having a conversation with someone who is aggressive, gets in your face, and talks too loudly, compared with someone who stands back, speaking quietly and calmly.

In loudspeakers, perspective is often the result of a peak or dip in the midrange (a peak is too much energy, a dip is too little). In fact, the midrange between 1kHz and 3kHz is called the *presence region* because it provides a sense of presence and immediacy. The harmonics of the human voice span the presence region; thus, the voice is greatly affected by a product's perspective.

Untrained listeners who can't specifically identify whether a reproduced musical presentation is forward or laid-back will feel a tension that usually translates into a desire to turn down the music if the presentation is too forward. Conversely, a laid-back presentation can make the music uninvolving rather than riveting.

Note that the terms associated with overall perspective can be used to describe specific aspects of the presentation (such as the treble) in addition to describing the overall perspective. If we say the treble is forward, we mean that it is overly prominent, sounding as if it is closer to the listener than the rest of the music.

The Treble

Good treble is essential to high-quality music reproduction. In fact, many otherwise excellent audio products fail to satisfy musically because of poor treble performance.

The treble characteristics we want to avoid are described by the terms *bright, tizzy, forward, aggressive, hard, brittle, edgy, dry, white, bleached, wiry, metallic, sterile, analytical, screechy*, and *grainy*. Treble problems are pervasive; look how many adjectives we use to describe them.

If a product has too much apparent treble, it overstates sounds that are already rich in high frequencies. Examples are overemphasized cymbals, excessive sibilance (*s* and *sh* sounds) in vocals, and violins that sound thin. A product with too much apparent treble is called *bright*. Brightness is a prominence in the treble region, primarily between 3kHz and 6kHz. Brightness can be caused by a rising frequency response in loudspeakers, or by poor electronic design. Many digital sources and solid-state amplifiers that measure as having a flat (accurate) frequency response nevertheless add prominence to the treble.

Tizzy describes too much upper treble (6kHz–10kHz), characterized as a *whitening* of the treble. Tizzy cymbals have an emphasis on the upper harmonics, the sizzle and air that rides over the main cymbal sound. Tizziness gives cymbals more of an *ssssss* than a *sssshhhh* sound.

Forward, if applied to treble, is very similar to *bright*; both describe too much treble. A forward treble, however, also tends to be dry, lacking space and air around it.

Many of the terms listed above have virtually identical meanings. *Hard, brittle*, and *metallic* all describe an unpleasant treble characteristic that reminds one of metal being struck. In fact, the unique harmonic structure created from the impact of metal on metal is very similar to the distortion introduced by a power amplifier when it is asked to play louder than it is capable of playing.

I find the sound of the alto saxophone to be a good gauge of hard, brittle, and metallic treble, particularly lower treble. If reproduced incorrectly, sax can take on a thin, reedy, very unpleasant tone. The antithesis of this sound is *rich, warm*, and *full*. When the sax's upper harmonics are reproduced with a metallic character, the whole instrument's sound collapses. Interestingly, the sound of the saxophone has the most complex harmonic structure of any instrument. It's no wonder that it is so revealing of treble problems.

White and *bleached* have meanings very similar to *bright*, but I associate them more with a thinness in the treble, often caused by a lack of energy (or what sounds like a lack of energy) in the upper midrange. With no supporting harmonic structure beneath it, the treble becomes threadbare and thin, much like an overexposed photograph. Cymbals should have a gong-like low-frequency component with a sheen over it. If cymbals sound like bursts of white noise (the sound you hear between radio stations), what you're probably hearing is a *white* or *bleached* sound.

A particularly annoying treble characteristic is *graininess*. Treble grain is a coarseness overlaying treble textures. I notice it most on solo violin, massed violins, flute, and female voice. On flute, treble grain is recognizable as a rough or fuzzy sound that seems to ride on top of the flute's dynamic envelope. (That is, the grain follows the flute's volume.) Grain makes violins sound as though they're being played with hacksaw blades rather than bows—a gross exaggeration, but one that conveys the idea of the coarse texture added by grain.

Treble grain can be of any texture, from very fine to coarse and rough. (Think of the difference between 400-grit and 80-grit sandpaper.) The more coarse the grain, the more objectionable it is. The preceding discussion of grain applies in even larger measure to midrange textures, which I'll discuss later.

Treble problems can foster the interesting perception that the treble isn't integrated into the music's harmonic tapestry, but is riding on top of it. The top end seems somehow separate from the music, not an integral part of the presentation. When this happens, we are aware of the treble as a distinct entity, not as just another aspect of the music. The treble should sound like an extension of the upper midrange rather than separate from it. If the treble calls attention to itself, be suspicious.

The most common sources of these problems are, in rough order of descending magnitude: tweeters in loudspeakers, overly reflective listening rooms, digital source components, preamplifiers, power amplifiers, cables, and dirty AC power sources.

So far, I've discussed only problems that emphasize treble. Some products tend to make the treble softer and less prominent than live music. This characteristic is often designed into the product, either to compensate for treble flaws in other components in the system, or to make the product sound more palatable. Deliberately softening the treble is the designer's shortcut; if he can't get the treble right, he just makes it less offensive by softening it.

The following terms, listed in order of increasing magnitude, describe good treble performance: *smooth, sweet, soft, silky, gentle, liquid,* and *lush*. When the treble becomes overly smooth, we say it is *romantic, rolled-off,* or *syrupy*. A treble described as "smooth, sweet, and silky" is being complimented; "rolled-off and syrupy" suggests that the component goes too far in treble smoothness, and is therefore *colored*.

A rolled-off and syrupy treble may be blessed relief after hearing bright, hard, and grainy treble, but it isn't musically satisfying in the long run. Such a presentation tends to become *bland, uninvolving, slow, thick, closed-in,* and *lacking detail*. All these terms describe the effects of a treble presentation that errs too far on the side of smoothness. The presentation will lack *life, air, openness, extension,* and a sense of *space* if the treble is too soft. The music sounds closed-in rather than being big and open.

Top-octave air describes a sense of almost unlimited treble extension, which fosters the impression of hearing the air in which the music exists. A slight treble rolloff (a reduction in the amount of treble) in loudspeakers, for example, can diminish top-octave air. Loss of top-octave air is also associated with an opaque soundstage.

The best treble presentation is one that sounds most like real music. It should have lots of energy—cymbals can, after all, sound quite aggressive in real life—yet not have a synthetic, grainy, or dry character. We don't hear these characteristics in live music; we shouldn't hear them in reproduced music. More important, the treble should sound like an integral part of the music, not a detached noise riding on top of it. If a component has a colored treble presentation, however, it is less musically objectionable if it errs on the side of smoothness rather than brightness.

The Midrange

J. Gordon Holt, *Stereophile* magazine founder and the father of subjective audio equipment evaluation, once wrote, "If the midrange isn't right, nothing else matters."

The midrange is important for several reasons. First, most of the musical energy is in the midrange, particularly the important lower harmonics of most instruments. Not only does this region contain most of the musical energy, but the human ear is much more sensitive to midrange and lower treble than to bass and upper treble. Specifically, the ear is most sensitive to sounds between about 800Hz and 3kHz, and to small changes in both low volume and frequency response within this band. The ear's threshold of hearing—i.e., the softest sound we can hear—is dramatically lower in the midband than at the frequency extremes. We've developed this additional midband acuity probably because the energy of most of the sounds we heard every day for thousands of years—the human voice, rustling leaves, the sounds of other animals—is concentrated in the midrange.

Midrange colorations can be extremely annoying. Loudspeakers with peaks and dips in the mids sound very unnatural; the midrange is absolutely the worst place for loudspeaker imperfections. Confining our discussion to loudspeakers for the moment, midrange colorations overlay the music with a common characteristic that emphasizes certain sounds. The male speaking voice is particularly revealing of midrange anomalies, which are often described by comparisons with vowel sounds. A particular coloration may impart an *aaww* sound; a coloration lower in frequency may emphasize *ooohhh* sounds; a higher-pitched coloration may sound like *eeeee*; another coloration might sound *hooty*.

Some midrange colorations can be likened to the sound of someone speaking through cupped hands. Try reading this sentence while cupping your hands around your mouth. Open and close your hands while listening to how the sound of your voice changes. That's the kind of midrange coloration we sometimes hear from loudspeakers—particularly mass-market ones.

In short, if recordings of male speaking voice sound monotonous, tiring, and resonant, it's probably the result of peaks and dips in the loudspeaker's frequency response. (These colorations are most apparent on male voice when listening to just one loudspeaker.)

Terms to describe poor midrange performance include *peaky, colored, chesty, boxy, nasal, congested, honky,* and *thick*. *Chesty* describes a lower-midrange coloration that makes vocalists sound as though they have colds. *Boxy* refers to the impression that the sound is coming out of a box instead of existing in open space. *Nasal* is usually associated with an excess of energy that spans a narrow frequency range, producing a sound similar to talking with your nose pinched. *Honky* is similar to nasal, but higher in frequency and spanning a wider frequency range.

Loudspeaker design has progressed so much in the past ten years that horrible midrange colorations are largely a thing of the past (at least in loudspeakers with high-end aspirations). Still, colorations persist—particularly in inexpensive loudspeakers—though they tend to be more subtle.

One type of lower-midrange coloration that still afflicts even moderately priced high-end loudspeakers is caused by vibration of the loudspeaker's cabinet. The loudspeaker will resonate at certain frequencies, launching acoustic energy from the cabinet when those frequencies are excited by certain notes in the music. Cabinet resonances are heard as certain notes "sticking out" or "jumping forward." This problem is clearly audible on solo piano; ascending or descending left-hand lines reveal that certain notes possess a character clearly different from the rest of the instrument's notes. These problem notes jump out because they have more energy; some of the sound is being produced by the loudspeaker cabinet, and not just by the loudspeaker's drive units (the cones that move back and forth to create sound). The same thing happens with musical instruments; the musician's expression "wolf tone" describes the same phenomenon. This problem also afflicts the upper bass. In Chapter 6 I talk in much greater detail about loudspeaker cabinet vibration.

As described previously under "Perspective," too much midrange energy can make the presentation seem forward and "in your face." A broad dip in the midrange response (too little midrange energy over a wide frequency span) can give an impression of greater distance between you and the presentation.

When choosing loudspeakers, be especially attuned to the midrange colorations described. What is a very minor—even barely noticeable—problem heard during a brief audition can turn into a major irritant over extended listening.

The preceding descriptions apply primarily to midrange problems introduced by loudspeakers. Expanding the discussion to include electronics (preamps and power amps) and source components (LP playback or a digital source) introduces different aspects of midrange performance that we should be aware of.

An important factor in midrange performance is how instrumental textures are reproduced. Texture is the physical impression of the instrument's sound—its fabric rather than its tone. The closest musical term for texture is timbre, defined by *Merriam Webster's Collegiate Dictionary, Tenth Edition* as "the quality given to a sound by its overtones; the quality of tone distinctive of a particular singing voice or instrument." Sonic artifacts added by electronics often affect instrumental and vocal textures.

The term *grainy,* introduced in the description of treble problems, also applies to the midrange. In fact, midrange grain can be more objectionable than treble grain. Midrange grain is characterized by a coarseness of instrumental and vocal textures; the instrument's texture is granular rather than smooth.

Midrange textures can also sound hard and brittle. Hard textures are apparent on massed voices; a choir sounds *glassy, shiny,* and *synthetic.* This problem gets worse as the choir's volume increases. At low levels, you may not hear these problems. But as the choir swells, the sound becomes hard and irritating. Piano is also very revealing of hard midrange textures, the higher notes sounding annoyingly *brittle.* When the midrange lacks these unpleasant artifacts, we say the textures are *liquid, smooth, sweet, velvety,* and *lush.*

A related midrange problem is *stridency.* I think of stridency as a combination of thinness (lack of warmth), hardness, and forwardness, all in the midrange. Strident vocals emphasize mouth noises and overlay the texture with a whitish grain. A saxophone can thus sound thin and reedy, though in a different way from that described as treble grain. A strident sax is hard, more forward in the mids, and granular in texture. Stridency can be caused by an apparent thinness in the lower midrange that makes the upper midrange too prominent. A product described as "strident" has been severely criticized.

Many of these midrange and treble problems are grouped together under the term *harshness.*

Other midrange characteristics affect such areas as clarity, transparency, and detail. These are discussed later, under more specific categories.

The Bass

Bass performance is the most misunderstood aspect of reproduced sound, among the general public and hi-fi buffs alike. The popular belief is that the more bass, the better. This is reflected in ads for "subwoofers" that promise "earthshaking bass" and the ability to "rattle pant legs and stun small animals." The ultimate expression of this perversity is boom trucks that have absurd amounts of extraordinarily bad bass reproduction.

But we want to know how the product reproduces *music,* not earthquakes. What matters to the music lover isn't *quantity* of bass, but the *quality* of that bass. We don't just want the physical feeling that bass provides; we want to hear subtlety and nuance. We want to hear precise pitch, lack of coloration, and the sharp attack of plucked acoustic bass. We want to hear every note and nuance in fast, intricate bass playing, not a muddled roar. If Ray Brown, Stanley Clarke, John Patitucci, Dave Holland, or Eddie Gomez is working out, we want to hear *exactly* what they're doing. In fact, if the bass is poorly reproduced, we'd rather not hear much bass at all.

Correct bass reproduction is essential to satisfying musical reproduction. Low frequencies constitute music's tonal foundation and rhythmic anchor. Unfortunately, bass is difficult to reproduce, whether by source components, power amplifiers, or—especially—loudspeakers and rooms.

Perhaps the most prevalent bass problem is lack of *pitch definition* or *articulation.* These two terms describe the ability to hear bass as individual notes, each having an attack, a decay, and a specific pitch (these terms are explained in Appendix A). You should hear the texture of the bass, whether it's the sonorous resonance of a bowed double bass or the unique character of a Fender Precision. Low frequencies contain a surprising amount of detail when reproduced correctly.

When the bass is reproduced without pitch definition and articulation, the low end degenerates into a dull roar underlying the music. You hear low-frequency content, but it isn't musically related to what's going on above it. You don't hear precise notes, but a blur of sound—the dynamic envelopes of individual instruments are completely lost. In music in which the bass plays an important rhythmic role—rock, electric blues, and some jazz—the bass guitar and kick drum seem to lag behind the rest of the music, putting a drag on the rhythm. Moreover, the kick drum's dynamic envelope (what gives it the sense of sudden impact) is buried in the bass guitar's sound, obscuring its musical contribution. These conditions are made worse by the common mid-fi affliction of too much bass.

Terms descriptive of this kind of bass include *muddy, thick, boomy, bloated, tubby, soft, fat, congested, loose,* and *slow.* Terms that describe excellent bass reproduction include *taut, quick, clean, articulate, agile, tight,* and *precise.* Good bass has been likened to a trampoline stretched taut; poor bass is a trampoline hanging slackly.

The amount of bass in the musical presentation is very important; if you hear too much, the music is overwhelmed. Excessive bass is a constant reminder that you're listening to reproduced music. This overabundance of bass is described as *heavy.*

If you hear too little bass, the presentation is *thin, lean, threadbare,* or *overdamped.* An overly lean presentation robs music of its rhythm and drive—the full, purring sound of bass guitar is missing, the depth and majesty of double bass or cello are gone, and the orchestra loses its sense of power. Thin bass makes a double bass sound like a cello, a cello like a viola. The rhythmically satisfying weight and impact of bass drum are reduced to shadows of their former power. Instruments' harmonics are emphasized in relation to the fundamentals, giving the impression of well-worn cloth that's lost its supporting structure. A thin or lean presentation lacks *warmth* and *body.* As described earlier in this chapter in the discussion of audio sins of commission and omission, an overly lean bass is preferable to a fat and boomy bass.

Two terms related to what I've just described about the quantity of bass are *extension* or *depth.* Extension is how deep the bass goes—not the bass and upper bass described by *lean* or *weighty,* but the very bottom end of the audible spectrum. This is the realm of kick drum and pipe organ. All but the very best systems *roll off* (reduce in volume) these lowermost frequencies. Fortunately, deep extension isn't a prerequisite to high-quality music reproduction. If the system has good bass down to about 35Hz, you don't feel that much is missing. Pipe-organ enthusiasts, however, will want deeper extension and are willing to pay for it. Reproducing the bottom octave correctly can be very expensive.

Just as colorations caused by peaks and dips in frequency response can make the midrange unnatural, bass can be colored by peaks and dips. Bass colorations create a monotonous, droning characteristic that quickly becomes tiring. In the most extreme example, *one-note bass,* the bass seems to have only one pitch. This impression is created by a large peak in the system's frequency response at a specific frequency. This pitch is then reproduced more loudly than other pitches. One-note bass is also described as being *thumpy*. Ironically, this undesirable condition is *maximized* in boom trucks. The playback system is tuned to put out all its energy at one frequency for maximum physical impact. The drivers of those vehicles don't seem to care that they're losing the wealth of musical information conveyed by the bass.

Many of the terms used to describe midrange colorations actually apply to the upper bass. *Chesty, thick,* and *congested* are all useful in describing colored bass reproduction. Bass lacking these colorations is called *smooth* or *clean*.

Much of music's dynamic power—the ability to convey wide differences between loud and soft—is contained in the bass. Though I'll discuss dynamics later in this section, bass dynamics bear special discussion—they are that important to satisfying music reproduction.

A system or component that has excellent bass dynamics will provide a sense of sudden impact and explosive power. Bass drum will jump out of the presentation with startling power. The dynamic envelope of acoustic or electric bass is accurately conveyed, allowing the music full rhythmic expression. We call these components *punchy,* and use the terms *impact* and *slam* to describe good bass dynamics.

A related aspect is *speed,* though, as applied to bass, "speed" is somewhat of a misnomer. Low frequencies inherently have slower attacks than higher frequencies, making the term technically incorrect. But the *musical* difference between "slow" and "fast" bass is profound. A product with fast, tight, punchy bass produces a much greater rhythmic involvement with the music. (This is examined in more detail later.)

Although reproducing the sudden attack of a bass drum is vital, equally important is a system's ability to reproduce a fast decay; i.e., how a note ends. The bass note shouldn't continue after a drum whack has stopped. Many loudspeakers store energy in their mechanical structures and radiate that energy slightly after the note itself. When this happens, the bass has *overhang,* a condition that makes kick drum, for example, sound *bloated* and *slow*. Music in which the drummer used double bass drums is particularly revealing of bass overhang. If the two drums merge into a single sound, overhang is probably to blame. You should hear the attack and decay of each drum as distinct entities. Components that don't adequately convey the sudden dynamic impact of low-frequency instruments rob music of its power and rhythmic drive.

Soundstaging

Soundstaging is the apparent physical size of the musical presentation. When you close your eyes in front of a good playback system, you can "see" the instrumentalists and singers before you, often existing within an acoustic space such as a concert hall. The soundstage has the physical properties of *width* and *depth*, producing a sense of great size and space in the listening room. Soundstaging overlaps with *imaging*, or the way instruments appear as objects hanging in three-dimensional space within the recorded acoustic. As mentioned previously in this chapter, a large and well-defined soundstage is most often heard when playing audiophile-grade recordings made in a real acoustic space such as a concert hall or church.

The most obvious descriptions of the soundstage are its physical dimensions—width and depth. You hear the musical presentation as existing beyond the left and right loudspeaker boundaries, and extending farther away from you than the wall behind the loudspeakers.

Of all the ways music reproduction is astounding, soundstaging is without question the most miraculous. Think about it: The two loudspeakers are driven by two-dimensional electrical signals that are nothing more than voltages that vary over time. From those two voltages, a huge, three-dimensional panorama unfolds before you. You don't hear the music as a flat canvas with individual instruments fused together; you hear the first violinist to the left front of the presentation, the oboe farther back and toward the center, the brass behind the basses on the right, and the tambourine behind all the other instruments at the very rear. The sound is made up of *individual objects* existing within a space, just as you would hear at a live performance. Moreover, you hear the oboe's timbre coming from the oboe's position, the violin's timbre coming from the violin's position, and the hall reverberation surrounding the instruments. The listening room vanishes, replaced by the vast space of the concert hall—all from two voltages.

A soundstage is created in the brain by the time and amplitude differences encoded in the two audio channels. When you hear instrumental images toward the rear right of the soundstage, the ear/brain is synthesizing those aural images by processing the slightly different information in the two signals arriving at your ears. Visual perception works the same way: there is no depth information present on your retinas; your brain extrapolates the appearance of depth from the differences between the two flat images.

Audio components vary greatly in their abilities to present these spatial aspects of music. Some products shrink soundstage width and shorten the impression of depth. Others reveal the glory of a fully developed soundstage. I find good soundstage performance crucial to satisfying musical reproduction. Unfortunately, many products destroy or degrade the subtle cues that provide soundstaging.

Terms descriptive of poor soundstage width are *narrow* and *constricted*—the music, squeezed together between the loudspeakers, does not envelop the listener. A soundstage lacking depth is called *flat, shallow*, or *foreshortened*. Ideally, the soundstage should maintain its width over its entire depth. A soundstage that narrows toward the presentation's rear robs the music of its size and space.

The illusion of soundstage depth is aided by resolution of low-level spatial cues such as hall reflections and reverberation. In particular, the reverberation decay after a loud climax followed by a rest helps define the acoustic space. The loud signal is like a flash of light in a dark room; the space is momentarily illuminated, allowing you to see its dimensions and characteristics.

To produce a realistic impression of a real instrument in a real acoustic space, the reverberation and hall sound must be distinct from the image itself. Better audio components place the image within, rather than superimposed on, the recorded acoustic. Poor-quality components don't resolve these spatial cues; they shorten soundstage depth, truncate reverberation decay, and fuse the reverberation into the instrumental images. When this happens, an audio system's ability to transport you to the original acoustic space is diminished.

Now that we've covered space and depth, let's discuss how the instrumental images appear within this space. Images should occupy a specific spatial position in the soundstage. The sound of the bassoon, for example, should appear to emanate from a specific point in space, not as a diffuse and borderless image. The same could be said for guitar, piano, sax, or any other instrument in any kind of music. The lead vocal should appear as a tight, compact, definable point in space exactly between the loudspeakers. Some products, particularly large loudspeakers, distort image size by making every instrument seem larger than life—a classical guitar suddenly sounds ten feet wide. A playback system should reveal somewhat correct image size, from a 60′-wide symphony orchestra to a solo violin. I say "somewhat" because it is impossible to re-create the correct spatial perspectives of such widely divergent sound sources through two loudspeakers spaced about 8′ apart. Although image size and placement are characteristics inherent in the recording, they are dramatically affected by components in the playback system.

Terms that describe a clearly defined soundstage are *focused, tight, delineated,* and sharp. *Image specificity* also describes tight image focus and pinpoint spatial accuracy. A poorly defined soundstage is described as *homogenized, blurred, confused, congested, thick,* and lacking *focus.*

A related issue is soundstage *layering*. This is the ability of a sound system to resolve front-to-back cues that present some images toward the soundstage front and at varying distances toward the soundstage rear. The greater the number of layers of depth gradation, the better. Poor components produce the impression of just a few layers of depth: you hear perhaps three or four discrete levels of distance within the soundstage. The best components produce a sense of distance along a continuum: very fine gradations of depth are clearly resolved. Again, the absence of these qualities constitutes subconscious cues that what you're hearing is artificial. When this important spatial information is revealed, however, you can more easily forget that you're not hearing the "real thing."

Bloom—the impression that individual instrumental images are surrounded by a halo of air—is often associated with soundstaging. Although image outlines may be clearly delineated, a soundstage with bloom has an additional sense of diffused air around the image. It is as though the instrument has a little space around it in which it can "breathe." Bloom gives the soundstage a more natural, open, and relaxed feeling.

The term *action,* coined by my colleague at *The Absolute Sound,* Jonathan Valin, describes the sense of bloom expanding outward into space from the instrument, and the way in which the instrument's dynamic envelope grows. Action is bloom with a dynamic component. Think of how a trumpet sounds in real life, and the way the sound seems to expand in space as the instrument gets louder. Some audio components express this characteristic better than others; those with a good sense of action sound more realistic, vibrant, and alive than those that lack action.

A product's soundstaging performance should be evaluated with a wide range of musical signals. Some products may throw a superb soundstage at low levels, only to have that collapse when the volume increases during musical climaxes. Listen for changes in the spatial perspective with signal level.

Some products produce a crystal-clear, *see-through* soundstage that allows the listener to hear all the way to the back of the hall. Such a *transparent* soundstage has a lifelike immediacy that makes every detail clearly audible. Conversely, an *opaque* soundstage is *thick* or *murky*, with less of an illusion of "seeing" into space. *Veiling* is often used to describe a lack of transparency.

Soundstaging is toward the top of my list of sonic priorities (after tonal balance and lack of grain). The ability to present the music as a collection of individual images surrounded by space, rather than as one big image, is very important to the creation of musical realism. A large contributing factor to musical involvement is the impression that the music exists independently of the loudspeakers—that is, when we hear music as existing in space rather than attached to the loudspeakers. A soundstage with space, depth, focus, layering, bloom, and transparency is nothing short of spectacular.

To evaluate soundstaging and hear the characteristics I've described, you must use recordings that contain these spatial cues. Studio recordings made with multiple microphones and overdubs rarely reveal the soundstaging characteristics described. Recordings made in real acoustic spaces with stereo microphone techniques and a pure signal path are essential to hearing all aspects of soundstaging. In short, soundstaging is provided by a combination of the recording and the playback system. If soundstaging cues aren't present in the recording, you'll never know how well or how poorly the component or system under evaluation reveals them. Most audiophile recordings are made with purist techniques (usually two microphones) that naturally capture the spatial information present during the original musical event..

A term related to soundstaging is *envelopment*, a word that comes from the field of home theater and multichannel sound. Envelopment describes the feeling of immersion in a soundfield created by front and side- or rear-positioned surround speakers. Well-done surround sound can wrap around the listener, creating an almost seamless continuity between the front and rear sounds. (Envelopment is described in more detail in Chapter 12.)

Finally, superb soundstaging is relatively fragile. You need to sit directly between the loudspeakers, and every component in the playback chain must be of high quality. Soundstaging is easily destroyed by low-quality components, a bad listening room, or poor loudspeaker placement. This isn't to say you have to spend a fortune to get good soundstaging; many very-low-cost products do it well, but it is more of a challenge to find those bargains. Chapters 14 and 15 contain many practical suggestions for achieving the best soundstaging possible from your system.

Dynamics

The *dynamic range* of an audio system isn't how loudly it will play, but rather the *difference* in level between the softest and loudest sounds that the system can reproduce. It is often specified technically as the difference between the component's noise level and its maximum output level. A symphony orchestra has a dynamic range of about 100 decibels, or dB (decibels are defined in Appendix A); a typical rock recording's dynamic range is about 10dB. In other words, comparatively speaking, the rock band is *always* loud; it has little dynamic range.

Dynamics are a very important part of music reproduction. They propel the music forward and involve the listener. Much of music's expression is conveyed by dynamic contrast, from *pp* (pianissimo) to *fff* (triple forte).

There are two distinct kinds of dynamics. *Macrodynamics* refers to the presentation's overall sense of slam, impact, and power—bass-drum whacks and orchestral crescendos, for examples. If the system has poor macrodynamics, we say the sound is *compressed* or *squashed*.

Microdynamics occur on a smaller scale. They don't produce a sense of impact, but are essential to providing realistic dynamic reproduction. Microdynamics describe the fine dynamic structure in music, from the attack of a triangle or other small percussion instruments in the back of the soundstage to the suddenness of a plucked string on an acoustic guitar. Neither sound is very loud in level, but both have dynamic structures that require agility and speed from the playback system.

Products with good dynamics—macro and micro—make the music come alive, allowing a vibrancy and life to emerge. Dynamic changes are an important vehicle of musical expression; the more you hear the musicians' intent, the greater the musical communication between performers and listener. Some otherwise excellent components fail to convey the broad range of dynamic contrast.

These characteristics are associated with *transient response*, a system's ability to quickly respond to an input signal. A transient is a short-lived sound, such as that made by percussion instruments. Transient response describes an audio system's ability to faithfully reproduce the quickness of transient signals. For example, a drum being struck produces a waveform with a very steep attack (the way the sound begins) and a fast decay (the way a sound stops). If any component in the playback system can't respond as quickly as the waveform changes, a distortion of the music's dynamic envelope occurs, and the steepness is slowed. Audio components described as *quick* or *fast* reproduce the suddenness of transient signals. A related concept is *transient coherence*, the impression that the system's transient response is the same across all frequencies. When this happens, bass transients seem lined up in time with midrange and treble transients. Many loudspeakers lack this transient coherence, sounding slower in the bass than in the rest of the spectrum.

But just because a component or system can reproduce loud and soft levels doesn't necessarily mean it has good dynamics. We're looking for more than a wide dynamic range. The system must be capable of expressing fine gradations of dynamics, not just loud and soft. As the music changes in level (which, except in most rock recordings, it's doing most of the time), you should hear loudness changes along a smooth continuum, not as abrupt jumps in levels.

Another aspect of dynamics we seek is the ability to play loudly without the sound becoming *congested*. Many products—particularly digital sources and loudspeakers—sound thick during musical climaxes. As the music gets louder, the sound becomes harder, timbre is obscured, and the soundstage degenerates into a confused mess. The word *congeal* aptly describes the homogenization of instrumental images within the soundstage. The nice sense of space and image placement heard at moderate volumes often collapses during loud passages. This produces a sense of *strain* on musical peaks that greatly detracts from your enjoyment. When a component doesn't exhibit these problems, we say it has *effortless* dynamics.

Detail

Detail refers to the small or low-level components of the musical presentation. The fine inner structure of an instrument's timbre is one kind of detail. The term is also associated with transient sounds (those with a sudden attack) at any level, such as those made by percussion instruments. A playback system with good resolution of detail will infuse music with that sense that there is simply more music happening.

Assembling a good-sounding music system or choosing between two components can often be a tradeoff between smoothness and the resolution of detail. Many audio components hype detail, giving transient signals an *etched* character. Etch is an unpleasant hardness of timbre on transients that emphasizes their prominence. Sure, you can hear all the information, but the presentation becomes too *aggressive, analytical,* and *fatiguing*: low-level information is brought up and thrust at you, and you feel a sense of relief when the music is turned down or off—not a good sign.

Components that err in the opposite direction don't have this etched and analytical quality, but neither do they resolve all the musical information in the recording. These components are described as overly *smooth,* or having *low resolution.* They tend to make music bland by removing parts of the signal needed for realistic reproduction. These kinds of components don't rivet your attention on the music; they are *uninvolving* and dull. You aren't offended by the presentation, as you are with an analytical system, but something is missing that you need for musical satisfaction.

Lack of detail is often an unintended consequence of removing treble brightness or grain. The design aspects that tame the aggressive treble also obliterate essential musical information. Such components are said to have low resolution. Some listeners prefer such low-resolution presentation to a more natural rendering. In fact, some products are actually designed to gloss over detail to make the products sound more "musical."

It is a rare product indeed that presents a full measure of musical detail without sounding etched. The best products will reveal all the low-level cues that make music interesting and riveting, but not in a way that results in listening fatigue—that sense of tiredness after a long listening session. The music playback system must walk the very fine line between resolution of real musical information and sounding analytical.

Pace, Rhythm, and Timing

The British audio reviewer Martin Colloms deserves the credit for identifying and defining important aspects of music reproduction he calls *pace, rhythm,* and *timing*. These terms refer to a system's ability to involve the listener in a physical way in the music's forward flow and drive. Pace is that quality that makes your body want to move with the beat, your foot to tap, and your head to bob—the feeling of being propelled forward on the music's rhythm. The sound of good pace and rhythm from an audio system can produce a sense of physical exhilaration.

Though the music's tempo doesn't change in an objective sense from component to component (one power amplifier to another, for example), the subjective impression of a timing difference can be profound. Some products put a drag on the rhythm, making the tempo seem slower or *sluggish.* Others are *upbeat,* conveying the music's rhythmic tautness and drive. In addition to apparently slowing the tempo, components with poor pace give the impression that the band is less tight and musically "up." The music has less vitality and energy, and the band is less locked-in to the groove.

Martin Colloms wrote this in his revelatory article on this overlooked aspect of music reproduction, published in the November 1992 issue of *Stereophile.*

While good rhythm is a key aspect of both live and reproduced music-making, it is not easy to analyze. It's as if the act of focusing on the details of a performance blinds one to the parameter in question. The subjective awareness of rhythm is a continuous event, registered at the whole-body level, and recognized in a state of conscious but relaxed awareness. Once you've learned that reproduced sound can impart that vital sense of music-making as an event, that the impression of an upbeat, involving drive can be reproduced again and again, you can't help but pursue this quality throughout your listening experience.

Any random energy, such as in a power amplifier's power supply or a loudspeaker's cabinet—even a loosening of the bolts that secure drivers to loudspeaker baffles—can degrade pace and timing. Rhythm and pace are more important in rock, jazz, blues, pop, than in classical music. Much of the feeling of propulsion comes from the bass drum and bass guitar working together.

Many listeners have heard examples of good pace or the lack of it without recognizing it as such. Instead, they may have felt an indefinable involvement in the music if the system had good timing, or a general apathy or boredom with the music if the system didn't convey pace and rhythm.

How do you listen for pace and rhythm? Forget about it; if you find yourself wanting to dance, the component probably has it.

Coherence

Coherence describes the impression that the music is integrated into a satisfying whole, and is not merely a collection of bass, midrange, and treble. The music's harmonic tapestry is integrally woven, not pieced together like a patchwork quilt.

Coherence also applies to the system's or component's dynamic performance, specifically the impression that all the transient edges are lined up with each other. This provides a sense of wholeness to the sound that we react to favorably. Coherence is more a feeling of musical "naturalness" than any specific characteristic.

A related term, coined by Harry Pearson, is *continuousness,* the impression that the music is harmonically whole across the spectrum.

Musicality

Finally, we get to the most important aspect of a system's presentation—*musicality.* Unlike the previous characteristics, musicality isn't any specific quality that you can listen for, but the overall musical satisfaction the system provides. Your sensitivity to musicality is destroyed when you focus on a certain aspect of the presentation; i.e., when you listen critically. Instead, musicality is the *gestalt*, the whole of your reaction to the reproduced sound. We also use the term *involvement* to describe this oneness with the music. A sure indication that a component or a system has musicality is when you sit down for an analytical listening session and minutes later find yourself immersed in the music and abandoning the critical listening session. This has happened many times to me as a reviewer, and is a good measure of the product's fundamental "rightness." Ultimately, musicality—not dissecting the sound—is what high-end audio is all about.

The following essay grapples with the conflict between analytical listening and musicality. It first appeared as an editorial in the November 1992 issue of *Stereophile*, and is reprinted with permission.

Between the Ears
by Robert Harley

Audiophiles constantly seek ways to improve the experience of hearing reproduced music. Preamps are upgraded, digital processors are compared, turntables are tweaked, loudspeaker cables are auditioned, dealers are visited, and, yes, magazines are read—all in the quest to get just a little closer to the music.

These pursuits have one thing in common: they are all attempts by physical means to enjoy music more. But there's another way of achieving that goal that is far more effective than any tweak, better than any component upgrade, and more fulfilling even than having *carte blanche* in the world's finest high-end store. And it's free.

I'm talking about what goes on *between* our ears when listening, not what's impinging on them. The ability—or lack of it—to clear the mind of distractions and let the music speak to us has a huge influence on how much we enjoy music. Have you ever wondered why, on the same system and recordings, there is a vast range of involvement in the music? The only variable is our state of mind.

Because audiophiles care about sound quality, we are often more susceptible to allowing interfering thoughts to get in the music's way. These thoughts are usually concerned with aspects of the *sound's* characteristics. Does the soundstage lack depth? Does the bass have enough extension? Is the treble grainy? How does my system compare to those described in magazines?

Unfortunately, this mode of thinking is perpetuated by high-end audio magazines. The descriptions of a product's sound—its specific performance attributes—are what make it into print, not the musical and emotional satisfaction to which the product contributes. The latter is ineffable: words cannot express the bond between listener and music that some products facilitate more than others. Consequently, we are left only with descriptions of specific characteristics that can leave the impression that being an audiophile is about dissection and critical commentary, and not about more closely connecting with the music's meaning.

A few months after I became an audio reviewer (and a much more critical listener), I underwent a kind of crisis; I found myself no longer enjoying music the way I once did. Listening became a chore, an occupational necessity, rather than the deeply moving experience that made me choose a career in audio. My dilemma was precipitated by the mistaken impression that whenever I heard music I had to have an opinion about the quality of its reproduction. Music became secondary to the sound. Music was merely an assemblage of

a specific component sounds when I haven't heard that product for a year or more. Because I record my sonic experiences when writing product reviews for *The Absolute Sound,* I have formed a capsule impression of that product's basic character. I don't remember every detail, of course, but the mental image may be something like "slightly etched treble, very transparent and spacious soundstage, lean bass, somewhat lacking in dynamics, analytical rather than smooth." I also associate a general value judgment with that component: would I want to listen to it for pleasure over a long period of time?

By writing down these impressions, you'll not only have a written record of your experience, but also a more tangible mental impression, one that you can use when making later comparisons with other components.

Critical Listening Setup Procedures

Now that we've explored the lexicon of critical listening, let's look at the procedures required to reach valid listening judgments. These procedures are controls on the process that help ensure that our listening impressions are valid. It's easy to be fooled if you don't take certain precautions.

The following procedures are the result of my experience evaluating products professionally, and reporting those evaluations to a sizable audience. In my profession, the need to be right about products is paramount. Not only will readers buy or not buy products based on my opinions, but the fortunes of high-end manufacturers are greatly affected by reviews, positive or negative. Accurately describing a product's sound, and reaching correct value judgments about that product, are huge responsibilities—to readers, manufacturers, and the truth. The techniques described are essential to reaching the right conclusions. I wouldn't even consider forming concrete opinions about products without these safeguards. Neither should you.

By using these controlled procedures, you can stop second-guessing what you hear and concentrate on the real musical differences between components. Removing the variables that may skew the listening impressions will allow you to have greater confidence that the differences you're hearing are real musical differences between products, not the results of improper setup. If you follow these procedures and hear differences, you can have confidence in your listening impressions.

As you go into the listening session, turn your mind into a blank slate. Forget about the brand names of the products you're about to audition, their reputation for certain sonic characteristics, the products' prices, and whom you may offend if you like product A better than product B. Be completely receptive to whatever your ears tell you. If your impressions match your preconceived ideas about what the product sounds like, suspect bias. But if you find that your listening impressions contradict your prejudices, you've developed the ability to listen without bias. Removing bias is essential to reaching accurate judgments.

The first rule of critical listening is that only one variable should be changed at one time. If comparing preamps, use the same source components, interconnects, AC power conditioner, power amp, cables, loudspeakers, music, and room. This is obviously impossible when comparing products at different dealers with different playback systems. But changing one variable at a time is an absolute requirement for arriving at valid opinions.

This is why I stress taking the competing products home over the weekend for comparison. Not only can you evaluate them under the same conditions, but you get to hear how they interact with the rest of your system.

Changing only one component at a time is vital; big differences (two different pairs of loudspeakers, for example) will often swamp small differences. You can't separate what one new component is doing from what the second new component is doing. The ultimate expression of this mistake is someone who hears a system for the first time and remarks on how good (or bad) the loudspeaker cables are. It's ridiculous; you can't isolate what the loudspeaker cable is doing within the context of a completely unknown system. An exception to this rule is when evaluating AC power cords. It is often more revealing to change all the AC cords in the system at once rather than one at a time.

The second rule is to match volume levels between components under audition. Level matching ensures that both products produce the same volume at the loudspeakers, and should become an integral part of critical listening sessions. All it takes is an inexpensive voltmeter and a few minutes of time. If you're comparing two digital sources, for example, match their output levels to within, at most, 0.2dB of each other, and preferably to within 0.1dB. You'll be well rewarded for those few minutes spent level-matching with more accurate value judgments about the products. (The section at the end of this chapter explains how to match levels.)

Level-matching is crucial; slight level differences between products can lead you to the wrong conclusion. As described in Appendix A, the ear's sensitivity to bass and treble increases disproportionately with volume. That is, you hear more bass and treble when the music is loud. If product A is played louder than product B, product A may sound brighter, more forward, more detailed, more dynamic, and have more bass. If product A has a soft treble, lacks detail, and has a lean bass balance, you may not know it if the levels aren't precisely matched. If you *can* distinguish product A's character with mismatched levels, the degrees to which the treble is soft, the detail lacking, the dynamics missing, and the bass thin are much harder to gauge. With level matching, you'll know precisely how *much* these characteristics are present in the product you're evaluating; you can forget about trying to mentally compensate for different levels and concentrate on characterizing the sonic differences.

With the same playback system and matched levels, you're ready to compare products. The typical method is called A/B, in which you hear product A, then immediately afterward the same music through product B. One problem with A/B listening is called "The A/A Paradox." This states that when hearing two identical presentations (A/A), you tend to hear differences between them. The A/A paradox happens because music has meaning; you perceive it and react to it a little differently each time you hear it. Moreover, if the music is unfamiliar, a second hearing allows you to hear more detail. These factors combine to fool you into hearing differences that aren't there.

Fortunately, overcoming the A/A paradox is easy. When comparing products, listen to A, then B, then A again. The differences you heard between A and B are often solidified after hearing A again. After selecting another piece of music for further comparison, make the order B/A/B. This technique of hearing three presentations (or five, A/B/A/B/A) confirms or refutes first impressions. Usually, a characteristic heard in the first comparison is more apparent in a second comparison.

Selecting the right *source material* (i.e., recordings) is important in fully exploring a product's character. If the source material doesn't contain certain cues, you won't know how well the product reproduces those cues. For example, if you're comparing two power amplifiers and use only dry studio recordings of pop music, you'll never know how good the amplifiers are at resolving soundstage depth and other spatial cues. Similarly, if you listen to nothing but chamber music in evaluating loudspeakers, you can't tell how the loudspeaker conveys bass dynamics, punch, and rhythmic intensity.

Establish a repertoire of familiar music for evaluating components, each recording selected for some sonic characteristic that will tell you how a component behaves. Remember, this is diagnostic listening, not listening for pleasure. A range of source material should include: a full-scale symphonic recording, chamber music, orchestra with choir, solo piano, popular vocal, rock, blues, or other music with electric bass and kick drum, jazz with acoustic bass, and music with cymbals. Some of the jazz and orchestral recordings should be naturally miked; that is, recorded in a way that captures the original spatial aspects of the music and concert hall. Audiophile record labels will often describe how the recordings were made, and even include a photograph of the hall and where the microphones were placed. Get to know these recordings intimately. By hearing the same recording played back on many different systems, you can quickly characterize a component's sonic signature.

Another factor that can lead you astray in critical listening is *absolute polarity* (discussed in Appendix A). If one product under comparison inverts absolute polarity and the other one doesn't, you may be led to the wrong conclusions; or, at the very least, be confused about which product is better. Many owner's manuals will state if the product inverts polarity. If you know that product A does not invert polarity while product B does, reverse the positive and negative connections on *both* loudspeakers: the red wire goes to the loudspeaker's black terminal, and the black wire goes to the loudspeaker's red terminal when auditioning product B. If your preamp has a polarity inversion switch, you can just throw it instead of swapping loudspeaker leads. Fortunately, very few components invert absolute polarity.

It's also easy to be misled by a colored playback system. Say you have loudspeakers that are bright, forward, and a little rough in the treble, and you're comparing two digital sources to determine which one to buy. Digital source A produces a sound through your system that is bright and forward in the treble. Digital source B renders a smoother treble and is more musical in your system.

Is source B inherently more musical? Not necessarily. It probably has colorations—an overly soft treble—that complement the loudspeaker's bright treble. Two products in the playback chain that have reciprocal flaws will often produce a musical result. And there's no reason not to buy source B if you intend on keeping your loudspeakers. But don't make the mistake of thinking source B is inherently better than source A. In fact, source A is probably more neutral and would be a much better choice with different, smoother loudspeakers. Although not ideal, choosing a product based on how it corrects other faults in your system is valid if the end result is a more musical presentation. Just recognize the situation when it occurs.

Single-Presentation Listening—What It's All About

The type of analytical listening I've described will reveal specific sonic attributes of components in relation to each other: CD transport A has deeper bass extension, smoother treble, and a deeper soundstage, while transport B has more inner detail, quicker transients, and a more focused soundstage. This kind of analytical listening will tell you a lot about the products' strengths and weaknesses, but it isn't the whole story. What's missing from this dissection is how each product makes you *feel* about the music. Does your mind wander, thinking about what you're going to do next? Or are you riveted in the listening seat, playing record after record well into the night?

This ability to express the music in a way that compels you to continue listening is called a product's *musicality* and *involvement*. Musicality and involvement are rarely revealed in A/B comparisons and can't be described in terms of specific sonic attributes. Instead, this fundamental characteristic of a product's ability to provide long-term musical satisfaction is discovered only when listening for pleasure, without level-matching, selecting music for its sonic rather than musical qualities, or analytical thinking. Dissecting the music to characterize specific sonic attributes, as in A/B comparisons, destroys our receptivity to the music's meaning. Involvement comes only when you consider music as a whole, not as a collection of sonic parts.

This is why all critical listening should include not only A/B/A comparisons, but also what's called "single-presentation listening" and listening just for pleasure. Single-presentation listening is listening to the product under evaluation over a long time period—days or weeks rather than hours—with an unconscious ear to how the product affects the music. Listening for sheer pleasure involves no thinking about the sound, only about the music. If, after a long session, you feel exhilarated and fulfilled, you can be confident in the product's ability to convey the music's expression and meaning. In fact, this is without question the most important indicator of quality in an audio component.

There is a general correlation between a product's sonic performance as revealed during analytical listening and its musicality as discovered during listening for pleasure. Specific sonic flaws—grainy treble, hard textures, flat soundstage—will often distract you from the music in ways that prevent deep involvement. The less the presentation sounds artificial, synthetic, or affronts your ears, the easier it is to forget the hardware and hear only the music. Indeed, making the hardware disappear is the Holy Grail of high-end audio.

Your evaluative listening should thus include a mix of analytical A/B/A comparisons with single-presentation listening (with an ear to specific sonic attributes) and pure listening for pleasure.

Critical Listening Summary

I've summarized the setup procedures for conducting critical listening that will provide valid, useful information about the product under evaluation:

- ***Use the same playback system for comparisons***
- ***Change only one component at a time***
- ***Match levels between products under audition***
- ***Listen to A/B/A, not A/B***
- ***Use very familiar music***
- ***Select music that reveals specific sonic characteristics; i.e., bass, midrange liquidity, dynamics, treble, and soundstaging***
- ***Listen to the product for pleasure over a long period of time***

Do I follow all these guidelines every time I listen critically? No. But I don't reach firm opinions about products unless I do. If you're listening to make a purchase decision, follow these guidelines religiously. Be suspicious of any listening impression received in poor circumstances. And don't reach firm opinions unless all these guidelines have been followed.

By applying these techniques, you'll be more likely to reach accurate value judgments, and consequently buy products that will provide the greatest long-term musical satisfaction.

Level Matching

Level-matching procedures are identical for all electronic components, whether a digital source, preamp, or power amp. You'll need a voltmeter (even a cheap one will work) and a test CD that contains pure tones at various frequencies. First, play music at a comfortable level through the system to set the volume. Then play the track on the test CD of 1kHz sinewave recorded at a low level. Measure the voltage across the loudspeaker terminals when the tone is playing, and note the voltage. Put a piece of masking tape on the preamp's face plate around the preamp's volume control and mark the position of the volume control.

After you switch components (power amp, digital source, etc.), play the 1kHz sinewave again and measure the voltage across the loudspeaker terminals. Adjust the preamp's volume control so that the voltage across the loudspeaker terminals is the same as what you measured for the first component. Mark the volume-control position on the masking tape. By setting the volume control at the first mark for product A and at the second mark for product B, you've got matched levels.

Some preamplifiers with detented (click stop) volume controls but continuous attenuation make level matching easier and more precise by removing the need for tape. Just count the number of steps necessary to achieve the same voltage across the loudspeaker terminals. Setting the volume control between detents is surprisingly accurate and repeatable.

If you're comparing preamps, you don't need masking tape—just find the volume setting on both preamps that produces the same voltage across the loudspeaker terminals when playing the test tone, and leave the volume controls in the same position.
Matching levels for phono cartridges and phono preamplifiers is identical, except that a test LP replaces the test CD.

Level-matching isn't required when comparing products that don't affect the volume. These include interconnects, loudspeaker cables, digital interconnects, AC line conditioners, AC power cords, vibration-damping devices, room treatments, and equipment stands.

Ideally, the playback level between products should be matched to within 0.2dB, and preferably to within 0.1dB. To calculate the dB difference between two measured voltages, you need only a calculator with a "log" button. Here's how to do it.

First, find the ratio of the two measured voltages by dividing one by the other. Then find the logarithm of that ratio by pressing the calculator's "log" button with the ratio in the display. Next, multiply the result by 20. The answer is the difference in dB between the two voltages.

For example, if you measured 2.82 volts (V) across the loudspeaker terminals with digital processor A, and 2.88V across the loudspeaker terminals with digital processor B after level matching, you first divide 2.88 by 2.82, which gives you the ratio of the two voltages (1.0213:1). Next, push the "log" button on the calculator with 1.0213 in the display, and multiply the result, 0.00914, by 20 to get the answer: the levels are matched to within 0.18dB, or close enough for meaningful comparisons.

4 Preamplifiers

Introduction

The preamplifier is the Grand Central Station of your hi-fi system. It receives signals from source components—disc players, tuners, DACs, music servers—and allows you to select which of these to send to the power amplifier for listening. In addition to allowing you to switch between sources, the preamplifier performs many other useful functions, such as amplifying the signal from your phono cartridge (in some preamplifiers), adjusting the balance between channels, and allowing you to set the volume level.. The preamplifier is the component you will most often use, touch, and adjust. It also has a large influence on the system's overall sound quality. (Note: preamplifiers are built into, or "integrated" with, integrated amplifiers and receivers, instead of being housed in a separate chassis.)

There are many types of preamplifiers, each with different capabilities and functions. Choosing the one best suited to your system requires you to define your needs. Listeners without a turntable, for example, won't need a preamplifier that amplifies the tiny signals from a phono cartridge. Others will need many inputs to multiple source components, and some will need multichannel capability. Let's survey the various preamplifiers and define some common preamplifier terms.

Line-Stage Preamplifier: Accepts only line-level (low-level) signals, which include every source component except a turntable. Line stages have become much more popular as listeners increasingly rely on digital sources rather than LPs as their main signal source. If you don't have a turntable, you need only a line-stage preamplifier.

Phono Preamplifier (also called a phono stage): Takes the very tiny signal from your phono cartridge and amplifies it to line level. It also performs RIAA equalization on the signal from the cartridge. RIAA equalization is a bass boost and treble cut that counteract the bass cut and treble boost applied in disc mastering, thus restoring flat response. Some phono preamplifiers offer other equalization curves besides RIAA (more on this later).

A phono stage can be an outboard stand-alone unit in its own chassis, or a circuit section within a full-function preamplifier. If you play records, you must have a phono stage, either as a separate component or as part of a full-function preamplifier.

Step-Up Transformer: Steps-up the voltage from a low-output moving-coil cartridge. The step-up transformer's output feeds a phono preamplifier.

Full-Function Preamplifier: Combines a phono stage with a line-stage preamplifier in one chassis.

Tubed Preamplifier: A tubed preamplifier uses vacuum tubes to amplify the audio signal.

Solid-State Preamplifier: A solid-state preamplifier uses transistors to amplify the audio signal.

Hybrid Preamplifier: A hybrid preamplifier uses a combination of tubes and transistors.

Audio/Video Controller: A device analogous to a preamplifier that includes video switching, multiple audio channels (typically eight), and surround-sound decoding such as Dolby Digital and DTS. (A/V controllers are described in Chapter 12, "Audio for Home Theater.")

Multichannel Preamplifier: A preamplifier with multiple audio channels (typically six) for playback of multichannel music. Differs from an A/V controller in that the multichannel preamplifier has no surround decoding, video switching, or other functions for film-soundtrack reproduction.

Digital Preamplifier: A preamplifier that accepts digital input signals (such as from a CD transport) and processes the audio signal in the digital domain. Digital preamplifiers usually include a digital-to-analog converter so that the digital preamplifier can drive a conventional power amplifier.

Passive Level Control: Sometimes erroneously called a passive preamp, the passive level control can replace a line-stage preamplifier in some situations. It is inserted in the signal path between a source component and the power amplifier.

Instead of amplifying the source, a passive level control merely attenuates (reduces) the signal level driving the power amplifier. It doesn't plug into the wall, and cannot amplify a signal as does an active line-stage preamplifier.

No Preamplifier: Some source components, notably disc players and DACs, have the ability to drive a power amplifier directly with no need for a preamplifier. The source component must have a variable output level so that you can adjust the volume. Some DACs, for example, have multiple digital inputs along with source switching, obviating the need for a preamplifier.

How to Choose a Preamplifier

Once you've decided on a line-stage, a full-function preamp, or separate line and phono stages (the last are generally more expensive), it's time to define your system requirements. The first is the number of inputs you'll need. Count your source components and be sure that the preamplifier has that number of inputs plus one (to accommodate future needs). Many preamplifiers have a tape loop that can be used as a line input if you don't have a recording device. A tape loop is a pair of input and output jacks for recording onto a recording device and receiving a signal from that recording device. When you press the Tape Monitor button on your preamplifier, you are routing the signal from the Tape Input jacks to your power amplifier for listening.

Most high-end preamps have few features—and for good reason. First, the less circuitry in the signal path, the purer the signal and the better the sound. Second, the preamp designer can usually put a fixed manufacturing budget into making a preamp that either sounds superb or has lots of features—but not both. Mass-market mid-fi equipment emphasizes vast arrays of features and buttons at the expense of sound quality. Don't be surprised to find very expensive preamps with almost no features; they were designed, first and foremost, for the best musical performance. Most high-end preamps don't even have tone (bass and treble) controls. Not only do tone controls electrically degrade the signal—and thus the musical performance—but the very idea of changing the signal is antithetical to the values of high-end audio. The signal should be reproduced with the least alteration possible. Tone controls are usually unnecessary in a high-quality system.

Another school of thought, however, holds that a playback system's goal isn't to perfectly reproduce what's on the recording, but to achieve the most enjoyable experience possible. If changing the tonal balance with tone controls enhances the pleasure of listening to music, use them. If you're comfortable with the latter philosophy, be aware that tone controls invariably degrade the preamplifier's sonic quality. (Some preamps with tone controls do allow you to switch them out of the circuit when they're not being used, a feature called "tone defeat.")

A similar debate rages over whether or not to include balance controls in a high-end preamp. A balance control lets you adjust the relative levels of the left and right channels. If the recording has slightly more signal in one channel than the other, the center image will appear to shift toward the louder channel, and the sense of soundstage layering may be reduced. A similar problem can occur if the listening room has more absorptive material on one side than the other, pulling the image off- center. A small adjustment of the balance control can correct these problems. Like tone controls, balance controls can slightly degrade a preamplifier's sonic performance. It isn't unusual to find a $25,000 state-of-the-art preamplifier with no tone or balance controls, and a $199 mass-market receiver with both of these features.

When making a purchasing decision, you should also consider the preamplifier's look and feel. Because the preamplifier is the component you'll interact with the most, ask yourself: Are the controls well laid out? Is the volume control easy to find in the dark? Does the preamp have a mute switch? A mute switch is handy when your listening is interrupted, for protecting the rest of your system when disconnecting cables, or for maintaining the same volume level when comparing two other components, such as digital sources or cables.

One convenience feature once found on only a few preamps but is now ubiquitous is remote control. The ability to adjust the volume (and sometimes absolute polarity) from the listening seat is a huge advantage. Remote volume control let's you set the volume exactly for the particular recording, enhancing the musical experience.

Balanced and Unbalanced Connections

Some preamplifiers have balanced inputs, balanced outputs, or both. A balanced signal is carried on a three-pin XLR connector rather than the conventional RCA plug. (Balanced and unbalanced connections are described in Chapter 11.) Fig.4-2 later in this chapter shows the rear panel of a preamplifier with balanced inputs and outputs on XLR jacks and unbalanced inputs and outputs on the familiar RCA jacks. If you have a balanced source component—usually a CD player or DAC—you should consider choosing a preamplifier with a balanced input. Nearly all source components with balanced outputs also have unbalanced outputs. You can use either output, but you may not get the best sound quality unless you use the balanced output option. (See Chapter 7, "Disc Players, Transports, and DACs," for the reason why.) In addition, not all "balanced" preamplifiers are created equal, as we'll see later in this chapter.

A preamp with unbalanced inputs and a balanced output can accept unbalanced signals but still drive a power amplifier through a balanced interconnect. If your power amplifier has balanced inputs, getting a balanced-output preamplifier is a good idea. You can listen to the system through both balanced and unbalanced lines and decide which sounds better. Some products sound better through balanced connections; others perform best with unbalanced lines. (A more technical description of balanced preamplifiers is included later in this chapter.)

Other Considerations in Choosing a Preamplifier

In addition to providing volume control and letting you select which source component you listen to, the preamplifier is a buffer between your source components and power amplifier. That is, the preamplifier acts as an intermediary, taking in signals from source components and conditioning those signals before sending them on to the power amplifier. Source components can easily drive a preamplifier, with the burden of driving a power amplifier through long cables falling on the preamp's shoulders. By buffering the signal, the preamplifier makes life easier for your source components and ensures good technical performance.

Some line-stage preamplifiers let you completely bypass all the active electronics (tubes or transistors) in the unit for purer sound. The preamp then becomes a passive level control, doing nothing more than controlling the system's volume. Although removing the active electronics from the signal path has certain sonic advantages, the preamplifier no longer acts as a buffer between source components and the power amplifier. The result can be poor technical performance (for reasons described later in this chapter) and degraded sound quality (primarily soft bass and loss of dynamics). Conversely, some systems will work just fine without the preamplifier acting as a buffer, and will benefit from bypassing the preamplifier's active electronics. Fortunately, you can try such a preamplifier in both modes (called active and passive) just by throwing a front-panel switch. When in the passive mode, the preamplifier suffers all the limitations of passive level controls, described later in this chapter. If you see a front-panel switch marked Bypass, you'll know what that means.

Another approach is the minimalist preamplifier that includes tubes or transistors that are always in the signal path, but provides no gain (amplification). This variety of preamplifier functions as a buffer between your source components and power amplifier, but cannot amplify signals. Some designers feel that these so-called unity gain preamplifiers sound better than conventional preamplifiers that amplify source signals. Your source components must put out enough voltage to adequately drive a power amplifier when using a unity-gain preamplifier.

Yet another technique uses a unity-gain buffer when the preamplifier's volume control is set to a low level, and kicks in a simple gain stage (amplifying circuit) only when the volume is turned up. You can see a commonality in these disparate approaches: keeping the preamplifier's signal path as short and clean as possible so that it has as little effect on the musical signal as possible. Preamplifiers using these techniques are not the norm; most preamplifiers always amplify and buffer the signal.

After you've decided on your functional needs, determine how much you can spend on a preamp using the guidelines in Chapter 2. Then put together a short list of preamps worthy of serious auditioning. Use dealer recommendations, read reviews in responsible audiophile magazines, and ask friends who have high-end systems. If possible, try the preamp in your own system before you buy.

The technical terms associated with preamplifiers (direct-coupled, Class-A, discrete, etc.) are defined later in this chapter under "How a Preamplifier Works."

Note that these recommendations for choosing the preamp best suited to your system also apply to integrated amplifiers. Think of an integrated amplifier as a preamplifier that also happens to have a power amplifier built-in. An integrated amplifier is nearly identical functionally to a preamplifier. The only difference is that, instead of connecting the preamplifier output to a power amplifier, you attach speaker cables to the integrated amplifier. (See Chapter 5 for more information on integrated amplifiers.)

Now that you know what features and functions you're looking for, it's time to begin listening.

What to Listen For

The preamplifier has a profound effect on the music system's overall performance. Because each of the source signals must go though the preamp, any colorations or unmusical characteristics it imposes will be constantly overlaid on the music. You can have superb source components, a top-notch power amplifier, and excellent loudspeakers, yet still have mediocre sound if the preamp isn't up to the standards set by the rest of your components. The preamplifier can establish the lowest performance level of your system; careful auditioning and wise product selection are crucial to building the best-sounding playback system for your budget.

A preamplifier's price doesn't always indicate its sonic quality; I know of one $2000 model that is musically superior to another preamp selling for nearly $6000. If you do your homework and choose carefully, you'll avoid paying too much for a poor-sounding product.

In addition to the usual listening procedures described in Chapter 3, preamplifiers offer several methods of sonic evaluation not possible with other components. We can therefore more precisely evaluate preamps and choose the best one for the money.

We'll start with the standard listening-evaluation techniques. First, the same musical selection can be played on the same system, alternating between two competing preamps. Be sure to match levels between the two preamps under audition. Listen for the presentation differences described in Chapter 3—particularly clarity, transparency, lack of grain, low-level detail, soundstaging, and a sense of ease. The better preamplifier will "get out of the way" of the music, imposing less of itself on the sonic presentation.

The most common sonic problems in preamplifiers are hardening of timbres, a thickening of the soundstage, and a reduction in low-level detail. Many preamps, particularly inexpensive solid-state units, overlay the midrange and treble with a steely, metallic hardness. These preamps can give the impression of more musical detail, but quickly become fatiguing. The treble becomes drier, more forward, and etched. Cymbals lose their sheen, instead sounding like bursts of white noise. Vocal sibilants (s and sh sounds) become objectionably prominent; violins become screechy and thin. The poor-sounding solid-state preamplifier emphasizes the brightness of the strings and diminishes the resonance of the violin's wooden body. Such preamps also reduce the saturation of tonal color, sounding thin and bleached.

The preamplifier that thickens the sound makes the soundstage more opaque. The transparent quality is gone, replaced by a murkiness that obscures low-level detail and reduces resolution. Instruments and voices no longer hang in a transparent, three-dimensional space. Instead, the presentation is thick, confused, congealed, and lacks clarity. Even some expensive models impose these characteristics on the music.

Beware of the preamplifier that tends to make all recordings sound similar in timbre, tonal balance, or spatial presentation. The best preamplifiers let you hear changes in the size of the concert hall, how closely the microphones were positioned, and can clearly resolve the differences between a Steinway piano and a Bösendorfer, among other such subtleties. In addition, listen for how well the preamp resolves the fine inner detail of instrumental timbres. Even the best preamplifiers shave off some of the very lowest level details.

Compare the preamps under audition to the very best preamp in the store. Listen for the qualities distinguishing the best preamp, and see if those characteristics are in the preamps you're considering. This will not only give you a reference point in selecting a preamplifier for yourself, but will sharpen your listening skills. The more experience you have listening to a variety of products, the better your ability to judge component quality.

A useful way of evaluating preamplifiers—one not possible with other components—is the bypass test. This technique compares the preamplifier to no preamplifier, revealing the preamplifier's editorial effect on the music. This most revealing of comparisons leaves the preamplifier's shortcomings nowhere to hide.

To conduct a bypass test, drive a power amplifier at a comfortable volume with a known high-quality preamplifier, or directly from a CD player or digital source that has an output level control. Using a voltmeter and a test CD with test tones, set the preamplifier under evaluation for unity gain—the volume position at which the input level is identical to the output level (usually a third to halfway up on the volume control). Put the preamplifier under evaluation in the signal path between the known preamp and the power amp. Compare the system's sound with and without the preamp in the playback chain. Did the preamp thicken the sound? Did the treble become dry and brittle with the preamp in the system? Was the sense of space, clarity, and transparency replaced by an opaque thickness that homogenized the individual instruments? Did the preamplifier impart a common sonic characteristic over varied recordings? Most important, was the music less involving? The bypass test provides a quick and accurate assessment of the type and amount of coloration imposed by a preamplifier.

If you're comparing two preamplifiers, use preamp A to drive the power amplifier; listen, then put preamp B in the signal path at unity gain between preamp A and the power amplifier. Listen for the difference between the presentations with and without preamp B in the signal path. Then reverse the roles of the two preamps, listening for preamp A's effect on the presentation. This technique, shown in Fig.4-1, quickly reveals exactly what each preamp does to the music. This method, however, doesn't provide listening conditions of the highest resolution; the colorations of the first preamplifier can obscure qualities in the second preamplifier. It also adds another run of interconnects and terminations to the signal path.

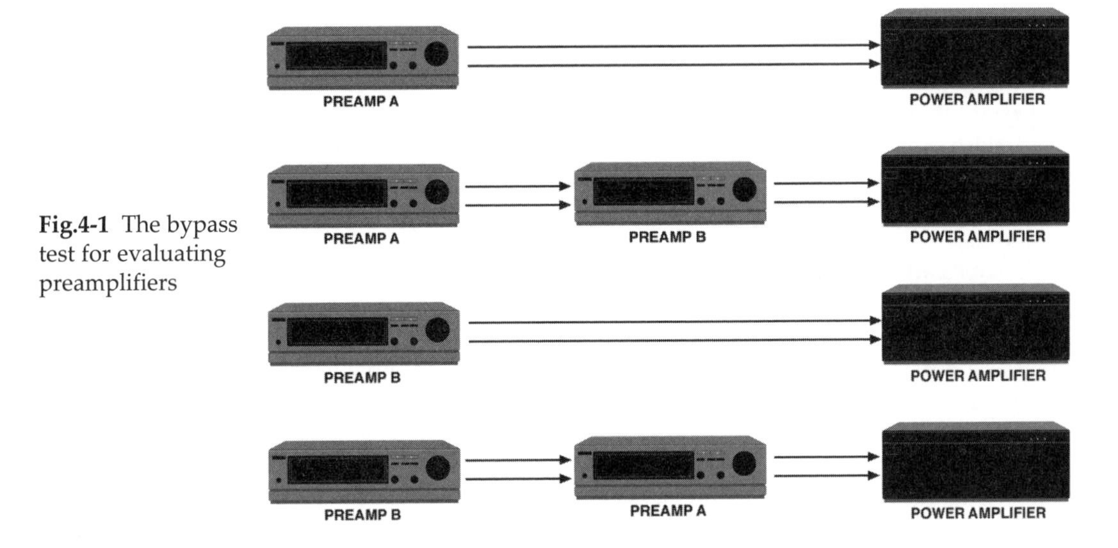

Fig.4-1 The bypass test for evaluating preamplifiers

You can also conduct a bypass test by substituting a passive level control for the preamplifier. Although passive level controls can reduce dynamic contrast and soften the bass in some systems, they rarely add grain and etch, or reduce soundstage depth and clarity, as do most active preamplifiers.

The increasing prevalence of digital source components with a variable output level along with the ability to drive a power amplifier directly provides a test bed for evaluating preamplifier transparency. This is especially true when listening to very high-resolution files (176.4kHz/24-bit) decoded through state-of-the-art DACs; the high-res material is extremely revealing of any colorations downstream. In my recent experience listening to such sources with and without a preamplifier in the signal path, I've found that even heroic (and expensive) preamplifiers noticeably degrade the sound. Timbres are not quite as pure, the soundstage slightly less expansive, and the very lowest levels of fine detail are obscured. If you have a DAC with a variable output level, I encourage you to perform the bypass test; it leaves preamplifier colorations nowhere to hide.

A less analytical method is to borrow the preamp from your dealer for a weekend and just listen to it. How much more exciting and involving is the music compared to using your existing preamp? Does the new preamp reveal musical information you hadn't heard before in familiar recordings? How much does the new preamp compel you to continue playing music? These are the best indicators of the product's ability to provide long-term musical satisfaction. Trust what your favorite music tells you about the preamplifier.

Tubes vs. Transistors

Of the many components that make up a hi-fi system, the preamplifier is the most likely to use vacuum tubes instead of solid-state devices (transistors). This is because preamplifiers handle only low-level signals, making tubes more practical and affordable there than in power amplifiers. Power-amplifier tubes are large, expensive, run hot, and require replacement. Further, one theory of audio design holds that if the system is to include tubes, they are best employed closest to the signal source—such as in the preamplifier.

Moreover, the qualities that have endeared many music lovers to the magic of tubes are much more affordable when used in a preamplifier. Tubed audio components are more expensive to build and maintain than solid-state units, but that cost differential is much lower in preamplifiers than in power amplifiers. If you want the special qualities of tubes but not their heat, greater expense, and higher maintenance, the tubed preamplifier is the way to go.

Tubes are often claimed to sound sweeter and warmer, portray greater soundstage dimensionality, and to have a more natural treble. Many solid-state preamps tend to make the treble dry, brittle, metallic, and etched. The result is steely-sounding strings (particularly violins), unnaturally emphasized vocal sibilants (s and sh sounds), and cymbals that sound like bursts of high-frequency noise rather than a delicate brass-like shimmer. Because these unpleasant artifacts can be introduced by many components (digital sources, power amplifiers, cables, tweeters), a natural-sounding tubed preamplifier can tend to ameliorate the system's tendency toward these amusical characteristics.

The popularity of tubed electronics has nothing to do with nostalgia and everything to do with sound quality. Tubes have several advantages in their favor as audio amplifying devices. First, the circuitry associated with a vacuum tube—the ancillary parts that make the tube work—is generally much simpler than a circuit using transistors. Second, the distortion tubes produce is very different from the distortion created by solid-state electronics. Harmonic distortion, introduced by all active electronics, adds spurious frequencies to the signal being amplified. For example, if a preamplifier is passing a 1kHz signal, the preamp will generate distortion at 2kHz (second harmonic), 3kHz (third harmonic), 4kHz (fourth harmonic), and so forth. Tubed circuits generally produce lower-order harmonic distortion components such as second and third harmonic, in contrast with the higher-order distortion (seventh and ninth) introduced by transistors. Lower-order distortion is much less musically objectionable than the upper-order harmonics. Moreover, a large amount of second-harmonic distortion (say, 5%) is harder to hear than a very small amount (say, 0.5%) of seventh-harmonic distortion. Tube proponents point to these facts as evidence of the superiority of tubed preamplifiers.

Be aware, however, that some tubed preamps are intentionally designed to sound very colored. Rather than present the music with the least added effect, tubed preamps often add significant amounts of euphonic coloration. This form of distortion can at first be pleasing to the ear, but represents a departure from the original signal. This type of "tubey" coloration is characterized by a soft treble, an overly laid-back and easygoing presentation, lack of detail, and a "syrupy" sound. Many audiophiles, in their attempts to avoid the worst characteristics of solid-state, turn to euphonically colored tubed preamps to make their systems listenable.

It is a far better approach, however, to make sure that each component in the chain is as transparent as possible. If this is achieved, there will be no need for "tubey" preamps. Ideally, the listener shouldn't be aware that she is listening to tubes; instead, she should be aware only of the music. Just as poor solid-state preamps color the sound by adding grain, treble hardness, and etch, the poor tube preamp will often err in the opposite direction, obscuring detail, adding false "bloom," and reducing resolution. Be equally aware of both forms of coloration.

It's a mistake to "fall in love" with either solid-state or tubes for the wrong reasons. The solid-state lover may think he is getting "more detail," and the tube aficionado may fall for the "lush sweetness." Both extremes are to be avoided in the pursuit of a truly musical playback system. The overly "sweet" preamplifier may become uninvolving over time because of its low resolution; the "detailed" and "revealing" solid-state unit may eventually become unmusical for the fatigue it produces in the listener. Not all tube and solid-state preamps can, however, be categorized so neatly. Many tubed preamps are extremely transparent and neutral, having very little effect on the sound.

Nor should you buy a preamp purely because it uses tubes. Certain circuits are better implemented with tubes, others with solid-state devices. There are no magic components or circuit designs that will ensure a product's musicality. Worthy—and unworthy—products have been made from both tubes and transistors. That's why some companies design tubed, solid-state, and hybrid tubed/solid-state preamplifiers. The designer picks the best device for the particular application.

Ideally, both tubed and transistor preamplifiers strive for musical perfection, but approach that goal from different directions. The gross colorations I've described are largely a thing of the past, or relegated to fringe products that appeal to a minority of unsophisticated listeners. Today's reality is that tubed and solid-state designs sound more and more alike every year. The best preamplifier will be a transparent window on the music—no matter what its design.

The best advice is to choose the preamp that has the least effect on the music; you'll get much more musical satisfaction from it for a longer period of time. And remember: The perfect tubed preamp and the perfect solid-state preamp would sound identical.

Tube Life and Replacement Options

The small-signal tubes found in preamplifiers (generally between two and eight tubes) need replacing after 1000 to 2000 hours of use. Any time the preamplifier is turned on, the hour meter is ticking. This fact, along with the heat produced by tubes, encourages many owners of tubed preamplifiers to turn their units off when not listening. Although this extends tube life, an electronic component doesn't sound its best until it's been on for a few hours. The best advice is to turn on the preamplifier an hour or so before a listening session, if possible. If you know you'll be listening at night after work, turn on the preamplifier in the morning and it will be at peak performance when you get home. Another alternative is to buy a mechanical AC timer (about $10) from a hardware store and set it to switch on the preamp two hours before you expect to come home. Solid-state preamplifiers, on the other hand, draw very little current, run cool, and don't need tube replacement—they can be left on continuously.

Replacement tubes vary in cost from $10 for an untested generic type to $100 for a premium-grade tube that has been thoroughly tested and selected for best electrical performance. You may also want to consider paying a premium for matched pairs of tubes that have the same gain (amount of amplification). If the preamplifier's channel balance (similarity of gain between left and right channels) is determined by the gain of separate tubes, it's worth the extra cost for matched pairs. This ensures that one channel won't be louder than the other.

Because tube quality has a big effect on the preamplifier's sound, high-quality tubes are worth the money. New Old Stock (NOS) tubes (tubes that are unused but that haven't been manufactured in a long time) carry the highest prices because of limited supply and high demand. Different brands of tubes equal in technical performance will also sound different, making choosing difficult. Your local dealer is the best guide in selecting the optimum replacement tubes for your preamplifier. Equipment reviews of tubed preamplifiers will sometimes include the reviewer's opinions about different tube brands used in a particular preamplifier. Changing tubes in an amplifier is called "tube rolling."

Fortunately for the tube aficionado, the supply of reliable, high-quality tubes has never been better. In fact, the increasing popularity of tubed audio products is partially due to the recent availability of great tubes from the former Soviet Union. Their technology was heavily reliant on vacuum tubes (which are still used in MIG fighter jets), and their factories produced excellent products. With the end of the cold war, this large supply of superb, reasonably priced tubes has fueled the recent growth of the tubed audio industry. The Chinese tube industry has also stepped up its output and quality in recent years.

The Line-Stage Preamplifier

A typical minimalist line-stage preamplifier is shown in Fig.4-2. This model, an Audio Research Reference 5, has six stereo pairs of inputs plus a dedicated "Processor" input. Signals input to the Processor input appear at the preamplifier's output with no gain or attenuation, and the volume control is bypassed. This feature allows the preamplifier to be used in conjunction with a multichannel AV controller as part of a dual-purpose music and home-theater system. (Chapter 12 includes a complete description of how a preamplifier and an AV controller work together in the same system.) A front-panel input selector switch selects which input is routed to the output. A mute switch shuts off the signal at the preamplifier output. (This is useful when changing records or disconnecting and connecting source components—engaging the mute switch prevents a loud thump from going through your system.) Note that this preamplifier has no balance control for adjusting the relative levels between left and right channels. Two sets of main output jacks are provided, along with a "record output" whose level is fixed regardless of the volume control setting. The front-panel controls include a switch that selects between balanced and unbalanced inputs, a mono switch, and polarity inversion.

Fig. 4-2 A line-stage preamplifier's front panel and rear-panel connections. (Courtesy Audio Research Corporation)

A pair of 12V input/output jacks allow the preamp to be turned on by another component, or to output 12V to turn on other components. This 12V "trigger" signal is widely used as a way of simplifying powering up or down an entire system. For example, a disc player's 12V trigger output would drive the preamplifier, automatically turning on the preamplifier when the disc player is turned on. The preamplifier's 12V trigger output, in turn, feeds the power amplifier's 12V trigger input, turning on the power amplifier when the preamplifier is turned on. This 12V trigger system allows you to turn on your entire system with one button push.

The Phono-Stage Preamplifier

The phono stage amplifies the very tiny signal (between a few tens of microvolts and a few millivolts) from the phono cartridge to a line-level (about one volt) signal. This line-level signal can then drive a line-stage preamplifier, just as any other source component would. A phono stage can be an integral part of a full-function preamplifier, an optional board that plugs into some line-stage preamplifiers, or an outboard unit in its own chassis. Outboard phono stages have no volume controls; they usually feed a line input on a line-stage preamplifier.

In the days before CD, virtually all preamplifiers included integral phono stages. In many of today's preamplifiers, however, a phono stage is an option (usually about $200–$800) for those listeners who play LPs. This arrangement reduces the preamplifier's price for those needing only a line-stage preamplifier. It's also less expensive to buy a preamplifier with an integral phono stage than buying separate line and phono stages. With one chassis, one power supply, one owner's manual, and one shipping carton, a full-featured preamp is less expensive. These integral phono stages can also offer exceptional performance.

Not all line stages accept a phono board; if you think you'll want to play records in the future, be certain the line-stage preamp you buy will accept a phono board. If the preamp's rear panel has an input marked Phono, a ground lug, and a front-panel input selection position marked Phono, you can be confident that the preamp has a built-in phono stage or will accept a phono board. Note, however, that some line-stage preamps have a line-level input marked Phono but no phono stage. Look on the rear panel for the ground lug—a post for connecting a ground wire between the preamplifier and the turntable. If the preamp has a ground lug, it probably has provisions for accepting phono signals.

Some phono stages are simple boxes with a pair of RCA inputs, a pair of RCA outputs, and a power cord. Others offer multiple inputs, remote control, a front-panel display of gain and cartridge loading (described later in this chapter), and balanced inputs or outputs (Figs.4-3 and 4-4).

Fig. 4-3 A simple, inexpensive phono stage. (Courtesy Pro-Ject and Sumiko)

Fig. 4-4 An elaborate phono stage with multiple inputs, variable cartridge loading, front-panel gain adjustment, balanced outputs, and remote control. (Courtesy Aesthetix)

RIAA Equalization

In addition to amplifying the cartridge's tiny output voltage, the phono stage performs RIAA equalization on the signal. RIAA stands for Recording Industry Association of America, the body that standardized the record and playback equalization characteristics. Specifically, phono-stage RIAA equalization boosts the bass and attenuates the treble during playback. This equalization counteracts the bass cut and treble boost applied to the signal when the record was cut. By combining exactly opposite curves in disc mastering and playback, a flat response is achieved. RIAA equalization curves are shown in Fig.4-5. The dotted trace is the curve applied when the lacquer master is cut; the solid trace is the curve applied in the phono preamplifier. The combination of these complementary curves is flat response.

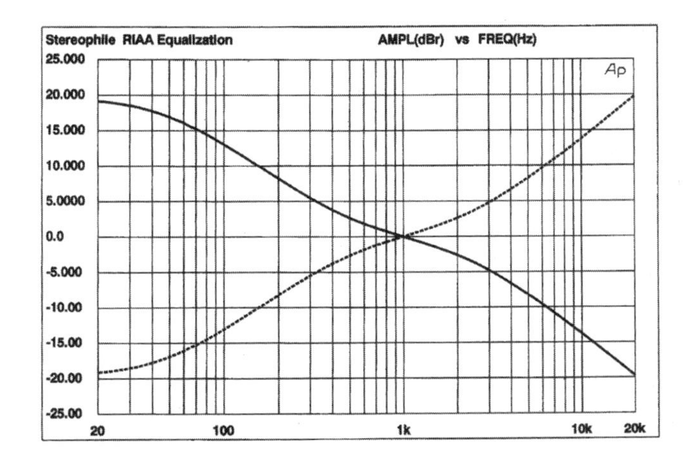

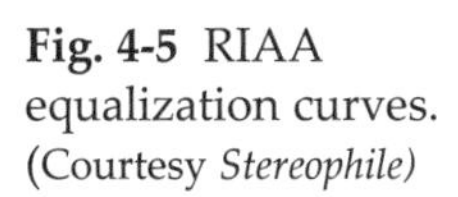

Fig. 4-5 RIAA equalization curves. (Courtesy *Stereophile*)

Attenuating bass and boosting treble when the disc is cut allows more signal to be cut into the record groove and increases playing time. Because bass takes up more room in the groove than does treble, reducing the amount of bass in the groove allows the grooves to be spaced closer together. In addition, the treble boost in disc cutting and subsequent treble cut on playback act as a noise-reduction system: attenuating treble on playback also attenuates record-surface noise. RIAA equalization is why you can't plug a line-level source into a preamplifier's phono input. Similarly, a phono cartridge can't be plugged into a preamplifier's line input.

One aspect of a phono preamp's technical performance is its RIAA accuracy. This measurement or specification indicates how closely the phono stage's RIAA equalization circuit matches the standard curve used in disc mastering. Typical performance is 20Hz–20kHz, ±0.5dB. That means the phono preamplifier will exhibit frequency-response variations (some frequencies are emphasized over others) of no more than 1dB (plus half a dB and minus half a dB) over the audioband of 20Hz to 20kHz. Because RIAA errors are essentially frequency-response errors, they can greatly affect the phono stage's sound. A positive RIAA error in the treble can make the phono stage bright and etched; a negative error in the treble can make the sound dull and lifeless. Examples of RIAA errors are shown in Chapter 16, "Specifications and Measurements."

Note that European audio equipment often has a low-frequency rolloff curve built into the phono stage's RIAA circuit. That is, low frequencies are intentionally reduced in volume. This prevents the very low frequencies (below about 30Hz) generated by warped records from getting into the rest of the system, but can also reduce bass extension on playback systems capable of reproducing frequencies down to 20Hz.

In the late 1990s, a few phono stages began appearing that offered alternate phono equalization curves on the premise that some older LPs might have been cut with an EQ curve other than RIAA. Before the RIAA standard was established in 1954, different record labels used their own equalization curves. If those recordings were played back with today's standard RIAA curve, the result would be a tonal imbalance, such as too much treble. Selectable equalization curves allow the listener to apply the correct equalization curve and hear a flat frequency response. These adjustable preamplifiers usually include a "Columbia" curve and a "Decca" curve in addition to RIAA.

If you don't listen to records made before 1954, selectable phono equalization curves is a moot point. All record labels converted to RIAA in the mid- to late-1950s; playing back an LP with a curve other than RIAA (which introduces greater treble rolloff) is a deviation from neutrality, but one that flatters the many overly bright records of that era. Some listeners, however, have suggested that proprietary equalization was used at some labels long after 1954, and that playing back those LPs with a "Columbia" or "Decca" curve results in more accurate sound.

Phono-Stage Gain

The amount of amplification provided by a phono stage (or any other amplifier) is called its gain. Gain is specified either in decibels (dB), or as a number expressing the ratio between input and output voltages. Phono stages have much more gain than line stages. Where a line stage may have 10 to 20dB of gain, a phono stage typically amplifies the signal by 40 to 75dB.

The amount of phono-stage gain required depends on the type of phono cartridge driving the phono preamplifier. Phono stages are of two varieties, each named for the type of cartridge with which it is designed to work. The first is the moving-magnet phono stage. Moving-magnet phono stages have their gain optimized to work with the relatively high output voltages from moving-magnet cartridges. Moving-magnet cartridges have high output voltages as cartridges go, on the order of two to eight millivolts (2–8mV). Consequently, they need less gain; a moving-magnet phono stage's gain is toward the lower end of the range, typically about 35dB.

Moving-coil cartridges have much lower output voltages due to their different method of generating a signal. Moving-coil output levels range from 0.15mV to 2.5mV. Consequently, they need more amplification (gain) in the phono preamplifier to reach line level than do moving-magnet cartridge signals. Moving-coil phono preamplifiers have about 35–60dB of gain. Note that moving-coil output voltages vary greatly with the cartridge design, with some so-called "high-output" models reaching moving-magnet levels.

Because of this wide variation in cartridge output level, a gain mismatch can occur between the cartridge and phono stage: The phono stage can have either too much or too little gain for a specific cartridge output voltage. If the phono preamp doesn't have enough gain, the volume control must be turned up very high for sufficient playback levels. This raises the noise floor (heard as a loud background hiss), often to the point of becoming objectionable. Conversely, a high-output cartridge can overload a moving-coil phono stage's input circuitry, causing distortion during loud musical passages. This condition is called input overload. A high-output cartridge driving a high-gain phono preamplifier can also make the preamplifier's volume control too insensitive. A moderate listening level may be achieved with the volume control barely cracked open; this makes small volume adjustments difficult. This latter condition may be accompanied by a distortion on musical peaks; the phono stage's excessive output level causes the linestage's input circuitry to overload on musical peaks.

Correctly matching the cartridge output voltage to the phono-stage gain avoids excessive noise and the possibility of input overload. A moving-coil cartridge specified at 0.18mV output needs about 75dB of gain. A typical moving-magnet output of 3mV should drive a phono stage that has about 50dB of gain. Some phono stages and full-function preamps have internal switches that adjust the gain between moving-magnet and moving-coil levels.

A pre-preamplifier is a small stand-alone component that can boost a moving-coil signal up to moving-magnet levels. If you have a moving-magnet phono input and upgrade to a moving-coil cartridge, you can add a pre-preamplifier instead of getting a new phono stage. Similarly, a step-up transformer increases a moving-coil's output voltage to a higher level. A transformer can improve the phono system's signal-to-noise ratio, or allow a moving-coil cartridge to drive a moving-magnet phono input.

High phono-stage gain carries the penalty of increased noise, particularly in all-tubed phono stages. Some tubed phono stages use a transistor (usually a FET, or Field Effect Transistor) in the first amplification stage, followed by subsequent tubed stages. This technique greatly increases the product's signal-to-noise ratio, but at the expense of adding a solid-state gain stage to the signal path. The amount of noise in a tubed phono stage is greatly influenced by the particular tubes in the unit. Replacing noisy tubes can greatly reduce the noise level. Although phono stages in general have poorer signal-to-noise specifications than other components, very-high-gain phono preamps can be objectionably noisy. All other factors being equal, the greater the gain, the higher the noise. Select a phono preamp with just enough gain for your cartridge.

For the mathematically inclined, here's a formula for precisely calculating the optimum phono-stage gain for a given cartridge output voltage. The formula is NdB = 20 log V1/V2. "NdB" is the number of decibels of gain, "V1" is 1 Volt (the desired output voltage from the phono stage), "V2" is you cartridge's output voltage, and "20 log" is 20 times the logarithm (base 10) to of the ratio between V1 and V2. We'll use an example of a cartridge with an output voltage of .7mV in our example. We first divide 0.0007 into 1 and get the ratio 1428. The logarithm of 1428 (you can use an on-line log calculator) is 3.155. We then multiply 3.155 by 20 to arrive at the optimum gain of 61dB.

Why is 1V the ideal output voltage from the phono stage? A cardinal rule of audio is that gain followed by attenuation degrades the signal-to-noise ratio. A typical power amplifier driving a loudspeaker of typical sensitivity produces a normal listening level when driven by a signal of about 1V. If the phono stage output level is 1V, the line stage between the phono stage and the power amplifier operates at unity gain (no gain or attenuation) and thus introduces the minimum amount of noise itself.

Cartridge Loading

Cartridge loading is the impedance and capacitance the phono cartridge "sees" when driving the phono input. Cartridge loading, specified in both impedance and capacitance, has a large effect on how the cartridge sounds, particularly moving-magnet types. Improper load-

only a continuously variable potentiometer (volume control), or a stepped knob that switches-in one of many discrete resistors.

Removing the capacitors, transistors (or vacuum tubes), circuit-board traces, and wiring found inside a line-stage preamplifier reduces the likelihood of degrading the signal passing through the passive level control. I've found passive controls to be extremely transparent, having very little effect on the signal passing through them. Only the best active line-stages have the passive level control's lack of coloration and degradation of the music. In addition, passive level controls are less expensive than active preamplifiers, making them attractive to the purist on a budget.

There are, however, many factors to consider when deciding if a passive level control will work in your system. Because the passive level control cannot amplify or buffer the source signal, the burden of driving the power amplifier and interconnects falls on the source component. I described a preamplifier's buffering function earlier, which is to act as an intermediary between source components and the power amplifier. Because of the active preamplifier's buffering function, source components have an easy time driving the preamplifier. The more difficult job of driving the power amplifier through interconnects falls on the shoulders of the active preamplifier.

Most source components are designed to drive the relatively high input impedance of preamplifiers (usually 47k ohms), not the low input impedance of passive level controls. The source component's output impedance must be added to the passive control's output impedance to find the total output impedance driving the interconnects and power amplifier. This high output impedance can cause high-frequency rolloff, particularly if high-capacitance interconnects run between the passive level control and the power amplifier. (A full explanation of how and why this high-frequency rolloff occurs is included in Appendix B.) Moreover, the passive level control's output impedance will vary with the amount of attenuation selected.

Passive level controls aren't the answer for all systems. They're often limited in functions, have only one or two inputs, and are tricky to match to the rest of your system. But replacing a colored line-stage preamplifier with a passive level control can greatly improve your system's sound. Listen to the passive level control in your system before you buy.

How a Preamplifier Works

A preamplifier consists of several stages. Each is an electronic circuit that performs a specific function, and can be pictured as a building block of an entire preamp.

The preamplifier's input stage is the first circuit element in the signal path. It acts as a buffer between the preamp's internal circuitry and the components driving it. That is, the input stage presents a high input impedance to source components driving the preamp, and a low output impedance to the next preamplifier stage. Impedance is, broadly speaking, the resistance to current flow. You'd intuitively think that more resistance to electrical current flow (a higher input impedance) in the preamplifier would put greater stress on the source components driving the preamp. The opposite is true: the higher the preamplifier's input impedance, the less current the source component must deliver.

Think of a garden hose connected to a faucet. The water pressure is the voltage from the source component, the hose is the interconnect cable, and your thumb and finger squeezing the hose are equivalent to the preamplifier's input impedance. The harder you squeeze the hose, the greater the resistance (impedance), reducing the amount of water flowing through it. If you let go of the hose altogether (a very low input impedance), the water pressure drops because the water supply is limited. In an audio system, some digital-source components with limited current output will distort (or have soft bass) when asked to drive a low-input-impedance preamplifier. The same digital source components will sound just fine when driving a high-input-impedance preamplifier. The preamplifier's high-impedance input stage is designed to make life easy for the source components driving the preamp.

The next circuit stage provides gain, amplifying the input signal. This is where the preamplifier actually boosts the signal level. (Although the input stage can have some gain.) The last stage is the output stage, the circuit block that serves as a buffer between the gain stage and the power amplifier connected to the preamp. The output stage has a high input impedance, low output impedance, and is designed to drive interconnects and a power-amplifier input. These circuit blocks are shown in Fig.4-6. Each of these stages can be made from transistors, vacuum tubes, integrated circuits (operational amplifiers, or op-amps), or a combination of these devices.

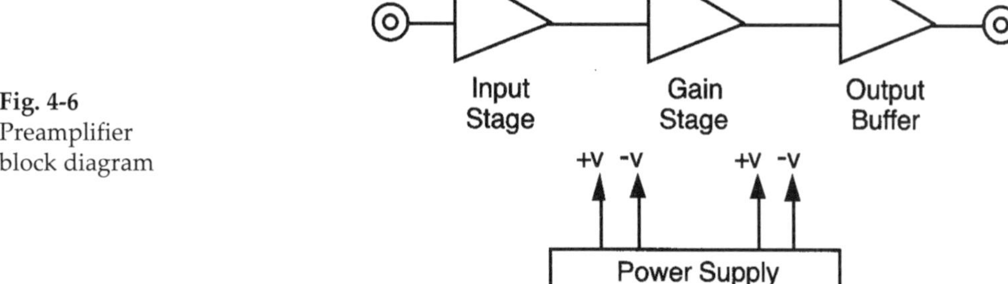

Fig. 4-6
Preamplifier block diagram

A preamplifier will also have a source-selection switch for selecting which source component is connected to your power amplifier for listening. This can be a series of relays controlled by electronic circuits, or simply a rotary mechanical switch.

The volume control is an important element of a preamplifier. Many preamps use a potentiometer, a continuously variable resistor in which a wiper attached to the preamp's volume knob moves in an arc around a resistive element. In preamps with remote control, this potentiometer (or "pot") is motor-driven. Another approach is to use a integrated circuit chip that provides digital control over the analog signal. The preamp's front-panel volume control (and remote) sends a digital code to the chip that selects which resistors of a resistor-ladder network the audio is routed through. The resistor values determine the amount of attenuation. This chip-based attenuator provides stepped levels rather than a continuous volume adjustment. The steps are typically half a dB in the range where the control is most often used, and as much as 3dB at the very bottom of the range. An improvement on this technique is the stepped attenuator built from discrete resistors rather than the resistors within a chip. The discrete resistors can be switched into the circuit directly by the rotary volume knob, or by transistors outside the signal path. In the latter case, the volume control is merely an optical encoder that controls which FETs, and thus which resistors, are in the audio signal path. Fig.4-7 shows a cost-no-object implementation of the discrete-resistor stepped attenuator. Note that the attenuator has four elements rather than two. This is because the preamplifier in which it is used is fully balanced—a distinction explained later in this chapter. A rarely used technique is to employ a transformer with many secondary taps, with the volume control selecting the transformer tap.

Fig. 4-7 A state-of-the-art switched-resistor attenuator. (Courtesy BAlabo and Pure Audio)

Most preamplifiers have separate miniature output stages to drive the tape output jacks. These output stages are nearly always of lower quality than the main output stage, but their sonic role is less crucial. Such products are said to have "independently buffered tape outputs."

If you plan to drive two power amplifiers from your preamplifier (as in a bi-amplified system), make sure that the preamplifier's output jacks are "independently buffered." This means that each output is driven by its own output stage rather than a single output stage driving both outputs.

A feature missing from many high-end preamplifiers is a headphone jack. This connection lets you plug headphones directly into the preamp and listen to music without turning on your power amplifier. Most of the headphone amplifiers in preamps that do include this feature are of compromised quality. The best way to listen to headphones is through a separate headphone amplifier that accepts the main output from your preamplifier. These products are described in Chapter 15. A few preamplifiers, however, offer a high-quality headphone amplifier. If headphone listening is important to you, look for such a model.

The preamplifier's power supply delivers a supply of direct current (DC) to power the preamplifier's circuits. To prevent this DC from appearing at the preamplifier output along with the audio signal, DC blocking capacitors are often placed between stages, or at the preamplifier output. Some preamps avoid capacitors in the signal path by using a DC servo circuit that also prevents DC from appearing at the output jacks. A preamplifier without capacitors in the audio signal path is said to be direct-coupled.

Class-A refers to a type of tube or transistor operation in which the tube or transistor is always biased on, meaning that it always conducts current. A discrete preamplifier is one that uses no integrated circuits (op-amps), only separate transistors or vacuum tubes. Discrete preamps are more expensive than op-amp–based designs. Virtually all state-of-the-art preamps are discrete Class-A designs. (Descriptions of Class-A operation are included in Appendix B and in Chapter 5, "Power and Integrated Amplifiers.")

Balanced and Unbalanced Preamplifiers

As described earlier in this chapter, preamplifiers can have balanced inputs, balanced outputs, or both. Although all balanced preamps have XLR jacks, not all balanced preamplifiers are created equal. Two preamps that have balanced inputs and outputs can be very different in how they treat the signal.

Most preamps accepting a balanced signal immediately convert it to an unbalanced signal, perform the usual preamplifier functions (provide gain and volume adjustment) on the unbalanced signal, then convert the signal back to balanced just before the main output. A preamp with balanced inputs and outputs, but unbalanced internal topology, often adds two active stages to the signal path: a differential amplifier at the input and the phase splitter at the output. A differential amplifier converts a balanced signal to unbalanced; a phase splitter converts an unbalanced signal to balanced. This kind of preamplifier is shown in block form in Fig.4-8.

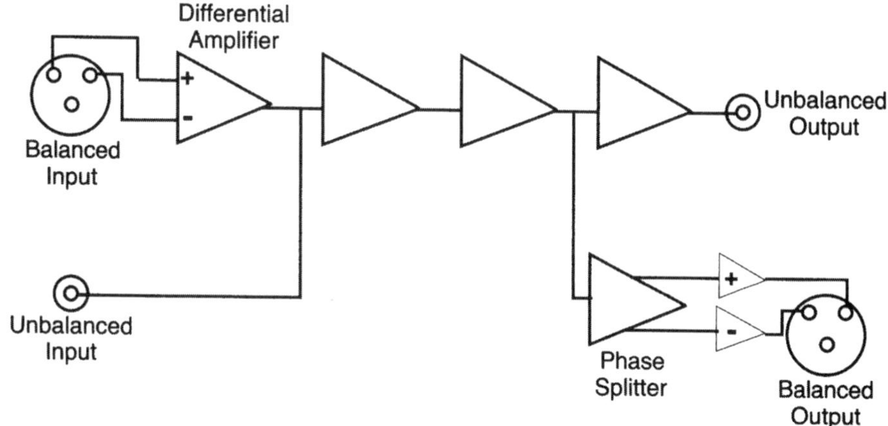

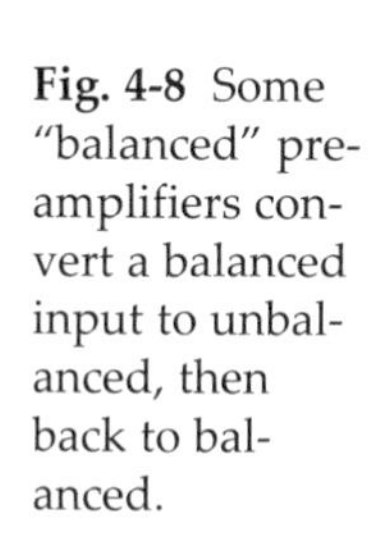

Fig. 4-8 Some "balanced" preamplifiers convert a balanced input to unbalanced, then back to balanced.

The preferred, but much more expensive, method is to keep the signal balanced throughout the preamplifier. This technique requires double the audio circuitry; each portion of the balanced signal is amplified separately. Moreover, very close tolerances between halves of the balanced signal are required. The two signal halves must have identical gain and noise characteristics. Further, the four-element volume control must maintain precise

Watt: A unit of electrical power. Power is the ability to do work; in this case, the ability of the amplifier to make a loudspeaker's diaphragm move.

Power Output: The maximum amount of power the amplifier can deliver to the loudspeaker, measured in watts.

Load: In power-amplifier terminology, a load is the loudspeaker(s) the power amplifier must drive.

Tubed: A tubed power amplifier uses vacuum tubes to amplify the audio signal.

Solid-State: A solid-state power amplifier uses transistors to amplify the audio signal.

Switching Amplifier (Class-D): A power amplifier in which the output transistors turn fully on or fully off. (Compare with a linear amplifier.)

Single-Ended Amplifier: A power amplifier in which the output device (a tube or transistor always amplifies the entire musical waveform. (Compare with "push-pull.")

Push-Pull: A power amplifier in which pairs of output devices (tubes or transistors) alternately "push" and "pull" current through the loudspeaker, with one output device amplifying the positive half of the waveform, another output device amplifying the negative half of the waveform. (Compare with "single-ended.")

Class A: A mode of amplifier operation in which the amplifying device (tube or transistor) amplifies the entire musical waveform.

Class A/B: A power amplifier that operates in Class A up to a small fraction of its output power, and then switches to Class B operation.

Hybrid: A power amplifier combining vacuum tubes and solid-state devices. The input and driver stages are usually tubed, the output stage is usually solid-state.

Bridging: Converting a stereo power amplifier into a monoblock power amplifier. Some power amplifiers have a rear-panel bridging switch. Also called "strapping" into mono.

Bi-amping: Driving a loudspeaker's midrange and treble units with one amplifier, the woofer with a second amplifier.

Digital Power Amplifier: An amplifier that takes in a digital, rather than an analog, audio signal. The digital signal is converted to a stream of pulses that turn on and off the output transistors.

How to Choose a Power Amplifier

Unlike most of the other components in your system, power amplifiers vary greatly in electrical performance. Consequently, choosing a power amplifier requires careful system matching for electrical compatibility, not just musical compatibility. While any digital source will function in a system (even though it may not be musically ideal), some power amplifiers just won't work well with certain loudspeakers on a technical level. Choosing a power amplifier is thus a technical, as well as aesthetic, decision that requires careful attention to system matching. We'll discuss these technical factors throughout this chapter.

Because a power amplifier's cost is often proportional to its output power, read this chapter's "How Much Power Do You Need?" section to select just the right power rating for your needs. Whatever your budget, the power amplifier should consume about 20 to 25% of your total system budget.

The first division of power amplifiers—a stereo unit or a pair of monoblocks—will be decided by your budget. Monoblocks generally start at about $2500 per pair. At this price level, a single stereo unit may make more sense; with only one chassis, power cord, and shipping carton, the manufacturer can put more of the manufacturing cost into better parts and performance. I advise against monoblocks if your amplifier budget is less than about $4000. There may be exceptions to this figure, but it nonetheless offers a broad guideline. Many excellent stereo units, for example, cost upward of $6000. A very popular price range for high-quality power amplifiers is $800–$2000, with the $2000 models sometimes offering musical performance close to that of the most expensive amplifiers.

Monoblocks generally perform better than a single stereo unit for several reasons. First, because the two amplifier channels are separate chassis, there is no chance of interaction between channels. Consequently, monoblocks typically have better soundstage performance than stereo units. Second, monoblocks have completely separate power supplies, even down to the power transformers: the left- and right-channel amplifier circuits don't have to share their electrical current source. This gives monoblocks the ability to provide more instantaneous current to the loudspeaker, all other factors being equal. Finally, most manufacturers put their cost-no-object efforts into monoblocks, which are often the flagships of their lines. If you want all-out performance and can afford them, monoblocks are the way to go.

Integrated Amplifiers

At the other end of the scale from monoblocks is the integrated amplifier, in which a preamplifier and a power amplifier are combined in the same chassis. Though the power output from integrated amplifiers is generally lower than that from separate power amplifiers, integrateds are much less expensive, and ideal for budget to moderately priced systems. True high-end integrated amplifiers start at about $300 and can cost as much as $15,000 for top models.

High-end integrated amplifiers have changed radically in the past few years. Once relegated to low-powered units from European manufacturers, with idiosyncratic operation and non-standard connectors, integrated amplifiers have finally come into their own. Leading high-end manufacturers have realized that an integrated amplifier makes sense for many music lovers. The cost and convenience advantages of an integrated amplifier are compelling: integrateds take up less space, are easier to connect, reduce the number of cables in your system, and can even offer the performance of separate components. Now that high-end manufacturers have taken the integrated amplifier seriously, they're putting their best technology and serious design efforts into their integrateds.

Consequently, many manufacturers have enjoyed booming sales of integrated amps in the $2000–$4000 price range that produce about 50–150Wpc (watts per channel) of power. Some manufacturers have even included a quality tuner with their integrated amplifier. Not so long ago, the term "high-end stereo receiver" was an oxymoron. Today, however, there's no reason why a receiver designed and built with the dedication given to separate components should offer anything but high-end musical performance.

These newer integrated amplifiers have also overcome one of the limitations of earlier designs: the inability to upgrade just the power amplifier or preamplifier section. Today's integrateds often include preamplifier-out jacks for connecting the integrated to a separate, more powerful amplifier. They also often have power-amplifier input jacks if you want to upgrade the preamplifier section. Higher-end integrated amplifiers feature a dual mono design in which the left and right audio channels are completely separate from each other, even down to the power transformers. These premium-quality integrateds also boast technologies found in upper-end preamplifiers, such as discrete-resistor stepped attenuators (described in the previous chapter), a discrete Class-A input stage, and fully balanced operation.

The digital age has driven a radical transformation of the integrated amplifier from a simple, no-frills product into a technological showcase. For example, today's integrated amplifiers often include a digital-to-analog converter with a USB input for connection to a computer-based music system. If you would like to plug a portable music player into your integrated amplifier, look for one with a front-panel 1/8" stereo jack. Some integrateds also offer an iPod docking port. If you're interested in the latter feature, there's a distinction you should know about: Some iPod docks tap into the iPod's analog signal, while others take the iPod's digital output and convert that digital signal to analog with the integrated amplifier's digital-to-analog converter. The latter approach replaces the iPod's compromised DAC and analog output stage with the integrated amplifier's superior circuitry.

Other useful features on an integrated amplifier include "gain offset" and "theater bypass." Gain offset allows you to attenuate the signal level on each input individually to compensate for the varying output levels of source components. This adjustment prevents large jumps in playback volume when switching between sources. Theater bypass is important if you plan on using the integrated amplifier as part of a home-theater system. This switch, or sometimes a dedicated input, sets the integrated amplifier at a fixed gain (the volume control is disabled) so that you calibrate the channel levels with an AV controller and maintain that calibration. Chapter 12 includes a complete description of how a stereo preamplifier or stereo integrated amplifier works with an AV controller in a home-theater system.

When choosing an integrated amplifier, combine the advice in Chapter 4 ("Preamplifiers") with the guidelines in the rest of this chapter. If your budget is under $5000 for amplification, seriously consider one of the new breed of high-end integrated amplifiers rather than separates.

How Much Power Do You Need?

The first question to answer when shopping for a power amplifier or integrated amplifier is how much output power you need. Power output, measured in watts into a specified loudspeaker impedance, varies from about 20Wpc in a very small integrated amplifier to about 1000Wpc. Most high-end power amplifiers put out between 80 and 250Wpc. Single-ended triode amplifiers, described later in this chapter, generally produce between 3Wpc and 20Wpc.

Choosing an appropriate amplifier power-output range for your loudspeakers, listening tastes, room, and budget is essential to getting the best sound for your money. If the amplifier is under-powered for your needs, you'll never hear the system at its full potential. The sound will be constricted, fatiguing, lack dynamics, and the music will have a sense of strain on climaxes. Conversely, if you spend too much of your budget on a bigger amplifier than you need, you may be shortchanging other components. Choosing just the right amplifier power is of paramount importance.

The amount of power needed varies greatly according to loudspeaker sensitivity, loudspeaker impedance, room size, room acoustics, and how loudly you like to play music. Loudspeaker sensitivity is by far the biggest determining factor in choosing an appropriate power output. Loudspeaker sensitivity specifies how high a sound-pressure level (SPL) the loudspeaker will produce when driven by a certain power input. A typical sensitivity specification will read "88dB SPL, 1W/1m." This means that the loudspeaker will produce an SPL of 88 decibels (dB) with one watt of input power when measured at a distance of one meter. Although 88dB is a moderate listening volume, a closer look at how power relates to listening level reveals that we need much more than one watt for music playback.

Each 3dB increase in sound-pressure level requires a doubling of amplifier output power. Thus, our loudspeaker with a sensitivity of 88dB at 1W would produce 91dB with 2W, 94dB with 4W, 97dB with 8W, and so on. For this loudspeaker to produce musical peaks of 109dB, we would need an amplifier with 128W of output power.

Now, say we had a loudspeaker rated at 91dB at 1W/1m—only 3dB more sensitive than the first loudspeaker. We can quickly see that we would need only half the amplifier power (64W) to produce the same volume of 109dB SPL. A loudspeaker with a sensitivity of 94dB would need just 32W to produce the same volume. The higher-sensitivity speaker simply converts more of the amplifier's power into sound.

This relationship between amplifier power output and loudspeaker sensitivity was inadvertently illustrated in an unusual demonstration more than 50 years ago. In 1948, loudspeaker pioneer Paul Klipsch conducted a demonstration of live vs. reproduced sound with a symphony orchestra and his Klipschorn loudspeakers. His amplifier power: 5W. The Klipschorns are so sensitive (an astounding 105dB SPL, 1W/1m) that they will produce very high volumes with very little amplifier power. Klipsch was attempting to show that his loudspeakers could closely mimic the tonal quality and loudness of a full symphony orchestra.

The other end of the speaker-sensitivity spectrum was illustrated by a demonstration I attended of an exotic new loudspeaker. During the demo, the music was so quiet that I could barely hear it. I looked at the power amplifiers—300Wpc monsters with large power meters—and was astonished to see that the power meters were nearly constantly pegged at full power. This unusual speaker converted only a minuscule amount of the amplifier's output power into sound.

impedance. I likened the situation to a water faucet and hose: the water pressure is voltage, the flow of water through the hose is electrical current, and squeezing the hose forms a resistance (impedance) to the flow. You can extend this analogy to a power amplifier driving a loudspeaker, with the loudspeaker's impedance forming the resistance in the hose. The lower the loudspeaker's impedance, the less the resistance to current flow from the amplifier, and the harder the amplifier must work to deliver current to the loudspeaker. As we'll see in Appendix A, if the impedance is halved (say, from 8 ohms to 4 ohms), the amplifier is asked to deliver double the current to the loudspeaker.

If the amplifier isn't up to the job, the musical result is strain or even distortion on musical peaks, weak bass, loss of dynamics, hardening of timbre, and a collapsing soundstage. In short, we can hear the amplifier give up as it runs out of power. Conversely, amplifiers that can continue increasing their output power as the impedance drops generally have very deep, extended, and powerful bass, virtually unlimited dynamics, a sense of ease and grace during musical peaks, and the ability to maintain correct timbre and soundstaging, even during loud passages. If you have relatively high-impedance loudspeakers with no severe impedance dips, you're much less likely to encounter sonic problems, even with modest power amplifiers; the loudspeaker simply demands less current from the power amplifier.

Amplifiers with high current capability (indicated by their ability to increase output power into low impedances) are often large and expensive. Their current capability comes from massive power transformers, huge power supplies, and lots of output transistors—all expensive items. Fig.5-1 shows such a power amplifier.

The massive power transformer and filter capacitors seen in the middle of the amplifier can supply large amounts of current to the circuits. The output stage has many output transistors working together for increased current capacity. These transistors are mounted to large heatsinks (the fins on the outside of an amplifier) to keep the transistors running relatively cool. Heatsinks transfer heat from the output transistors to the amplifier exterior so that it can be dissipated into the air.

In the "Specifications and Measurements" section on power amplifiers in Chapter 16, we'll look at how to evaluate a power amplifier's ability to deliver current into low-impedance loads. You can also figure out for yourself which amplifiers are good at driving low-impedance speakers by converting their output-power ratings at 8 ohms and 4 ohms into dBW. The less the difference in the dBW ratings into 8 and 4 ohms, the better the amplifier's ability to deliver current.

Keep in mind, however, that not all systems require large power amplifiers. If you have sensitive loudspeakers with a fairly high impedance, the loudspeaker's current demands are vastly lower. Consequently, smaller amplifiers work just fine. As we'll see in the discussion of single-ended triode amplifiers later in this chapter, amplifiers with as little as 3Wpc and very limited ability to deliver current can sound highly musical.

What to Look For when Comparing Power Ratings

When comparing amplifier power ratings, make sure the specified power is continuous or RMS rather than peak. Some manufacturers will claim a power output of 200W, for example, but not specify whether that power output is available only during transient musical events such as drum beats, or if the amplifier can deliver that power continuously into a

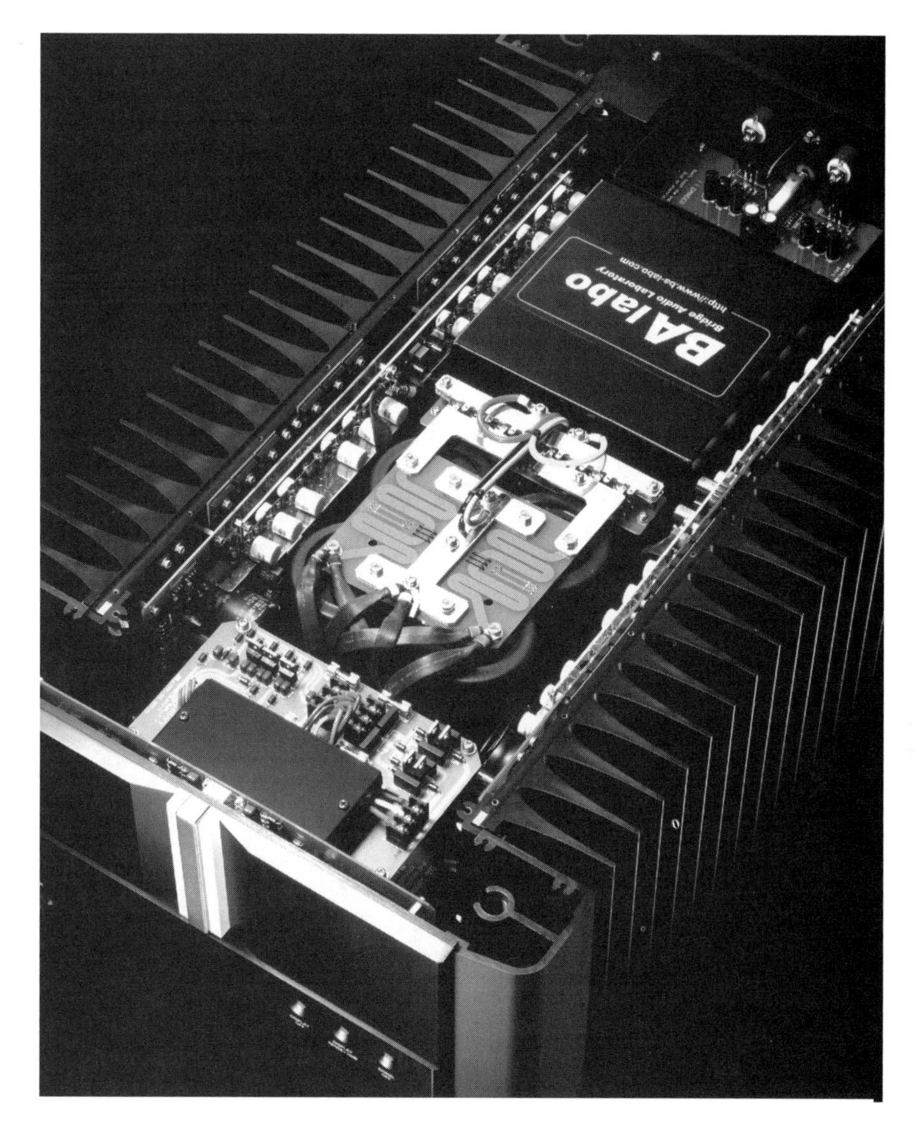

Fig. 5-1 Inside a high-current power amplifier (Courtesy BAlabo and Pure Audio)

load. RMS stands for "Root Mean Square," a mathematical calculation expressing the effective, or average, power output. Very few amplifiers are, however, specified by peak power.

Another way manufacturers exaggerate power ratings is by not specifying the power bandwidth. This term describes the frequency range over which a power amplifier can deliver its power. A power amplifier delivering 200W at 1kHz is far less powerful than one that can deliver 200W over the full audio bandwidth of 20Hz–20kHz. You'll often see mass-market audio/video receivers with power-output ratings specified only at 1kHz, or from 50Hz to 20kHz. Further, stereo power amplifiers can deliver more power with only one channel driven—look for the words "both channels driven." The maximum power output should also be specified at a certain distortion level.

You can see the potential for misleading power-amplifier output claims. The abuses were so bad at one time that the Federal Trade Commission (FTC) stepped in to regulate power claims—the only example of an audio specification being regulated by a governmental body. The FTC mandate for power ratings requires that the power rating be continuous (not peak), that the load impedance and bandwidth be specified, and that the Total Harmonic Distortion (THD) be given at full power and measured over the audio bandwidth. You may see a power specification that reads "50Wpc continuous (or RMS) power into 8 ohms, both channels driven, 20Hz–20kHz, with less than 0.1% THD." A power specification including all these conditions is called an "FTC power rating." Some manufacturers no longer adhere to the FTC-mandated power ratings, figuring that the issue has blown over and is no longer enforced. You see fudged power ratings on mass-market audio/video receivers that must now power five or seven loudspeakers rather than two, and in single-ended triode amplifiers that cannot meet the stringent FTC power-output specification requirements.

If you're amplifier-shopping for low-impedance loudspeakers, look at the power-output specifications into 4 ohms. Make sure you see the words "continuous" or "average" in the power rating. See if the bandwidth and distortion are specified. These figures don't tell us what we need to know about the amplifier's musical qualities, but nevertheless indicate good technical performance.

Why Amplifier Power Isn't Everything

We've seen how loudspeaker sensitivity greatly affects how much amplifier power you need, and how power amplifiers with the same 8 ohm power rating can differ radically in their abilities to drive loudspeakers. Now let's look at some other factors influencing how much amplifier power you need.

The first is room size. The bigger the room, the more amplifier power you'll need. A rough guide suggests that quadrupling the room volume requires a doubling of amplifier power to achieve the same sound-pressure level. How acoustically reflective or absorptive your listening room is will also affect the best size of amplifier for your system. If we put the same-sensitivity loudspeakers in two rooms of the same size, one room acoustically dead (absorptive) and the other acoustically live (reflective), we would need roughly double the amplifier power to achieve the same sound-pressure level in the dead room as in the live room.

Finally, how loudly you listen to music greatly affects how much amplifier power you need. Chamber music played softly requires much less amplifier power than rock or orchestral music played loudly. The relationships between loudspeaker sensitivity, room size, room acoustics, and amplifier power are shown in Fig.5-2.

We can see that a low-sensitivity loudspeaker, driven by orchestral music in the large, acoustically dead room of someone who likes high playback levels, may require hundreds of times the amplifier power needed by someone listening to chamber music at moderate listening levels through high-sensitivity loudspeakers in a small, live room. A 20Wpc amplifier may satisfy the second listener; the first listener may need 750Wpc.

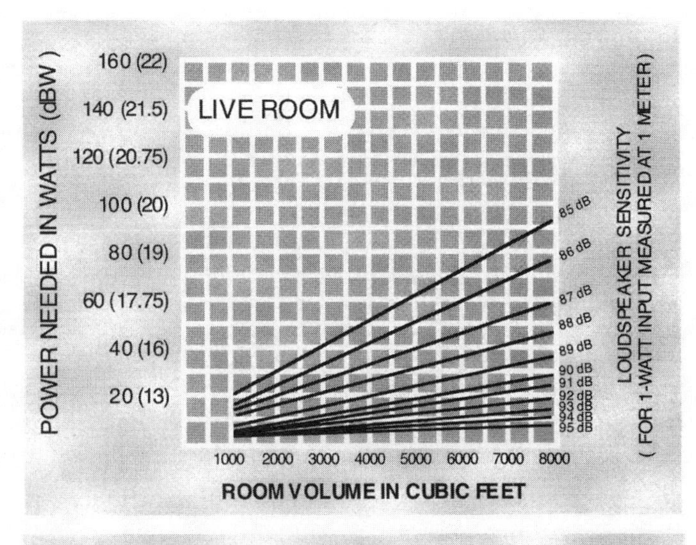

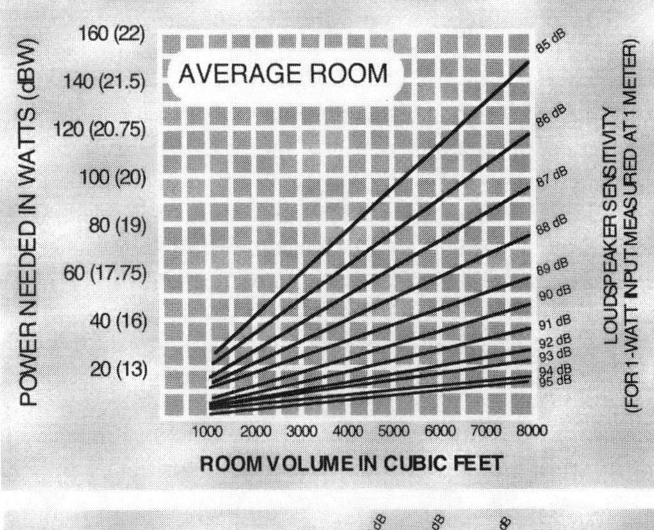

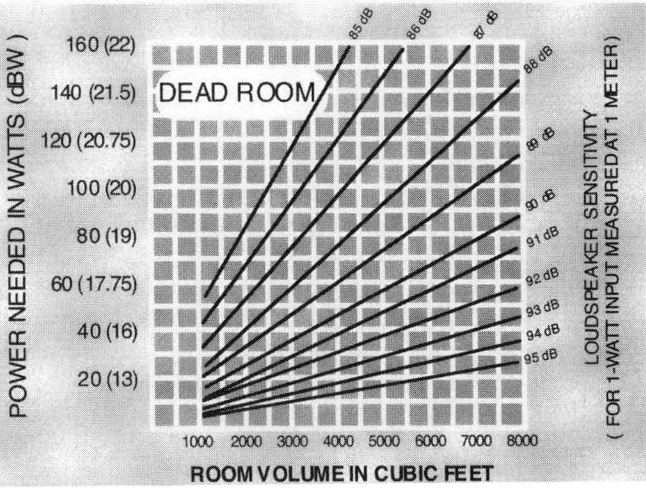

Fig. 5-2 The relationship between loudspeaker sensitivity, room size, room acoustics, and power-amplifier output power (after *High Fidelity*)

Other Power-Amplifier Considerations

Tubes vs. Transistors

When transistor amplifiers were introduced in the 1960s, it appeared that the days of the vacuum tube were over. Transistors were smaller, lighter, and cheaper than tubes, ran cooler, and produced more output power. If that wasn't enough, transistor amplifiers didn't need an output transformer, a component that added considerably to the amplifier's size, weight, and cost. All the audio manufacturers of the day scrapped their tubed amplifiers in favor of transistor units virtually overnight.

But many music lovers found the sound of these newfangled "solid-state" amplifiers unlistenable. They likened the sound to that of a pocket transistor radio, only louder. Unfortunately, those perceptive audiophiles couldn't buy a new tubed amplifier after the transistor's introduction.

The modern tubed amplifier was created in 1970 by William Zane Johnson of Audio Research Corporation. As did many music lovers, he found the sound of tubed amplifiers more musical. Johnson's demonstration of a tubed amplifier at a 1970 hi-fi show prompted one show-goer to remark, "You've just set the audio industry back 10 years!" But instead of being a setback, Johnson's amplifier began a renaissance in tubed equipment that is still going strong more than 40 years later.

The perennial tubes-vs.-transistors debate arises when you're faced with choosing a power amplifier. Tubed units can offer stunning musical performance, but they have their drawbacks. Here are the advantages and disadvantages of tubed power amplifiers. (For the moment, I'll confine my observations to conventional "push-pull" tubed amplifiers, not the exotic single-ended triode varieties described later in this chapter.)

Tubed power amplifiers are more expensive than their similarly powered solid-state counterparts. The cost of tubes, transformers, and more extensive power supplies all make owning a tubed power amplifier a more expensive proposition than owning a solid-state unit. More-over, the tubes will need replacing every few years, further adding to the real cost of ownership. A set of sixteen EL34 output tubes, for example, can cost from $200 to $400.

In terms of bass performance, tubed power amplifiers can't compete with good solid-state units. Tubes have less control in the bass, making the presentation less punchy, taut, and extended. Further, tubed power amplifiers often have limited current delivery into low-impedance loads, making them a poor choice for current-hungry loudspeakers. Tubes also require monthly biasing (small adjustments made with a screwdriver) to maintain top performance. Biasing is very easy, but some music lovers would prefer not to have to think about performing routine maintenance on their music-playback systems.

Power-amp tubes can fail suddenly, sometimes in smoke and (momentary) flames. Such dramatic failure is rare, however; I've used tubed amplifiers for decades and have had two tubes fail in that time, both of them uneventfully.

Finally, there is the possibility of small children or pets burning themselves on hot tubes. Many, but not all, tubed power amplifiers have exposed output tubes. If you have small children, consider a tubed amplifier that is surrounded by a ventilated metal cage.

Given these drawbacks, why would anyone want to own a tubed power amplifier? It's simple: tubes can sound magical. When matched with an appropriate loudspeaker, tubed power amplifiers offer unequaled musicality, in my experience. Even small, moderately priced tubed amplifiers have more than a taste of tube magic.

Many important aspects of music reproduction seem to come naturally to tubed amplifiers. They generally have superb presentation of instrumental timbre, smooth and unfatiguing treble, and spectacular soundstaging. The hard, brittle, edgy midrange and treble presentation of many solid-state amplifiers is contrasted with the purity of timbre and sense of ease conveyed by a good tubed amp. Music has a warmth, ease, and natural musicality when reproduced by many tubed designs. The soundstage has an expansive quality, with a sense of bloom around instrumental images. This isn't to say there aren't good-sounding solid-state power amplifiers, only that tubes seem to more consistently deliver the musical goods. A tubed amplifier's softer bass is often willingly tolerated for its magical midrange, treble, and sometimes holographic soundstaging.

Beware, however, of the tubed power amplifier that sounds lush on every recording; the amplifier is probably imposing its own sonic signature on the music. Some tubed amplifiers have euphonic colorations, or departures from accuracy that are initially pleasing to the ear. Listen for a change in timbre, spatial presentation, or bass quality with different recordings. These characteristics should change with the recording; if they don't, it's likely that the power amplifier is imposing a common sonic signature on the music. Ideally, the power amplifier will reveal exactly what's on the recording without adding any characteristics of its own.

As good as tubed amplifiers can sound, solid-state amplifiers have some decided sonic and technical advantages. For example, tubed units are no match for solid-state amplifiers in bass performance. Transistor power amplifiers have tighter, deeper, and much more solid bass than tubed units. The feeling of bass tautness, kick, extension, and power are all better conveyed by solid-state amplifiers, regardless of how good the tubed amplifier is. Speaking technically, solid-state amplifiers can deliver more current to low-impedance loudspeakers, making them a better choice for such loads.

No one but you can decide if a tubed power amplifier is ideal for your system. I strongly suggest, however, that you audition at least one tubed amplifier before making a purchasing decision. You may get hooked.

Balanced Inputs

If your preamp has balanced outputs, you may want to consider a power amplifier with balanced inputs. Most power amps with balanced inputs also provide unbalanced inputs, allowing you to compare these two connection methods before deciding which one to leave in your system. Some preamplifier/power-amplifier combinations sound better via balanced connection, others via the unbalanced jacks. The best way to discover which method is better is by listening to both. (See Chapter 4 for a discussion of why some preamplifiers may sound better from their unbalanced outputs.) Borrow a pair of interconnects from your dealer so you can make this comparison. If you have a power amplifier with both balanced and unbalanced inputs, but use the unbalanced inputs, be sure you have a shorting plug in the amplifier's balanced input jacks. This is a small U-shaped plug that shorts pin 3 of the balanced input connector to ground, preventing noise from getting into the amplifier.

Bridging

Some stereo power amplifiers can be "bridged" to function as monoblocks. Bridging configures a stereo amplifier to function as a more powerful single-channel amplifier. The amplifier will have a switch on the rear panel to convert the amplifier to bridged operation. Note that two bridged amplifiers are needed for stereo. If you have a stereo amplifier that can be bridged and you want more power, simply buy a second, identical amplifier and bridge the two for more total power. In theory, bridging results in a four-fold increase in output power. That's because bridging doubles the amplifier's maximum output voltage, and, according to Ohm's Law, quadruples the power. In practice, however, bridging roughly doubles an amplifier's 4 ohm power rating, due to the amplifier's current-output limitations.

Bridging changes the amplifier's internal connections so that one channel amplifies the positive half of the waveform and the other channel amplifies the negative half. The loudspeaker is connected as the "bridge" between the two amplifier channels instead of between one channel's output and ground.

Bridging is most beneficial when the power amplifiers are asked to drive low-sensitivity, high-impedance loudspeakers (8 ohms nominal). High-impedance speakers are driven more by voltage than by current. Conversely, low-impedance speakers demand more current from the power amplifier. Bridging doubles the amplifier's maximum output voltage, but quadruples its maximum current output (because two amplifier channels are now driving one loudspeaker). Moreover, connecting a 4 ohm speaker to a bridged power amplifier causes the amplifier to "see" a 2-ohm load, further stressing the amplifier's current capacity. The result can be amplifier overheating, which will either damage the amplifier or cause its protection circuit to activate and shut down the amplifier while music is playing.

Bi-Amping

Any loudspeaker with two sets of input terminals is a candidate for bi-amping. In a bi-amped setup, two power amplifiers are used to drive one loudspeaker. One amplifier drives the woofer, the other drives the midrange and tweeter.

Let's first consider passive bi-amping. The preamplifier output is split into two left-channel signals and two right-channel signals. (Many preamps have two stereo output pairs for this purpose.) The left-channel preamp output drives the left-channel inputs on both power amps. The right-channel preamp output drives the right-channel inputs on both power amps. One power amp drives the bass input on both loudspeakers, the other power amp drives the midrange and tweeter input on both loudspeakers. Note that the loudspeaker's crossover still operates; both amplifiers receive and amplify a full-bandwidth signal. This is called passive bi-amping because the frequency division into bass and midrange/treble is performed by passive components (capacitors, resistors, and inductors) in the loudspeaker's crossover.

In active bi-amping, rather than allow the loudspeaker's crossover to split the frequency spectrum, an electronic crossover between the preamp and power amps does the job. Active bi-amping divides the frequency spectrum before the power amplifiers, not after, as in passive bi-amping. The losses and problems inherent in any passive crossover are eliminated by active bi-amping. Moreover, each power amplifier receives only the portion of the frequency band it will amplify. Active bi-amping is more efficient than passive bi-amping, but requires bypassing the loudspeaker crossovers and adding an electronic crossover. Bi-amping not only increases the total power delivered to the loudspeaker, but allows each amplifier to work over a narrower frequency range. We can thus choose a bass amplifier for its superior bottom-end performance regardless of how it sounds in the mids and treble, and a midrange/treble amplifier without concern for its bass. An ideal combination may include a large solid-state amplifier for the bass and a sweet-sounding tubed amplifier for the top end. The small tubed amplifier is relieved of the burden of driving the bass, allowing it to do what it does best. Bi-amping is one way of getting the best of both the tubed and solid-state worlds.

Another advantage of (active) bi-amping is that the woofer amplifier can be connected directly to the woofer's voice coil rather than through a passive crossover network, particularly the series inductor found in most woofer crossovers. This provides much tighter damping (the ability of the amplifier to control woofer motion) and results in tighter, deeper bass reproduction.

When bi-amping is chosen, the two amplifiers' gains (their amounts of amplification) must be exactly matched. If you're using identical amplifiers, the gain will be inherently matched. But with different amplifiers, a gain mismatch will produce too much or too little bass. An amplifier can be padded down at the input with an attenuator to reduce the input level and match its output to the second amplifier. If you're using an active external crossover, it may include independent output-level controls to balance the levels between the woofer and midrange/tweeter amplifiers.

Bi-amping can be tricky. I don't recommend it for most systems unless you fully understand what's happening, or unless your dealer helps you. One serious drawback is that replacing the speaker's internal crossover with an external one changes the crossover frequencies and characteristics the loudspeaker designer decided were optimum for that loudspeaker's drivers. Many loudspeaker designers don't take kindly to someone second-guessing their crossover characteristics.

A few manufacturers—Linn Products, for example—have designed easy bi-amping upgradeability into their products; bi-amping is just a matter of installing active crossover boards in the company's power amplifiers, adding a second amplifier, and removing the crossovers inside the loudspeaker. In this case, the loudspeaker works optimally with the external crossover.

Incidentally, adding an active subwoofer (one with an integral power amplifier—see Chapter 6) automatically makes your system bi-amped. The subwoofer's internal amplifier drives the low bass, and your main amplifier handles only the upper bass, mids, and treble. Installing an active subwoofer is one way of getting the benefits of bi-amping without changing amplifiers or adding another electronic crossover. Loudspeakers that come with a dedicated woofer amplifier, such as those employing servo-driven woofers are also inherently bi-amplified.

The benefits of bi-amping can be extended to tri-amping, or three amplifiers driving one loudspeaker.

What to Listen For

How can you tell if the power amplifier you're considering will work well with your loudspeakers? Simple: Borrow the amplifier from your dealer for the weekend and listen to it. This is the best way of not only assessing its musical qualities, but determining how well it drives your loudspeakers. In addition, listening to the power amplifier at home will let you hear if the product's sonic signature complements the rest of your system.

The next best choice is if the dealer sells the same loudspeakers you own, and allows you to audition the combination in the store. If neither of these options is practical, consider bringing your loudspeakers into the store for a final audition.

All the sonic and musical characteristics described in Chapter 3 apply to power amplifiers. However, some sonic characteristics are more influenced by the power amplifier than by other components.

The first thing to listen for is whether the amplifier is driving the loudspeakers adequately. The most obvious indicator is bass performance. If the bass is soggy, slow, or lacks punch, the amplifier probably isn't up to the job of driving your loudspeakers. Weak bass is a sure indication that the amplifier is underpowered for a particular pair of speakers. Other telltale signs that the amplifier is running out of current include loss of dynamics, a sense of strain on musical peaks, hardening of timbre, reduced pace and rhythm, and soundstage collapse or congestion. Let's look at each of these individually.

First, play the system at a moderate volume. Select music with a wide dynamic range—either full orchestral music with a loud climax accompanied by bass drum, or music with a powerful rhythmic drive from bass guitar and kickdrum working together. Audiophile recordings typically have much wider dynamic ranges than general-release discs, making them a better source for evaluation. Music that has been highly compressed to play over the radio on a 3" car speaker will tell you less about what the system is doing dynamically.

After you've become used to the sound at a moderate level, increase the volume—you want to push the amplifier to find its limits. Does the bass seem to give out when you turn it up, or does the amplifier keep on delivering? Listen to the dynamic impact of kickdrum on a recording with lots of bottom-end punch. It should maintain its tightness, punch, quickness, and depth at high volume. If it starts to sound soggy, slow, or loses its power, you've gone beyond the amplifier's comfortable operating point. After a while, you can get a feel for when the amplifier gets into trouble. Is the sound strained on peaks, or effortless? NOTE: When performing this experiment, be sure not to overdrive your loudspeakers. Turn down the volume at the first sign of loudspeaker overload (distortion or a popping sound).

Compare the amplifier's sound at high and low volumes. Listen for brass instruments becoming hard and edgy at high volumes. See if the soundstage degenerates into a confused mess during climaxes. Does the bass drum lose its power and impact? An excellent power amplifier operating near its maximum power capability will preserve the senses of space, depth, and focus, while maintaining liquid instrumental timbre. Moreover, adequate power will produce a sense of ease; lack of power often creates listener fatigue. Music is much more enjoyable through a power amplifier with plenty of reserve power.

All the problems I've just described are largely the result of the amplifier running out of current. Just where this happens is a function of the amplifier's output power, its ability to deliver current into the loudspeaker, the loudspeaker's sensitivity and impedance, the room size, and how loudly you expect your hi-fi system to play. Even when not pushed to its maximum output, a more powerful amplifier will often have a greater sense of ease, grace, and dynamics than a less powerful amplifier.

Assuming that the power amplifier can drive your loudspeakers, let's evaluate its musical characteristics.

First, listen for its overall perspective. Determine if it's forward or laid-back, then decide if the amplifier is too "up-front" or too recessed for your system or tastes. A forward perspective seems to put you in Row C of the concert hall; a laid-back perspective puts you in Row W. Forwardness gives the impression of presence and immediacy, but can quickly become fatiguing. A presentation that's too laid-back is less exciting and compelling. If the amplifier's overall perspective isn't right, nothing else it does well may matter.

Next, listen for brightness, hardness, treble grain, and treble etch. If you feel a sense of relief when the volume is turned down, you can be pretty sure that the treble is at fault. Listen to midrange timbre, particularly in voice, violin, piano, and flute. Midrange textures should be sweet and liquid, and devoid of grain, hardness, or a glassy edge. Many power amplifiers—particularly inexpensive solid-state units—make the treble dry, forward, and unpleasant. If you hear even a hint of these characteristics in a few hours' listening, rest assured that they will only grow more annoying over time. Many moderately priced amplifiers don't add these unmusical characteristics—keep shopping.

Power amplifiers vary greatly in their ability to throw a convincing soundstage. With naturally miked recordings, the soundstage should be deep, transparent, and focused, the loudspeakers and listening room seeming to disappear. Instruments should hang in space within the recorded acoustic. The presentation should provide a transparent, picture-window view into the music. If the sound is flat, veiled, or homogenized, you haven't found the right amplifier.

Bass performance is also largely influenced by the power amplifier. It should sound taut, deep, quick, dynamic, and effortless. The music should be propelled forward, involving your body in its rhythmic drive. Kickdrum should have depth, power, and a feeling of suddenness. The bass guitar and kickdrum should provide a tight, solid, and powerful musical foundation.

The best way to find out which power amplifiers sound good and are well-made is by conferring with a trusted dealer and reading product reviews in reputable high-end magazines. Responsibly written reviews can save you from spending time and trouble on unworthy contenders, and point you toward products that offer promise. Thorough reviews will also include insightful commentary on the amplifier's technical performance. If the amplifier has technical characteristics that make it questionable or unsuitable for a particular type of loudspeaker, the review will reveal this fact. A review's technical section may state that, "Although the XYZ power amplifier sounded terrific with the Dominator 1 loudspeakers, its high output impedance and limited current delivery make it a poor choice with low-sensitivity loudspeakers (below 86dB/1W/1m), or those with an impedance below 5 ohms." This kind of commentary can point you toward the best amplifiers for your loudspeakers.

In short, read lots of product reviews before drawing up a list of candidates to audition. Restrict candidates to those that fall within your budget, and take the few contenders home for an evening's listen. And don't be in a hurry to buy the first amplifier you audition—listen to several before choosing. Your effort and patience will be rewarded with a more musical amplifier, and one better suited to your system.

A Survey of Amplifier Types

So far I've described mainstream power amplifiers that most audiophiles are likely to buy. But a number of variations on the basic design are a significant force in the amplifier marketplace. These include the single-ended triode amplifier, the single-ended solid-state amplifier, switching power amplifiers, and digital amplifiers.

Single-Ended Triode Amplifiers

One of the most interesting trends in high-end audio over the past 20 years has been the return of the single-ended triode power amplifier. The single-ended triode (SET) amplifier was the first audio amplifier ever developed, dating back to Lee De Forest's patent of the triode vacuum tube in 1907 and his triode amplifier patent of 1912. SET amplifiers generally deliver very low power, sometimes just a few watts per channel.

You heard right: Large numbers of audiophiles are flocking to replace their modern power amplifiers with amplifiers based on 100-year-old technology. Have the past 100 years of amplifier development been a complete waste of time? A surprising number of music lovers and audio designers think so.

The movement back to SET amplifiers began in Japan in the 1970s, specifically with designer Nobu Shishido, who combined SET amplifiers with high-sensitivity horn-loaded loudspeakers. Many who heard SET amplifiers were startled by their goosebump-raising musical immediacy and ability to make the music "jump" out of the loudspeakers. Thus began the rage for SET amplifiers in Japan, which is about 10 years ahead of the SET trend in the United States. You can't open a high-end audio magazine today without seeing lots of ads for very-low-powered single-ended triode amplifiers. The SET enthusiast's mantra, coined by reviewer Dick Olsher, is, "If the first watt of amplifier power doesn't sound good, why would you want 199 more of them?"

One of the early audio-amplifier triodes, the Western Electric 300B, was suddenly in such demand that audiophiles were paying as much as $500 for a single tube. This shocking fact prompted Western Electric to start producing the 300B again. If you told Western Electric executives of the 1980s that in 1997 they would put the 300B back into production, they'd have thought you were crazy. (A single-ended triode power amplifier using the 300B output tube is shown in Fig.5-3 The 300B is the bulbous tube in the middle.)

The triode is the simplest of all vacuum tubes; its glass envelope encloses just three electrical elements rather than the five elements in the more common (and modern) pentode tube. Triodes have much less output power than pentodes, but more benign distortion characteristics. Virtually all modern tubed amplifiers before the SET comeback used pentodes. In a single-ended triode amplifier, the triode configured so that it always amplifies the entire audio signal. That's what "single-ended" means. Virtually all other power amplifiers are "push-pull," meaning that opposing pairs of tubes (or transistors) alternately "push" and "pull" current through the loudspeaker. (Later in this chapter we'll look more closely at amplifier output-stage topology and class of operation.) SET proponents believe that because the triode amplifies the entire waveform, SET amplifiers offer the ultimate in sound quality and musicality. Moreover, SET amplifiers have no need for a circuit called a phase splitter, making them even simpler. Note that a single-ended tubed amplifier can use more

Fig. 5-3 A single-ended triode monoblock amplifier based on the 300B tube (Courtesy Cary Audio)

than one output tube; what makes it single-ended is that the tubes are configured in such a way that they always conduct current throughout the entire musical waveform.

In addition, SET circuits are extremely simple and often use very little or no negative feedback. Negative feedback is taking some of the amplifier's output signal and feeding it back to the input. Such feedback lowers distortion, but many feel that any feedback is detrimental to amplifier musicality.

On the test bench, SET amplifiers have laughably bad technical performance. They typically produce fewer than 25Wpc of output power, and have extremely high distortion—as much as 10% Total Harmonic Distortion (THD) at the amplifier's rated output. Although most SET amplifiers use a single triode output tube, additional triodes can be combined to produce more output power. Some SET amplifiers, however, put out just 3Wpc. In fact, there's a kind of cult around SET amplifiers that strives for lower and lower output power. These SET enthusiasts believe that the lower the output power, the better the sound. One SET designer told a reviewer, in all seriousness, "If you liked my 9W amplifier, wait until you hear my 3W model."

In addition to low output power and high distortion, SET amplifiers have a very high output impedance as amplifiers go: on the order of 1.5–3 ohms. This is contrasted with the 0.1 ohm output impedance of most solid-state amplifiers, and the 0.8 ohms of many push-pull tubed designs. Because a loudspeaker's impedance isn't constant with frequency, the SET amplifier's high output impedance reacts with the loudspeaker's impedance variations to produce changes in frequency response. That is, the SET amplifier will have a different frequency response with every loudspeaker it drives. These variations can range from just 0.1dB with some loudspeakers that have a fairly constant impedance, to as much as 5dB with other loudspeakers. The SET amp's sound will therefore be highly dependent on the loudspeaker with which it is matched.

Despite these technical drawbacks, my listening experience with SET amplifiers suggests that this ancient technology has many musical merits. SET amps have a certain presence and immediacy of musical communication that's hard to describe. It's as though the musicians aren't as far removed from here-and-now reality as they are with push-pull amplifiers. SET amps also have a wonderful liquidity and purity of timbre that is completely devoid of grain, hardness, and other artifacts of push-pull amplifiers. When I listen to SET amplifiers (with the right loudspeakers), it's as though the musicians have come alive and are playing in the listening room for me. There's a directness of musical expression that's impossible to put into words, but is immediately understood by anyone who has listened for themselves. You must hear a SET firsthand to know what the fuss is about; no description can convey how they sound.

When auditioning a SET amplifier, it's easy to be seduced by the midrange. That's because SET amplifiers work best in the midband, and less well at the frequency extremes of bass and treble. If the SET demo is being run for your benefit, be sure to listen to a wide variety of music, not just small-scale music or unaccompanied voice—these will accentuate the SET's strengths and hide its weaknesses.

The importance of matching a SET amplifier to the right loudspeaker cannot be overemphasized. With a low-sensitivity speaker, the SET will produce very little sound, have soft bass, and reproduce almost no dynamic contrast. The ideal loudspeaker for a SET amplifier has high sensitivity (higher than 93dB/1W/1m), high impedance (nominal 8 ohms or higher), and no impedance dips (a minimum impedance of 6 ohms or higher). Such a speaker will produce lots of sound for a small amount of input power, and require very little current. (Remember the example earlier of Paul Klipsch's high-sensitivity speaker that produced concert-hall levels with 5W?) There's been a resurgence in high-sensitivity speakers that has paralleled the popularity of SET amplifiers. Some loudspeakers designed for SET amplifiers have sensitivities of more than 100dB, which enable them to produce satisfying listening levels with 5Wpc. SET amplifiers are often coupled with horn-loaded loudspeakers, which have extremely high sensitivity but, in my experience, often introduce unacceptable levels of coloration.

The popularity and unmistakable sound quality of SET amplifiers pose a serious dilemma: how can an amplifier that performs so poorly by every "objective" measure produce such an involving musical experience? How can 100-year-old technology eclipse, in many ways, amplifiers designed in the 21st century? What aren't we measuring in SET amplifiers that would reflect their musical magic? Why do conventional measurements fail so dismally at quantifying what's right in SET amplifiers? Do SET amplifiers sound so good *because* of their high distortion, or *despite* it? As of yet, no one has the answers to these questions.

Single-Ended Solid-State Amplifiers

Single-ended amplifiers aren't confined to those using ancient vacuum-tube technology. Transistors can also be configured to amplify the entire musical waveform. A solid-state, single-ended amplifier is shown in Fig.5-4. Note the large heatsinks required to dissipate the additional heat produced by Class-A operation.

Single-ended solid-state amplifiers have better technical performance than single-ended triode amps, with a lower output impedance, more power, and the ability to drive a

Fig. 5-4 A single-ended solid-state power amplifier (Courtesy Pass Laboratories)

wider range of loudspeakers. They share many of the benefits of SET amps, particularly the very simple signal path, lack of crossover distortion, and greater linearity. Although single-ended solid-state amplifiers generally produce less power than their push-pull counterparts, they generally have much more output power than single-ended tubed units. Nonetheless, it's a mistake to equate single-ended solid-state with single-ended tube amplifiers: there are so many other design variables that single-ended solid-state and single-ended tubed amplifiers should be considered completely different animals.

Switching (Class D) Power Amplifiers

If single-ended triode amplifiers represent a return to fundamental technology, the switching power amplifier may represent the future of audio amplification. Switching amplifiers, also called Class D amplifiers, have been gaining in popularity due to their small size, low weight, high efficiency, and low cost. You can hold some 250Wpc switching amplifiers in the palm of an outstretched hand. Many of them dissipate so little heat that they can be housed in an enclosed equipment cabinet—something you'd never do with a conventional amplifier (also called a linear amplifier). That's because a linear amplifier is typically about 15% efficient, meaning that it converts only about 15% of the power it draws from the wall into the electrical signal that drives the loudspeakers. The other 85% is dissipated as heat. By contrast, a switching amplifier is as much as 90% efficient, converting just 10% of its power draw into heat. Fig.5-5 shows a switching power amplifier. This monoblock delivers 175W into 8 ohms and 335W into 4 ohms. It weighs just eight pounds and measures 8.5" x 14" x 1.8".

Switching amplifiers are sometimes erroneously called "digital" amplifiers, but that appellation is reserved for a special type of switching amplifier described later in this chapter.

At the low-end of the audio spectrum, switching amplifiers are ubiquitous in home-theater-in-a-box units and as integral subwoofer amplifiers. A home-theater-in-a-box may

Fig. 5-5 This eight-pound switching power amplifier can deliver 175W into 8 ohms. (Courtesy NuForce)

need to power six loudspeakers from a DVD player-sized chassis—all for a few hundred dollars. Such a unit can output perhaps 300 watts (50Wpc x 6), yet run cool enough to be placed in a cabinet. In this application, the advantages of a switching amplifier are undeniable. But are switching amplifiers suitable for high-end systems?

Before tackling that question, let's first look at how a switching amplifier works. In a conventional linear amplifier, the output transistors amplify a continuously variable analog signal—the musical waveform. The current flow through the output transistors (or tubes) is continuously variable, and a direct analog of the musical waveform. In a switching amplifier, the analog input signal is converted into a series of on and off pulses. These pulses are fed to the output transistors, which turn the transistors fully on or fully off. When the transistors are turned on, they conduct the DC supply voltage to the loudspeaker. When turned off, no voltage is connected to the loudspeaker. The audio information is contained in the durations of these on-off cycles. The train of pulses amplified by the transistors is smoothed by a filter to recover the musical waveform and remove the switching noise. Because the signal amplitude is contained in the width of the pulses, switching amplifiers are also called pulse-width modulation (PWM) amplifiers. In fact, the Direct Stream Digital encoding system used in SACD is nearly identical to the pulse-width modulation scheme in Class D power amplifiers. (The graphic in Chapter 7 showing how Direct Stream Digital's pulses represent an analog waveform is most illustrative.)

One major drawback of switching amplifiers is the need for an extremely clean power supply. Remember that the transistors switch the DC supply directly to the loudspeaker. Consequently, any noise or ripple (a small amount of AC on the DC voltage) will appear at the loudspeaker's input terminals. Moreover, the output transistors may not turn on and off at precisely the right time, introducing distortion. Finally, switching amplifiers generate large amounts of switching noise that must be filtered by large inductors and capacitors at the amplifier output. In practice, the sound quality of switching amplifiers seems to be highly dependant on the environment, the loudspeaker, and the loudspeaker cables. A switching amplifier that sounds reasonably good in one system might be unlistenable in another.

Nonetheless, some successful high-end amplifiers employ switching technology. The field is relatively new, and manufacturers are finding ways to get good sound from switching amplifiers. A few of the high-end switching amplifiers I've heard sounded excellent, suggesting that switching technology may have a future in products other than car stereo and home-theaters-in-a-box.

Digital Amplifiers

A related amplifier technology uses a switching output stage, but accepts digital, rather than analog, input signals. These "digital" amplifiers take in the pulse-code modulation (PCM) signal from a CD transport, music server, or other source and convert those audio data to a pulse-width modulated signal. This PWM signal drives the output transistors, just as in a switching amplifier. The difference between a switching amplifier and a digital amplifier is that the digital amplifier accepts digital data rather than an analog signal.

This difference might not seem that great at first glance, but consider the signal path of a conventional digital-playback chain driving a switching power amplifier. In your CD player, data read from the disc go through a digital filter and are converted to analog with a DAC; the DAC's current output is converted to a voltage with a current-to-voltage converter; the signal is low-pass filtered and then amplified/buffered in the CD player's analog-output stage. This analog output signal travels down interconnects to a preamplifier with its several stages of amplification, volume control, and output buffer. The preamp's output then travels down another pair of interconnects to the power amplifier, which typically employs an input stage, a driver stage, and the switching output stage. In addition to the D/A conversion, that's typically six or seven active amplification stages before the signal gets to the power amplifier's output stage.

To reiterate the contrast with a true digital amplifier, PCM data are converted by DSP into the pulse-width modulation signal that drives the output transistors. That's it. There are no analog gain stages between the PCM data and your loudspeakers. The signal stays in the digital domain until the switching output stage, which, by its nature, acts as a digital-to-analog converter in concert with the output filter. The volume is adjusted in DSP. Digital amplifiers are usually not just power amplifiers, but also include a range of inputs, source selection, and volume control, effectively giving them the functional capabilities of an integrated amplifier. Fig.5-6 shows a digital amplifier, and Fig.5-7 is a block diagram comparing the signal path of this amplifier with a conventional system architecture.

Fig. 5-6 A true digital amplifier accepts digital input signals for conversion to a pulse-width modulated signal. (Courtesy NAD Electronics)

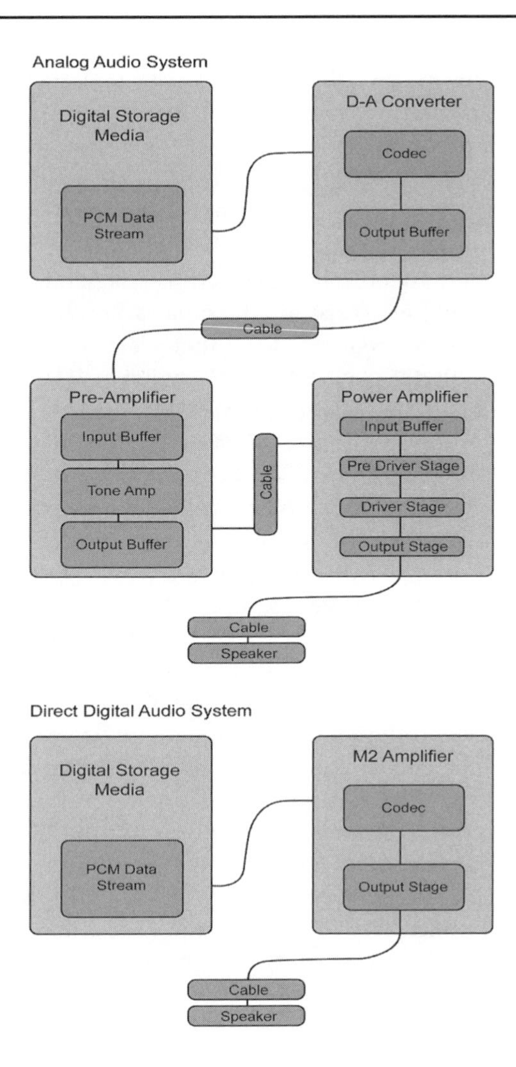

Fig. 5-7 Block diagram comparing a conventional analog audio system architecture with a direct-digital amplifier architecture (Courtesy NAD Electronics)

Output Stage Topology and Class of Operation

We've seen that the output stage of a power amplifier can be configured either as single-ended or push-pull. To reiterate, in a singled-ended amplifier the output devices (tubes or transistors) are configured so that they always amplify the entire musical waveform. The single-ended amplifier cannot operate in any other way—that's the very definition of "single-ended." In a push-pull output stage, pairs of opposing devices are arranged so that they work alternately—one device "pushes" current through the loudspeaker and then other device "pulls" current through the loudspeaker. Multiple pairs of output devices can be grouped together to increase the power output, called parallel push-pull.

That's a description of an amplifier's output-stage topology—how the amplifying devices are configured. Now let's look at a separate but related factor—the amplifier's class of operation.

Class of operation is how a given output stage is driven. The two main classes of operation are Class A and Class B. In a Class A amplifier, the output stage (single-ended or push-pull) amplifies the entire musical waveform. In a Class B amplifier, opposing pairs of transistors (or tubes) are operated so that one transistor in the pair amplifies the positive half of the waveform and the opposing transistor amplifies the negative half of the waveform.

Class of operation is easily confused with single-ended and push-pull output-stage topologies. But there's an important distinction. A single-ended amplifier always operates in Class A. But a push-pull amplifier can be operated in either Class A or Class B. It might seem like a push-pull Class A amplifier is a contradiction in terms, but it's not. In a push-pull Class A amplifier, opposing pairs of transistors are driven in such a way that current flows through both transistors throughout the entire musical waveform. All the output devices participate in amplifying through the full cycle of the audio signal. One device pulls current through the loudspeaker and the other pushes current, but both are always turned on and conducting current.

By contrast, in a Class B push-pull output stage, one transistor amplifies the signal during the positive-going portion of the signal, and the other amplifies the negative-going signal half. When one is working, the other is resting (and getting some needed cooling). The two factors are the output-stage topology—how the output devices are configured—and class of operation—how those output devices are driven. Here are four statements that may clarify the matter:

- Single-ended *topology* always *operates* in Class A
- Class A *operation* can be implemented in a single-ended or push-pull *topology*
- Push-pull *topology* can be *operated* in Class A or Class B
- Class B *operation* is always implemented by a push-pull *topology*

Let's move on from these distinctions to the more practical aspects of amplifier class of operation.

Most power amplifiers are called Class A/B power amplifiers because they operate in Class A at very low power outputs, then default to Class B operation at higher power outputs. A 100Wpc amplifier may put out 5W of Class A power, and then switch to Class B above that level. Even the heftiest Class A/B amplifiers can put out only a small portion of their rated power in Class A. A typical value is about 1 or 2%. Although this may not sound like much, an amplifier may be running at just a couple of watts at low listening levels with high-sensitivity speakers.

How much of an amplifier's power output is Class A is determined by the amount of bias applied to the output transistors. Bias is a DC current that flows through the output stage at idle. The higher the bias, the more current flows through the transistors when no signal is present. More bias results in more Class A output power. Class B operation has no bias current; Class A/B has moderate bias current; Class A has high bias current. The designer can keep increasing the bias in a push-pull output stage until all of its output power capability is delivered in Class A mode. The amplifier's power output rating would be the point at which the amplifier leaves Class A operation and begins operating in Class B. The limiting factors to increasing the bias current are the ability of the transistors to handle the greatly increased current flow through them, the power supply to keep up with the transistors' current demands, and the heatsinking to dissipate the considerable heat caused by the high bias current.

To give you an idea of the demands placed on the output stage, power supply, and heatsinking by an output stage biased for Class A operation, let's compare two products with identical output stages that are biased completely differently. The two products are the Pass Labs INT-150, a 150Wpc Class A/B integrated amplifier and the Pass Labs INT-30A, a 30Wpc pure Class A integrated amplifier. The INT-150's push-pull output stage is biased so that it produces 10Wpc of Class A power before switching into Class B to deliver its full rated output of 150Wpc. The amplifier can double its output power to 300Wpc into 4 ohms. The INT-30A is exactly the same amplifier, employing the same power supply, an identical number and type of output transistors, and the same heatsinks as the INT-150, but is rated at just 30Wpc. The difference is that the INT-30A's 30Wpc are pure Class A watts. The amplifier biased into Class A delivers just one fifth the power output of its Class A/B counterpart.

As you can see, Class A operation is enormously inefficient. A Class A amplifier converts nearly all the power it draws from the wall outlet into heat, and consumes just as much power at idle as when it is operating at its maximum output power. Moreover, a Class A amplifier is much more expensive to build on a "watts per dollar" basis than its Class A/B counterpart.

So why would designers go to the trouble and expense of creating Class A amplifiers, and why would consumers pay such a huge premium for "Class A watts" over "Class A/B watts"?

Class A has many theoretical and practical advantages. For starters, Class B and A/B amplifiers suffer from crossover distortion, a discontinuity at the zero-crossing point where one transistor in the opposing pair "hands off" the signal to the other transistor in the pair. A waveform discontinuity can occur at this transition, and is lessened as bias current is introduced and increased in value. Crossover distortion can't occur in Class A operation because each transistor amplifies the entire audio waveform, not just half of it. Second, the large thermal hardware capacity required by Class A has the advantage of keeping the output transistors in better thermal stability (a more constant temperature). This makes their operating characteristics more uniform, and less subject to changes resulting from the signal characteristics the transistors are amplifying. For example, if the transistors have just delivered a surge of current to the loudspeakers, they won't behave differently—and thus sound different—immediately afterward because they are momentarily hotter. Third, increasing the bias current so that an amplifier produces more Class A power not only reduces harmonic distortion, but more importantly, changes the nature of the harmonic distortion. As the bias is increased, the upper-order harmonics (everything above the third harmonic) that are most sonically harmful are most dramatically reduced in amplitude, leaving the predominant distortion component the more sonically benign second and third harmonics.

Class A power amplifiers can sound extremely good, with a sweetness and liquidity that set them apart from Class A/B amps. In my experience (I've been listening to a pair of 100W Class A monoblocks for the past 18 months as of this writing), Class A amplifiers have many of the virtues of a tubed amplifier but without the tube amp's technical limitations. This isn't to say that a Class A amplifier mimics the sonic character of tubes, but rather that Class A avoids many of the characteristic nonlinear distortions of Class A/B solid-state amplifiers.

How a Power Amplifier Works

It's intuitive to conceptualize a power amplifier as a device that simply makes the input signal bigger. A more accurate visualization is of a device that pulls 120V, 60Hz alternating current (AC) from your wall outlet through the amplifier's power transformer, converts the AC into direct current (DC), stores that energy in large capacitors (the can-shaped objects), and then allows the tiny audio signal at the amplifier input to modulate the stored energy as electrical current that is driven through the loudspeakers by the power amplifier's output transistors or tubes.

This way of thinking about a power amplifier produces the realization that a power amplifier's power supply (the section that converts the AC from the wall outlet to the DC that supplies the output transistors) is actually in the audio signal path. The current that ultimately drives the cones in your loudspeakers back and forth comes from the wall outlet via the amplifier's power supply. That's one reason a power amplifier's power supply is so important—and why good power amplifiers tend to have large and heavy power supplies. At its most basic level, a power amplifier consists of a power supply, an input stage, a driver stage, and an output stage. The input and driver stages prepare the signal for the output stage, which is where transistors or tubes drive electrical current through the loudspeaker. Let's look at each of these sections in more detail.

The Power Supply

The power supply converts 60Hz, 120V AC (Alternating Current) from the wall outlet into the DC the audio circuits need. Power-amplifier power supplies differ greatly from those for other audio components. (A close reading of Appendix B will be helpful in understanding the rest of this explanation.) Power-amplifier supplies must be able to provide large amounts of current, operate at much higher voltages, and have much larger storage capacities. Parts sizes and ratings are thus much higher in a power amplifier than in the power supply of a preamp or disc player.

The first element of the power supply is the power transformer, which steps down the 120V AC line voltage to a lower value, perhaps 80V. You can get a pretty good idea of how powerful the amplifier is by the power transformer's size and rating. Transformers are rated in "VA" or "kVA" (x 1000); i.e., voltage (V) multiplied by the maximum current in amperes (A) they can supply. A transformer with a VA rating of 1kVA can deliver 8.34 amperes of current (1000 divided by the AC line voltage of 120V) to the amplifier's circuits (assuming the transformer isn't stepping down the voltage). Because most power transformers step down the voltage to 80V, a 1kVA transformer could deliver 12A.

Note that the amplifier's overall continuous output current for both channels combined can never exceed the power transformer's current rating. The amplifier can, however, deliver short bursts of higher current (such as on musical transients like bass-drum whacks) because of the energy stored in the capacitors—provided the output transistors can handle the current. One very large power amplifier's transformer, with a VA rating of 4.5kVA, weighs 85 of the amplifier's total 147 pounds. A power amplifier of more moderate size may have a power transformer with a rating of 300VA.

switchable option also lets you select the best output mode for your loudspeakers. High-sensitivity speakers may work just fine when driven by the amplifier in triode mode; lower-sensitivity models may require the higher output power provided by pentode operation. Another type of tubed output stage, called ultra-linear, was developed in 1951 by David Hafler and Herbert Keroes. Ultra-linear operation is a method of connecting a pentode so that it operates more like a triode. Specifically, some of the plate voltage is returned to the pentode's screen grid. Ultra-linear operation became the dominant circuit topology in tubed amplifiers for decades because it offered the output power of pentodes with the distortion characteristics of triodes.

Some tubed amplifiers, particularly the single-ended variety, let you vary the amount of feedback. You can thus dial in the sound you want, and decide which musical tradeoffs are acceptable. Reducing the amount of feedback often results in purer timbre, but at the expense of bass tautness.

Note that tubed amplifiers don't increase their power outputs as the load impedance drops, as do solid-state units. The output transformer acts as a current source rather than as a voltage source, decreasing the available output voltage when the load impedance drops. This is one reason why tubed amplifiers have less tight and extended bass than solid-state power amplifiers.

Finally, tubed power-amplifier power supplies are different from their solid-state counterparts. The voltage at the output-stage supply rail is many hundreds of volts, not ±60V to ±100V as in a solid-state amplifier's power supplies. Because of this very high voltage, the reservoir capacitors generally store vastly more energy than the reservoir caps in a solid-state amp. This so-called "B+" voltage is very high—the capacitors can deliver lots of current through your body, and can even kill you. Energy storage is measured in Joules, and is a function of the capacitance in Farads (a unit of stored charge) multiplied by the voltage squared. Given the same capacitance (same number of microfarads) but ten times the voltage, power supplies for a tubed amplifier have much greater energy storage than a solid-state amplifier. This factor may account for tube amplifiers' subjectively greater loudness for the same rated power compared to a solid-state amplifier.

Tubed amplifiers must also supply the tube filaments (the part that glows) with a low DC voltage. In low-signal-level tubes, the number beginning the tube's designation (6DJ8, 12AX7, etc.) indicates the filament voltage, usually 6V or 12V. Filaments are also called by their more descriptive name, "heaters."

From a design standpoint, tubed power amplifiers have several inherent advantages. First, they're simple. Comparing the schematics of solid-state and tubed amplifiers reveals many more parts in the transistor design, most of them in the signal path. Fewer parts in the signal path generally results in better sound. When overdriven, tubes soft-clip, meaning that the waveform gently rounds off rather than flattening immediately. Moreover, a tubed amplifier in clipping produces primarily second- and third-harmonic distortion, which is fairly benign sonically. A transistor amplifier driven into clipping immediately produces a whole series of very objectionable upper-order odd harmonics (fifth, seventh, ninth). In fact, 10% second-harmonic distortion is less annoying than 0.5% seventh-harmonic distortion.

6 Loudspeakers

Introduction

Of all the components in your audio system, the loudspeaker's job is by far the most difficult. The loudspeaker is expected to reproduce the sound of a pipe organ, the human voice, and a violin through the same electromechanical device—all at the same levels of believability, and all at the same time. The tonal range of virtually every instrument in the orchestra is to be reproduced from a relatively tiny box. This frequency span of 10 octaves represents a sound-wavelength difference of 60 feet in the bass to about half an inch in the treble.

It's no wonder that loudspeaker designers spend their lives battling the laws of physics to produce musical and practical loudspeakers. Unlike other high-end designers who create a variety of products, the loudspeaker designer is singular in focus, dedicated in intent, and deeply committed to the unique blend of science and art that is loudspeaker design.

Although even the best loudspeakers can't convince us that we're hearing live music, they nonetheless are miraculous in what they can do. Think about this: a pair of loudspeakers converts two two-dimensional electrical signals into a three-dimensional "soundspace" spread out before the listener. Instruments seem to exist as objects in space; we hear the violin here, the brass over there, and the percussion behind the other instruments. A vocalist appears as a palpable, tangible image exactly between the two loudspeakers. The front of the listening room seems to disappear, replaced by the music. It's so easy to close your eyes and be transported into the musical event.

To achieve this experience in your home, however, you must carefully choose the best loudspeakers from among the literally thousands of models on the market. As we'll see, choosing loudspeakers is a challenging job.

Another mistake is to drive high-quality loudspeakers with poor amplification or source components. The high-quality loudspeakers will resolve much more information than lesser loudspeakers—including imperfections in the electronics and source components. All too many audiophiles drive great loudspeakers with mediocre source components and never realize their loudspeakers' potential. Match the loudspeakers' quality to that of the rest of your system. (Use the guidelines in Chapter 2 to set a loudspeaker budget within the context of the cost of your entire system.)

3) Musical Preferences and Listening Habits

If the perfect loudspeaker existed, it would work equally well for chamber music and heavy metal. But because the perfect loudspeaker remains a mythical beast, musical preferences must play a part in choosing a loudspeaker. If you listen mostly to small-scale classical music, choral works, or classical guitar, a minimonitor would probably be your best choice.

Conversely, rock listeners need the dynamics, low-frequency extension, and bass power of a large full-range system. Different loudspeakers have strengths and weaknesses in different areas; by matching the loudspeaker to your listening tastes, you'll get the best performance in the areas that matter most to you.

Other Guidelines in Choosing Loudspeakers

In addition to these specific recommendations, there are some general guidelines you should follow in order to get the most loudspeaker for your money.

First, buy from a specialty audio retailer who can properly demonstrate the loudspeaker, advise you on system matching, and tell you the pros and cons of each candidate. Many high-end audio dealers will let you try the loudspeaker in your home with your own electronics and music before you buy.

Take advantage of the dealer's knowledge—and reward him with the sale. It's not only unfair to the dealer to use his or her expensive showroom and knowledgeable salespeople to find out which product to buy, and then look for the loudspeaker elsewhere at a lower price; it also prevents you from establishing a mutually beneficial relationship with him or her.

In general, loudspeakers made by companies that make only loudspeakers are better than those from companies who also make a full line of electronics. Loudspeaker design may be an afterthought to the electronics manufacturer—something to fill out the line. Conversely, many high-end loudspeaker companies have an almost obsessive dedication to the art of loudspeaker design. Their products' superior performance often reflects this commitment. There are, however, a few companies that produce a full line of products, including loudspeakers, that work well with each other.

Don't buy a loudspeaker based on technical claims. Some products claiming superiority in one aspect of their performance may overlook other, more important aspects. Loudspeaker design requires a balanced approach, not reliance on some new "wonder" technology that may have been invented by the loudspeaker manufacturer's marketing department. Forget about the technical hype and listen to how the loudspeaker reproduces music. You'll hear whether or not the loudspeaker is any good.

Don't base your loudspeaker purchases on brand loyalty or longevity. Many well-known and respected names in loudspeaker design of 20 years ago are no longer competitive. Such a company may still produce loudspeakers, but its recent products' inferior performance only throws into relief the extent of the manufacturer's decline. The brands the general public thinks represent the state of the art are actually among the worst-sounding loudspeakers available. These companies were either bought by multinational business conglomerates who didn't care about quality and just wanted to exploit the brand name, or the company has forsaken high performance for mass-market sales.

The general public also believes that the larger the loudspeaker and the more drivers it has, the better it is. Given the same retail price, there is often an inverse relationship between size/driver count and sonic performance. A good two-way loudspeaker—one that splits the frequency spectrum into two parts for reproduction by a woofer and a tweeter—with a 6" woofer/midrange and a tweeter in a small cabinet is likely to be vastly better than a similarly priced four-way in a large, floorstanding enclosure. Two high-quality drivers are much better than four mediocre ones. Further, the larger the cabinet, the more difficult and expensive it is to make it free from vibrations that degrade the sound. The four-way speaker's more extensive crossover will require more parts; the two-way can use just a few higher-quality crossover parts. The large loudspeaker will probably be unlistenable; the small two-way may be superbly musical.

If both of these loudspeakers were shown in a catalog and offered at the same price, however, the large, inferior system would outsell the high-quality two-way by at least 10 times. The perceived value of more hardware for the same money is much higher.

The bottom line: You can't tell anything about a loudspeaker until you listen to it. In the next section, we'll examine common problems in loudspeakers and how to choose one that provides the highest level of musical performance.

Finding the Right Loudspeaker—Before You Buy

You've done your homework, read reviews, and narrowed down your list of candidate loudspeakers based on the criteria described earlier—you know what you want. Now it's time to go out and listen. This is a crucial part of shopping for a loudspeaker, and one that should be approached carefully. Rather than buying a pair of speakers on your first visit to a dealer, consider this initial audition to be simply the next step. Don't be in a hurry to buy the first loudspeaker you like. Even if it sounds very good to you, you won't know how good it is until you've auditioned several products.

Audition the loudspeaker with a wide range of familiar recordings of your own choosing. Remember that a dealer's strategic selection of music can highlight a loudspeaker's best qualities and conceal its weaknesses—after all, his job is to present his products in the best light. Further, auditioning with only audiophile-quality recordings won't tell you much about how the loudspeaker will perform with the music you'll be playing at home, most of which was likely not recorded to high audiophile standards. Still, audiophile recordings are excellent for discovering specific performance aspects of a loudspeaker. The music selected for auditioning should therefore be a combination of your favorite music, and diagnostic recordings chosen to reveal different aspects of the loudspeaker's performance. When listening to your favorite music, forget about specific sonic characteristics and pay attention to how much you're enjoying the sound. Shift into the analytical mode only when playing the diagnostic recordings. Characterize the sound quality according to the sonic criteria described in Chapter 3, and later in this chapter.

Visit the dealer when business is slow so you can spend at least an hour with the loudspeaker. Some loudspeakers are appealing at first, and then lose their luster as their flaws begin to emerge over time. The time to lose patience with the speakers is in the dealer's showroom, not a week after you've bought them. And don't try to audition more than two sets of loudspeakers in a single dealer visit. If you must choose between three models, select between the first two on one visit, then return to compare the winner of the first audition with the third contender. You should listen to each candidate as long as you want (within reason) to be sure you're making the right purchasing decision.

Some loudspeakers have different tonal balances at different listening heights. Be sure to audition the loudspeaker at the same listening height as your listening chair at home. A typical listening height is 36", measured from the floor to your ears. Further, some loudspeakers with first-order crossovers sound different if you sit too close to them. When in the showroom, move back and forth a few feet to be certain the loudspeaker will sound the same as it should at your listening distance at home.

Make sure the loudspeakers are driven by electronics and source components of comparable quality to your components. It's easy to become infatuated with a delicious sound in a dealer's showroom, only to be disappointed when you connect the loudspeakers to less good electronics. Ideally, you should drive the loudspeakers under audition with the same level of power amp as you have at home, or as you intend to buy with the loudspeakers.

Of course, the best way to audition loudspeakers is in your own home—you're under no pressure, you can listen for as long as you like, and you can hear how the loudspeaker performs with your electronics and in your listening room. Home audition removes much of the guesswork from choosing a loudspeaker. But because it's impractical to take every contender home, and because many dealers will not allow this, save your home auditioning for only those loudspeakers you are seriously considering.

What to Listen For

There are several common flaws in loudspeaker performance that you should listen for. Though some of these flaws are unavoidable in the lower price ranges, a loudspeaker exhibiting too many of them should be quickly passed over.

Listen for thick, slow, and tubby bass. One of the most annoying characteristics of poor loudspeakers is colored, peaky, and pitchless bass. You should hear distinct pitches in bass notes, not a low-frequency, "one-note" growling under the music. Male speaking voice is a good test for upper-bass colorations; it shouldn't have an excessive or unnatural chesty sound.

Certain bass notes shouldn't sound louder than others. Listen to solo piano with descending or ascending lines played evenly in the instrument's left-hand, or lower, registers. Each note should be even in tone and volume, and clearly articulated. If one note sounds different from the others, it's an indication that the loudspeaker may have a problem at that frequency.

The bottom end should be tight, clean, and "quick." When it comes to bass, quality is more important than quantity. Poor-quality bass is a constant reminder that the music is being artificially reproduced, making it that much harder to hear only the music and not the loudspeakers. The paradigm of what bass should not sound like is a "boom truck." Those car stereos are designed for maximum output at a single frequency, not articulate and tuneful bass. Unfortunately, more bass is generally an indicator of worse bass performance in low- to moderately-priced loudspeakers. A lean, tight, and articulate bass is preferable in the long run to the plodding boominess that characterizes inferior loudspeakers.

Listen to kickdrum and bass guitar working together. You should hear the bass drum's dynamic envelope through the bass guitar. The drum should lock in rhythmically rather than seem to lag slightly behind the bass guitar. A loudspeaker that gets this wrong dilutes rhythmic power, making the rhythm sound sluggish, even slower. But when you listen to a loudspeaker that gets this right, you'll find your foot tapping and hear a more "upbeat" and involving quality to the music.

Midrange coloration is a particularly annoying problem with some loudspeakers. Fortunately, coloration levels are vastly lower in today's loudspeakers than they were even since the previous edition of this book (2004). Still, there are lots of colored loudspeakers out there. These can be identified by their "cupped hands" coloration on vocals, a nasal quality, or an emphasis on certain vowel sounds. A problem a little higher in frequency is manifested as a "clangy" piano sound. A good loudspeaker will present vocals as pure, open, and seeming to exist independently of the loudspeakers. Midrange problems will also make the music sound as though it is coming out of boxes rather than existing in space.

Poor treble performance is characterized by grainy or dirty sound to violins, cymbals, and vocal sibilants (s and sh sounds). Cymbals should not splash across the soundstage, sounding like bursts of undifferentiated white noise. Instead, the treble should be integrated with the rest of the music and not call attention to itself. The treble shouldn't sound hard and metallic; instead, cymbals should have some delicacy, texture, and pitch. If you find that a pair of speakers is making you aware of the treble as a separate component of the music, keep looking.

Another thing to listen for in loudspeakers is their ability to play loudly without congestion. The sounds of some loudspeakers will be fine at low levels, but will congeal and produce a giant roar when pushed to high volumes. Listen to orchestral music with crescendos—the sound should not collapse and coarsen during loud, complex passages.

Finally, the loudspeakers should "disappear" into the soundstage. A good pair of loudspeakers will unfold the music in space before you, giving no clue that the sound is coming from two boxes placed at opposite sides of the room. Singers should be heard as pinpoint, palpable images directly between the loudspeakers (if that's how they've been recorded). The sonic image of an instrument should not "pull" to one side or another when the instrument moves between registers. The music should sound open and transparent, not thick, murky, or opaque. Overall, the less you're aware of the loudspeakers themselves, the better.

Some loudspeakers with less-than-high-end aspirations have colorations intentionally designed into them. The bass is made to be big and fat, the treble excessively bright to give the illusion of "clarity." Such speakers are usually extremely sensitive, so that they'll play loudly in comparisons made without level matching. These loudspeakers may impress the unwary in a two-minute demonstration, but will become extremely annoying not long after you've brought them home. You're unlikely to find such products in a true high-end audio store.

Finally, the surest sign that a loudspeaker will provide long-term musical satisfaction at home is if, during the audition, you find yourself greatly enjoying the music and not thinking about loudspeakers at all.

The flaws described here are only the most obvious loudspeaker problems; a full description of what to listen for in reproduced music in general is found in Chapter 3.

Loudspeaker Types and How They Work

Many mechanisms for making air move in response to an electrical signal have been tried over the years. Three methods of creating sound work well enough—and are practical enough—to be used in commercially available products. These are the dynamic driver, the ribbon transducer, and the electrostatic panel. A loudspeaker using dynamic drivers is often called a box loudspeaker because the drivers are mounted in a box-like enclosure or cabinet. Ribbon and electrostatic loudspeakers are called planar loudspeakers because they're usually mounted in flat, open panels. The term transducer describes any device that converts energy from one form to another. A loudspeaker is a transducer because it converts electrical energy into sound.

The Dynamic Driver

The most popular loudspeaker technology is without question the dynamic driver. Loudspeakers using dynamic drivers are identifiable by their familiar cones and domes. The dynamic driver's popularity is due to its many advantages: wide dynamic range, high power handling, high sensitivity, relatively simple design, and ruggedness. Dynamic drivers are also called point-source transducers because the sound is produced from a point in space.

Dynamic loudspeaker systems use a combination of different-sized dynamic drivers. The low frequencies are reproduced by a cone woofer. High frequencies are generated by a tweeter, usually employing a small metal or fabric dome. Some loudspeakers use a third dynamic driver, the midrange, to reproduce frequencies in the middle of the audio band.

Despite the very different designs of these drivers, they all operate on the same principle (Fig.6-1). First, the simplified explanation: Electrical current from the power amplifier flows through the driver's voice coil. This current flow sets up a magnetic field around the voice coil that expands and collapses at the same frequency as the audio signal. The voice coil is suspended in a permanent magnetic field generated by magnets in the driver. This permanent magnetic field interacts with the magnetic field generated by current flow through the voice coil, alternately pushing and pulling the voice coil back and forth. Because the voice coil is attached to the driver's cone, this magnetic interaction pulls the cone back and forth, producing sound.

On a more technical level, the voice coil is a length of wire wound around a thin cylinder called the voice-coil former. The former is attached to the diaphragm (a cone or dome). Electrical current from the amplifier flows through the voice coil, which is mounted in a permanent magnetic field whose magnetic lines of flux cross the gap between two permanent magnets. According to the "right-hand rule" of physics, the circular flow of current

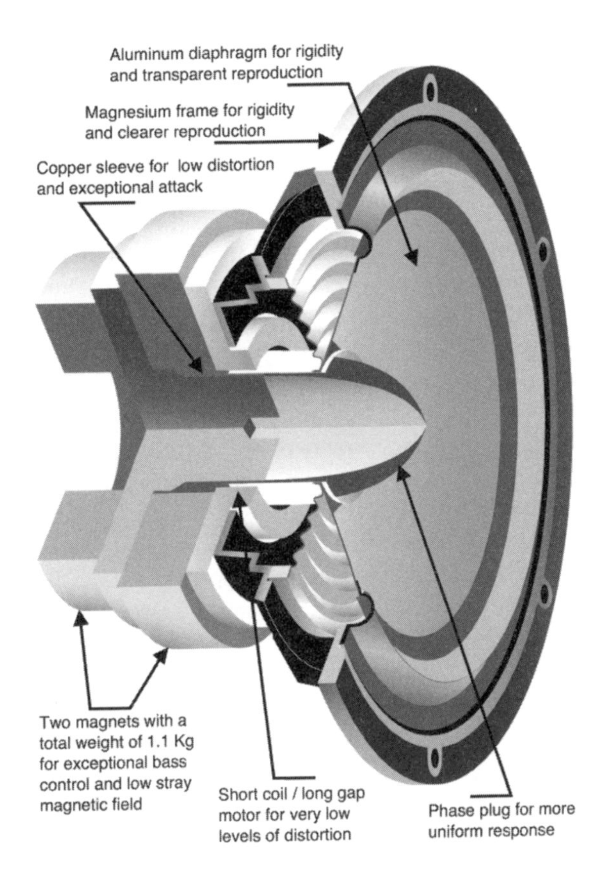

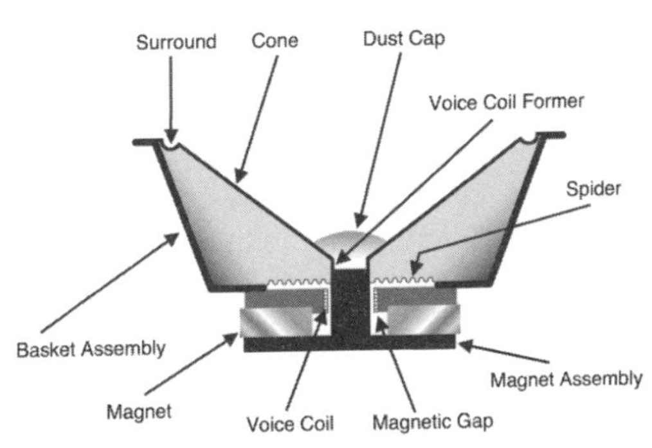

Fig. 6-1
Dynamic driver cutaway and cross section (Courtesy Thiel Audio)

through the voice-coil windings generates magnetic forces that are directed along the voice coil's axis. The interaction between the fluctuating field of the voice coil and the fixed magnetic field in the gap produces axial forces that move the voice coil back and forth, carrying the diaphragm with it. The faster the audio signal alternates, the faster the diaphragm moves, and the higher the frequency of sound produced. Dynamic drivers are also called moving-coil drivers, for obvious reasons.

Other elements of the dynamic driver include a spider that suspends the voice coil in place as it moves back and forth. The basket is a cast- or stamped-metal structure holding the entire assembly together. (Cast baskets are generally found in higher-quality loudspeakers, stamped baskets in budget models.) A ring of compliant rubber material, called the surround, attaches the cone to the basket rim. The surround allows the cone to move back and forth while still attached to the basket. The maximum distance the cone moves back and forth is called its excursion.

Common cone materials include paper, paper impregnated with a stiffening agent, a form of plastic such as polypropylene, metal (titanium, for example), or exotic materials such as carbon-fiber, Kevlar (the material used in bullet-proof vests), and proprietary composites. Designers use these materials, and sometimes a sandwich of different materials, to prevent a form of distortion called breakup. Breakup occurs when the cone material flexes instead of moving as a perfect piston. Because the cone is driven at a small area at the inside (an area the size of the voice coil), the cone tends to flex, which produces non-linear distortion. Stiff cone materials help prevent breakup. Although all dynamic drivers exhibit breakup at a certain frequency, the competent loudspeaker designer makes sure that a driver is never driven by frequencies that would cause breakup. For example, if a woofer has its first breakup mode at 4kHz, the designer would probably operate the driver only up to about 2kHz, well away from the breakup frequency.

The cone should be lightweight as well as stiff. A lighter cone has less inertia, allowing it to respond faster to transient signals and stop faster after the drive signal has ceased. Think of a bass-drum whack. A large and heavy cone may not be able to move fast enough to reproduce the sound's attack, diminishing the drum's dynamic impact. Similarly, once the drum whack is over, the heavy cone's mass will want to continue moving. Loudspeaker designers therefore search for cone materials that combine high stiffness with low mass. Many of the advances made in loudspeaker design of the past 15 years have been the result of materials research that has yielded lighter yet stiffer diaphragms.

Tweeters work on the same principle, but typically use a 1" dome instead of a cone. Common dome materials include plastic, woven fiber coated with a rubbery material, titanium, aluminum and aluminum alloys, and gold-plated aluminum. A recent trend has been toward making tweeter diaphragms from materials such as beryllium and even diamond. These materials, once found only in a company's expensive flagship models, have found their way into less costly products. Unlike cone drivers, which are driven at the cone's apex, dome diaphragms are driven at the dome's outer perimeter. Most dome tweeters use Ferrofluid, a liquid cooling agent, to remove heat from the tweeter's small voice coil. The first breakup mode of well-designed modern dome tweeters is above 25kHz, well out of the audible range.

Midrange drivers are smaller versions of the cone woofer. Some, however, use dome diaphragms instead of cones.

Dynamic Compression

When a dynamic driver plays loudly, it must dissipate a tremendous amount of heat from its voice coil. As the voice coil heats up, its electrical resistance increases, reducing the amount of current flow through the voice coil. Because it is current flow through the voice coil that causes the cone to move back and forth, increased voice-coil resistance reduces the amount of sound produced by the driver. In other words, you can keep turning up the volume, but the speaker reaches a point where it won't play any louder.

This phenomenon, called dynamic compression, obviously alters musical dynamics. Specifically, loud musical peaks aren't reproduced quite as loudly as they should be. An electroacoustic mechanism is changing the score's dynamic markings; fff may be reproduced as ff, for example.

Although this distortion of the music's dynamics is a departure from accuracy, dynamic compression introduces a worse form of distortion. The multiple drivers in a three-way dynamic loudspeaker will have different levels at which they begin to compress. If the tweeter exhibits dynamic compression at a lower volume than the midrange and woofer, the sound will be slightly duller on musical peaks. Conversely, if the woofer and midrange compress at a lower level than the tweeter, the sound will grow brighter as it gets louder. In essence, the speaker's tonal balance changes as a function of the music's dynamics. You may not hear this phenomenon directly as a change in tonal balance, but this form of distortion creates an impression of strain on musical peaks that may momentarily draw our attention away from the music. It's just one more cue to the brain that we're not hearing live music. Loudspeaker designers combat dynamic compression by creating drivers with oversized voice coils, physical structures that naturally cool the voice coil, and through innovative voice-coil wire shape and winding techniques. Higher-sensitivity drivers are less prone to dynamic compression, all other factors being equal.

There's a related mechanism by which drivers can distort music. When a driver is producing a loud sound, the ends of its voice coil can momentarily leave the magnetic field in which the coil sits. This robs the driver of its ability to faithfully move back and forth in response to the drive signal, again changing the music's dynamics and creating distortion. This problem is addressed by designing drivers with a very short voice coil and a very long magnetic gap. Although a short voice coil is more susceptible to dynamic compression, it doesn't introduce this form of non-linear distortion. Loudspeaker designers must balance these trade-offs.

Finally, a dynamic driver can "bottom out," heard as a popping or cracking sound from the woofer. The popping sound is caused by the voice-coil former (the bobbin around which the voice coil is wound) hitting the back of the magnet structure. If you ever hear this sound, immediately turn down the volume.

Problems with Dynamic Drivers

Of all the components in your system, the loudspeaker is the most likely to develop trouble. The most common problem is a blown driver, usually the tweeter—it simply stops working. Tweeters are often destroyed by too much current flowing through their voice coils. The tweeter can't dissipate the resultant heat quickly enough, and its voice coil is burned open.

Another common cause of speaker failure is the buzzing sound produced by a loose voice coil. Too much current through the voice coil melts the glue holding the coil to the former (the bobbin around which wire is wound to make the coil). This loosens turns of wire, which then rub against the magnet to cause the buzzing sound.

Driver mounting bolts can loosen over time and degrade a loudspeaker's performance by allowing the entire driver to vibrate rather than just its diaphragm. Gently tightening these bolts from time to time—particularly when a loudspeaker is new—can improve its sound. Be careful not to overtighten and strip the bolts.

The Electromagnetic Dynamic Driver

The dynamic driver I've described eventually runs up against the laws of physics. Specifically, the magnetic field strength generated by the fixed magnets is limited, which in turn places restrictions on the cone weight, how low in frequency the driver will play, and how sensitive the driver is. A heavy cone goes lower in frequency (all other factors being equal), but requires greater magnetic-field strength surrounding the voice coil to drive it. These limitations are particularly acute in large woofers.

A solution to this physics problem is the electromagnetic driver in which the driver's fixed magnets are replaced with a large coil that functions as an electromagnet. The coil is driven with direct current from an outboard power supply that plugs into an AC outlet. This current flow through the coil creates the magnetic field against which the voice-coil–generated magnetic field pushes and pulls. The electro-magnet produces a magnetic field strength in the gap (the area in which the voice coil sits) nearly double that of a conventionally driven woofer. Consequently, the electromagnetic woofer can be heavier (giving it a lower resonant frequency) yet simultaneously more efficient. Moreover, the woofer's bass output can be adjusted by varying the current through the electromagnetic coil. This is accomplished via a control on the outboard supply that drives current through the electromagnetic coil. One can thus adjust the electromagnetic woofer's bass output to better integrate the system into a variety of listening rooms.

An example of the electromagnetic driver is the woofer in the Focal Grande Utopia EM loudspeaker. The system's woofer has a very high sensitivity (97dB for 1W) but very low resonance (24Hz). In other words, the woofer delivers lots of very low bass with very little input power. The price of this performance is the need for the outboard supply that has to be plugged into an AC outlet, along with the sheer weight of the woofer. The EM's 16" woofer weighs 63 pounds, 48 of which is the electromagnetic coil.

The Planar Magnetic Transducer

The next popular driver technology is the planar magnetic transducer, also known as a ribbon driver. Although the term "ribbon" and "planar magnetic" are often used interchangeably, a true ribbon driver is actually a sub-class of the planar magnetic driver. Let's look at a true ribbon first.

Instead of using a cone attached to a voice coil suspended in a magnetic field, a ribbon driver uses a strip of material (usually aluminum) as a diaphragm suspended between the north-south poles of two magnets (see Fig.6-2). The ribbon is often pleated for additional strength. The audio signal travels through the electrically conductive ribbon, creating a magnetic field around the ribbon that interacts with the permanent magnetic field. This causes the ribbon to move back and forth, creating sound. In effect, the ribbon functions as both the voice coil and the diaphragm. The ribbon can be thought of as the voice coil stretched out over the ribbon's length.

In all other planar magnetic transducers, a flat or slightly curved diaphragm is driven by an electromagnetic conductor. This conductor, which is bonded to the back of the diaphragm, is analogous to a dynamic driver's voice coil, here stretched out in straight-line segments. In most designs, the diaphragm is a sheet of plastic with the electrical conductors

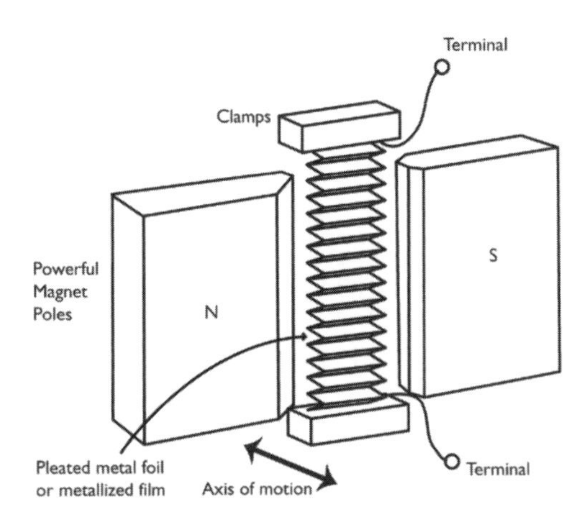

Fig. 6-2
A true ribbon driver (After Martin Colloms, *High Performance Loudspeakers)*

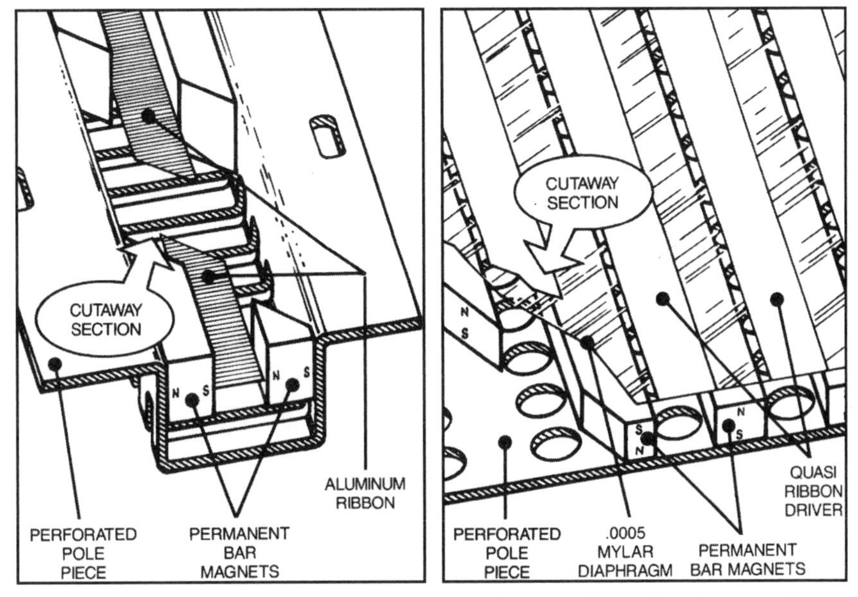

Fig. 6-3
Cross-section comparison of true ribbon (left) and quasi-ribbon (right) drivers (Courtesy Magnepan)

bonded to its surface. The flat metal conductor provides the driving force, but it occupies only a portion of the diaphragm area. Such drivers are also called quasi-ribbon transducers. Fig.6-3 shows the difference between a true ribbon and a quasi-ribbon driver.

A planar driver is a true ribbon only if the diaphragm is conductive and the audio signal flows directly through the diaphragm, rather than through conductors bonded to a diaphragm, as in quasi-ribbon drivers. (Despite this semantic distinction, I'll use the term "ribbon" throughout the rest of this section, with the understanding that it covers both true ribbons and quasi-ribbon drivers.)

Ribbon drivers like the one in Figs.6-2 and 6-3 are called line-source transducers because they produce sound over a line rather than from a point, as does a dynamic loudspeaker. Moreover, a ribbon's radiation pattern changes dramatically with frequency. At low frequencies, when the ribbon's length is short compared to the wavelength of sound, the ribbon will act as a point source and produce sound in a sphere around the ribbon—just like a point-source woofer. As the frequency increases and the wavelength of sound approaches the ribbon's dimensions, the radiation pattern narrows until it looks more like a cylinder around the ribbon than a sphere. At very high frequencies, the ribbon radiates horizontally but not vertically. This can be an advantage in the listening room: the listener hears more direct sound from the loudspeaker and less reflection from the sidewalls and ceiling. Reduced wall reflections aid in soundstaging: Pinpoint imaging and the ability to project a concert hall's acoustic signature are hallmarks of good ribbon loudspeakers.

The main technical advantage of a ribbon over a moving-coil driver is the ribbon's vastly lower mass. Instead of using a heavy cone, voice coil, and voice-coil former to move air, the only thing moving in a ribbon is a very thin strip of aluminum. A ribbon tweeter may have one quarter the mass and 10 times the radiating area of a dome tweeter's diaphragm. Low mass is a high design goal: the diaphragm can respond more quickly to transient signals. In addition, a low-mass diaphragm will stop moving immediately after the input signal has ceased. The ribbon starts and stops faster than a dynamic driver, allowing it to more faithfully reproduce transient musical information.

The ribbon driver is usually mounted in a flat, open-air panel that radiates sound to the rear as well as to the front. A loudspeaker that radiates sound to the front and rear is called a dipole, which means "two poles." Fig.6-4 shows the radiation patterns of a point-source loudspeaker (left) and a dipolar loudspeaker.

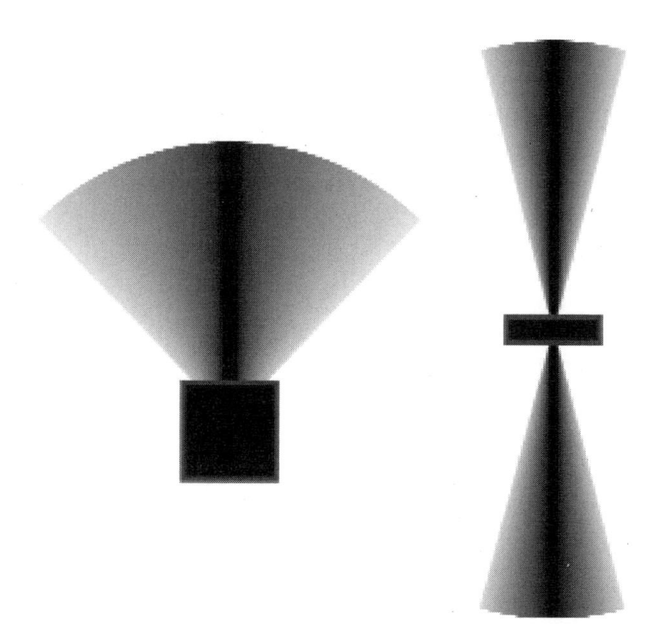

Fig. 6-4
A point-source loudspeaker (left) radiates sound in one direction; a dipole loudspeaker (right) radiates sound equally to the front and rear.

Another great advantage enjoyed by ribbons is the lack of a box or cabinet. As we'll see in the section of this chapter on loudspeaker enclosures, the enclosure can greatly degrade a loudspeaker's performance. Not having to compensate for an enclosure makes it easier for a ribbon loudspeaker to achieve stunning clarity and lifelike musical timbres.

A full-range quasi-ribbon loudspeaker is illustrated in Fig.6-5. The large panel extends the system's bass response: when the average panel dimension approaches half the wavelength, front-to-rear cancellation reduces bass output. Consequently, the larger the panel, the deeper the low-frequency extension.

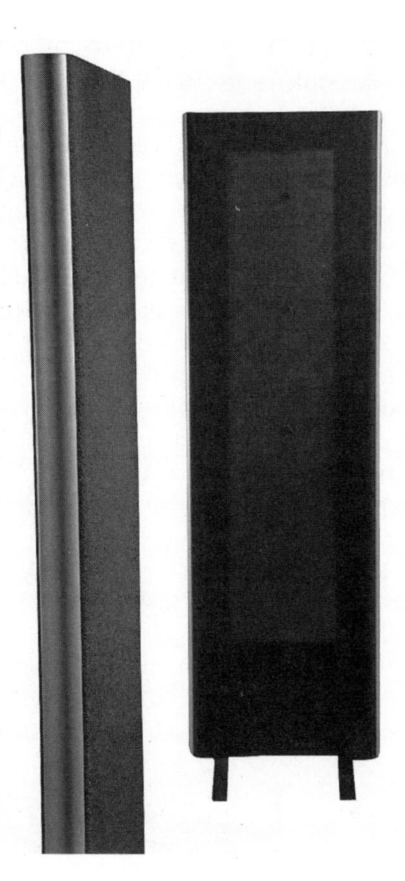

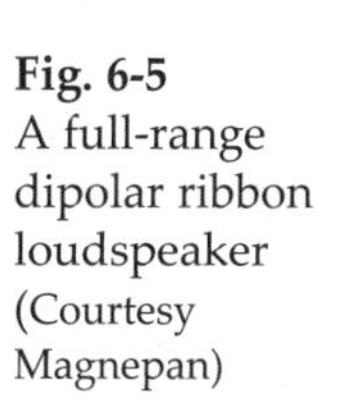

Fig. 6-5
A full-range dipolar ribbon loudspeaker (Courtesy Magnepan)

Ribbon loudspeakers are characterized by a remarkable ability to produce extremely clean and quick transients—such as those of plucked acoustic guitar strings or percussion instruments. The sound seems to start and stop suddenly, just as one hears from live instruments. Ribbons sound vivid and immediate without being etched or excessively bright. In addition, the sound has an openness, clarity, and transparency often unmatched by dynamic drivers. (Incidentally, these qualities are shared by ribbon microphones.) Finally, the ribbon's dipolar nature produces a huge sense of space, air, and soundstage depth (provided this spatial information was captured in the recording). Some argue, however, that this sense of depth is artificially produced by ribbon loudspeakers, rather than being a *re*production of the actual recording.

Despite their often stunning sound quality, ribbon drivers have several disadvantages. The first is low sensitivity; it takes lots of amplifier power to drive them. Second, ribbons inherently have a very low impedance, often a fraction of an ohm. Most ribbon drivers therefore have an impedance matching transformer in the crossover to present a higher impedance to the power amplifier. Design of the transformer is therefore crucial to prevent it from degrading sound quality.

From a practical standpoint, ribbon-based loudspeakers are more difficult to position in a room. Small variations in placement can greatly change the sound, due primarily to their dipolar radiating pattern. This dipolar pattern requires that the ribbon loudspeakers be placed well away from the rear wall, and that the rear wall be acoustically benign.

Low-profile, ribbon-based loudspeakers with the ribbon top at the same height as the listener's ears will have a radically different treble balance if the listener moves up or down a few inches. That's because ribbon loudspeakers have very narrow vertical dispersion, meaning they radiate very little sound above and below the ribbon at high frequencies. If you sit too high or listen while standing, you'll hear less treble. Some ribbon loudspeakers have a tilt adjustment that allows you to set the correct treble balance for a particular listening height.

Ribbons also have a resonant frequency that, if excited, produces the horrible sound of crinkling aluminum foil. Consequently, the ribbon must be used within strict frequency-band limits. In addition, ribbon drivers are tensioned at the factory for optimum performance. If under too much tension, a ribbon will produce less sound; if under too little tension, a ribbon can produce distortion that sounds like the music is "breaking up." This is most noticeable on piano; the transient leading edges sound "shattered" and distorted. A sudden increase in ambient temperature can cause a ribbon driver to lose some of its tension and introduce the distortion described. If you hear this sound from your ribbon loudspeakers, contact the manufacturer for advice. The solution may be as simple as turning a few tensioning bolts half a turn.

Ribbon drivers don't necessarily have to be long and thin. Variations on ribbon technology have produced drivers having many of the desirable characteristics of ribbons but few of the disadvantages.

Finally, some loudspeakers use a combination of dynamic and ribbon transducers to take advantage of both technologies. These so-called hybrid loudspeakers typically use a dynamic woofer in an enclosure to reproduce bass, and a ribbon midrange/tweeter (Fig.6-6).

Fig. 6-6
A dynamic-ribbon hybrid loudspeaker employing a dynamic woofer and a ribbon tweeter (Courtesy Volent)

The hybrid technique brings the advantages of ribbon drivers to a lower price level (ribbon woofers are big and expensive), and exploits the advantages of each technology while avoiding the drawbacks. The challenge in such a hybrid system is to achieve a seamless transition between the dynamic woofer and the ribbon tweeter, with no audible discontinuity between the drivers.

The Heil Air-Motion Transformer

Designed in the early 1970s by Dr. Oskar Heil, the Heil Air-Motion Transformer operates in a completely different way than dynamic or planar drivers. Rather than causing a diaphragm to move back and forth like a piston, the AMT's pleated polyethylene membrane alternately squeezes and expands the air in front of it in response to the audio signal. A conductor is bonded to the diaphragm, and then suspended in a magnetic field. Although ribbons are often pleated for strength, the AMT's pleats are much deeper and are responsible for the squeezing effect. The "Air-Motion Transformer" name comes from the fact that air is squeezed out of the pleats at five times the speed of the diaphragm's motion. This squeezing effect, and the accompanying acceleration of the air, is analogous to the rapid expulsion of a seed from a piece of fruit that's gently squeezed. The seed's motion is many times faster than the force applied.

The AMT has excellent transient response due to the diaphragm's low mass and very small range of motion, as well as high acoustic output. A 1" wide AMT produces the same acoustic output as an 8" round driver.

A number of companies make a version of the Heil Air-Motion Transformer now that the patents have expired.

The Electrostatic Driver

Like the ribbon transducer, an electrostatic driver uses a thin membrane to make air move. But that's where the similarities end. While both dynamic and ribbon loudspeakers are electromagnetic transducers—they operate by electrically induced magnetic interaction—the electrostatic loudspeaker operates on the completely different principle of electrostatic interaction.

No discussion of electrostatic loudspeakers would be complete without mentioning the classic electrostatic loudspeaker, the Quad ESL-57, created in 1957 by Peter Walker. The ESL-57 revolutionized the standard for transparency upon its introduction, and still holds it own more than 50 years later. Many listeners' first exposure to high-quality audio was through an ESL-57. A large number of contemporary loudspeaker designers still have a pair of ESL-57s on hand as a reference. The ESL-57 doesn't have much low bass, won't play very loudly, and produces a very narrow sweet spot, but when operated within its limitations, it's magical.

In an electrostatic loudspeaker (sometimes called an ESL), a thin moveable membrane—usually made of transparent Mylar—is stretched between two static elements called stators (Fig.6-7). The membrane is charged to a very high voltage with respect to the stators. The audio signal is applied to the stators, which create electrostatic fields around them that vary in response to the audio signal. The varying electrostatic fields generated around the stators interact with the membrane's fixed electrostatic field, pushing and pulling the membrane to produce sound. One stator pulls the membrane, the other pushes it. This illustration also shows a dynamic woofer as part of a hybrid dynamic/electrostatic system. (A full-range electrostatic loudspeaker is illustrated in Fig.6-8.)

The voltages involved in an electrostatic loudspeaker are very high. The polarizing voltage applied to the diaphragm may be as high as 10,000 volts (10kV). In addition, the audio signal is stepped up from several tens of volts to several thousand volts by a step-up

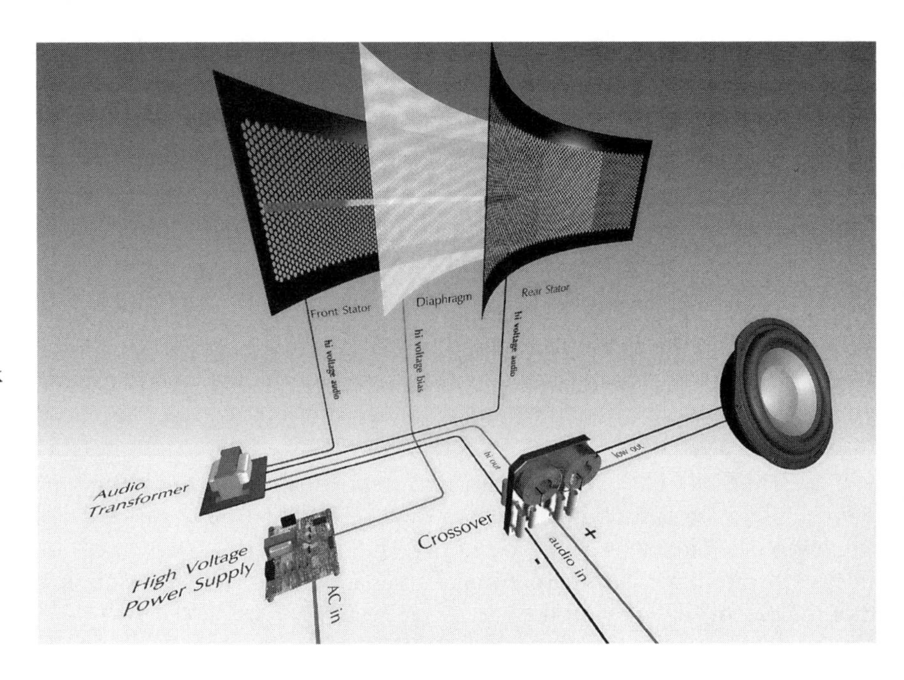

Fig. 6-7
An electrostatic loudspeaker employs two stators separated by a membrane that is pulled back and forth by the stators. (Courtesy MartinLogan)

Fig. 6-8
A full-range electrostatic loudspeaker (Courtesy MartinLogan)

transformer inside the electrostatic loudspeaker. These high voltages are necessary to produce the electrostatic fields around the diaphragm and stators.

To prevent arcing—the electrical charge jumping between elements—the stators are often coated with an insulating material. Still, if an electrostatic loudspeaker is overdriven, the electrostatic field strips free electrons from the oxygen in the air, making it ionized; this provides a conductive path for the electrical charge. Large diaphragm excursions— i.e., a loud playing level—put the diaphragm closer to the stators and also encourage arcing. Arcing can destroy electrostatic panels by punching small holes in the membrane. Arcing is a greater problem in humid climates than in dry climates because moisture makes the air between the stators more electrically conductive.

Electrostatic panels are often divided into several smaller panels to reduce the effects of diaphragm resonances. Some panels are curved to reduce the lobing effect (uneven radiation pattern) at high frequencies. Lobing occurs when the wavelength of sound is small compared to the diaphragm. Lobing is responsible for electrostatics' uneven high-frequency dispersion pattern, which *Stereophile* magazine founder J. Gordon Holt has dubbed the vertical venetian-blind effect, in which the tonal balance changes rapidly and repeatedly as you move your head from side to side.

Electrostatic panels are of even lighter weight than planar magnetic transducers. Unlike the ribbon driver, in which the diaphragm carries the audio signal current, the electrostatic diaphragm need not carry the audio signal. The diaphragm can therefore be very thin, often less than 0.001". Such a low mass allows the diaphragm to start and stop very quickly, improving transient response. And because the electrostatic panel is driven uniformly over its entire area, the panel is less prone to breakup. Both the electromagnetic planar loudspeaker (a ribbon) and the electrostatic planar loudspeaker enjoy the benefits of limited dispersion, which means less reflected sound arriving at the listening position. Like ribbon loudspeakers, electrostatic loudspeakers also have no enclosure to degrade the sound. Electrostatic loudspeakers also inherently have a dipolar radiation pattern. Because the diaphragm is mounted in an open panel, the electrostatic driver produces as much sound to the rear as to the front. Finally, the electrostatic loudspeaker's huge surface area confers an advantage in reproducing the correct size of instrumental images.

In the debit column, electrostatic loudspeakers must be plugged into an AC outlet to generate the polarizing voltage. Because the electrostatic is naturally a dipolar radiator, room placement is more crucial to achieving good sound. The electrostatic loudspeaker needs to be placed well out into the room and away from the rear wall to realize a fully developed soundstage. Electrostatics also tend to be insensitive, requiring large power amplifiers. The load impedance they present to the amplifier is also more reactive than that of dynamic loudspeakers, further taxing the power amp. (Reactance is described later in this chapter.) Nor will they play as loudly as dynamic loudspeakers; electrostatics aren't noted for their dynamic impact, power, or deep bass. Instead, electrostatics excel in transparency, delicacy, transient response, resolution of detail, stunning imaging, and overall musical coherence.

Electrostatic loudspeakers can be augmented with separate dynamic woofers or a subwoofer to extend the low-frequency response and provide some dynamic impact. Other electrostatics achieve the same result in a more convenient package: dynamic woofers in enclosures mated to the electrostatic panels. Some of these designs achieve the best qualities of both the dynamic driver and electrostatic panel. (An example of a hybrid electrostatic/dynamic loudspeaker is illustrated in Fig.6-9.) Audition such hybrid speakers carefully; they

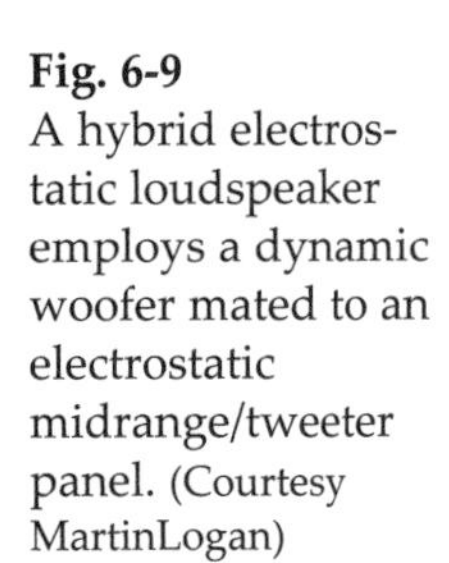

Fig. 6-9
A hybrid electrostatic loudspeaker employs a dynamic woofer mated to an electrostatic midrange/tweeter panel. (Courtesy MartinLogan)

sometimes exhibit an audible discontinuity at the transition frequency at which the woofer leaves off and the electrostatic panel takes over. Listen, for example, for a change in a piano's timbre, bloom, projection, and image size as it is played in different registers. Acoustic bass in jazz is also a good test of woofer/panel discontinuity in dynamic/electrostatic hybrid loudspeakers.

One great benefit of full-range ribbons and full-range electrostatics is the absence of a crossover; the diaphragm is driven by the entire audio signal. This prevents any discontinuities in the sound as different frequencies are reproduced by different drivers. In addition, removing the resistors, capacitors, and inductors found in crossovers greatly increases the full-range planar's transparency and harmonic accuracy. Even hybrid planars put the crossover frequency between the dynamic woofer and the planar panel very low (below 800Hz, a frequency nearly an octave above middle A), so there's no discontinuity between drivers through most of the audible spectrum.

Finally, the large diaphragms of electrostatic and ribbon drivers are gently driven over their entire surface areas, rather than forcefully over the relatively small voice-coil area of a dynamic driver. This high force over a small area contributes to the dynamic driver's breakup described earlier, a phenomenon less likely to occur with large planar diaphragms.

The Dipolar Radiation Patterns of Ribbons and Electrostatics

Because planar loudspeakers (ribbons and electrostatics) are mounted in an open frame rather than an enclosed box, they radiate sound equally from the front and back. As we saw earlier, the term dipolar describes this radiation pattern. This is contrasted with point source loudspeakers, whose drivers are mounted on the front of a box. Point-source loudspeakers are usually associated with dynamic drivers, but any type of driver in an enclosed cabinet qualifies as a point-source loudspeaker.

The dipolar radiation patterns of ribbons and electrostatics make using them very different from point-source loudspeakers. Because they launch just as much energy to the rear as they do to the front, positioning a dipole is more crucial, particularly their distance from the rear wall. Dipoles need a significant space behind them to work well. In addition, the rear wall's acoustic properties have a much greater influence over the sound. A few attempts have been made to absorb a ribbon's rear energy inside an enclosure to make placement easier, but these have been largely unsuccessful.

Ribbons and electrostatics also tend to have much narrower radiation patterns than point-source loudspeakers. That is, they disperse less sound to their sides. Consequently, the listener hears much more of the loudspeaker's direct sound and much less sound reflected from the floor, ceiling, and sidewalls. This is particularly true in the midrange and treble, where the radiation pattern becomes cylindrical. The result is often a greatly improved transparency and more natural timbre. (The effects of these reflections are discussed in Chapter 14.) The negative effects of this tendency for planar loudspeakers to "beam" at high frequencies is a disturbing shift in the tonal balance with small motions of the head from side to side—the "vertical venetian-blind effect" described earlier. This tendency can be minimized with a curved diaphragm (seen in the earlier photographs of an electrostatic loudspeaker) or by electrically segmenting the diaphragm.

Bipolar and Omnidirectional Loudspeakers

We've seen that a diaphragm mounted in an open panel will naturally radiate sound equally to the front and rear. These front and rear waves are out of phase with each other. That is, as the diaphragm moves forward, it creates a positive pressure wave in front of the diaphragm, and a negative pressure wave (called a rarefaction) behind the diaphragm.

Some loudspeaker designers see the advantages of front- and rear-launched sound, but believe that the front and rear waves should be in phase with each other. Because this is impossible to achieve with a diaphragm mounted in an open panel, these designers employ a second set of dynamic drivers on the rear of the loudspeaker enclosure. This design, called a bipolar loudspeaker, has a radiation pattern similar to that of a dipole, but the front and rear waves are now in phase. Other designers add just a rear-firing tweeter, which they contend produces a smoother treble response at the listening position and better reproduces a sense of spaciousness and ambience. To recap, in a dipolar loudspeaker the front and rear waves are out-of-phase with each other. In a bipolar loudspeaker the front and rear waves are in-phase.

A bipolar loudspeaker produces a radiation pattern slightly different from that of a dipole. The dipole's out-of-phase front and rear waves cancel at the sides of the speaker, narrowing their radiation pattern. A bipolar speaker's in-phase waves don't exhibit this cancellation, and thus have wider dispersion.

Advocates of dipoles and bipoles believe that reflecting sound from the wall behind the speakers increases the sense of musical realism because we hear such reflected sound in live music. In addition, point-source speakers (those firing in one direction) exhibit frequency-response changes off-axis (to the speaker's sides). The sound reflected from the sidewalls has a different spectral balance compared with the direct sound from the loudspeaker. This confuses the ear/brain, and provides yet another clue that the sound isn't real. Because dipole and bipole loudspeakers distribute a more uniform spectral balance around the room, aficionados of such designs believe that they produce a more musically natural sound.

An extension of this idea is the omnidirectional loudspeaker that produces a spherical dispersion pattern; that is, it radiates sound in all directions. Proponents claims that omnidirectional loudspeakers produce a ratio of direct to reflected sound that more closely matches that of live music. Omnidirectional loudspeakers produce a large and expansive soundstage, but not "pinpoint" imaging. Fig.6-10 shows an omnidirectional loudspeaker, the MBL X-Treme. Note also that this design employs a novel driver technology called a Radialstrahler in which "petals" of metal are driven by magnets located in the top and bottom anchors holding the petals so that they flex, creating sound.

Fig. 6-10
An omnidirectional loudspeaker radiates sound in all directions.
(Courtesy MBL)

Horn Drivers

One of the first loudspeaker technologies, the horn, has enjoyed a resurgence in popularity in the last 20 years. The horn loudspeaker employs a small dynamic driver mounted in the throat (narrow end) of a horn structure, which more efficiently couples the driver's diaphragm to the air (Fig.6-11).

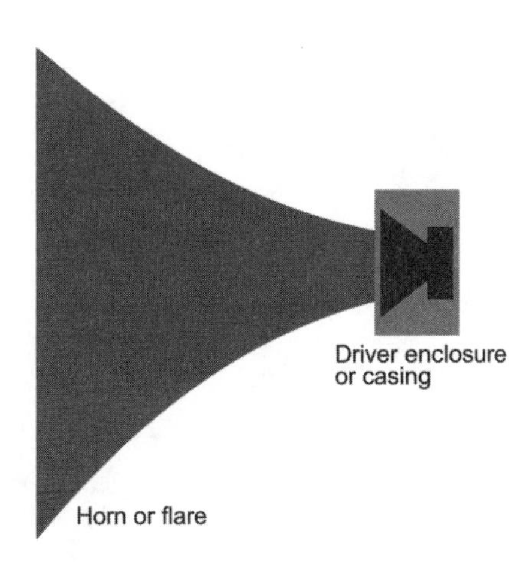

Fig. 6-11 Horn-loading a driver more efficiently couples the driver's diaphragm to the air, increasing sensitivity.

Any type of driver can be horn-loaded, but it is uncommon to find true-horn loading in woofers because of the enormous size required. The lower the frequency at which the horn is designed to operate, the larger the horn must be. Consequently, horn loading is generally confined to midrange drivers and tweeters. Full-range horn-based systems commonly use a conventionally loaded woofer to reproduce bass.

Horn-loading a driver confers many performance benefits. First, horns have extraordinarily high sensitivity, and can be driven to very high volume with just a few watts of power. It's not unusual for a horn-loaded system to have a sensitivity of 100dB or more. This brings us to the second benefit of horns: they can be driven by very small power amplifiers, often those with just 10Wpc (or less). Third, the excursion (back and forth motion) of the cone or dome diaphragm can be an order of magnitude less than in a direct-radiating driver. This allows the driver to operate in the linear range of its motion at all times, greatly reducing distortion. The electrical and magnetic forces involved in moving a horn-loaded diaphragm back and forth are about one-tenth those of a direct-radiating driver. As a result of these attributes, horn-loaded systems have state-of-the-art dynamics and lifelike recreation of music's transient nature. The sharp attack of a snare drum, for example, is reproduced with a stunning sense of realism. This quality, known as "jump factor," gives horn-loaded loudspeakers a lifelike presentation unmatched by any other loudspeaker technology.

Now the bad news. Horn-loading often introduces large tonal colorations. Try this experiment: read the previous sentence, and then read it again, this time with your hands cupped around your mouth to make a horn. For many listeners, the horn-loaded system's colorations (particularly through the midrange) are a deal-breaker, outweighing all of the horn's advantages.

There are, however, a few horn-loaded systems that do not sound colored in the slightest. These systems are very large, exotic, and extremely expensive, the result of the extraordinary manufacturing required. The horn-loaded system in Fig.6-12 is an example. In this five-way system, three of the horns are machined out of solid blocks of aircraft-grade aluminum; the top horn, which handles lower-midrange frequencies, is constructed from thick aluminum reinforced with 56 hand-welded ribs. Note the lower-midrange horn's massive size, necessary to maintain the horn-loading effect down to 120Hz, the frequency at which it is crossed over to the conventionally loaded woofer. If a horn's mouth isn't large enough relative to the wavelength being reproduced, the horn's natural acoustic amplification effect diminishes. At this point, the driver operates like a conventionally loaded driver and sensitivity drops off rapidly.

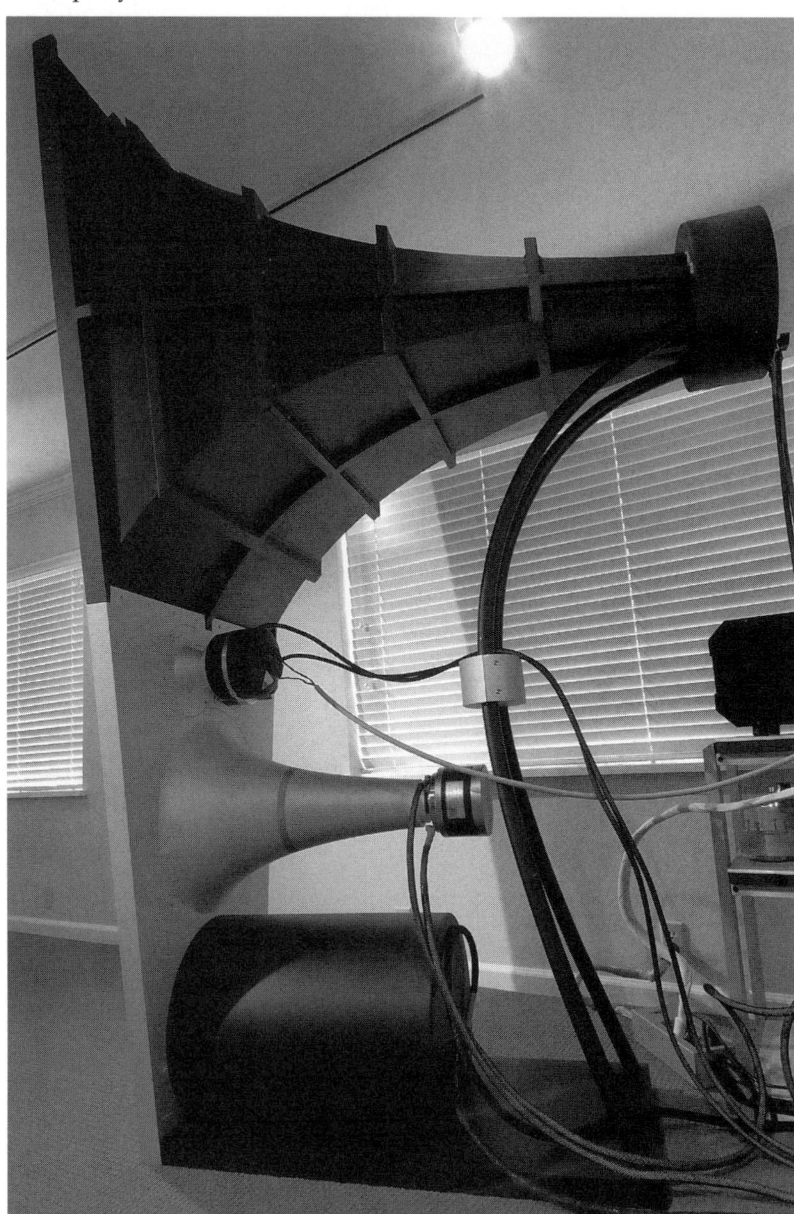

Fig. 6-12
A massive horn loudspeaker
(Courtesy Magico)

For the technically minded I've included a more in-depth description of how a horn structure acts as a natural acoustic amplifier. The horn loudspeaker operates on the same principle as a megaphone. Just as with a megaphone, the sound source (the driver in the case of the horn-loaded speaker) is mounted in the narrow end of a tube that exhibits an exponential flair. This narrow throat restricts air motion and thus presents a high acoustical impedance to the driver diaphragm. Consequently, the diaphragm can develop very high pressures in the throat with very little excursion. As this high-pressure wavefront travels down the horn, the horn's exponentially increasing size gradually converts this high-pressure into increasingly lower pressure until it reaches the horn's mouth. At the mouth, the waveform is of lower pressure but higher amplitude. The horn's shape essentially converts the high-pressure, low-displacement wavefront over a small area (in the throat) to low pressure but high-amplitude wavefront to a larger area (at the mouth). The horn has been likened to an acoustic impedance-matching transformer that more efficiently couples the driver diaphragm to the air.

A related technology is the compression driver, a dynamic transducer used in horn loudspeakers. The compression driver gets its name from the fact that it uses a diaphragm to "compresses" the air through an aperture that is very small relative to the diaphragm's size. When a compression driver is mounted in the throat of a horn, the combination is capable of extremely high sensitivity—as high as 115dB SPL for 1W of power. The compression driver employs massive magnets relative to the diaphragm size and consequently, is quite heavy. Compression drivers that are hand-built by Japanese artisans can cost as much as $15,000 each. You can see such compression drivers in Fig.6-12 earlier; they are the small cylindrical structures at the throats of the three horns.

Waveguides

A tweeter is sometimes mounted in a recessed flair in the baffle, a structure called a *waveguide*. The waveguide tailors the tweeter's dispersion so that it more closely matches the midrange driver's dispersion at the crossover frequency. This technique avoids a discontinuity at the midrange-to-tweeter transition that could be audible.

To understand how a waveguide works, you need to know that at the lower end of its frequency range, a midrange driver has very wide dispersion. At higher frequencies (shorter wavelengths relative to the driver size), the midrange driver's dispersion becomes narrower. Similarly, the tweeter's dispersion is wide at the low end of its frequency range (because the wavelengths are long relative to the driver size) and becomes narrower as the frequency it is reproducing increases. The waveguide physically restricts the tweeter's dispersion so that it more closely matches the midrange driver's dispersion at the crossover point. A waveguide also has a slight horn-loading effect (described earlier in this chapter), increasing the tweeter's sensitivity. That is, a tweeter in a waveguide requires less amplifier power to produce the same sound-pressure level. This has the added benefit of reducing the stress on the tweeter; the diaphragm doesn't have to move as far, and the voice coil doesn't have to handle as much current. These benefits give the loudspeaker designer the option of setting a lower crossover frequency between the midrange driver and the tweeter. The waveguide's shape and depth determine the degree of this horn-loading effect. Taken to an extreme, the waveguide can introduce the problems of horns described earlier in this chapter, which is why waveguides are generally shallow and have a gentle flair, as seen in Fig.6-13.

Fig. 6-13 A tweeter mounted in a waveguide. (Courtesy Revel)

Loudspeaker Enclosures

The enclosure in which a set of drivers is mounted has an enormous effect on the loudspeaker's reproduced sound quality. In fact, the enclosure is as important as the drive units themselves. Designers have many choices in enclosure design, all of which affect the reproduced sound.

In addition to the very different enclosure types described in this section, any speaker cabinet will vibrate slightly and change the sound. The ideal enclosure would produce no sound of its own, and not interfere with the sound produced by the drivers. Inevitably, however, some of the energy produced by the drive units puts the enclosure into motion. This enclosure vibration turns the speaker cabinet into a sound source of its own, which colors the music.

One of the factors that makes current high-end loudspeakers so much better than mass-produced products is the extreme lengths to which some manufacturers go to prevent the enclosure from vibrating. Mass-market manufacturers generally skimp on the enclosure because, to the uninformed consumer, it adds very little to the product's perceived value.

Let's look at why a loudspeaker needs an enclosure, and survey the most popular enclosure types.

An enclosure around a woofer is required to prevent the woofer's rear-radiated wave from combining with the woofer's front-radiated wave. When the woofer moves forward, a positive pressure wave is launched at the front, and a rarefaction (negative pressure) wave of equal intensity is launched to the rear. Because low frequencies can refract (bend) around small objects such as a woofer, the rear wave will combine with the front wave and cancel the sound, resulting in reduced bass output. Putting an enclosure around the woofer prevents this cancellation and allows the loudspeaker to generate low frequencies. Although this phenomenon occurs with planar loudspeakers, their large size lowers the frequency at which this cancellation occurs. That's why full-range planar loudspeakers don't go very low in the bass, and are often augmented by a woofer in a traditional enclosure. In addition to preventing this front-to-back cancellation, the enclosure should also present the optimum environment for the drivers, particularly the woofer. This is called loading. Various loading techniques have evolved to optimize the way the woofer operates in the cabinet. The most common techniques are the infinite baffle (also called a sealed cabinet, or air suspension in smaller enclosures), and the reflex or ported enclosure. Less common loading techniques are the transmission line, Isobarik, and finite baffle designs.

These different loading techniques affect the speaker's bass response. A woofer in an enclosure will go only so low in the bass; as the frequency drops below a certain point, so does the woofer's output. The point at which the woofer's output is down by 3dB is called the low-frequency cutoff point. Below this frequency, the woofer produces less and less sound as the signal driving the woofer decreases in frequency—a phenomenon called rolloff. Rolloff is why not every speaker can reproduce 20Hz bass. The loading technique chosen by the loudspeaker designer greatly affects the woofer's low-frequency cutoff point and how steeply this rolloff occurs.

Let's look at the advantages and drawbacks of the most popular loading techniques.

Infinite Baffle Loading

The infinite baffle simply seals the enclosure around the driver rear to prevent the two waves from meeting. In theory, such an enclosure approximates a baffle (the surface on which the drivers are mounted) of "infinite" size; in fact, it wraps around the driver. The air inside the sealed enclosure acts as a spring, compressing when the woofer moves in and creating some resistance to woofer motion. If this resistance to woofer motion by the compressed air is great enough, the infinite baffle can be called air suspension (we can dispense with the technical details of where this transition occurs). Although all air suspension loudspeakers are infinite baffle, not all infinite baffles are air suspension. Nonetheless, all sealed-enclosure speakers have a low-frequency rolloff slope of 12dB per octave; this means that, one octave below the system's resonant frequency, the output will be reduced by 12dB. This rolloff is relatively gradual, meaning you'll still hear some lower bass.

Reflex Loading

I mentioned earlier that a woofer in a box produces just as much sound inside the box as outside. The bass-reflex enclosure exploits this fact, channeling some of this bass energy out of the cabinet and into the listening room. You can instantly spot a reflex-loaded loudspeaker by the hole, or vent, on the front or back of the enclosure. Reflex-loaded loudspeakers are also called vented designs.

Reflex loading extends the low-frequency cutoff point by venting some of the energy inside the enclosure to the outside. That is, a reflex-loaded woofer will go lower in the bass than one in a sealed enclosure, other factors being equal. The designer can't, however, just cut a hole in the cabinet. Instead, a duct inside the cabinet delays the woofer's rear-firing wave so that it emerges in-phase with the front-firing wave. The rear wave channeled outside the cabinet by the port thus reinforces, rather than cancels, the front wave. A loading material such as long-hair wool or Fiberglas batting is sometimes put in the duct to smooth any peaks in the sound emerging from the port.

Reflex loading has three main advantages. First, it increases a loudspeaker's maximum acoustic output level—it will play louder. Second, it can make a loudspeaker more sensitive—it needs less amplifier power to achieve the same volume. Third, it can lower a loudspeaker's cutoff frequency—the bass goes deeper. Note that these benefits are not available simultaneously; the acoustic gain provided by reflex loading can be used either to increase a loudspeaker's sensitivity or to extend its cutoff frequency, but not both.

Although a reflex system's low-frequency cutoff point is lower than that of a sealed system (assuming the same woofer and enclosure volume), the reflex system's bass rolls off at a much faster rate. Specifically, a reflex system rolls off at 24dB/octave compared to the sealed enclosure's 12dB/octave rolloff. If all other factors are equal, the reflex-loaded system maintains flat bass response down to a lower frequency, but then the bass output drops off more quickly than it does in a sealed system. A comparison of low-frequency cutoff points and rolloff slopes is illustrated in Fig.6-14.

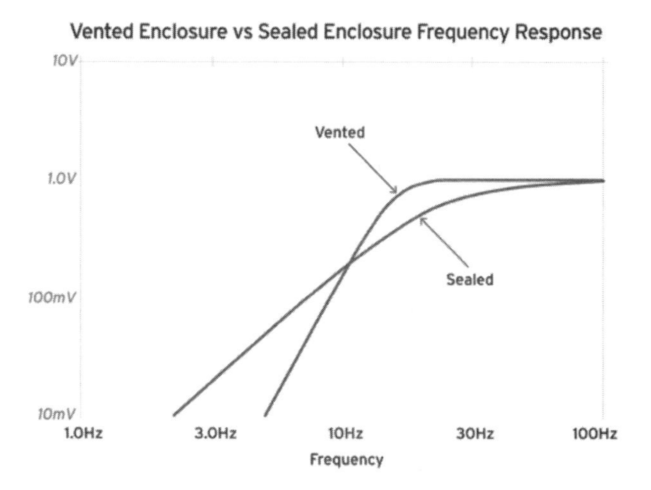

Fig. 6-14 Frequency-response characteristics of vented and sealed enclosures

Subjectively, the sealed loudspeaker's higher cutoff frequency and gradual rolloff provide a more satisfying feeling of bass fullness than the reflex system's lower cutoff frequency and steeper rolloff. Some designers have found that a loudspeaker's –10dB point is the most reliable indicator of a loudspeaker's subjective bass fullness and extension because it takes into account not only the low-frequency cutoff point, but also the steepness of the rolloff.

Reflex loading is more often used in small enclosures, which trade low-frequency extension and sensitivity for compactness. It's easier to get more bass from a small reflex enclosure than from a sealed one. However, a small reflex enclosure can still achieve some of the extension and sensitivity of a larger cabinet.

Some controversy exists over the merits of reflex loading. Some designers believe that reflex systems are fundamentally wrong. Reflex systems in general have gotten a bad name, both because of inherent limitations and poor implementation in certain products. Small, inexpensive loudspeakers are often reflex-loaded to increase their sensitivity.

Ineptly implemented reflex systems tend to be boomy, slow, and poorly defined in the bass. In addition, you can sometimes hear the sound emerging from the port, a phenomenon perfectly (and onomato-poetically) described by the term chuffing. Critics also point to the reflex system's poor transient performance compared with a sealed loudspeaker. Specifically, a woofer in a vented enclosure will tend to keep moving after the drive signal has stopped, as shown in Fig.6-15. This difference in transient response can be manifested as sluggish sound in, for example, a kickdrum. A sealed enclosure has better transient response and better bass definition, but at the cost of lower sensitivity and less deep bass extension. Nonetheless, many very-high-quality loudspeakers have been based on ported designs.

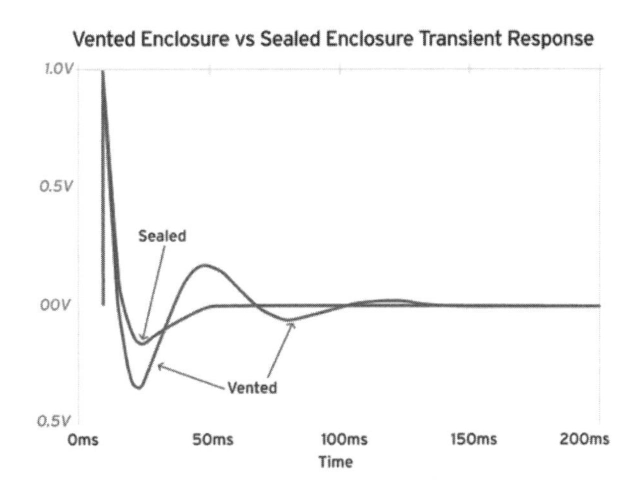

Fig. 6-15 Transient-response comparison between vented and sealed enclosures

Passive Radiators

A variation on the reflex system is the passive radiator, also called an auxiliary bass radiator, or ABR. This is usually a flat diaphragm with no voice coil or magnet structure, that cannot produce sound on its own. Instead, the diaphragm covers what would have been the port in the reflex system, and moves in response to varying air pressure inside the cabinet caused by the woofer's motion. The passive radiator smoothes out any response peaks, and eliminates a ported system's wind noise (the "chuffing" mentioned earlier) and port resonances.

Transmission-Line Loading

Another loading technique is the transmission line. In a transmission line, the rear wave from the woofer is channeled through a folded labyrinth or duct—the transmission line—filled with absorbent material such as wool. The end of the transmission line appears at the outside of the enclosure, just as in a reflex system. The acoustic energy traveling down the transmission line is absorbed, with the goal of no acoustic output at the end of the transmission line. In theory, the woofer's rear wave is dissipated in the line. This isn't the case in practice: most transmission lines have some output at the end of the line, causing them to function partially as reflex systems. Because the line must be at least as long as the lowest wavelength to be absorbed, transmission-line enclosures are usually very large. Proponents of transmission lines argue that the springiness of sealed enclosures is not the ideal environment for the woofer. The transmission line removes the spring action of the sealed enclosure without the disadvantages of reflex designs.

Transmission-line loudspeakers have extremely deep extension and the ability to deliver very loud and very clean bass with an effortless quality not heard from other loading designs. They aren't as popular as their sound quality would suggest they should be, mostly because of their large size.

British loudspeaker manufacturer Bowers and Wilkins (B&W) has developed a variation on transmission-line loading the company calls Nautilus. Each driver in a Nautilus system is mounted in a tapered or coiled (thus the Nautilus name) enclosure that absorbs the driver's rear wave. The rear wave cannot interact with the drivers, nor can it form standing waves inside the enclosure. An exploded view of a B&W loudspeaker using Nautilus technology is shown in Fig.6-16. Note also the spherical midrange enclosure, which reduces cabinet diffraction. This "boxless" approach to loudspeaker design is an attempt to reduce the enclosure's contribution to the reproduced sound.

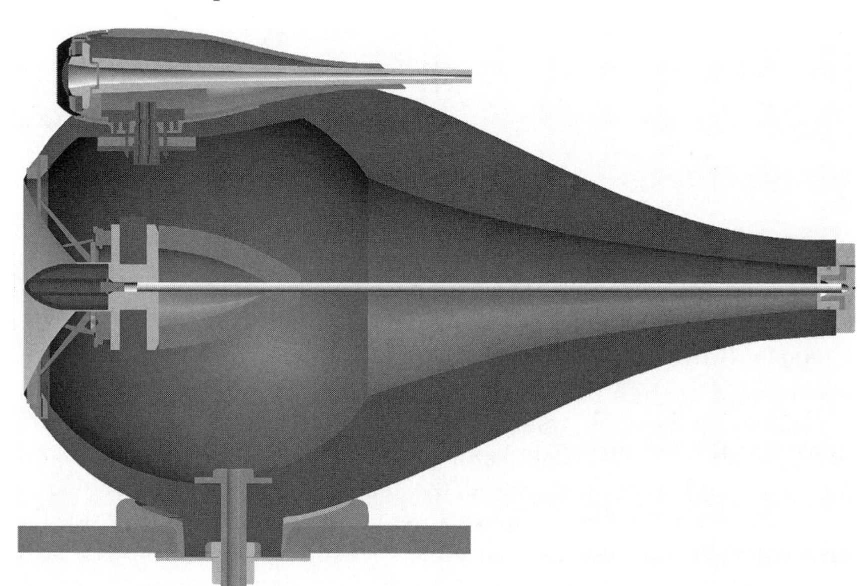

Fig. 6-16 B&W's Nautilus technology absorbs the drivers' rear-firing energy via a computer-optimized tapered-tube enclosure. (Courtesy Bowers & Wilkins Loudspeakers)

Isobarik Loading

A fourth type of loading, used primarily by Linn Products, is called Isobarik (Linn's trade name), or by the more descriptive generic term constant pressure chamber. (See Fig.6-17.) In the Isobarik system, a second woofer is mounted directly behind the first woofer and driven in parallel with it. As the front woofer moves back, so does the second; this maintains a constant pressure inside the chamber separating the two woofers. This technique offers deeper low-frequency extension, higher power handling, greater linearity, and reduced standing-wave reflections inside the enclosure. Isobarik loading reduces sensitivity because the amplifier must drive two woofers, although only one produces acoustic output. Two woofers mounted in an Isobarik configuration can be modeled as a single woofer whose greater mass and compliance deliver useful deep bass in an enclosure half the usual size.

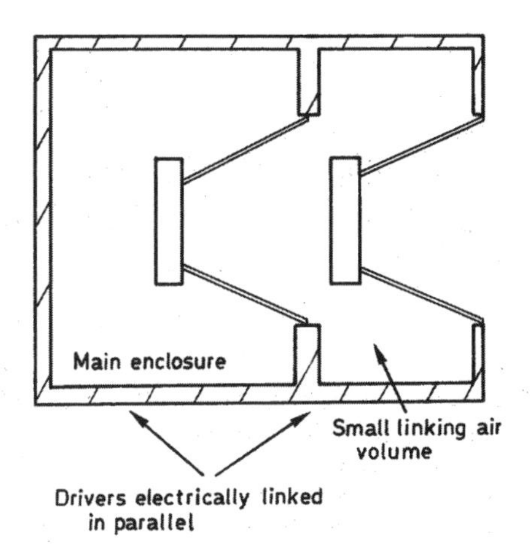

Fig. 6-17
Isobarik loading
(Courtesy Martin Colloms, *High Performance Loudspeakers)*

Energy Multiplied Bandpass

Another dual-driver loading cleverly exploits driver resonance to work to the designer's advantage. The technology, found in subwoofers made by James Loudspeakers, is called Energy Multiplied Bandpass by that company. In the EMB-1000 subwoofer (the first product to use EMB technology), a 10" woofer is hidden inside the cabinet behind the 10" woofer you see at the outside of the cabinet. This internal woofer is driven by the subwoofer's internal amplifier. The internal woofer's acoustic output is acoustically coupled to the passive external woofer to produce sound.

To understand this approach, you need to know that the lower the frequency the woofer is reproducing, the farther the woofer's cone must move. For example, a woofer must move 16 times farther at 25Hz than at 100Hz to produce the same sound-pressure level. Such high excursion creates distortion as the woofer's voice coil moves out of the magnetic gap, making its movement non-linear. In other words, as frequency decreases, distortion increases.

Back to the EMB technique. The external passive woofer is not driven by an electrical signal, but by air compression in the chamber between it and the driven woofer behind it. The passive woofer's resonant frequency is 35Hz, which causes its acoustic output to rise by 12.5dB at that frequency. To maintain flat response, the signal driving the internal woofer is equalized with a 12.5dB dip at 35Hz, which perfectly complements the passive woofer's resonant peak to achieve flat response. Because of this equalization dip in the signal driving the internal woofer, the woofer isn't driven as hard in the low frequencies, lowering its excursion and, with it, distortion. In this system, the driven woofer moves only about one-tenth as far at 35Hz as it would in a conventional system. The passive driver moves much farther than the internal woofer (thus the "multiplying" factor claimed in the technique's name). The result is lower distortion, greater output in the band where bass energy is concentrated, and increased dynamic headroom. Fig.6-18 is a cutaway illustration of this technique.

Fig. 6-18 In the Energy Multiplied Bandpass technique, an active, interior woofer is acoustically coupled to an exterior passive woofer. (Courtesy James Loudspeakers)

The Finite Baffle

The only examples of the finite-baffle, or open-panel, loading technique are planar loudspeakers such as the full-range ribbon system illustrated in Fig.6-5 earlier. The finite baffle has a rolloff of only 6dB/octave until the driver's free-air resonance frequency, then drops at the rate of 18dB/octave.

System Q

Just as a struck bell produces a certain pitch, a woofer in an enclosure will naturally resonate at some frequency. The nature of that resonance is an important characteristic of the loudspeaker, and one that greatly influences its sound. The term Q, for "quality factor," is a unitless number that expresses how a woofer resonates in an enclosure.

Specifically, a loudspeaker's Q equals the resonant peak's center frequency divided by the peak's bandwidth. A woofer that "rings" (resonates) over a very narrow frequency band is said to have a higher Q than a woofer that resonates less severely over a wider band of frequencies. The steeper the resonance, the higher the Q.

The woofer has its own resonant Q, which is modified by the enclosure's Q. These resonances combine and interact to reach the system Q, which usually falls between 0.7 and 1.5, as shown in Fig.6-19. A Q of less than 1 is considered overdamped, while a Q of more than 1 is underdamped. You'll sometimes hear a loudspeaker described as having subjectively "underdamped bass," which means the bass is full and warm but lacks tightness. Technically, these terms refer to the system's anechoic response (the speaker's response in a reflection-free room), specifically whether the response is up or down at the resonant frequency. A "critically damped" system having a Q of 0.5 provides perfect transient response, with no detectable overhang. That is, the woofer stops moving the instant the drive signal stops. The higher the Q, the more the woofer rings.

Subjectively, an underdamped alignment has lots of bass but lacks tightness, has poor pitch definition, and tends to produce "one-note" bass. An overdamped alignment produces a very tight, clean, but decidedly lean bass response. An overdamped loudspeaker has less bass, but that bass is of higher quality than the bass from an underdamped system. Overdamped speakers tend to satisfy intellectually by resolving more detail in the bass, but often lack the bass weight and power that viscerally involve your whole body in the music. Most loudspeaker designers aim for a Q of about 0.7 to reach a compromise between extended bass response (down only 3dB at resonance) and good transient response (very slight overhang). Some designers maintain that a Q of 0.5 is ideal, and that a higher Q produces bass of poorer quality.

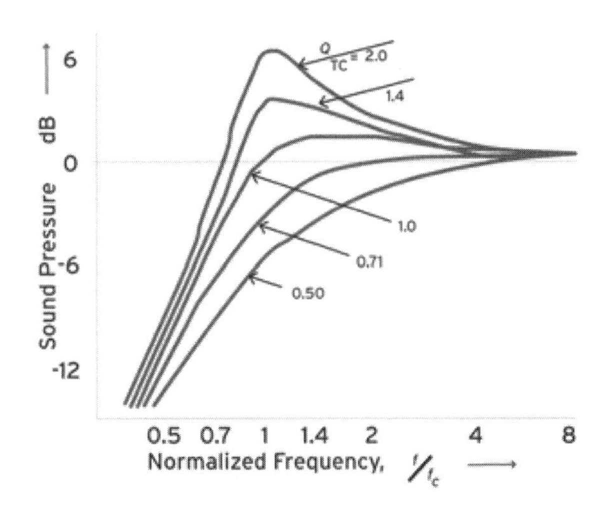

Fig. 6-19
A comparison of system Qs (After Martin Colloms, *High Performance Loudspeakers)*

Mass-market loudspeakers are virtually always underdamped (high Q) so that the unwary will be impressed by the loudspeaker's big bottom end. An example of absurdly high Q is the "boom truck" that produces a big bass impact but fails to resolve pitch, dynamic nuances, or any semblance of musical detail. That boom you hear is the woofer resonating in its enclosure at a specific frequency—the antithesis of what we want in a high-end loudspeaker.

Powered and Servo-Driven Woofers

Loudspeaker designers are increasingly choosing to include a power amplifier within the loudspeaker to drive the woofer cone. In such designs, the external power amplifier is relieved of the burden of driving the woofer, and the loudspeaker designer has more control over how the woofer behaves. For example, equalization can be applied to extend the bass response.

In the most notable and successful example of a powered-woofer system, the Vandersteen Model 5A and Model 7, the bass is rolled off by an external crossover before it reaches the loudspeaker, and a complementary boost restores flat response to the signal driving the woofer. This technique eliminates crossover parts (specifically, the large series inductor) between the internal amplifier and the woofer cone, conferring a significant improvement in the amplifier's ability to control woofer motion, which, in my experience, translates to tighter and better-defined bass. This loudspeaker also employs an 11-band equalizer to minimize room-induced frequency-response variations below 120Hz. As you can see, a loudspeaker employing an integral power amplifier has some decided advantages.

Some designers don't try to juggle these laws of physics to produce the most musically satisfying compromise in bass performance; instead, they take brute-force control of woofer movement with the servo-driven woofer. A servo woofer system consists of a woofer with an accelerometer attached to the voice coil, and a dedicated woofer power amplifier. An accelerometer is a device that converts motion into an electrical signal. The accelerometer sends a signal back to the woofer amplifier, telling the woofer amplifier how the woofer cone is moving. The woofer amplifier compares the drive signal to the cone's motion; any difference is a form of distortion. The woofer amplifier can then change the signal driving the woofer so that the woofer cone behaves optimally.

For example, the inertia in a woofer would cause it to continue moving after a bass-drum whack. The woofer servo amplifier would see the cone moving and instantly stop the cone motion. In fact, if you try to gently press in the cone of a servo-driven woofer with the amplifier connected without playing music, you'll find that the cone doesn't move. Because there's no drive signal, the servo amplifier knows the cone shouldn't be moving and locks it in place.

A servo-driven woofer doesn't follow the same rules as a woofer in an enclosure with no servo drive. Instead of rolling off in the bass below the woofer's resonant frequency, a servo-driven woofer can have flat response down to very low frequencies. The woofer's natural rolloff is counteracted by equalization in the servo amplifier that forces the woofer to keep going lower and lower in frequency. The servo amplifier simply pumps more and more current through the woofer's voice coil as the frequency gets lower, which increases the woofer's back-and-forth excursion to produce more sound at low frequencies. The lower the frequency, the farther the woofer cone must move back and forth to maintain the same volume.

This puts enormous stress on both the woofer and the servo amplifier. The woofer's voice coil must be able to handle the large amount of current flow, and the woofer cone is susceptible to flexing because it is driven over an area the size of the voice coil—the outside area of the cone may not move exactly in step with the area near the voice coil.

Servo systems therefore must be designed to withstand such rigors. Servo amplifiers are often large and powerful, with huge power transformers and many output transistors. The woofer cone may be made of a trilaminate material that sandwiches a damping compound between two metal diaphragms. This structure ensures that the cone moves as a perfect piston, even when driven hard. Oversized voice coils provide high power handling, and the woofer's structure accommodates high excursion.

A loudspeaker using servo-driven woofers may use many woofers operating together. An array of woofers reduces the contribution each must make to the sound, reducing the demands placed on individual woofers. Not every woofer in such an array will have an accelerometer; one accelerometer may serve four woofers, the servo amplifier assuming that the other three woofers are behaving identically.

Servo-driven woofers are characterized by a sense of slam and impact in the bottom end. There's a feeling of suddenness to bass transients that sounds closer to live music.

Moreover, the bass stops just as quickly, with no overhang, slowness, or blurring of the music's dynamic structure. Kickdrum has a crispness and punch rather than sounding like a dull thud. I also find that servo-driven woofers have superb rendering of pitch and precise articulation of each note. Servo-driven woofers can also produce deep bass from a small enclosure.

Servo-driven woofer systems are uncommon because they are more expensive than "cones-in-a-box" loudspeakers, and require specialized expertise to get them to work properly. Keep in mind that any loudspeaker system that uses a dedicated woofer amplifier and a separate amplifier to drive the midrange and tweeter enjoys all the advantages conferred by bi-amping, such as greater ability to play loudly without strain.

Enclosure Resonances

A casual acquaintance once tried to impress me with how great his brother's hi-fi system was by describing how water in a glass placed on top of the loudspeaker would splash out when the system played loudly. This person held the mistaken impression that the ability to make the water fly from the glass was an indicator of the system's power and quality.

Ironically, his description told me immediately that this loudspeaker system was of poor quality. That's because low-quality loudspeaker systems have thin, vibration-prone cabinets that color the sound. The more a loudspeaker cabinet vibrates, the worse it is. Any speaker system that will splash water out of a glass must sound dreadful. We're about to see why.

When excited by the sound from the driver (primarily the woofer), the enclosure resonates at its natural resonant frequencies. Some of the woofer's back-and-forth motion is imparted to the cabinet, either mechanically through the coupling of the driver to the enclosure, or acoustically by the sound inside the enclosure. This causes the enclosure panels to move back and forth, producing sound. The enclosure thus becomes an acoustic source: we hear music not only from the drivers, but also from the enclosure.

Enclosure vibrations produce sound largely over one or more narrow bands of frequencies centered on the panel's resonant frequencies. The loudspeaker has greater acoustical output at those frequencies. Consequently, cabinet resonances change the timbres of instruments and voices. With a recording of double-bass or piano, you can easily hear cabinet resonances as changes in timbre at certain notes. Some musical instruments exhibit a similar phenomenon, producing "wolf tones."

Enclosure resonances not only color the sound spectrally (changing instrumental timbre), they smear the time relationships in music. The enclosure stores energy and releases that energy slowly over time. When the next note is sounded, the cabinet is still producing energy from the previous note. Loudspeakers with severe cabinet resonances produce smeared, blurred bass instead of a taut, quick, clean, and articulate foundation for the music. Enclosure vibration also affects the midrange by reducing clarity and smearing transient information.

The acoustic output of a vibrating surface such as a loudspeaker enclosure panel is a function of the excursion of the panel and the panel's surface area. Because loudspeaker cabinet panels are relatively large, it doesn't take much back-and-forth motion to produce sound. A large panel excited enough that you can feel it vibrating when you play music will color the sound of that loudspeaker. That's why smaller loudspeakers often sound better than similarly priced but larger loudspeakers; it's much harder (and more expensive) to keep a large cabinet from vibrating.

Cabinet vibration is why the same loudspeaker sounds different when placed on different stands, or with different coupling materials between the speaker and stand. Cones, isolation feet, and other such accessories all cause the speaker cabinet to have a different resonant signature. I noted with interest a paper presented at an Audio Engineering Society convention that explored how much cabinet resonance was audible. Before the researchers could begin the experiment, however, they discovered that the material on which the loudspeaker was resting had an enormous effect on how the cabinet vibrated. For consistency in the experiment, they hung each speaker under test in a rope harness.

Enclosure resonances can be measured and displayed in a way that lets you intuitively see how bad the resonances are, and at what frequency they occur. (This technique is discussed in Chapter 16 under loudspeaker measurements.) An easier way of judging an enclosure's inertness is the "knuckle-rap" test. Simply knock on the enclosure and listen to the resulting sound. An enclosure relatively free of resonances will produce a dull thud; a poorly damped enclosure will generate a hollow, ringing tone. You can also drive the loudspeaker with a swept sinewave (a special test signal found on some test CDs consisting of a pure tone that gradually and continuously increases in frequency) with your hand on the enclosure. When the test signal's frequency momentarily matches the enclosure's resonant frequency, the enclosure will vibrate and you will hear a change in the tone's sound. The less the enclosure vibrates and the fewer the frequencies that cause the enclosure to vibrate, the better.

Designers reduce enclosure resonance by constructing cabinets from thick, vibration-resistant material. Generally, the thicker the material, the better. Most loudspeakers use 3/4" Medium Density Fiberboard (MDF). MDF 1" thick is better; some manufacturers use 3/4" on the side panels and top, and 1" or 2" MDF for the front baffle, which is more prone to vibration. Exotic materials and construction techniques are also used to combat cabinet resonances. These include making the front baffle from Corian (a synthetic material used in kitchen countertops), aluminum, or carbon fiber. One loudspeaker designer molded a baffle from a mixture of concrete and special polymers that weighed more than 100 pounds.

Cross-braces inside the enclosure reduce the area of unsupported panels and make the cabinet more rigid. Some braces are called "figure-8" braces because their four large holes form a figure-8 pattern. These braces are strategically placed for maximum effectiveness. The importance of cabinet bracing can be seen in Fig.6-20; note the elaborate aluminum internal structure whose sole purpose is to prevent enclosure vibration.

The modern trend toward building inert loudspeaker enclosures was started by Wilson Audio in 1986 with the WATT loudspeaker. This small two-way loudspeaker employed lead panels fitted in hollowed-out sections of the cabinet walls. Today's Wilson loudspeakers are made from a combination of materials Wilson calls "X" material and "M" material, both of which are resin-based ceramics that machine like metal.

Much of the improvement in loudspeaker systems over the past 15 years can be attributed to a greater understanding of the importance of reducing cabinet vibration.

Fig. 6-20 Some loudspeakers feature extraordinary construction designed to prevent the enclosure from vibrating. (Courtesy Magico)

Enclosure Shapes

The enclosure can also degrade a loudspeaker's performance by creating diffraction from the cabinet edges, grill frame, and even the drivers' mounting bolts. Diffraction is a re-radiation of energy when the sound encounters a discontinuity in the cabinet, such as at the enclosure edge. (Diffraction is illustrated conceptually in Fig. 6-21.) Diffracted energy combines with the direct sound to produce ripples (i.e., colorations) in the frequency response. Rounded baffles, recessed drivers and mounting bolts, and low-profile grille frames all help to reduce diffraction. The midrange enclosure shown earlier in Fig.6-16 (B&W's Nautilus technology) is an example of diffraction reduction taken to an extreme.

Some loudspeaker enclosures are tilted back to align the drivers in time. This tilt aligns the acoustic centers of all the drivers so that their outputs arrive at the listener at the same time (see Fig.6-22). Although many loudspeaker manufacturers use the words "time-aligned," the term is a trademark owned by loudspeaker designer Ed Long.

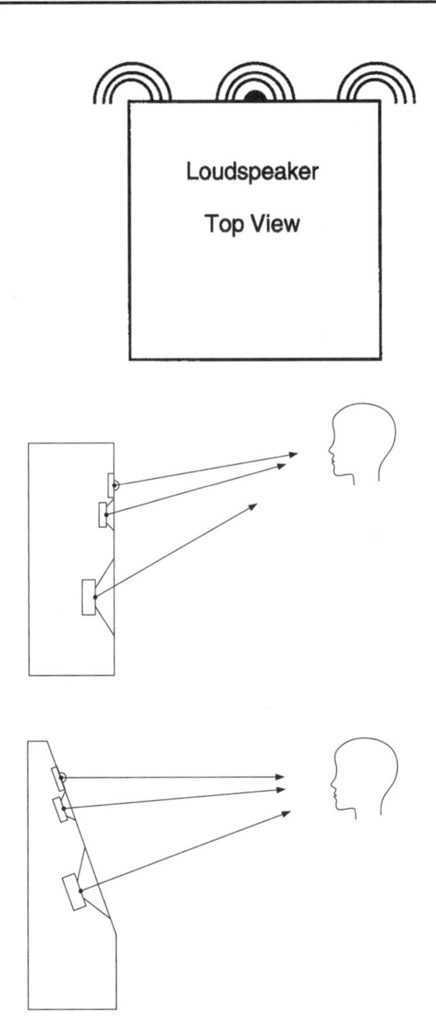

Fig. 6-21
Diffraction causes re-radiation of the sound from cabinet edges.

Fig. 6-22
A sloped baffle causes the sound from each driver to reach the listener simultaneously. (Courtesy Thiel Audio)

Crossovers

A loudspeaker crossover is an electronic circuit inside the loudspeaker that separates the frequency spectrum into different ranges and sends each frequency range to the appropriate drive unit: bass to the woofer, midband frequencies to the midrange, and treble to the tweeter (in a three-way loudspeaker).

A crossover is made up of capacitors, inductors, and resistors. These elements selectively filter the full-bandwidth signal driving the loudspeaker, creating the appropriate filter characteristics for the particular drivers used in the loudspeaker. A crossover is usually mounted on the loudspeaker's inside rear panel or bottom.

A crossover is described by its cutoff frequency and slope. The cutoff frequency is the frequency at which the transition from one drive unit to the next occurs—between the woofer and midrange, for example. The crossover's slope refers to the rolloff's steepness. A slope's steepness describes how rapidly the response is attenuated above or below the cutoff frequency. For example, a first-order crossover has a slope of 6dB/octave, meaning that the signal to the drive unit is halved (a reduction of 6dB) one octave above the cutoff frequency.

If the woofer crossover circuit produces a cutoff frequency of 1kHz, the signal will be rolled off (attenuated, or reduced in level) by 6dB one octave higher, at 2kHz. In other words, the woofer will receive energy at 2kHz, but that energy will be reduced in level by 6dB. A first-order filter producing a 6dB/octave slope is the slowest and most gentle rolloff used.

The next-steeper filter is the second-order crossover, which produces a rolloff of 12dB/octave. Using the preceding example, a woofer crossed over at 1kHz would still receive energy at 2kHz, but that energy would be reduced by 12dB (one quarter the amplitude) at 2kHz. A third-order crossover has a slope of 18dB/octave, and a fourth-order crossover produces the very steep slope of 24dB/octave. Using the previous example, the fourth-order filter would still pass 2kHz to the woofer, but the amplitude would be down by 24dB (1/16th the amplitude). Fig.6-23 compares crossover slopes.

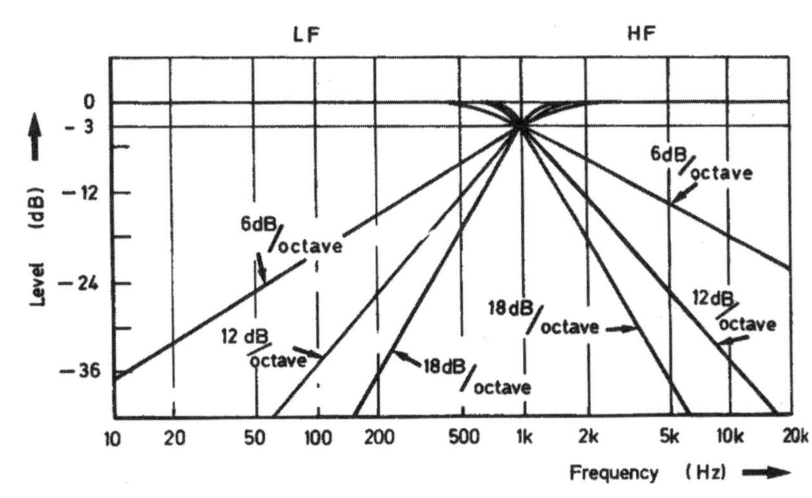

Fig. 6-23
A comparison of crossover slopes (Courtesy Martin Colloms, *High Performance Loudspeakers*)

Typical crossover points for a two-way loudspeaker are between 1kHz and 2.5kHz. A three-way system may have crossover frequencies of 800Hz and 3kHz. The woofer reproduces frequencies up to 800Hz, the midrange driver handles the band between 800Hz and 3kHz, and the tweeter reproduces frequencies above 3kHz.

Note that the actual acoustic crossover slope—the drive unit's acoustical output—may be different from the electrical slope produced by the crossover. If a drive unit is operated close to its own rolloff frequency, this inherent rolloff is added to the electrical rolloff. For example, a woofer rolling off naturally at 1kHz at 6dB/octave will produce an acoustic rolloff of 12dB/octave when crossed over at 1kHz with an electrical 6dB/octave crossover.

The crossover slope is a major design decision. A shallow first-order slope produces the least phase error in the crossover region between the two drive units. The two drive units operate in-phase at both the crossover point and outside the crossover region. That is, the sounds produced by the various drivers are in step with each other in time. A fourth-order filter, by contrast, can cause a phase lag between drivers; the individual drivers' output don't occur at exactly the same time. Loudspeakers with second-order crossovers usually have their tweeters wired in inverse polarity to correct for the crossover's phase lag and provide correct acoustic polarity.

Simple first-order crossovers have inherently better time behavior than more complex crossovers. First-order crossovers, however, produce considerable overlap between drive units. There is a relatively wide band over which both the woofer and the midrange driver produce the same frequency, and over which the midrange and tweeter overlap. This can cause dips in the frequency response if the listener's ears aren't exactly the same distance from each drive unit. The outputs of the two drivers can combine, causing destructive interference and irregular amplitude response. The listener should also sit farther away from a loudspeaker with first-order crossovers to give the sounds from the individual drive units more time to integrate.

In addition, first-order crossovers can allow the tweeter to be overdriven by midrange frequencies in the octaves below the nominal cutoff frequency. A tweeter's excursion quadruples with each halving of frequency, moving four times farther at 1kHz than at 2kHz, given the same drive level. At 500Hz its motion quadruples again, if it isn't rolled off by the tweeter's fundamental resonance. Although a 6dB/octave crossover slope halves the drive signal with each halving of frequency, that's not enough to offset the tweeter's natural quadrupling of excursion as the frequency is halved. The result can be reduced power handling by the system as the drivers are driven to their excursion limits at a lower sound-pressure level. A tweeter driven by excessive midband energy can also introduce distortion, which is manifested as a hard and metallic sound. Proponents of high-order crossovers suggest that any phase advantages of first-order slopes are more than offset by the unwanted overlap between drivers, and the concomitant potential for driver overload. A fourth-order crossover provides a sharper delineation between the frequency ranges of each pair of drivers.

A factor in choosing crossover points is the individual drivers' directivity —their dispersion of sound with frequency. As the frequency a driver is producing increases, the dispersion pattern becomes narrower. In other words, bass is omnidirectional and high frequencies tend to beam. Driver size and crossover points are chosen so that there is no discontinuity in directivity between drivers. This is why you don't see a 10″ woofer mated to a 1″ tweeter—the discontinuity between the two drivers' dispersion patterns would be audible.

Some loudspeakers can be connected to a power amp with two runs of loudspeaker cable, a technique called bi-wiring. A loudspeaker with provision for bi-wiring has separate high-pass (tweeter) and low-pass (woofer) crossover sections, along with two pairs of input terminals. You should always take advantage of this feature by running two pairs of loudspeaker cables: Bi-wiring can significantly improve a loudspeaker's sound. (Bi-wiring is discussed in more detail in Chapter 11.)

Crossovers vary greatly in the quality of parts used. Budget loudspeakers will likely use iron-core inductors rather than the preferable air-core types. Air-core inductors are larger and more expensive than iron-core coils, but are immune to a phenomenon called saturation, which introduces distortion. Similarly, lower-cost loudspeakers use electrolytic capacitors rather than the more expensive plastic-film types. Electrolytic capacitors are sometimes bypassed with small-value polypropylene or other similar high-quality capacitors. Bypassing is putting a smaller-value, higher-quality capacitor in parallel with the larger-value capacitor. Some of the sonic virtues of high-quality capacitors are retained, accompanied by the cost and size advantages of an electrolytic capacitor. Many ultra-high-end loudspeakers use very expensive gold and silver foil capacitors that can cost as much as $300 each. Finally, some loudspeakers are wired internally with high-end cable rather than generic cable.

The Crossoverless Coaxial Driver

Thiel Audio created unique driver that combines a dome tweeter within a midrange cone in the same drive unit. Although such "coaxial" loudspeakers aren't new, Thiel's innovation is that the driver requires no electrical crossover. The tweeter's physical mounting serves as a mechanical crossover, keeping midrange frequencies out of the tweeter. Getting rid of the tweeter crossover confers significant advantages in sound quality, and the coaxial alignment perfectly time-aligns the two drivers.

Digital Loudspeakers

The capacitors, inductors, and resistors that make up a loudspeaker crossover are far from perfect. Not only do they exhibit variations in value that affect performance, these components split up the frequency spectrum in a relatively crude way. As mentioned, higher-order crossovers produce better division of the frequency spectrum, but their phase response suffers. Moreover, a significant amount of amplifier power is wasted in the crossover.

The widespread availability of digital audio sources provides an opportunity to remove traditional crossovers from the signal path. If we can divide the frequency spectrum digitally—that is, by performing mathematical computations on the audio signal—we can create just about any crossover characteristics we want, with none of the problems of capacitors, inductors, and resistors.

These "digital loudspeakers" accept a digital input signal and implement the crossover in the digital domain with digital signal-processing (DSP) chips. Instead of subjecting the audio signal to resistors, capacitors, and inductors, DSP crossovers separate the frequency spectrum by performing mathematical processing on the digital audio data. DSP crossovers can have perfect time behavior, any slope and frequency the designer wants without regard for component limitations or tolerances, and employ equalization to the individual drive units.

Fig.6-24 is a side view of a combination pictorial/block diagram of a digital loudspeaker. The speaker accepts a digital input signal from a disc player, music server, or other digital source. DSP chips inside the speaker split the frequency spectrum into bass, midrange, and treble. These chips can also equalize and delay signals to produce nearly perfect acoustic behavior at the drive units' output. Each of the three digital signals (bass, midrange, treble) is then converted into an analog signal with it's own digital-to-analog converter (DAC). The DAC output feeds a power amplifier specially designed to power the particular drive unit used in the digital loudspeaker. This aspect of a digital loudspeaker confers a large advantage: the power amplifiers are designed to drive a known load. The power amplifiers amplify an analog signal, just like a conventional power amp, then drive conventional cone loudspeakers.

Digital loudspeakers offer the designer the ability to equalize the signal driving the woofer to achieve deeper bass extension for a given enclosure volume than is possible in a conventional loudspeaker. Earlier in this chapter we saw how a woofer in an enclosure begins rolling off just below its resonant frequency, with the resonant frequency and rolloff slope determined by the acoustical loading. In a digital loudspeaker, the signal driving the

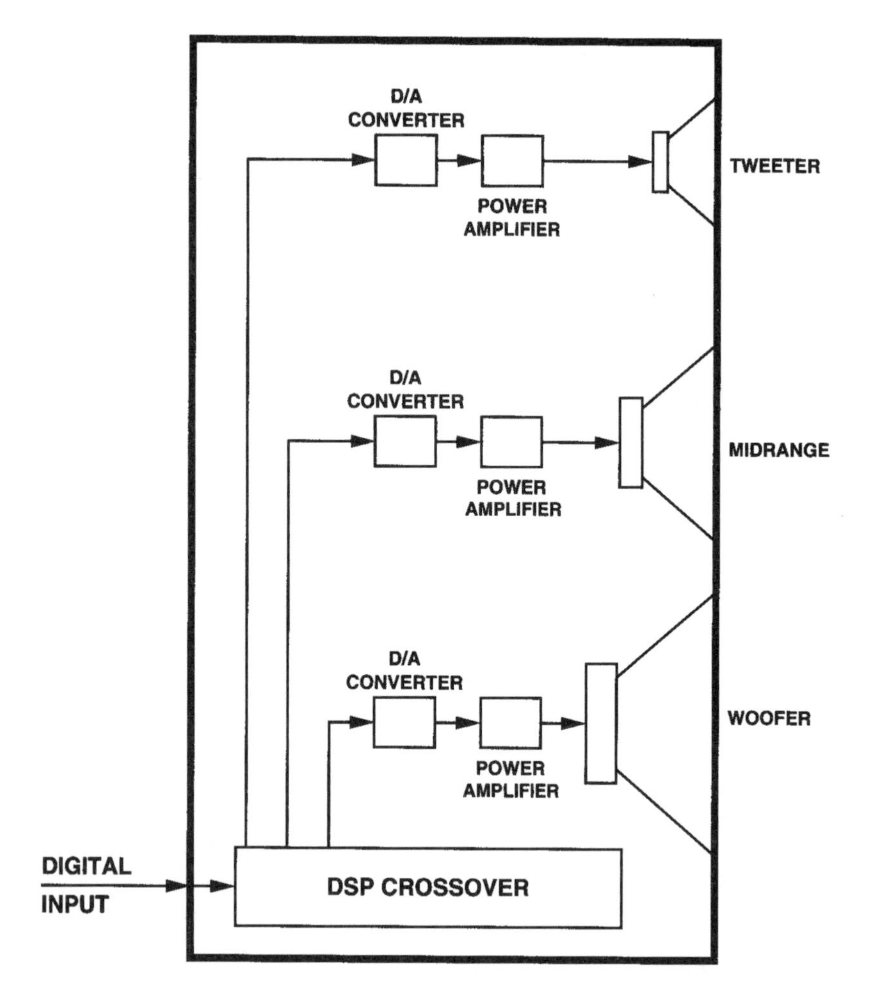

Fig. 6-24 A digital loudspeaker incorporates separate digital-to-analog converters and power amplifiers for each driver.

woofer can be boosted to compensate for this rolloff, extending the low-frequency response. It is possible to achieve amazingly deep bass extension from a moderately sized enclosure with this technique.

In addition to employing virtually perfect crossovers, digital loudspeakers also simplify your system. You need only a digital source and a pair of loudspeakers for a complete music system. This eliminates monoblock power amplifiers on the floor and long runs of garden-hose speaker cable. Moreover, the digital nature of the system provides for remote control of the loudspeaker. You can adjust the tonal balance, correct for different listening heights, and even delay one loudspeaker if your room doesn't permit symmetrical speaker placement. The digital speaker will also have a volume readout, which makes setting the correct playback level for different music more convenient. When considering the price of digital loudspeakers, keep in mind that you don't need to buy an outboard digital-to-analog converter or power amplifiers.

A digital loudspeaker also "knows" the sound-pressure level the loudspeaker is producing and can add a continuously variable loudness compensation. You may have seen the "loudness" button on some audio products, which engages a circuit that boosts bass and treble at low volumes to compensate for the ear's reduced sensitivity to bass and treble at low volume. (See Appendix A for a complete explanation.) Although loudness compensation is, in theory, a good idea, loudness controls are eschewed by audiophiles because they are crude and are effective at only one precise volume setting. A digital loudspeaker, by contrast, can apply exactly the right amount of loudness compensation at any volume setting. The correct amount of bass and treble boost for a given sound-pressure level is precisely known from decades of psychoacoustic research.

Another advantage of a loudspeaker under DSP control is the ability to protect the drivers from damage; the DSP knows the driver's limitations and can impose a maximum excursion limit.

Digital loudspeakers were pioneered by Meridian Audio, which has been producing them since 1993. Meridian's first models accepted the 44.1kHz, 16-bit datastream from a CD transport, but newer models have been upgraded to accept high-resolution digital audio with sampling rates up to 192kHz and word lengths up to 24 bits. Newer models also feature vastly more powerful digital signal processing engines that deliver better performance than earlier models.

Subwoofers

A subwoofer is a loudspeaker that produces low frequencies that augment and extend the bass output of a full-range loudspeaker system. The term "subwoofer" is grossly misused to describe any low-frequency driver system enclosed in a separate cabinet. But "subwoofer" actually means "below the woofer," and should be reserved for those products that extend bass response to below 20Hz. A low-frequency driver in an enclosure extending to 40Hz and used with small satellite speakers is more properly called a woofer.

You'll also see full-range speakers with a built-in "subwoofer" powered by its own amplifier. Most of these products actually employ woofers that are simply driven by an integral power amplifier. Such a design relieves your main amplifier of the burden of driving the woofer, but requires that the loudspeakers be plugged into an AC outlet.

Subwoofers come in two varieties: passive and active. A passive subwoofer is just a woofer or woofers in an enclosure that must be driven by an external amplifier. In one variation of the passive subwoofer, the same stereo amplifier driving the main loudspeakers also powers the subwoofer. In this least desirable method of connecting a subwoofer, the full-range output from a power amp is input to the subwoofer, and a crossover in the subwoofer removes low frequencies from the signal and outputs the filtered signal to the main loudspeakers. This technique puts an additional crossover in the signal path to filter speaker-level signals.

A better way of driving the subwoofer is with an electronic crossover and separate power amp. This method separates the bass from the signal driving the main loudspeakers at line-level, which is much less harmful to the signal than speaker-level filtering. Moreover, adding a separate power amp for the subwoofer greatly increases the system's dynamic range and frees the main loudspeaker amplifier from the burden of driving the subwoofer. Adding a line-level crossover and power amp turns the passive subwoofer into an active subwoofer, and also makes the system bi-amplified.

A self-contained active subwoofer combines a subwoofer with a line-level crossover and power amp in one cabinet, eliminating the need for separate boxes and amplifiers. Such a subwoofer has line-level inputs (which are fed from the preamplifier), line-level outputs (which drive the power amp), and a volume control for the subwoofer level. The line-level output is filtered to roll off low-frequency energy to the main loudspeakers. This crossover frequency is often adjustable on subwoofers to allow you to select the frequency that provides the best integration with the main loudspeakers (more on this later).

A useful subwoofer feature is an automatic equalization circuit that measures the frequency response of your system in your room and tailor's the subwoofer's output to provide flatter response. You connect a supplied calibration microphone to the subwoofer, push a button, and the subwoofer emits a series of tones that are picked up by the microphone and analyzed by a circuit in the subwoofer. The subwoofer then equalizes its output, applying a boost to certain frequencies and a cut to others so that the resultant output is as flat as possible.

Adding an actively powered subwoofer to your system can greatly increase its dynamic range, bass extension, midrange clarity, and ability to play louder without strain. The additional amplifier power and low-frequency driver allow the system to reproduce musical peaks at higher levels. Moreover, removing low frequencies from the signal driving the main loudspeakers lets the main loudspeakers play louder because they don't have to reproduce low frequencies. The midrange often becomes clearer because the woofer cone isn't furiously moving back and forth trying to reproduce low bass.

Now for the bad news. More often than not, subwoofers degrade a playback system's musical performance. Either the subwoofer is poorly engineered (many are), set up incorrectly, or, as is increasingly common, designed for reproducing explosions in a home-theater system, not resolving musical subtleties.

Let's first look at the theoretical problems of subwoofers. First, most subwoofers—passive or active—add electronics to the signal path. The active subwoofer's internal crossover may not be of the highest quality. Even well-executed crossovers can still degrade the purity of very-high-quality source components, preamplifiers, and power amps. This drawback can be avoided by running the main loudspeakers full-range (no rolloff), but you lose the dynamic advantages and additional midrange clarity conferred by keeping low frequencies out of the main loudspeakers.

Second, the subwoofer's bass quality may be poor. The subwoofer may move lots of air and provide deep extension, but a poorly designed subwoofer often adds a booming thumpiness to the low end. Rather than increasing your ability to hear what's going on in the bass, a subwoofer often obscures musical information.

Third, a subwoofer can fail to integrate musically with the main loudspeakers. Very low frequencies reproduced by the subwoofer can sound different from the midbass produced by the main loudspeakers. The result is an extremely distracting discontinuity in the musical fabric. This discontinuity is manifested as a change in the sound of, for example, acoustic bass in different registers. Ascending and descending bass lines should flow past the crossover point with no perceptible change in timbre or dynamics.

Another factor that can make integrating a subwoofer difficult is matching a slow and heavy subwoofer to taut, lean, articulate main loudspeakers. Put another way, the sound from an underdamped subwoofer won't integrate very well with that from an overdamped loudspeaker.

Fourth, subwoofers often trade tight control, pitch resolution, and lack of overhang for greater sensitivity or deeper extension. This is particularly true of subwoofers designed for home theater. Consequently, many subwoofers sound bloated, slow, and lacking in detail.

Finally, a subwoofer can fill the listening room with lots of low-frequency energy, exciting room-resonance modes that may not have been that bothersome without the subwoofer. Placement is therefore crucial—you can't just put a subwoofer anywhere and expect musical results. This problem of room-mode excitation can be ameliorated by two subwoofers; they excite different room modes, substantially smoothing out the room's low-frequency response.

All of these problems are exacerbated by most people's tendency to set subwoofer levels way too high. The reasoning is that if you pay good money for something, you want to hear what it does. But if you're aware of the subwoofer's presence, either its level is set too high, it isn't positioned correctly, or the subwoofer has been poorly designed. The highest compliment one can pay a subwoofer is that its contribution can't be heard directly. It should blend seamlessly into the musical fabric, not call attention to itself.

It's a rare subwoofer that doesn't degrade the signal driving the main loudspeakers, integrates well with main loudspeakers, has tight and controlled bass, and can improve a playback system. You should therefore approach the purchase of a subwoofer with great caution. I can think of only a few true subwoofers worth owning. Further, most well-engineered, full-range loudspeakers go deep enough in the bass for most listeners. Very little program material requires 16Hz extension; most analog recordings have very little energy below about 30Hz.

Subwoofer Technical Overview

Designers of subwoofers face the same low-frequency loading choices that designers of full-range loudspeakers do. The woofer or woofers can be loaded into an infinite baffle (sealed box), reflex enclosure (ported), or a transmission line. Transmission-line subwoofers are rare because of the large cabinet required for such a design. Instead, the sealed-box and reflex are the most popular. Some subwoofers use servo drive (explained earlier in this chapter).

The simplest subwoofer is a single driver—usually a 10" or 12" unit—in a reflex enclosure. The choice of a reflex design keeps the cabinet at a moderate size and maintains reasonable sensitivity. A larger enclosure extends the subwoofer's bass response, but is more expensive to manufacture, particularly if the enclosure is properly constructed (i.e., well-braced).

The sealed-box subwoofer is less common due to its inherently higher cutoff frequency. A sealed subwoofer would have to be enormous to achieve sub-20Hz extension; most sealed-box units are actually woofers, not subwoofers. They trade extension for higher-quality bass: better control, faster transient response, and excellent pitch definition, for example.

Some small sealed subwoofers can, however, achieve sub-20Hz extension because the signal driving them has been equalized to produce flat response. The equalization applies a boost to the low bass that counteracts the woofer's natural rolloff, producing flat response far below the woofer's natural cutoff frequency. Equalization puts enormous stress on the woofer cone because it forces the woofer to handle a large amount of power and have high excursion at very low frequencies. Consequently, a woofer designed for use in an equalized system will have a large magnet and huge voice coil. If an equalized subwoofer uses a separate, rather than integral, power amplifier and equalization circuit, be aware that the amplifier will work correctly only with the subwoofer for which it was designed. Similarly, a subwoofer designed to be driven by an equalized signal will work correctly only when driven by its matching amplifier.

Some subwoofers are mounted in bandpass enclosures that limit the system's bandwidth with both electrical and acoustical filtering. In a bandpass enclosure (see Fig.6-25), the woofer drives a chamber that is vented through a port or slot. The woofer is often loaded in a smaller, rear chamber. This version is called a second-order bandpass alignment. Porting the smaller chamber results in a fourth-order alignment, producing steeper filter slopes. Advantages of bandpass enclosures include greater low-frequency extension, lower distortion, greater sensitivity, and higher power handling.

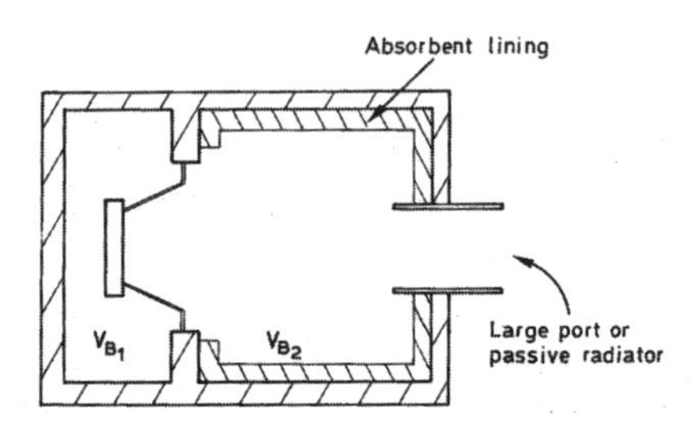

Fig. 6-25
A bandpass enclosure
(Courtesy Martin Colloms, *High-Performance Loudspeakers*)

Subwoofers usually offer a choice of crossover frequencies, either at fixed points (typically 80Hz, 120Hz, and 180Hz) or with a continuously variable knob. A higher crossover frequency would be selected for a minimonitor than for a full-range loudspeaker. Generally, the lower the crossover frequency, the better; the main loudspeaker's bass is probably of higher quality than the subwoofer's, and a low crossover frequency moves any crossover discontinuity lower in frequency, where it will be less audible. In addition, a low crossover frequency ensures that you won't be able to locate the sound source of the low bass. A subwoofer reproducing frequencies above 100Hz can be "localized"—i.e., the location of the source of the bass can be detected—which is musically distracting. A crossover frequency that is too low will, however, burden small loudspeakers with excessive bass and reduce the system's power handling and maximum listening level. I've used a subwoofer with a massive full-range loudspeaker system, setting the crossover frequency at just 30Hz. The subwoofer provided just a little deeper extension and power in the bottom octave.

Another variable in subwoofer crossovers is the slope. Most use second-order (12dB/octave) or higher filters. Ideally, the crossover frequency and slope would be tailored to the particular loudspeaker used with the subwoofer. But because the subwoofer manufacturer doesn't know which loudspeakers will be used with the subwoofer, these parameters are compromised for good performance with a variety of loudspeakers.

Some subwoofers have a "delay" control that lets you time-align the subwoofer's wavefront with that of the main loudspeakers. You can accomplish the same thing by moving the subwoofer forward or backward in relation to the main loudspeakers—a much less practical approach than the delay control.

If the subwoofer has a delay control, there's a simple trick for aligning the subwoofer with the main loudspeakers. Drive the system with a pure tone at exactly the crossover frequency. Many test CDs include a full range of test tones. Driving the system with a pure tone at the crossover frequency causes the main loudspeakers and the subwoofer to reproduce the same signal. Now invert the polarity of the main loudspeakers relative to the subwoofer by reversing the red and black leads going to both loudspeakers. Sit in the listening chair and have an assistant slowly vary the delay control until you hear the least bass. Return the loudspeaker leads to their former (correct) polarity. The delay control is now set optimally. Here's why: When the main loudspeakers' and subwoofer's wavefronts are 180° out of phase with each other, the greatest cancellation (the least sound heard) will occur. That's because as the subwoofer's cone moves outward, the main speakers' cones are moving in, canceling each other. When the loudspeaker leads are returned to the correct position (removing the 180° phase shift), the subwoofer and loudspeaker outputs are maximally in-phase. Any time lag between the main speakers and subwoofer has been eliminated.

When positioning a subwoofer, follow the guidelines on loudspeaker placement outlined in Chapter 14. If the subwoofer is between the loudspeakers, it should be located behind them to avoid interfering with the soundstage created by the main loudspeakers.

Loudspeaker Stands

Small loudspeakers should be mounted on stands for best performance. In fact, the stands' quality can greatly affect the reproduced sound. Flimsy, lightweight stands should be avoided in favor of solid, rigid models. The stands should include spikes on their bottoms to better couple the stand and loudspeaker to the floor. Some loudspeaker stands can be filled with sand or lead shot for mass loading, making them more inert and less prone to vibration. A great loudspeaker on a poor stand will suffer significantly degraded performance. Plan to spend several hundred dollars on stands. When comparing a floor-standing loudspeaker to one requiring stand-mounting, include the cost of the stands in your budgeting.

The interface between loudspeaker and stand also deserves attention. Spikes, cones, and other isolation devices (see Chapter 15) can allow the loudspeaker to perform at its best. In a series of experiments attempting to quantify the effects of different materials installed between stand and loudspeaker, an accelerometer was attached to the loudspeaker's side panel and its vibration measured with the different interface materials. The loudspeaker's resonant "signature" was changed considerably with the stand, and again with the material or device between the stand and loudspeaker. Most effective was a small ball of Bostik Blue-Tak placed at each corner of the speaker stand. This very effective yet inexpensive interface is a sticky, gum-like material available at hardware stores. Best of all, a lifetime supply costs about $2.

When setting up speaker stands, be sure that each of the stands' three or four spikes contacts the floor, and that the stand can't rock back and forth. You can also improve the coupling of the speaker to the stand, and the stand to the floor, by putting a heavy object on top of the speaker. Commercial products, made from a heavy granular material in a canvas bag, are available for this purpose. The bags also help damp enclosure resonances.

7 Disc Players, Transports, and DACs

Introduction

Digital source components are any products or combinations of products that play music from a digital medium. A digital front-end can be as simple as a CD player or as complex as a combination of a disc player, Internet-connected music server, outboard digital-to-analog converter, and separate master clock. Other digital source components include CD/SACD players, Blu-ray Disc players, and universal players that support all disc formats. Whatever the configuration, your system's digital front-end is a vital link in the audio reproduction chain, and the source you'll probably spend the most time listening to.

No area of audio has experienced greater technological change over the past few years than the digital arena. We've seen a revolution in how we access, store, and play digitally encoded music. Just a few years ago, digital audio was limited to listening to an optical disc (CD or SACD). Today, we can download high-resolution music files over the Internet, store those files on hard drives, and access our music via a software interface and mouse—sometimes wirelessly. Our entire music libraries can fit on a single hard-disk drive. The idea of purchasing a physical format (a disc), taking that disc out of a package, and putting it into a player seems to some music lovers like a quaint anachronism. In 2009 the president of Polygram Records expressed the view that the vast majority of the CDs sold are played just once—to transfer the music into some form of computer-based music system. Another sign of the times was the decision by Linn Products in early 2010 to cease making CD players and instead focus on their DS (Digital Streaming) devices that download music from the Internet.

Nonetheless, CD and SACD remain important formats for the audiophile. Optical discs might be on their way out, but they still have a few more years of useful service left. Moreover, many of the sonic and technical criteria for choosing a disc player as a digital front-end apply equally well to music servers and other newer technologies.

Before we look in depth at how to choose digital components let's define some terms.

CD Player: A CD player is a self-contained component that plays CDs. The CD player's analog line-level output feeds a line-level input on your preamplifier. Every CD player includes, in the same chassis, a transport mechanism (which spins and reads the CD) and a digital-to-analog converter (or DAC).

CD/SACD Player: A CD/SACD player is a self-contained component that will play conventional CDs as well as Super Audio Compact Discs.

Universal Disc Player: A machine that plays multiple 120mm disc formats, including CD, SACD, DVD-Video, DVD-Audio, and Blu-ray Disc.

Music Server: A device that stores digital audio on hard-disc drives or in solid-state memory. (Music servers are covered in detail in the next chapter.)

Digital-to-Analog Converter: Also called a "digital processor" or "D/A converter" or "DAC," the digital-to-analog converter is a component that receives digital audio data from a disc transport, music server, or other source, and converts it to an analog signal. The DAC has a digital input and analog output, the latter feeding one of your preamplifier's line-level inputs. Some disc players have a digital input that enables them to function as a DAC.

Disc Transport: The disc transport reads the digitally encoded information from a CD or other optical medium and sends it to the D/A converter for conversion to analog. Unlike a CD player, a transport has no DAC built into it. CD-only transports are becoming increasingly rare.

High-Resolution Digital Audio: Digital audio with a sampling rate of 88.2kHz or higher and 20-bit word length or greater. Contrasted with standard-resolution digital audio of 44.1kHz sampling with 16-bit word length.

Digital Output: A jack on digital source components that provides access to the digital datastream. A digital output allows a player to send digital data to a separate DAC or other device. A digital output could be in the S/PDIF, AES/EBU, FireWire, USB, or HDMI formats.

S/PDIF Interface: S/PDIF stands for "Sony/Philips Digital Interface Format," after the two companies that invented the compact disc. The S/PDIF interface is a standard format for transmitting digital audio, primarily between a source component and a DAC. The S/PDIF signal can be transmitted via a variety of interface types, such as optical or coaxial (described in detail later in this chapter). Nearly all consumer digital audio products—transports, DACs, and digital recorders—use the S/PDIF interface. A professional version of S/PDIF, called AES/EBU (Audio Engineering Society/European Broadcast Union), is sometimes included in consumer digital audio products.

USB Interface: The USB (Universal Serial Bus) was originally developed for the computer industry but has become a digital-audio interface with the proliferation of computer-based music servers.

HDMI: The High-Definition Digital Multimedia Interface can carry multichannel high-resolution digital audio along with high-definition video in a single cable. HDMI is the standard for connecting home-theater components, and also has some applications in high-end audio.

How to Choose a Digital Source—Overview

The first consideration in choosing digital source components is deciding which formats to invest in. For many music lovers a CD player will suffice, particularly those who don't want to boot-up a computer in order to listen to music. Others will opt for a CD/SACD player to take advantage of SACD's superior sound quality. Although SACD is generally considered a "failure" in the marketplace, it remains a robust format for audiophiles, particularly those who enjoy classical music. The SACD catalog abounds with outstanding performances in terrific sound. Moreover, SACD is *the* format for high-resolution multichannel digital audio; no other medium, whether packaged or downloaded, delivers high-res multichannel with nearly as many titles as SACD. Many record companies still regularly release new titles in multichannel SACD. If you enjoy classical music in multichannel sound, SACD is the format of choice.

The DVD-Audio format, once positioned as the successor to the compact disc, is essentially dead. Still, some universal-disc players offer DVD-A compatibility. The Blu-ray Disc format has the technical capacity to serve as a high-resolution multichannel music-only format, but so far the world's record companies have ignored its potential as a music carrier. They are instead focused on moving away from packaged physical media to electronic download. Nonetheless, the Blu-ray Disc format has much to offer the audiophile; it can deliver concerts and musical performances in high-definition video along with high-resolution multichannel audio that has the same quality as the studio master.

A major decision is whether to play music from discs or buy a music server and load it with your music library. We'll explore music servers in depth in the next chapter, but an overview of their strengths and weaknesses might help point you in the right direction. On the plus side of the ledger, a music server gives you instant access to your music library by showing you your entire collection on a computer screen or hand-held device such as an iPhone, iTouch, or iPad. Playing a particular piece of music is as simple as clicking a mouse button or tapping a touchscreen, allowing you to spend more time listening and less time finding the CD case and loading a disc in a player. A music server also allows you to program a large number of tracks in advance for uninterrupted listening. Some servers will assemble a long playlist based on you telling it what kind of music you'd like to hear. Living with a music server fundamentally changes your relationship to your music library; I've found myself listening deeper into my catalog as more titles are brought to my attention. In addition, music servers open up a vast new world of music via Internet downloads. You can add music to your library without leaving your listening chair. Anyone who has used an iPod and iTunes knows the power of a music server. And downloading to a music server is the best way to access high-resolution digital audio. Rather than waiting for record companies to agree on a physical format for high-res digital, music servers and Internet delivery bypass physical formats entirely.

And now for the downside. Unless you opt for an expensive turnkey system, you must go through the hassle of installing software and hardware, configuring the system, getting it to work correctly, and going through the learning curve to make the system play music. A do-it-yourself music server requires patience and above-average technical experience with personal computers. Moreover, some music lovers find that turning on a computer in order to enjoy music is anathema. Those who want the convenience of a server without the hassles can choose a turnkey system that also offers an outstanding user interface, but such systems are expensive. We'll look at such a system in the next chapter.

Finally, some audiophiles will opt for a music server along with a CD/SACD player either to take advantage of the SACD catalog, SACD's multichannel aspects, or both. Music servers generally don't support storing the Direct Stream Digital audio format that is the basis of SACD.

When choosing digital source products, keep in mind that of all the components in an audio system, the digital front-end is the most likely to be left behind by technology's inexorable progress. This is particularly true of music servers, which at their heart are computers. We all know how quickly computer technology becomes obsolete. In addition, computer-based audio is in its formative years; consequently, some of today's technology tends to be cumbersome, and aimed at those with technical expertise.

Whatever format or formats you choose for your music system, those digital bits must be converted into an analog signal. And it's the digital-to-analog conversion stage that has the greatest influence on sound quality. Without a great-sounding DAC (either a separate component or integral to a disc player), it doesn't matter what formats you've chosen or how sophisticated your music server is. The DAC is where the rubber meets the road, sonically speaking, and will exert an enormous influence on how much you enjoy the music you play back through it. "What to Listen For" later in this chapter offers a complete description of how to evaluate the sonic performance of any digital source.

Should You Buy a CD Player, Universal Disc Player, or Transport and DAC?

For those of you who will play CDs only, and have no interest in SACD or music servers, you can choose between a CD player and a CD transport/DAC combination. During most of the 1990s, all the cutting-edge work in digital was in separates. That situation has changed; designers are now using the same level of parts and bringing the same level of design quality to CD players that they once reserved for separates. These premium parts and techniques include large and well-regulated power supplies, high-quality DAC chips, and the designer's best analog circuits. Moreover, many innovative design techniques are starting to appear first in CD players. The CD player is no longer automatically the entry-level product in the manufacturer's line; instead, it's become worthy of a manufacturer's best efforts.

The CD player enjoys two significant advantages over separate transports and DACs: 1) By combining the transport and DAC in one chassis, with one power supply, one front panel, one shipping carton, and one AC cord, the manufacturer can put more of the manufacturing budget into better sound; and 2) A CD player has no need of a sonically degrading digital interface (cable) between its transport and DAC, which means that it might have better sound. (The interface's detrimental effect on musical performance is described later in this chapter.)

A CD player also makes your life much simpler than owning separates. Rather than requiring two chassis in your rack, two power cords, and a digital interconnect, the CD player lets you focus on the music rather than on the equipment.

Many excellent CD players are available for under $1000, including some legitimate high-end machines that sell for as little as $300. This is, however, the lower limit of true high-end CD players. Below this level you enter the realm of mass-market products designed for maximum features and minimum manufacturing cost, not for musical performance.

If you opt for a CD player, look for one that has a coaxial digital output on an RCA jack. This will let you use the CD player as a transport if you upgrade to a separate DAC in the future. Notice that I said a *coaxial* digital output on an RCA jack. Nearly all recently made mass-market CD players use a much inferior optical connector called TosLink. If you drive an outboard digital processor with TosLink, you won't be getting the sound you paid for. Insist on a CD player with a coaxial digital output. That way, you can be assured of a clear upgrade path in the future.

Another option is the CD player that is also compatible with SACD. About 15% of the CD players on the market also offer SACD playback. Such players are often a bit more expensive, but allow you to enjoy the large library of great-sounding SACD titles. Some of these CD/SACD players have one other advantage; they convert PCM data read from the CD to the Direct Stream Digital (DSD) format used in SACD, and then convert that DSD signal to analog. The conversion of DSD to analog is much simpler than PCM-to-analog conversion, resulting in better sound. In every CD/SACD player I've heard that offers the option of PCM-to-DSD conversion, I've preferred the sound with the conversion. (See "DSD Encoding on SACD" later in this chapter for the technical background on why DSD conversion might sound better.)

You can think of a CD/SACD player as simply a CD player with expanded capabilities. That is, the combination player isn't that different from a CD-only machine. But a universal disc player that is compatible with DVD-Video, DVD-Audio, and Blu-ray Disc as well as CD and SACD is a radically different animal. Unlike the CD/SACD player, the universal machine includes complex video-processing circuitry, and it must include a fundamentally different clock frequency that is based on 48kHz sampling rate, not CD's 44.1kHz sampling rate. (The DSD encoding system of SACD is based on an integer multiple of CD's 44.1kHz and thus doesn't require a completely different clock.) The addition of video circuitry and clocks of different frequency greatly increases the machine's complexity and potential for sonic degradation. Moreover, the more formats a machine can play, the greater the royalties the manufacturer must pay. This limits the amount the designer can spend on internal parts that affect sound quality, potentially compromising performance. Consequently, universal players rarely offer the best sound quality—on any format—for a given price level. If most of your listening is to CD, make sure that the universal player has acceptable sound quality when playing CDs. It would be a mistake to compromise performance of your primary medium just to gain compatibility with DVD and Blu-ray. If you are on a budget, want the simplicity of a single digital source component in your rack, or have an audio system that also functions as a home theater, a universal player will suit your needs. For those with more ambitious systems, a separate CD-based digital front-end combined with a separate universal-disc player may be more appropriate.

How to Choose a Digital Source—Features and Specs

The wide range of features and specifications available in digital source components affects not just sound quality but also how the products function in your system. That is, your overall system architecture can be greatly affected by the digital components' features and capabilities. A good example is a CD player with a variable output that allows you to drive a power amplifier directly without the need for a preamplifier.

When shopping for a digital front-end, remember that technical performance is secondary to sound quality. Manufacturers will often tout their products on the basis of some technical innovation, or because it uses the latest parts. Although interesting, such descriptions don't tell you how the product sounds. Many technical factors influence a component's musical performance; which parts it uses is only one of these. Don't buy a product just because it has a particular DAC or digital filter—many digital processors with excellent parts pedigrees just don't measure up in the listening room. Listen to the product and decide for yourself if it sounds good. Just as you wouldn't consider buying an amplifier based on how little THD it has, so you shouldn't choose a digital component because it features parts used successfully in highly regarded products.

Another claim you should consider with suspicion is a processor's clock-jitter specification. Some manufacturers make unverifiable jitter claims, sometimes even picking a jitter number out of the air. Not only are there no standards for jitter specifications, but jitter is also very difficult to measure—a condition that has led to the current confusion in the marketplace. When shopping for digital components, forget the marketing hype—just listen.

Let's look at the features and specs you should consider when choosing digital source components.

Disc Player and DAC Features and Specs

The first decision when choosing a disc player is which formats you intend to play, a topic covered earlier in this chapter. After you've decided on a CD player, CD/SACD machine, or universal-disc player, you'll need to consider which features and functions are important to you. Some of today's CD players sport a host of features and capabilities unimaginable to the format's creators.

A basic CD player will have one pair of analog outputs on RCA jacks, and usually a coaxial digital output on an RCA jack and/or an optical digital output on a TosLink jack. The digital output allows you to upgrade your digital front end by turning the disc player into a transport that feeds an outboard DAC.

The outboard DAC converts a digital datastream from a transport, music server, or other source to an analog signal that is fed to your preamplifier. Outboard DACs range in price from $200 to $20,000, but many good-sounding units can be had for under $1000. The most basic DAC has one digital input on an RCA jack or USB port and one pair of unbalanced analog outputs. More complex DACs have multiple digital inputs, digital outputs, balanced analog outputs, polarity-inversion switching, and sometimes even a volume control. Fig.7-1 shows examples of DACs of varying complexity.

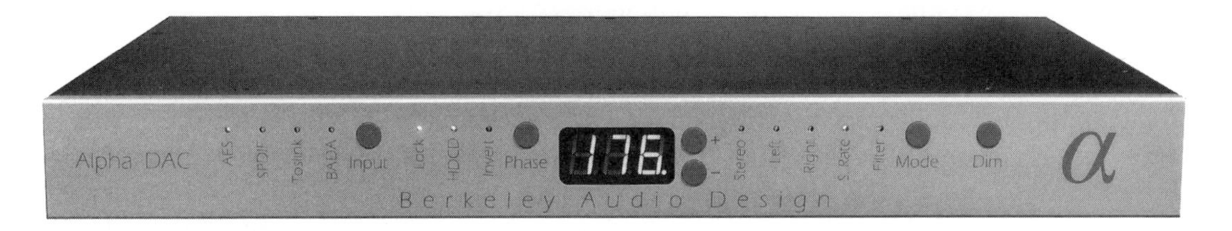

Fig. 7-1 A DAC can be a simple component with no controls (top), or an elaborate, feature-rich product (bottom). (Courtesy High-Resolution Technologies [top] and Berkeley Audio Design [bottom])

I've combined disc player and DAC features and specifications because the two component categories share so many similarities. Indeed, a disc player is simply a DAC with an integral disc-transport mechanism.

Let's take a closer look at the features and options available on disc players and DACs.

Volume Control: For those who listen only to digital sources (no analog tuner or turntable), the disc player or DAC with a volume control is ideal. This feature allows you to drive a power amplifier directly, with no need for a traditional analog preamplifier. This saves money—on both the preamp and an additional pair of interconnects—and has the potential to sound better because fewer electronic components are in the signal path.

A related feature is called *source switching*. A disc player or DAC with source switching has multiple digital inputs, allowing you to listen to different digital sources through the DAC. A few products even include one or two analog inputs in addition to digital ones, completely obviating the need for a traditional preamplifier in most systems. It is unlikely, however, that the preamplifier section built into a DAC can match the sonic performance of a stand-alone preamplifier.

The volume control in disc players and DACs can be implemented in the analog or the digital domain. That is, the analog audio signal can be put through a standard volume-control knob as is found on a preamplifier, or the volume can be adjusted by performing mathematical operations on the digital data representing the music. Before deciding on a disc player or DAC with volume control, you should know the tradeoffs inherent in each approach.

An analog volume control is a mechanical device that subjects the audio signal to a varying electrical resistance. That resistance, and the contacts between the resistance and parts inside the volume control, can slightly degrade the signal—no volume control is perfectly transparent Moreover, the volume control can introduce small channel-balance errors at certain volume settings. For example, when the volume is turned very low, the left channel may be half a dB louder than the right. This situation could reverse as the volume is turned up due to manufacturing tolerances in the volume control.

A digital volume control doesn't suffer from these problems but has its own drawbacks. Each 6dB reduction in volume from the maximum setting throws away one bit of resolution. A low volume setting (say, 30dB of attenuation) is equivalent to discarding five bits. If you had true 20-bit resolution in your D/A converter, you'd be listening to 15-bit audio instead of 20-bit. The lower the volume setting, the greater the loss in resolution. This shortcoming of digital volume controls is much less significant in today's products than in those of ten years ago. As computing horsepower has increased, designers can implement digital volume controls with greater precision. When implemented correctly, a digital volume control can be sonically benign at most listening levels. My current reference DAC has digital volume control and I hear no loss of resolution at typical playback volume.

Number and Type of Inputs: Most DACs offer a range of input connections. All provide S/PDIF input on an RCA jack, or sometimes on a BNC jack. A BNC jack is technically superior in this application, but very few sources provide BNC output (Fig.7-2 shows a range of digital inputs including BNC). The next most common input type is AES/EBU, recognizable from the three-pin XLR jack. AES/EBU is a professional version of S/PDIF that's carried on a balanced cable. Virtually all DACs have an optical input on a TosLink jack for compatibility, but TosLink is sonically inferior to the other input types. A very few DACs have ST-Type optical connection, an interface popular in the 1990s that seems to be making a limited comeback. Another interface of the 1990s that has begun appearing on DACs and transports is the I^2S (pronounced "I squared S") that transmits the audio data and clock on separate conductors for lower jitter. Finally, many of today's DACs provide a USB input for connection to a PC-based music server. Another computer-industry interface, FireWire (technically called IEEE1394), is found on some DACs.

When choosing a DAC, be sure that it has the type of inputs that match your digital source components. These interfaces are described in detail later in this chapter.

Fig. 7-2 A DAC's rear panel showing balanced and unbalanced outputs, along with a variety of input types. (Courtesy Berkeley Audio Design)

Decoding of High-Resolution Bitstreams: Virtually all DACs today (and the DAC sections within disc players) can decode digital bitstreams with resolution beyond CD's 44.1kHz sampling frequency and 16-bit word length. Some DACs are limited to 96kHz/24-bits; others can decode any sampling frequency up to 192kHz and word length up to 24-bits. The first generation of USB DACs was limited to 96kHz because of the available chipsets; later models can accommodate up to 192kHz, the highest frequency of any source material.

Ability to Read WAV Files: Some CD players and transports can read WAV files, whether standard resolution or high-resolution. A WAV file is a raw audio-data format commonly used in music servers. WAV files are typically high-resolution and supplied on a DVD disc. WAV files are usually read into a music server for playback from the server's hard drive. If you don't have a music server and want to play high-resolution WAV files from disc, be sure that the disc player you buy has this capability.

Upsampling: Nearly all of today's disc players and DACs offer upsampling, a technique that converts a bitstream of one resolution to a higher resolution. For example, an upsampling DAC will take in 44.1kHz/16-bit data and upsample it to 192kHz/24-bit for conversion by the digital-to-analog converter chips. No new information is created in this process, but it usually improves sound quality.

Clock Input: Some disc players and DACs have an input that allows them to be clocked by an external box that contains a precision clock. This arrangement replaces the clock within the disc player or DAC with one of ultra-high precision. This feature is only found on more expensive products.

DSD Decoding: Some DACs can accept the Direct Stream Digital (DSD) bitstream from an SACD player that has a DSD digital output and convert the DSD signal to analog. DSD is the encoding format used in Super Audio Compact Disc. The interface format is FireWire (IEEE1394), explained later in this chapter. Some DACs give the user the choice of converting PCM to DSD; the DSD bitstream is converted to analog rather than the PCM signal, which many listeners find to be an advantage.

HDCD Decoding: Some disc players and DACs can decode CDs recorded using the High Definition Compatible Digital (HDCD) process. HDCD is explained later in this chapter.

Balanced Outputs: Some disc players and DACs have balanced outputs on XLR jacks in addition to unbalanced outputs on RCA jacks. If you have a fully balanced preamplifier, consider a DAC with balanced output jacks.

As explained in Chapter 11, a balanced signal is carried on three conductors: a ground wire and two signal-carrying conductors. The two signals are identical but opposite in polarity. That is, they are mirror images of each other. When one is swinging positive, the other is swinging negative. One of the phases of a balanced signal is called the "+" phase, the other the "–" phase.

Designers can use one of two very different techniques to create this balanced signal. The easiest way is to simply put a phase splitter in the DAC's analog output stage. A phase splitter is an electronic circuit that generates a second signal of opposite polarity to the input signal. These two signals are then connected to an XLR jack on the processor's rear panel. This technique, shown in block form in Fig.7-3, is the inexpensive and easy way to create a balanced signal.

Some disc players and DACs, however, use a much more elaborate scheme. Rather than create the balanced signal in the analog domain with an additional active device (the phase splitter), these products create a balanced signal in the digital domain *before* digital-to-analog conversion. The left and right audio-channel digital datastreams are split into a balanced signal (left +, left –, right +, right –), which is processed and converted to analog separately. This technique requires nearly double the circuitry, with four digital-to-analog converter chips and four sets of analog output electronics. A digitally balanced DAC is shown in block form in Fig.7-4.

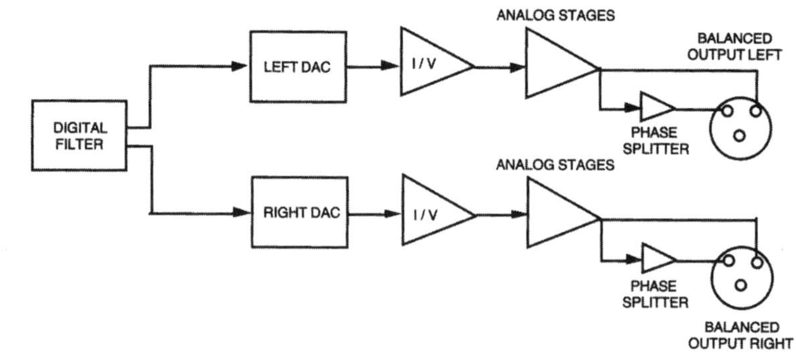

Fig. 7-3 A "balanced" DAC can be created simply by adding a phase splitter to the analog output stage.

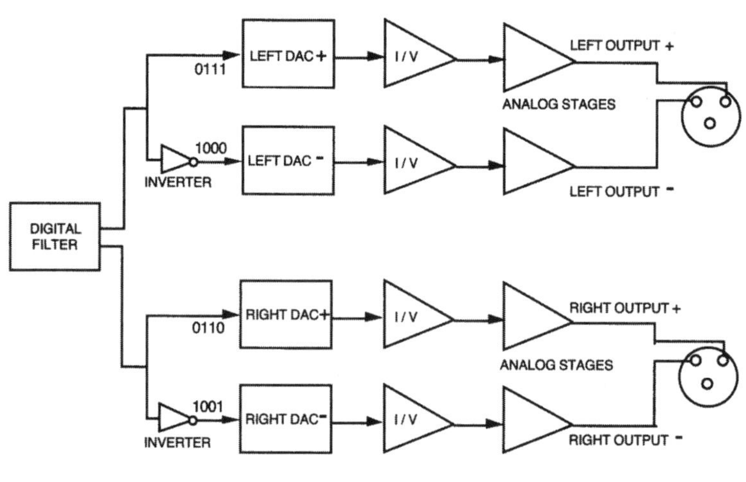

Fig. 7-4 A truly balanced DAC splits the signal into "+" and "–" phases in the digital domain, and then converts the signal to analog with two converters per channel.

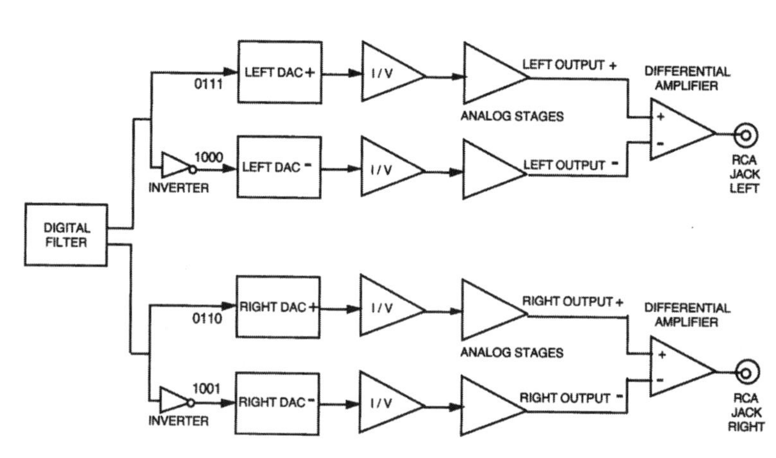

Fig. 7-5a A balanced DAC's unbalanced output can be generated by combining the "+" and "–" phases of the balanced signal.

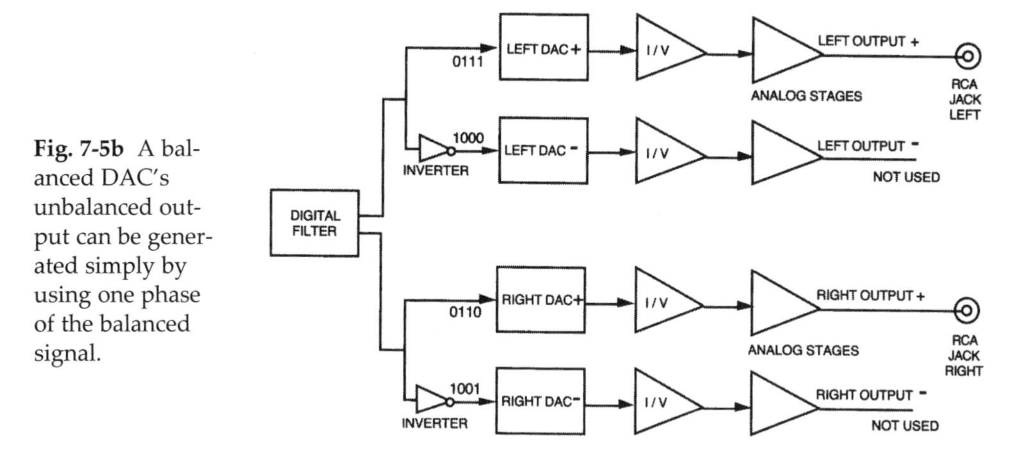

Fig. 7-5b A balanced DAC's unbalanced output can be generated simply by using one phase of the balanced signal.

This method is a lot more expensive than a simple phase splitter in the analog stage, but has many advantages. When the balanced signal is eventually summed (in the preamplifier or power amplifier), any noise or distortion common to both channels will be rejected. This phenomenon, called common-mode rejection, is described in Chapter 11, "Cables and Interconnects." Any artifacts or spurious signals introduced by the DACs will thus cancel. Moreover, balanced DACs yield a 3dB increase in signal-to-noise ratio. Finally, a fully balanced DAC doesn't add another active device (the phase splitter) to the signal path.

A final consideration with fully balanced DACs is how the unbalanced signal is derived. Some products merely connect half the balanced circuit (the left- and right-channel + signals) to the left and right rear-panel RCA jacks. Others sum the + and – signals with a differential amplifier (a circuit that converts a balanced signal to an unbalanced one) to drive the unbalanced RCA output jacks. The latter approach provides the benefits of balanced DACs even for those users who don't use the balanced XLR outputs. Conversely, taking half the balanced signal at the unbalanced output obviates the advantages of fully balanced DACs when using the unbalanced outputs. A comparison of these two techniques is shown in Fig.7-5a and Fig.7-5b.

Digital Output on a DAC: This feature allows you to send the digital signal to a digital recording device. A digital output is an important feature if you own a digital recording device—without it, you'll need to unplug the digital signal going into your digital processor and run it to the digital recorder every time you want to record through the recorder's digital input.

Polarity Inversion Switch: This front-panel switch inverts the absolute polarity of the signal to match its original polarity. To find correct absolute polarity, simply throw the switch back and forth to find the position that sounds the best (if there is an audible difference). A polarity-inversion switch is often marked "180°" to indicate polarity reversal.

Advanced Disc-Player Features: Two Examples

I mentioned earlier in this chapter that some of today's CD players include features unimaginable to the format's creators. Let's take a look at two examples of advanced disc player's feature set and expanded capabilities. The first product under consideration, the Esoteric SA-50, plays CDs and SACDs in its integral transport mechanism. It has digital inputs for decoding external sources such as a music server. These inputs include USB, coaxial, and TosLink, with front-panel switching between digital inputs. The analog output level is variable, allowing the player to drive a power amplifier directly without the need for a traditional preamplifier in the system. The device has an external clock input for connection to an optional high-precision outboard clock. PCM data read from a CD, or appearing at one of the digital inputs, can be upsampled to high-resolution PCM (192kHz) or to the Direct Stream Digital format before conversion to analog.

This player's features allow you to have a single box that serves as a disc player for both CD and SACD, a preamplifier with digital inputs and source switching, and a digital-to-analog converter for your music server—all with upsampling in a variety of user-selectable formats. Moreover, it has the capacity to be upgraded by adding an external high-precision clock that synchronizes the player with an external digital source. The SA-50 and its rear panel are shown in Fig.7-6.

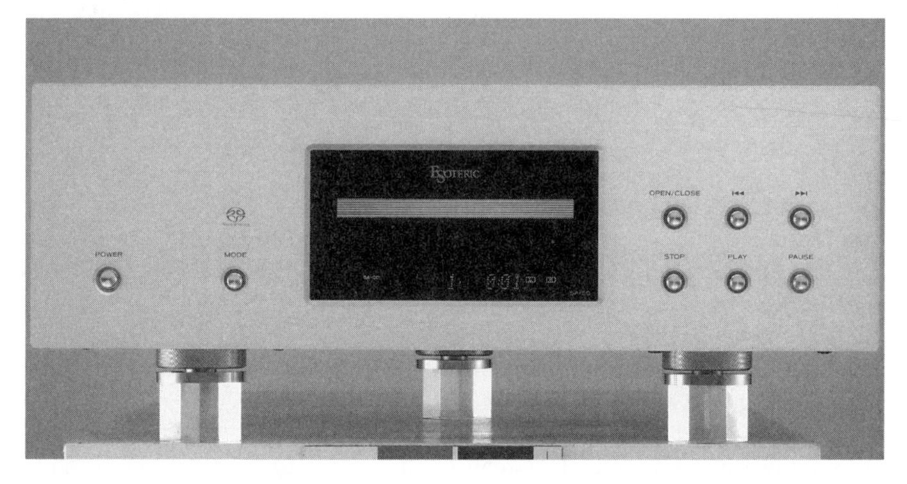

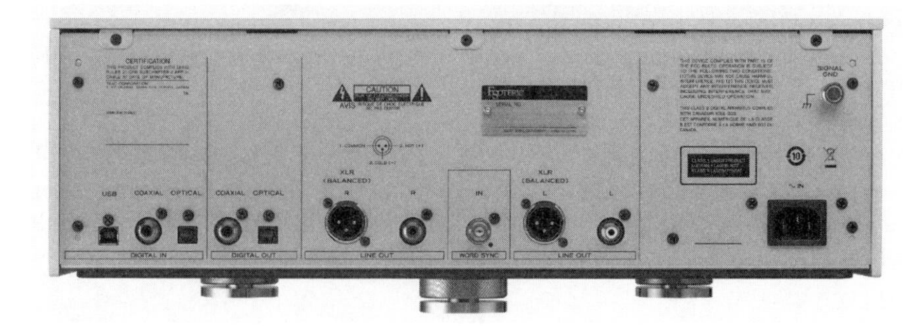

Fig. 7-6 A new generation of digital products combines multi-format disc playback with DAC and preamplifier functions. (Courtesy Esoteric)

Our second example is the PS Audio Perfect Wave Transport, a new breed of disc player that doesn't fall neatly into either the "disc player" category or the "music server" category (Figs.7-7 and 7-8). Rather, it's a new type of component that bridges the gap between a pure disc player and a music server. Moreover, it offers several advanced digital technologies not found in conventional disc players. I'll describe this product in detail in the next chapter, but give you an overview here. This device is a CD transport that incorporates a CD-ROM transport mechanism that will re-read the disc multiple times if necessary until the data on the disc are retrieved with no bit errors. It can accomplish this feat because the data from the disc are read into a buffer memory that can take in data in chunks, and then clock the data out with high precision. This technique isolates the transport's digital-output signal from the disc playback mechanism for better sound. That is, you listen to the output of the data buffer, not the output from the transport mechanism.

Fig. 7-7 Today's advanced transports offer music-server functionality. (right)

Fig. 7-8 This transport's rear panel along with the matching DAC. (Figs. 7-7 and 7-8 Courtesy PS Audio)

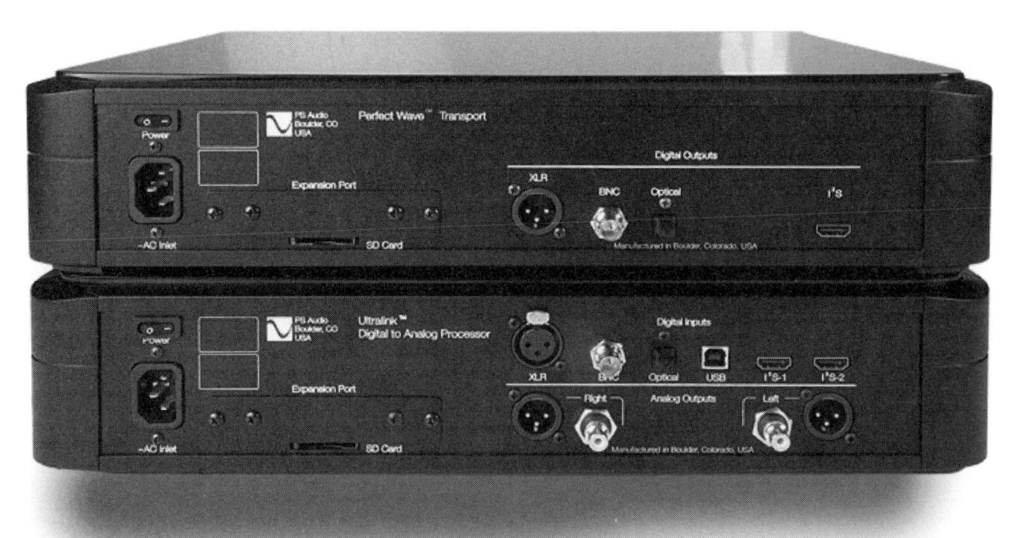

The product has a variety of digital connections in addition to S/PDIF. You can connect a network attached storage (NAS) hard drive to the transport and rip CDs to the NAS. You can play music from either the disc transport or as files from the NAS, with an iTouch or similar device as the user interface (the transport also has a color front-panel touch-screen). The transport provides an I²S output (described later in this chapter) for connection to the I^2S input on the companion DAC. The I^2S interface eliminates the jitter problem that plagues the S/PDIF interface. In addition, the transport and companion DAC can be augmented with a third device that accesses music from the Internet for storage on the NAS. The device's operating software can be updated via a rear-panel slot for an SD memory card. I consider this product more of a music server than a disc player, but included this description here to give you an idea of how the traditional disc player is being transformed by new technologies.

Transport Features and Specs

The CD transport can be thought of as a CD player without a digital-to-analog converter. Instead of generating an analog audio signal as a CD player does, the transport outputs a digital datastream that must be converted to analog by another component, the digital processor, also known as a DAC . A transport-and-processor combination is essentially a CD player in two boxes. Fig.7-9 shows the three digital front-end connection methods: a CD player, a transport/digital processor combination, and a CD player used as a transport.

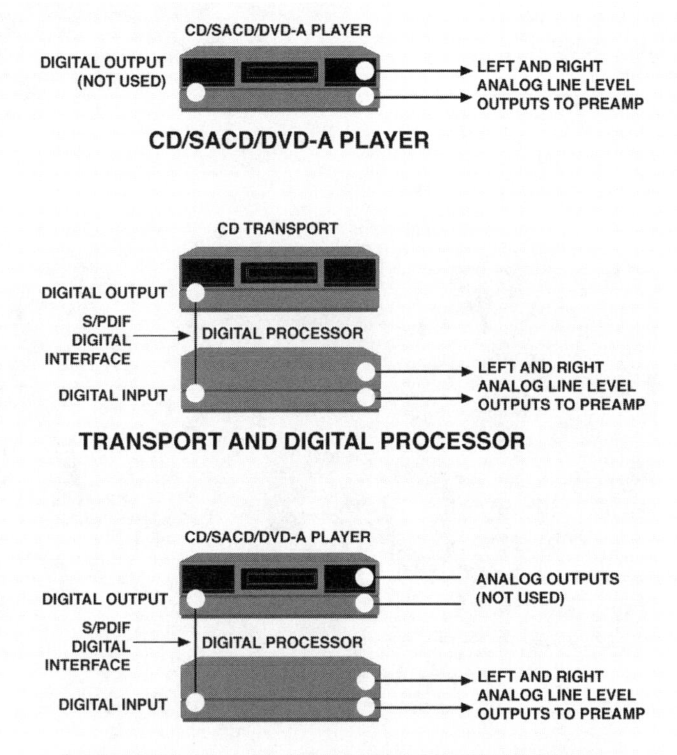

Fig. 7-9 Comparison of different digital front-end connections.

When choosing a transport (or disc player), pay close attention to the machine's ease of use, display, remote control, and overall user interface. Of all the components in your system, you'll probably interact the most with your transport (unless you own a music server). Some have small buttons bunched together, making it hard to find the one you want. Others have large, clearly marked buttons. A particularly useful feature is a back-lit remote control in which the buttons illuminate when pressed. This feature allows you to easily find the button you want in the dark. Look for a remote with an intuitive layout, and with larger buttons for the most frequently used commands. Although ergonomics take second place to sound quality, consider a transport's ease of use when deciding whether or not to buy it.

Another factor to consider is the transport's loading method. Some are top-loading, in which a lid or top opens to accept the disc. The second method is drawer-loading, with a drawer that recesses into the front panel. A top-loading transport usually requires the top shelf of an equipment rack, while a drawer-loading unit can fit on a lower shelf.

Transports have a variety of output connections. The three main types are coaxial (RCA jack), TosLink optical, and AES/EBU (XLR connector). A few also offer a BNC output jack. Virtually all high-end transports at least have coaxial output, and some offer all four types. Some manufacturers have begun to offer digital output on I^2S, as well as on proprietary interfaces.

The best way to audition a transport or disc player used as a transport is to listen to several models in your price range, using the same DAC that you'll be driving with the transport. This will allow you to get the best musical match, since how the transport sounds is affected by the DAC the transport is driving. That's because all DACs respond differently to transport jitter. Transport jitter is caused by timing variations in the digital datastream output from the transport. This datastream jitter is either passed along to the DAC's clock (where it degrades the sound), or is rejected by the DAC and thus is less sonically detrimental. Consequently, transports make much more of an audible difference with some DACs than with others. If you choose a DAC that's relatively immune to transport jitter, you can spend less on a CD transport and still get great sound.

The interface between transport and processor will also affect the sound. There are noticeable sonic differences not only between types of interface (coaxial, AES/EBU, and TosLink), but among cables in the same interface family. Indeed, two coaxial cables may sound almost as different as two transports.

Incidentally, evaluating digital transports and digital interconnects is much easier than comparing other components because the levels are automatically and precisely matched. All transports and interconnects will produce the same listening volume when driving the same DAC—the transport or interconnect doesn't change the ones and zeros in the digital code.

Sonic differences between digital transports are almost certainly solely the result of jitter in their S/PDIF outputs. Recovering the correct ones and zeros from the disc is relatively straightforward; the digital output from a transport is an identical copy, bit for bit, of the source data. The timing of those bits, however, can greatly affect playback quality. Fortunately, today's DACs have better jitter rejection than products made ten years ago, and are less affected by transport and interface jitter.

When evaluating digital transports, listen for the same artifacts we talked about in judging digital front-ends in general. Transports vary greatly in their low-frequency dynamics, treble smoothness, overall perspective, and soundstaging. Choose one that complements your DAC and playback system.

What to Listen For

Perhaps more than any other components except loudspeakers, digital sources come in the most "flavors." That is, their sonic and musical characteristics vary greatly between brands and models. This variability has its drawbacks ("Which one is right?"), but also offers the music lover the chance to select one that best complements his playback system's characteristics and suits his musical tastes. The different types of musical presentations heard in disc players and outboard DACs tend to reflect their designers' musical priorities. If the designer's parts budget—or skill—is limited, certain areas of musical reproduction will be poorer than others. The trick is to find the digital source that, in the context of your system, excels in the areas you find most important musically.

Selecting a digital source specifically tailored for the rest of your playback system can sometimes ameliorate some of the playback system's shortcomings. For example, don't choose a bright-sounding DAC for a system that is already on the bright side of reality. Instead, you may want to select a DAC whose main attribute is a smooth, nonfatiguing treble. Each digital product has its particular strengths and weaknesses. Only by careful auditioning—preferably in your own system—can you choose the product best for you.

To illustrate this, I've invented two hypothetical listeners—each with different systems and tastes—and two hypothetical DACs. I've used a DAC in the example, but disc players could be easily substituted. Although the following discussion could apply to all audio components, it is particularly true of digital components. Not only are there wide variations in sonic characteristics between DACs, but a poor-sounding DAC at the front-end of a superb system will ruin the overall performance.

Listener A likes classical music, particularly early music, Baroque, and choral performances. She rarely listens to full-scale orchestral works, and never plays rock, jazz, or pop. Her system uses inexpensive solid-state electronics and somewhat bright loudspeakers; the combination gives her a detailed, forward, and somewhat aggressive treble.

Listener B wouldn't know a cello from a clarinet, preferring instead electric blues, rock, and pop. He likes to feel the power of a kickdrum and bass guitar working together to drive the rhythm. His system is a little soft in the treble, and not as dynamic as he'd like.

Now, let's look at the sonic differences between two inexpensive and similarly priced DACs and see how each would—or wouldn't—fit in the two systems.

DAC #1 has terrific bass: tight, deep, driving, and rhythmically exciting. Unfortunately, its treble is a little etched, grainy, and overly prominent. DAC #2's best characteristics are its sweet, silky-smooth treble. The DAC has a complete lack of hardness, grain, etch, and fatigue. Its weakness, however, is a soft bass and limited dynamics. It doesn't have the driving punch and dynamic impact on drums of DAC #1.

I think you can guess which DAC would be best for each system and listener. DAC #1 would only exacerbate the brightness Listener A's system already has. Moreover, the additional grain would be more objectionable on violins and voices. DAC #2, however, would tend to soften the treble presentation in Listener A's system, providing much-needed relief from its relentless treble. Moreover, the sonic qualities of DAC #1—dynamics and tight bass—are less important musically to Listener A.

Conversely, Listener B would be better off with DAC #1. Not only would DAC #1's better dynamics and tighter bass better serve the kind of music Listener B prefers, but his system could use a little more sparkle in the treble and punch in the bass.

Which DAC is "better"? Ask Listener A after she's auditioned both products in her system; she'll think DAC #2 is greatly superior, and wonder how anyone could like DAC #1. But Listener B will find her choice lacking rhythmic power, treble detail, and dynamic impact. To him, there's no comparison: DAC #1 is the better product.

Though exaggerated for clarity, this example shows how personal taste, musical preference, and system matching can greatly influence which digital products are best for you. The only way to make the right purchasing decision is to audition the products for yourself. Use product reviews in magazines to narrow your choice of what to audition. Read reviewers' descriptions of a particular product and see if the type of sonic presentation described is what you're looking for. But don't buy a product solely on the basis of a product review—a reviewer's system and musical tastes may be very different from yours. You could be Listener A and be reading a review written by someone with Listener B's system and tastes.

Use reviews as guides in pointing you to products you might want to audition yourself, not as absolute truth. You're going to spend many hours with your decision, so listen carefully before you buy—it's well worth the investment in time. Moreover, the more products you evaluate and the more careful your listening, the sharper your listening skills will become.

It's important to realize that the specific sonic signatures described in the example are much more pronounced at lower price levels. Two "perfect" DACs would sound identical. At the very highest levels of digital playback, the sonic tradeoffs are much less acute—the best products have fewer shortcomings, making them ideal for all types of music.

Still, a significant factor in how good any DAC or CD player sounds is the designer's technical skill and musical sensitivity. Given the same parts, two designers of different talents will produce two very different-sounding products. Consequently, it's possible to find skillfully designed but inexpensive products that outperform more expensive products from less talented designers.

Higher-priced products are not necessarily better. Don't get stuck in a specific budget and audition products only within a narrow price range. If an inexpensive product has received a rave review from a reviewer you've grown to trust, and the sonic description matches your taste, audition it—you could save yourself a lot of money. If you decide not to buy the product, at least you've added to your listening database, and can compare your impressions with those of the reviewer.

In addition to determining which digital products let you enjoy music more, there are specific sonic attributes you should listen for that contribute to a good-sounding digital front-end. How high a priority you place on each of these characteristics is a matter of personal taste.

In the following sections, I've outlined the musical and sonic qualities I look for in digital playback.

The first quality I listen for in characterizing how a digital component sounds is its overall perspective. Is it laid-back, smooth, and unaggressive? Or is it forward, bright, and "in my face"? Does the product make me want to "lean into" the music and "open my ears" wider to hear the music's subtlety? Or do my ears tense up and try to shut out some of the sound? Am I relaxed or agitated?

A digital product's overall perspective is a fundamental characteristic that defines that product's ability to provide long-term musical satisfaction. If you feel assaulted by the music, you'll tend to listen less often and for shorter sessions. If the product's fundamental musical perspective is flawed, it doesn't matter what else it does right.

Key words in product reviews that describe an easy-to-listen-to digital product include ease, smooth, laid-back, sweet, polite, and unaggressive. Descriptions of bright, vivid, etched, forward, aggressive, analytical, immediate, and incisive all point toward the opposite type of presentation.

There is a fundamental conflict between these extremes of presentation. DACs that are smooth, laid-back, and polite may not actively offend, but they often lack detail and resolution. An absence of aggressiveness is often achieved at the expense of obscuring low-level musical information. This missing musical information could be the inner detail in an instrument's timbre that makes the instrument sound more lifelike. It could be the sharp transient attack of percussion instruments; a slight rounding of the attack gives the impression of smoothness but doesn't accurately convey the sound's dynamic structure. Consequently, very smooth-sounding digital products often have lower resolution than more forward ones.

The other extreme is the digital product that is "ruthlessly revealing" of the music's every detail. Rather than smoothing transients, these products hype them. In a side-by-side comparison, a ruthlessly revealing product will appear to present much more detail and musical information. It will sound more upbeat and exciting, and will appeal to some listeners. Such a presentation, however, quickly becomes fatiguing. The listener feels a sense of relief when the music is turned down—or off. The worst thing a product can do is make you want to turn down the volume, or stop listening altogether.

This conflict between lacking detail and over-emphasizing it can be resolved by buying a higher-quality (and often correspondingly higher-priced) processor. I've found a few models that can present all the music, yet are completely unaggressive and un-fatiguing. This is a rare virtue, and one that I find musically important. The digital front-end must walk a fine line between resolving real musical information and sounding etched and analytical.

Digital reproduction also has a tendency to homogenize individual instruments within the soundstage. This tendency to blur the distinctions between individual instruments occurs on two levels: the instruments' unique timbral signatures and specific locations within the soundstage.

On the first level, digital products can overlay music with a common synthetic character that diffuses the unique textures of different instruments and buries the subtle tonal differences between them—the music sounds as if it is being played by one big instrument rather than many individual ones. There is a "sameness" to instrumental textures that prevents their individual characteristics from being heard.

The second way in which digital playback can diffuse the separateness of individual instruments is by presenting images as flat "cardboard cutouts" pasted on top of each other. The instruments aren't surrounded by an envelope of air and space, the soundstage is flat and congested, and you can't clearly hear where one image ends and the next begins. Instead of separate and distinct objects (instruments and voices) hanging in three-dimensional space, the listener hears a synthetic continuum of sound. Good digital playback should present a collection of individual images hanging in three-dimensional space, with the unique tonal colors of each instrument intact and a sense of space and air between the instrumental images. This is easy for analog to accomplish, but quite difficult for digital. A recording with excellent portrayal of timbre and space will help you identify which digital products preserve these characteristics.

Another important quality in digital playback is soundstage transparency. This is the impression that the space in which the music is presented is crystal-clear, open, and transparent. (The opposite of this is thick, congested, and opaque.) Soundstage transparency is analogous to looking at a city skyline on a perfectly clear day: Just as smog or haze will reduce the buildings' immediacy, vibrancy, and visible detail, so too will soundstage opacity detract from the musical presentation.

I've focused on these aspects of the sonic presentation for evaluating digital products because they are the most common flaws in digitally reproduced music. You should also listen for other aspects—treble grain, rhythm, dynamics—described in Chapter 3.

Beyond these specifics, a good question to ask yourself is, "How long can I listen without wanting to turn the music down—or off?" Conversely, the desire—or even compulsion—to bring out CD after CD is the sign of a good digital front-end. Some components just won't let you turn off your system; others make you want to do something else. This ability to musically engage the listener is the essence of high-end audio. It should be the highest criterion when judging digital front-ends.

A Closer Look at Digital Interfaces

At first glance it would seem that the method of getting the digital datastream from a source to a DAC would have no effect on sound quality provided that the interface doesn't introduce errors. That is, if the ones and zeros are the same at the receiving end as at the transmitting end, the sound should be the same. It's a trivial matter to prove that digital interfaces don't introduce bit errors, but they nonetheless exhibit an analog-like variability in sound quality. Different interface types sound different from one another, and different brands of digital interconnects all exhibit their own sonic flavors. To some, the reports by audiophiles that digital cables don't all sound the same is the height of audiophile lunacy. According to the conventional wisdom, if the ones and zeros are the same, the sound must be the same. After all, a computer program runs identically whether from a computer's internal hard drive or from an external drive connected by a USB cable. As described later in this chapter ("Jitter in the Digital Interface") the analog-like variability in digital interfaces and cables is the result of jitter, or timing imprecision.

The S/PDIF Digital Interface

The S/PDIF interface is a method of transmitting digital audio from one component to another. For example, a digital transport's S/PDIF output is carried down a digital interconnect to the S/PDIF input on a digital processor. As described earlier, transports and DACs come fitted with a variety of connection methods—AES/EBU, coaxial, TosLink, or BNC. All are based on the S/PDIF (Sony/Philips Digital Interface Format) interface standard. Note that all transmit the same S/PDIF signal, but in different ways.

Digital interfaces can be divided into two categories: electrical and optical. In an electrical interconnect, electrons carry the signal down copper or silver wire. An optical interface transmits light down a plastic or glass tube.

Fig.7-10 shows the three common interconnect types: coaxial, TosLink, and AES/EBU. Coaxial and AES/EBU are both electrical; TosLink is optical. (The USB interface , which is covered in the next chapter, does not use the S/PDIF transmission standard.)

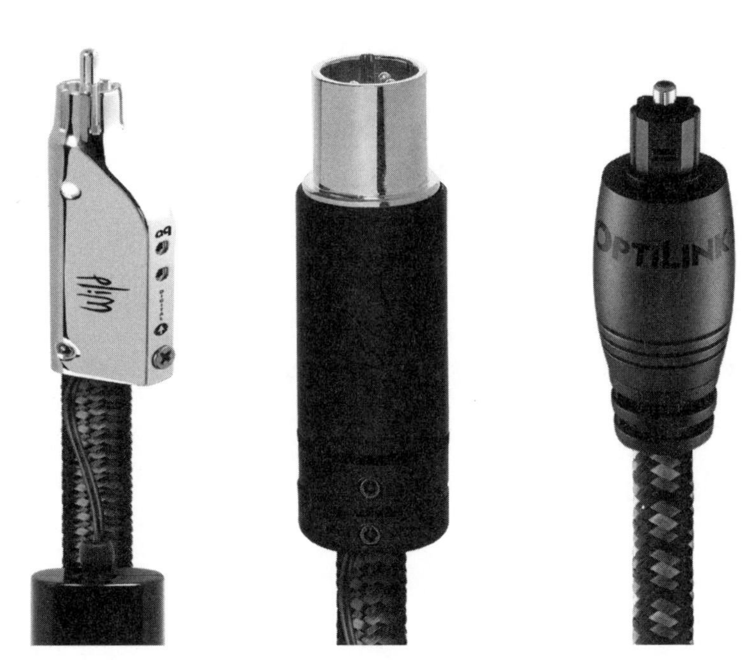

Fig. 7-10
Comparison of different digital interface terminations. From left to right: coaxial S/PDIF on RCA jack, AES/EBU on XLR jack, TosLink optical. (Courtesy AudioQuest)

The most common type of connection, coaxial, is carried on an RCA cable. This is the electrical connection found on virtually all CD transports, most good CD players, and other consumer digital audio products such as digital preamplifiers, room-correction systems, and digital recorders. A variation on the coaxial connection is the BNC cable, which is used in only a few products. Though BNC is better than RCA—mechanically, electrically, and sonically—it never caught on.

TosLink is the low-cost optical interface promoted by mass-market audio manufacturers as an alternative to coaxial connection. TosLink, a trademarked name of the Toshiba Corporation, is more properly called "EIAJ Optical," after the Electronics Industries Association of Japan. The major electronics companies had two good reasons for trying to convert the world to TosLink. First, TosLink jacks and cables are cheaper to make than coaxial jacks and cables. Second, TosLink connection makes it easier for the components to meet FCC (Federal Communications Commission) requirements for radiated noise. An electrical signal traveling down copper (such as the S/PDIF signal in a coax connection) throws off RF (radio frequency) noise that could interfere with radio and television transmission. The FCC will simply ban products that don't meet its criteria for radiated noise. Because TosLink sends the signal as light down glass or plastic fiber, it produces no radiated noise.

The ST-Type optical interface, developed by AT&T for telecommunications, was popular in the mid-1990s, but is rarely found on contemporary products. ST-Type connection transmits the optical signal down wide-bandwidth glass fiber instead of TosLink's plastic light path (some higher-quality TosLink cables also use glass). ST's locking bayonet connector ensures a good junction between the cable and optical transmitter and receiver. Although ST-Type is generally considered a very good-sounding interface, it is no longer common on high-end products.

Many products offer the AES/EBU interface, which is carried on a balanced line terminated with three-pin XLR connectors. Of the three conductors in a balanced signal, one is ground, one is the digital signal, and the third is the digital signal inverted. AES/EBU benefits from all the advantages of balanced lines (described in Chapter 11), and is transmitted at 5V, compared to S/PDIF's 0.5V. AES/EBU is usually the interface of choice unless the products offer a proprietary interface

Virtually all transports and processors have RCA jacks, and most include TosLink for compatibility. AES/EBU is usually only found on the more expensive components, or offered as an option on mid-priced digital equipment. TosLink is by far the worst interface, mechanically (the physical connection between cable and jack), electrically (it has the lowest bandwidth), and sonically. TosLink connection tends to blur the separation between individual instrumental images, adds a layer of grunge over instrumental textures, softens the bass, and doesn't have the same sense of black silence between notes. Better results can be achieved with a high-end glass-fiber TosLink cable, but my advice is to forget about TosLink unless you have a laserdisc player equipped only with TosLink output.

Some digital front-ends will benefit from an optical interface because there is no ground connection between the transport and processor, as exists with an electrical interconnect. Sonically degrading high-frequency noise on a transport's ground can contaminate the digital processor's ground through an electrical connection. Because an optical interface transmits the signal without an electrical connection, there is no chance for coupling ground noise between the two components.

In theory, the electrical interfaces will work best because they have the widest bandwidth. The transmission of digital data from a source to a digital processor should be over the widest possible bandwidth link for low jitter. TosLink has a bandwidth of 6MHz (6,000,000Hz), and electrical interfaces have a potential 500MHz bandwidth (if implemented correctly).

The best way to choose an interface is by listening to the various types. When comparing electrical and optical interfaces, be sure to disconnect the electrical interface from the processor when listening to the optical interface. The advantage of keeping the transport ground isolated from the processor ground is lost if the electrical cable is still connected between the transport and processor, even though it may not be active.

A primary reason for poor-sounding transports and interfaces in electrical connections is an impedance mismatch between the transport's output impedance, the cable's characteristic impedance, and/or the digital processor's input impedance. An impedance mismatch causes reflections of the signal in the cable, introducing jitter in the datastream. The impedance specification is 75 ohms (±5%) for S/PDIF and 110 ohms (±20%) for AES/EBU. Manufacturers of digital products should strictly adhere to these impedance standards, although many do not.

Finally, it should be noted that recent improvements in DACs have largely reduced the sonic differences between interface types, and between specific models of interface cables. Today's DACs are much better at rejecting jitter, and some even employ technologies to completely isolate interface jitter from the processor's clock. Keep in mind that the greater the differences you hear between interfaces, the worse the digital DAC is at rejecting interface-induced jitter.

The I²S Interface

The I²S interface (pronounced "I squared S") was originally developed by Philips as a means of transporting audio data and clocks between chips inside a CD player. A few high-end manufacturers have tapped into this bus to link separate components to each other.

The interface is usually implemented on a so-called 13W3 cable, which was originally designed for the computer industry for transmitting component video and control signals. It consists of three coaxial conductors and five twisted-pair lines bundled in an outer jacket. Two of the conductors carry the master clock in balanced mode. The twisted-pair lines carry the bit clock, word clock, channel-status data, and audio data. It's possible to carry I²S on other cable types, including RJ-45 and HDMI (High-Definition Digital Multimedia Interface).

A high-end variation is called I²S Enhanced, with two levels of implementation. Level 2 is the basic configuration in which the clock is generated by the transmitting device. In a Level 1 implementation, the clock is generated by the receiving device (the DAC), forcing the source to lock to this clock. Level 2 provides for lower jitter, all other factors being equal. These levels are compatible with each other. That is, a source with a Level 2 I²S Enhanced interface will work with a DAC fitted with a Level 1 implementation. You won't achieve Level 1 performance, but the system will work automatically without setting switches or otherwise telling the system which level you're using. If you upgrade to a Level 1 source, the interface will automatically recognize the change and provide Level 1 performance.

There was a movement in the early 1990s by high-end manufacturers to make I²S a standard, but only a few companies offered the interface. I²S fell out of favor to such a degree that I removed references to it in the third edition of this book. It's interesting to see I²S return.

Jitter in the Digital Interface

Jitter describes timing inaccuracies in the digital-to-analog conversion process that degrade sound quality. A clock in the digital processor controls *when* the digital samples that represent the music are converted to analog. If this clock isn't perfectly precise, sound quality is degraded.

Think of the bass drum in a marching band as a D/A-converter clock, and of the band members as the digital samples representing the music. Each row of musicians is one audio sample. On each drum beat, the musicians take one step forward in unison. Now consider what would happen if the drummer started to get sloppy, leaving too much time between some beats and not enough time between others. Without a precise timing reference, the distance between each row of musicians would no longer be uniform.

In digital audio reproduction, each clock pulse triggers the conversion of a digital sample to analog form. If these clock pulses aren't perfectly uniform—the very definition of jitter—we hear the timing imprecision as reduced musicality. Jitter causes the samples representing the music to be put back together with slightly staggered spacing, creating distortion in the audio waveform. Just as the marching-band musicians can't step together if the drummer isn't perfectly precise, digital audio samples can't be correctly reconstructed into music if the clock is jittered. To give you an idea of how precise the clock must be, timing errors of as little as 10 picoseconds (0.00000000001 second)—about the time it takes light to travel a tenth of an inch—are audible. (A more technical explanation of clock jitter appears in Appendix C.)

The S/PDIF interface is fundamentally flawed in that the clock is carried within the audio data (it's actually more technically correct to say that the audio data are embedded in the clock). This forces the receiving device to lock to that clock (with a phase-lock loop, or PLL) and generate a new clock based on the incoming clock. PLLs are not perfect devices; they tend to track and pass along variations in the incoming signal rather than rejecting them. The clock "recovered" by the PLL becomes the timing reference for the critical digital-to-analog conversion stage. This is the point where jitter matters—in the clock that controls *when* each digital sample is converted to an analog value in the DAC chip. If that clock isn't perfectly precise, the digital samples that represent the analog signal's amplitude are reconstructed with slightly staggered spacing between samples, resulting in tiny amplitude errors in the analog waveform. Such micro-amplitude errors don't exist in nature, which perhaps explains whey our brains are highly sensitive to the phenomena. This is why transports and interfaces affect sound quality.

Jitter is manifested as a reduction in the sense of space (the recorded acoustic sounds smaller and drier), less impression of three-dimensional objects (musical instruments) existing in space, less liquid instrumental textures, and a softening of the bass.

This interface-induced jitter isn't random in nature; it is directly correlated with the audio signal being transmitted. In other words, the clock is modulated by the music carried by the interface. The recovered clock is jittered at exactly the same frequency as the audio signal traveling down the interface. Using a jitter-analysis instrument, it's possible to listen to only the jitter component of the S/PDIF signal. Just from hearing the jitter, you can readily discern the piece of music being transmitted. This audio-correlated jitter ends up at the DAC's word clock, where it degrades sound quality.

Because a CD player has no need for an S/PDIF interface, it can theoretically have lower jitter than a separate digital source and DAC. One of the main sources of jitter is the digital interface that connects a CD transport to a DAC. Because the interface carries the digital audio data and a clock in the same cable, jitter can be introduced. By eliminating the interface, CD players have potentially lower jitter than separates.

Outboard Clocks

In the early 2000s, several manufacturers introduced a new breed of product that attempted to reduce jitter and its harmful effects. These jitter-reduction devices were inserted in the digital signal path between a digital transport (or any other digital source) and a DAC. In theory, the jitter-reduction device takes in a jittered S/PDIF digital signal and outputs a less jittered digital signal.

These devices exhibited various degrees of efficacy. One fundamental problem was that the de-jittered signal was output on an S/PDIF interface to delivery to the DAC. This interface promptly re-introduced jitter into the signal.

Jitter reduction devices fell out of favor after a few years, but manufacturers never stopped looking for solutions to the jitter problem. The modern approach is to use a high-precision clock housed in a separate chassis. When used with a transport and DAC, the clock's outputs supply both the transport and DAC with a precise clock (Fig.7-11). The critical digital-to-analog conversion stage is no longer controlled by a clock recovered from the S/PDIF interface. Moreover, the outboard clock can be extremely precise. When used with a CD player, the clock's output becomes the master clock for the entire player, with the transport mechanism locked to the high-precision clock.

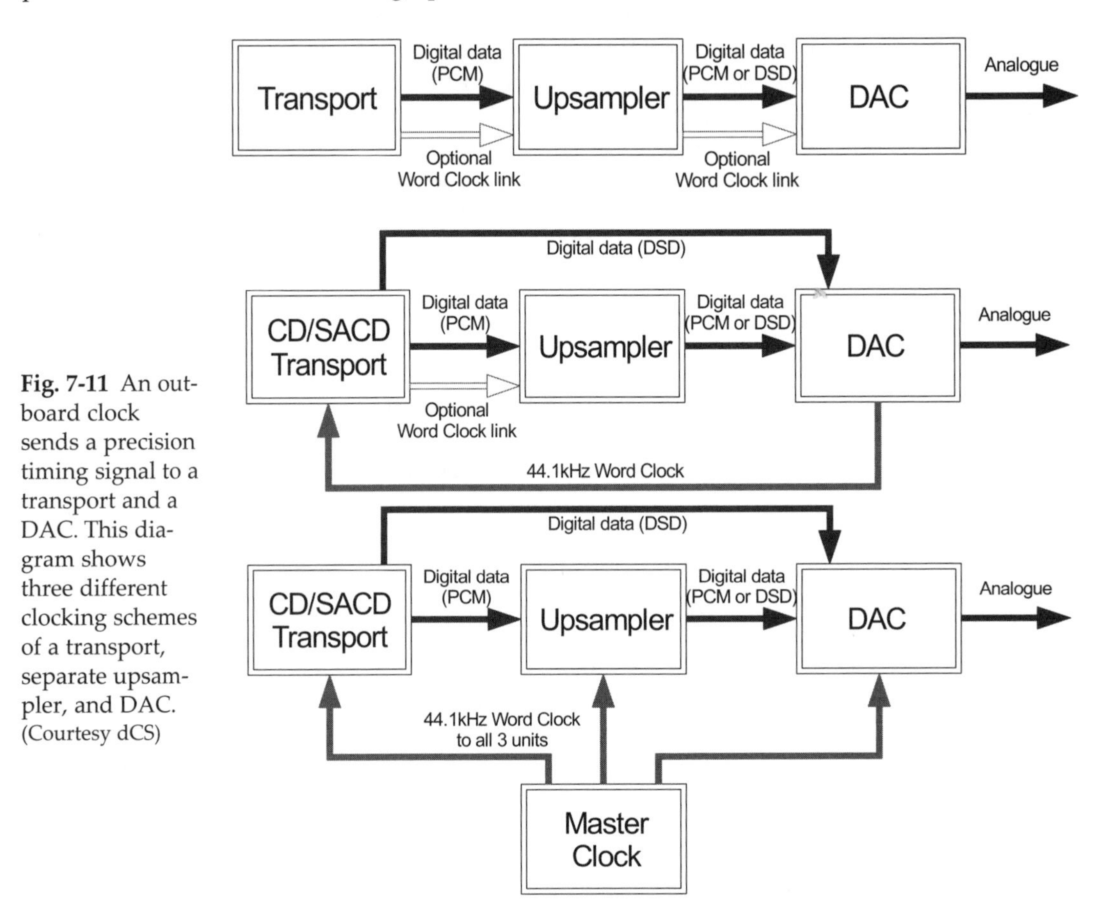

Fig. 7-11 An outboard clock sends a precision timing signal to a transport and a DAC. This diagram shows three different clocking schemes of a transport, separate upsampler, and DAC. (Courtesy dCS)

It's a fascinating exercise to listen to the sonic effects of an outboard clock by turning the clock on and off. When turned off, the CD player or DAC reverts to its own internal clock. It is thus possible to hear the effects of jitter with a simple button push. Here's an excerpt from my review of the dCS Puccini CD/SACD player and U-Clock USB Converter and Clock, first published in *The Absolute Sound*, Issue 200 (February, 2010).

After getting a general impression of the Puccini itself, I engaged the U-Clock. One little front-panel button push vaulted what was already a spectacular sound into entirely new territory. The U-Clock snapped images into sharp(er) focus, increasing the sense of clarity, precision, and definition I had enjoyed from the Puccini alone. The heightened focus had a profound effect on the sense of instruments existing within an acoustic. Without the U-Clock, reverberation tended to be connected to the image itself, as though the image and the hall were merely variations on the same sonic cloth. With the U-Clock, the instrumental image was presented as a clearly defined object existing within an acoustic space rather than simply fused to it. The instrument and the surrounding acoustic were presented in a closer facsimile to what we hear it in life.

That was just the beginning of the U-Clock's magic. The Puccini's reproduction of timbre, which already had a bell-like clarity, was taken to a new level by the U-Clock. Timbres had greater palpability and realism, partly the result of less grain and edge (which was already very low) and partly because of greater resolution of textural detail. Similarly, the U-Clock made the Puccini's reproduction of transient information even more lifelike. The leading edges of piano attacks, for example, had a trace of edge that vanished with the U-Clock engaged. Listen, for example, to the wonderful new recording of Vassily Primakov performing Chopin Mazurkas on Bridge Records. The U-Clock made the piano more lifelike in transient attack, in richness of tone color, and particularly, in the sense of space surrounding the instrument. I pulled out this CD as a diagnostic tool to listen for specific sonic attributes of the U-Clock but immediately forgot about the sound and listened to the entire disc, completely captivated by the compositions and Primakov's expressive performance. Such an experience is always the sign of a great component.

Perhaps the most ambitious clock is the Esoteric G-0Rb, a $16,000 device that employs at its heart the element rubidium, which resonates at a precise frequency to create the timing reference. The G-0Rb is in essence an atomic clock designed solely to reduce jitter in digital audio reproduction. This might seem like overkill, but listening to an Esoteric transport/DAC combination with and without the clock was revelatory. The G-0Rb had a profound influence on the sound quality, both in specific performance attributes as well as in overall musical involvement.

Asynchronous Sample-Rate Conversion

Asynchronous sample rate conversion is not to be confused with asynchronous-mode USB connection; the two are completely unrelated. Asynchronous sample-rate conversion refers to a technique in which digital audio data are converted to a different sample rate. The term "asynchronous" means the input and output frequencies can run independently of each other and need not be related or locked together; the input frequency can be varied and the output frequency will remain the same. First used in professional applications where multiple sampling rates were common (44.1kHz and 48kHz, for example), asynchronous sample-rate conversion has found its way into consumer audio products.

Asynchronous sample-rate conversion has the ability to remove jitter in the incoming signal. In theory, different transports, sources, and interfaces are all sonically identical with an asynchronous sample-rate converter chip in the signal path (the chip is usually located inside the DAC). The sample-rate converter chip is essentially an oversampling digital filter that interpolates new samples. This means the output data are not identical to the input data, even when the input and output sampling frequencies are identical. Obviously, this is cause for concern; we want perfect bit-for-bit accuracy between the original master recording and signal being converted to analog. Asynchronous sample-rate conversion outputs a signal with perfect timing, but with samples of the wrong value. It converts errors in the time domain (jitter) to errors in the amplitude domain.

High-Resolution Digital Interfaces

Record companies are terrified by the prospect of consumers having access to high-resolution digital bitstreams, since those bitstreams can be copied and distributed without sonic degradation. Consequently, DVD-A and SACD players generally do not provide high-resolution digital outputs, forcing listeners to use the player's analog outputs. Although DVD-A and SACD players include digital outputs, the signal has been downsampled to 44.1kHz with 16-bit resolution, rendering the output signal identical to that of a CD player's output, regardless of the disc type being played.

Some players, however, employ a proprietary high-resolution digital output that allows the company's DVD-A or SACD player to connect digitally with the company's other products. An example is the Meridian High-Resolution (MHR) Smart Link, which features proprietary encryption to prevent illegal copying of the datastream. Meridian's DVD-A players and controllers output high-resolution, multichannel digital audio on the MHR Smart Link, which can be received and decoded by Meridian's other products, notably its digital loudspeakers. The MHR interface also carries information about the disc content, such as whether the disc is stereo or multichannel, and the audio encoding format. This information allows the MHR-equipped controller to automatically invoke the correct decoding and DSP modes.

Some Sony SACD players offer high-resolution, multichannel digital output on an optical i.LINK connector, which can connect to some of Sony's other products, as well as outboard high-end digital-to-analog converters featuring i.LINK input. i.LINK is Sony's tradename for the IEEE1394 interface (also known as FireWire). This high-resolution digital output is copy protected by the Digital Transmission Content Protection (DTCP) system.

High-Resolution Digital Audio: Why 44.1kHz Sampling and 16-bit Quantization Aren't Enough

As explained in Appendix C, the Nyquist Theorem tells us that if the sampling frequency is more than twice as high as the highest audio frequency we want to encode, sampling is a lossless process. That is, it perfectly encodes frequencies at just under half the sampling rate. CD's 44.1kHz was chosen because 44.1kHz is slightly higher than twice 20kHz, the highest frequency humans can reportedly hear. (But why such an odd number? Early digital recorders were based on video formats such as ¾" U-Matic tape, and 44.1kHz sampling happened to fit mathematically with the 30 frame-per-second rate of black and white video.)

So if humans can hear only up to 20kHz (at best), and CD's sampling rate of 44.1kHz perfectly preserves a 20kHz sine wave, why does 96kHz sampling sound more open and transparent, and deliver greater transient detail?

When high sampling rates became available in professional recorders in the mid-to-late 1990s, a debate raged over whether the sonic improvements were attributable to the preservation and reproduction of information higher in frequency than 20kHz, or some other factor. That "other factor" turned out to be the relaxation in digital-filter requirements for higher sampling rates.

The digital filter in a CD player or outboard DAC removes all energy above half the sampling frequency. Unfortunately, digital filters introduce distortion in the time domain by smearing transients. Think of a snare drum that starts and stops very suddenly. Digital filters take some of that sudden energy and spread it out over time. Oddly, this energy-smearing occurs not just after the transient musical signal, but before it as well (Fig.7-12). Such distortion is also called ringing because the filter "rings" like a bell when "struck" by the music signal. The ringing that occurs before the musical transient is called "pre-ringing." The steeper the filter, the longer the time period before and after the signal the energy is smeared. In addition, steep filters introduce phase shift and amplitude-response ripples in the passband. (The range of frequencies passed by the digital filter is called the passband. The frequency range stopped by the digital filter is called the stop-band. The frequency range between the passband and the stopband is called the transition band.) This ringing (particularly pre-ringing) is largely responsible for the hardness and glare overlaying instrumental timbres and soundstage flattening that we associate with standard-resolution digital audio.

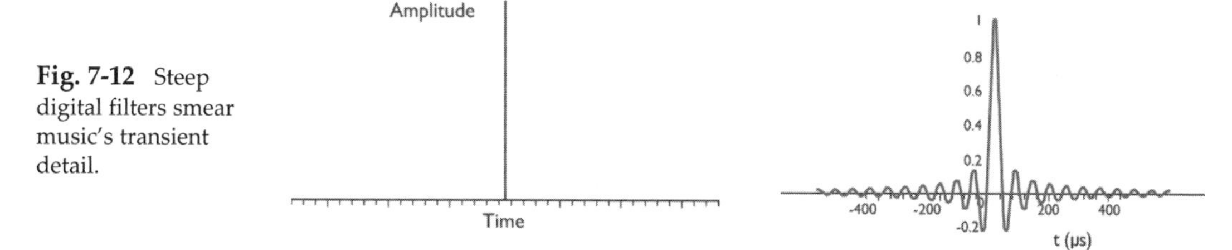

Fig. 7-12 Steep digital filters smear music's transient detail.

The problems caused by steep digital filters right next to the audio passband are sometimes exacerbated by the low quality of the filter. It's possible to design filters that reduce these problems by employing greater mathematical precision, but they are too costly to implement except in the most expensive products. And because high-end companies are often reliant on mass-market semiconductor manufacturers for their digital filters, even expensive digital components don't always include high-performance digital filters.

The steeper (more rapid) the filter's rolloff, the greater the ringing and energy-smear. Digital filters used in decoding 44.1kHz audio are steep indeed; they must pass the entire audioband up to 20kHz, yet attenuate the signal by a whopping 120dB at 22.05kHz (half the sampling frequency). For comparison, analog filters used in loudspeaker crossovers are much more gradual, attenuating the signal by about 12dB over a full octave in a typical loudspeaker.

A filter designed for a 96kHz-sampled system, however, can begin attenuating at a much higher frequency, and have a far gentler slope. The stopband for such a filter begins at 48kHz; the transition band is much wider; and the frequency at which the filter begins attenuating can be moved higher, away from the audio passband. The result is less distortion (amplitude-response ripple, phase shift, and energy smearing) within the audio passband, and thus improved sound quality. For an academic treatise on this subject, see "A Suggested Explanation for (some of) the Audible Differences Between High-Sample Rate and Conventional Sample Rate Audio Material" by Mike Story of Data Conversion Systems (dCS) at www. dcsltd.co.uk/papers.htm.

Although the relaxation of digital filter requirements is certainly a factor in 96kHz's improved sound quality, some respected researchers also suggest that even though we can't hear sine waves above 20kHz, we can nevertheless perceive the greater steepness—that is, speed—of transient musical information that an ultra-wideband digital audio system provides. The wider the bandwidth of a system, the faster the events (sudden changes in amplitude) it can preserve.

Consider a subwoofer, a very low-bandwidth device, reproducing the sound of a triangle. The subwoofer removes only the very-fast-changing components of the signal (the steep transient of the attack), but reproduces the rest of the signal. Similarly, a 20kHz-bandwidth audio system filters out the steepest transients of musical signals, perhaps introducing a kind of distortion. Some believe that the steeper transients allowed by a wide-bandwidth digital audio system improve the sense of space because the ear/brain uses transient information in localizing sound sources.

The filters I've described are so-called "linear phase" filters, which have been standard in CD players and DACs since the beginning of digital audio. Recently, however, designers have begun creating "minimum phase" filters that sacrifice flat frequency response (they exhibit a slight upper-treble rolloff) for reduced ringing (better time-domain performance). These filters have become more popular as high-end designers increasingly create their own filters by writing custom software that controls a general-purpose digital signal processing (DSP) chip. Often, the listener can select between filters by pushing a button on the front panel of the CD player or DAC.

In 2009, Meridian Audio introduced its so-called "apodising" digital filter that not only doesn't introduce pre-ringing of its own, but can also remove pre-ringing added to the signal by processing earlier in the chain, primarily in the analog-to-digital converter. (The term "apodising" comes from optics and radio astronomy. Sharp edges at the boundaries of optical lenses or radio dishes create ripples in the diffraction pattern, analogous to the ringing in digital filters. In radio telescopes, the contribution from the outer edge of the disc is attenuated to reduce this effect, a process called "apodisation.")

The pre-ringing that exists on nearly all the CDs in our collections is just as sonically detrimental as the pre-ringing added by the CD player's linear-phase filter. In fact, this pre-ringing is largely responsible for "CD sound," that combination of flatness of the soundstage and hardness of timbres. Meridian's apodising filter, first used in the groundbreaking 808.2 Reference Signature CD player, greatly improved the sound of CDs, especially older discs that sounded particularly hard and flat.

16 Bits, 20 Bits, and 24 Bits

Although the causes of the superior sound quality delivered by increasing the sampling rate are debatable, the benefits of the increase in word length from 16 to 18, 20, and even 24 bits are not in dispute. As I mentioned earlier, word length is the number of bits used to encode the audio signal's amplitude at each sample. Assigning a number—called a word—to represent the audio signal's amplitude is called quantization. The word length determines the system's resolution, dynamic range, distortion, and signal-to-noise ratio. We also call the word length "resolution," as in the phrase "20-bit resolution." The greater the number of bits per word, the higher the resolution.

Think of a digital clock with only two digits; we could discern only between the hours, with no resolution of minutes. If we add a third digit (to indicate tens of minutes), we can know the time with ten-minute precision. Adding a fourth digit (minutes) allows us to know the time with one-minute accuracy. A fifth digit to indicate tens of seconds further increases the clock's precision. Similarly, the more bits in a digital-audio word, the finer the precision with which we can encode the audio signal's amplitude.

The CD format uses 16-bit quantization, which provides 98dB of dynamic range. Although this number may sound high in theory, it's not enough in practice. Consider an engineer setting the recording levels for a symphony recording. He must adjust the signal driving the analog-to-digital converter so that at the end of the fourth movement when everyone is playing full-bore, the levels never exceed full-scale digital—the maximum amplitude that can be encoded. At full-scale, all 16 bits are used. Now think about a quiet flute passage in the slow movement; the low signal level is encoded with perhaps four or five bits, which offers very low resolution.

Consequently, the resolution of a digital audio signal isn't defined by the maximum number of bits available, but by the number of bits being used at any given moment. The lower the signal level, the lower the precision with which the signal is encoded. And if the recording engineer leaves 6dB of "headroom" to account for unexpectedly loud peaks, he's effectively throwing away one full bit of resolution.

The advent of 20- and 24-bit digital audio not only expands the dynamic range, but also increases the resolution of low-level detail. This low-level detail can be fine nuances of an instrument's timbre, which enhance the sense of realism. It can also be subtle spatial cues such as discrete acoustic reflections and reverberation decay, which the ear interprets as a more convincing soundstage.

Longer word lengths also contribute to better sound because the post-processing common in the recording or mastering studio (mixing, equalization, signal processing) is performed with much greater mathematical precision. Moreover, noise added by these processes is spread out over a wider bandwidth, which makes the noise less audible. The combination of higher sampling rate and longer word length results in greatly improved sound quality.

A Caveat About High-Resolution Digital Audio

It's important to realize that a fast sampling rate such as 176.4kHz and a 24-bit word length don't guarantee good sound. Rather, those specifications simply provide the *opportunity* for higher quality audio. Implementation is everything. A well-designed 44.1kHz/16-bit system will outperform a poorly conceived and executed "high-resolution" system.

Second, high-resolution analog-to-digital and digital-to-analog conversion is vastly more exacting than standard-resolution conversion. In a 16-bit system, the least significant bit of the 16-bit word represents one part in 65,536; in a 24-bit system the least significant bit represents one part in more than 67 million. Many DACs don't even deliver real 16-bit resolution, never mind full 24-bit resolution. High-res digital places demands not only on the DAC chip's low-level performance, but also on the analog output stage, clocking accuracy (jitter), power supply purity, and every other subsystem. The temperature variations within components that might marginally affect the performance of a 16-bit system can introduce gross errors in a 24-bit converter.

Third, the big numbers associated with high-resolution digital audio (96kHz, 176.4kHz, 20-bit, 24-bit) are often marketing-driven. The product might not deliver "96/24," but consumers often think that higher numbers automatically translates to better sound.

Although state-of-the-art digital delivers 24 bits, and most digital-to-analog converters have 24 "rungs" in their resistor ladders, those last four bits generally do not have sufficient precision to encode musical information. This is a limitation of the analog-to-digital and digital-to-analog converters. The additional 4 bits in a 24-bit DAC chip compared with a 20-bit chip are cynically called "marketing bits"; they serve no technical function. Although the DAC might have 24 "rungs" on its resistor-ladder network, those last four (or more) rungs are rarely capable of delivering real information.

Finally, a "high-resolution" datastream doesn't do you any good if the signal has previously been subjected to a standard-resolution A/D or sampling-rate conversion. Once the music has been encoded at 44.1kHz (or 48kHz) and 16 bits anywhere in the chain, it has been irreparably compromised. This is an important consideration when buying "high-res" downloads. Did they really go back and re-master the material from the original source elements in high-res, or simply upconvert the existing CD-quality datastream and call it "high-res"? And what about those albums tracked or mixed on the 48kHz/16-bit digital records prevalent in the late 1980s and early 1990s? This topic is discussed in more detail in the following chapter.

I raise these issues for two reasons: listeners should be aware that high-resolution specs are not a silver bullet for digital's flaws; and that high-resolution done right is rare.

How to Get High-Resolution Digital Audio

High-res digital is available through a variety of formats, although the number of titles is rather limited (but increasing daily). The easiest way of getting high-res into your system is the SACD format. We'll describe SACD in detail later in this chapter, but you should know that it delivers much-better-than-CD quality. A single SACD disc can include a six-channel presentation, a two-channel stereo mix, as well as a CD-compatible version of the music. If you are reluctant to download music from the Internet, SACD is a fabulous format with a significant catalog available, particularly in classical music.

The DVD-Audio format can also deliver multichannel high-resolution digital audio, but the number of titles is extremely limited. Moreover, DVD-A has very little industry support; it is essentially a dead format.

Another disc-based high-res media is Blu-ray Disc. This optical disc can encode up to eight channels of high-resolution digital audio along with high-definition video. The Blu-ray Disc specification calls for an audio-only carrier, but so far only a few specialty labels have released music in this format.

A decidedly non-standardized method of delivering high-resolution digital audio to consumers is the format pioneered by audiophile label Reference Recordings that stores WAV files on a DVD-R disc. In this format, called HRx by Reference Recordings, the original high-resolution master files are recorded on a blank DVD as WAV files. The catch is that you'll need a disc player capable of playing WAV files directly, or a music server to which the WAV files are ripped, along with an outboard DAC capable of decoding 176.4kHz/24-bit datastreams. You buy the HRx disc ($45 per title), load it onto your hard drive, and play it back using software such as MediaMonkey. You can't play HRx files directly from the disc except on a few players; they must first be transferred to a music server. A few other audiophile record labels have also released material as WAV files on DVD, but the format has limited appeal because of the high disc cost and requirement of a computer.

Finally, an increasing number of titles are available via download. These files are generally limited to 96kHz/24-bit, but that's sufficient for high-quality audio when done right. The next chapter on music servers goes into this topic in more detail.

Super Audio CD (SACD)

The compact disc format is limited in musical performance by its sampling rate of 44.1kHz and 16-bit quantization. When an audio signal is converted to digital for release on CD, the analog-to-digital converter "samples" (looks at) the audio waveform 44,100 times each second, in a process called sampling. The converter assigns a number to each sample; this number represents the waveform's amplitude at the sample point, a process called quantization. The CD format uses 16-bit quantization, meaning that each sample's amplitude is represented by a 16-bit binary (ones and zeros) number.

Sampling and quantization are the cornerstones of digital audio; sampling preserves the time information in an audio signal, quantization preserves the amplitude information. The faster the sampling rate, the wider the range of audio frequencies that can be preserved. The more bits in the quantization "word," the greater the dynamic range, the lower the noise, and the greater the resolution. These fundamentals of digital audio are described in much greater detail (with graphics) in Appendix C.

When the compact disc was developed in the late 1970s, 44.1kHz sampling and 16-bit quantization were chosen to achieve 74 minutes of playing time on a 120mm disc. The D/A converters of the day were limited to 14-bit resolution. Some argued at the time that a sampling rate of 32kHz with 14-bit quantization were sufficient for CD, but fortunately, the standard was established at 44.1kHz and 16 bits. These specifications are, however, not sufficient to encode all the musical information humans can hear. Nonetheless, the compact disc has been a massive commercial success, with more than 1.5 billion players sold.

Super Audio CD provides 74 minutes of high-quality digital audio in 2-channel and 6-channel modes. SACD also offers the possibility (but not the requirement) of a "hybrid" disc containing two layers on the same disc for backward compatibility with CD. One layer contains conventional 16-bit/44.1kHz audio and the second layer holds the high resolution multichannel version.

The first few SACD titles were from Sony Music and a few audiophile labels; as of late 2010, more than 6800 SACD titles are available from 120 different labels (up from 1500 titles in 2004). The ultimate source for information on SACD titles is the website www.sa-ced.net.

Hybrid SACD

A key element of SACD is backward compatibility with the world's more than 1.5 billion CD players. The advantages of backward compatibility are manifold: a single inventory for record companies and music retailers and the ability to play the same disc in a home SACD player and a CD-based car stereo. Moreover, consumers needn't make a conscious decision to stop buying music in one format (CD) and start buying into a new format. The transition from CD to SACD can be gradual and painless, particularly since SACDs don't cost any more than CDs. Indeed, some consumers buying a hybrid SACD may be unaware that the disc is not a conventional CD. The hybrid SACD re-release of the Rolling Stones catalog in 2003 did not include on the packaging any clue that the discs were hybrid SACD. This strategy avoided the possibility of the discs being relegated to the "new formats" or "audiophile" sections of record stores, or of confusing consumers.

A hybrid SACD looks just like a conventional CD, but contains two layers of information rather than one. The lower layer contains 16-bit/44.1kHz information that can be read by any CD player. The upper layer contains high-resolution digital audio information (called HD, for High Density; see Fig.7-13).

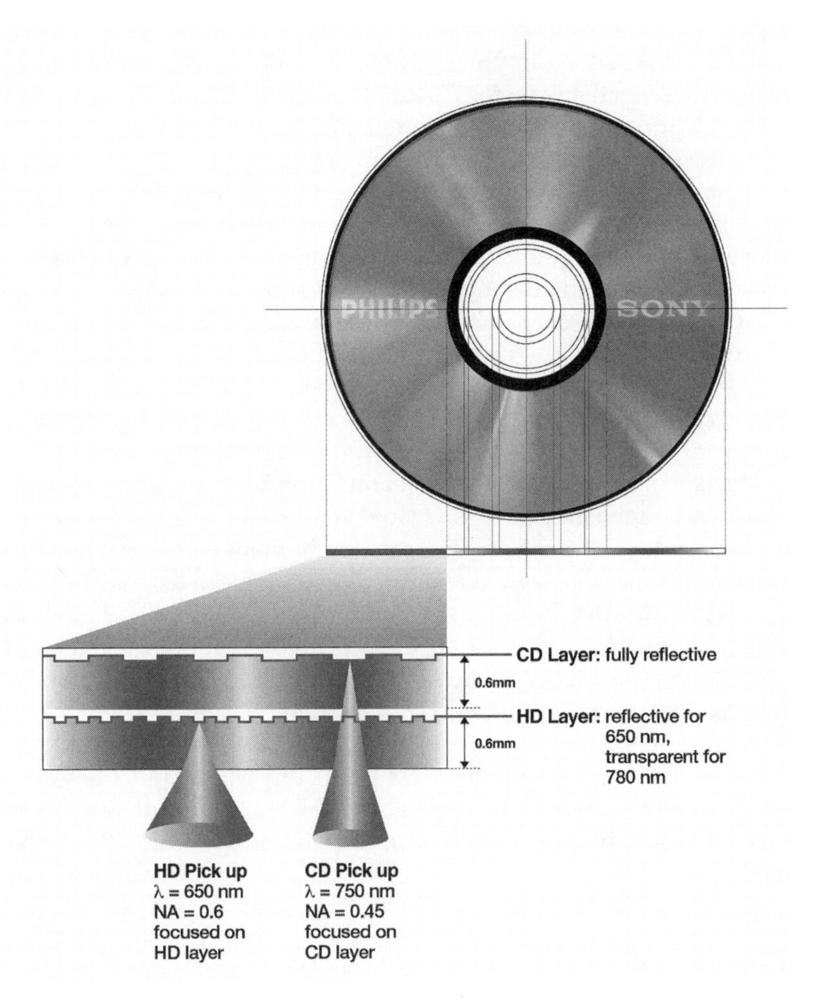

Fig. 7-13 A Super Audio CD contains a high-density layer and a conventional CD layer on the same disc. (Courtesy Sony Corporation)

Fig.7-14 shows the SACD's physical structure. It is, in essence, a CD with an additional layer of information-carrying pits embedded in its playing surface. This second, higher-density layer is semi-transparent to the 780-nanometer wavelength of CD laser light, but reflective to the shorter 650-nanometer wavelength of the SACD player's laser. The CD

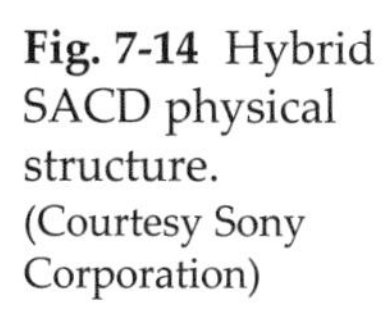

Fig. 7-14 Hybrid SACD physical structure. (Courtesy Sony Corporation)

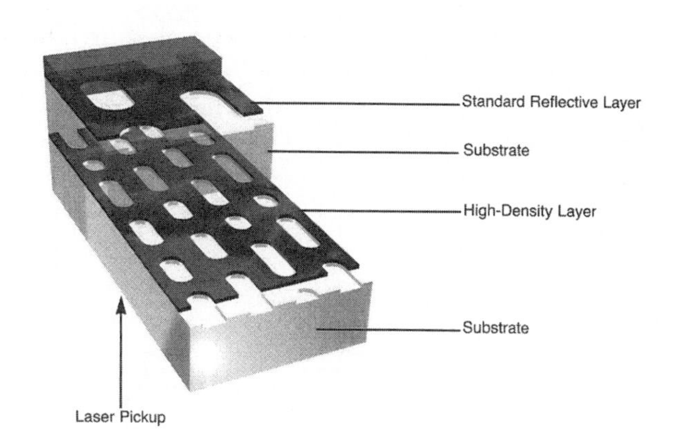

player's laser focuses all the way through the disc; the SACD player's laser focuses halfway through the disc to the HD layer. A switch on SACD players allows the user to switch between layers, and to compare the CD and high-resolution layers. The CD-quality layer is sometimes called the Red Book layer, after the color of the cover of the official specification describing the compact disc.

More information is contained in the second layer because the information-carrying pits are about half the size of a CD's pits, spaced closer together, and the information is recovered with a shorter wavelength playback laser. This high-density layer contains a 2-channel mix and, at the artist's discretion, a 6-channel version of the music, along with the potential for text, graphics, and limited video information. You can select between stereo and multichannel playback from the same disc, depending on your tastes and the number of channels in your system.

SACD also contains strong anti-copy and anti-piracy provisions, issues that are dear to the hearts of record-company executives. First, a digital code is embedded in the disc that allows the SACD player to read the disc. If this code, which is difficult to find and duplicate, is missing, the player simply ejects the disc immediately after it is inserted. Second, the SACD can carry an image faintly visible to the naked eye, similar to a watermark on paper. The image is created by slightly varying the width of the information-carrying pits, which changes the diffraction pattern given off by the disc. If the disc is pirated (as occurs on a massive, government-sanctioned scale in some parts of the world), this image disappears. Pirated discs are thus instantly recognizable.

Direct Stream Digital (DSD) Encoding on SACD

Sony and Philips developed for SACD an entirely different method of digitally representing music, called Direct Stream Digital (DSD). Instead of sampling the musical waveform at a relatively slow 44,100 times per second and assigning a 16-bit binary number to represent the amplitude at sample time (the parameters of CD), DSD samples the musical waveform at a lightning-fast 2.8224 million times per second. Each "sample" is, however, only one bit. If the waveform's amplitude is increasing, a binary "one" is recorded. If the waveform's amplitude is decreasing, a binary "zero" is recorded. An audio signal at full-scale positive is represented by all ones; a signal at full-scale negative is represented by all zeros. The zero point is represented by alternating ones and zeros, as shown in Fig.7-15. The musical information is contained in the width of the pulses, which is why this type of encoding is sometimes called Pulse Width Modulation (PWM). The data rate of DSD is exactly four times that of 16-bit/44.1kHz encoding (44,100 x 16 x 4 = 2,822,400 bits per second).

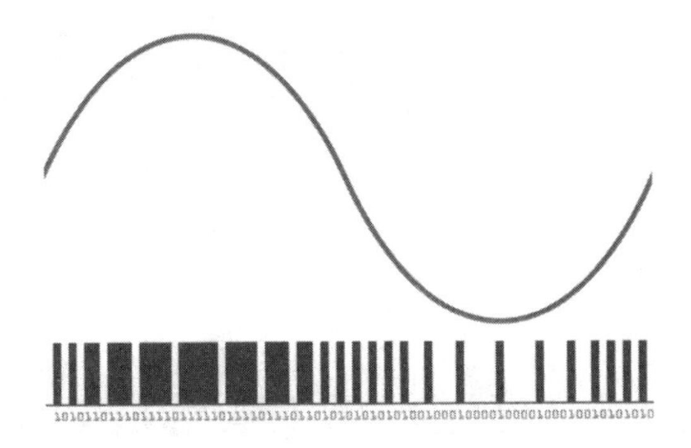

Fig. 7-15 The DSD bitstream is a train of single-bit pulses. (Courtesy Sony Corporation)

Interestingly, the pulse train generated by DSD encoding "looks" remarkably analog-like. That is, you can look at the pulse train and get an idea of what the analog waveform looked like. Moreover, converting this pulse train to analog can be as simple as putting the pulses through an analog low-pass filter.

In practice, the DSD-encoded signal is too noisy to convert directly to analog. Because this noise is distributed over a very wide bandwidth, a technique called noise shaping shifts the noise away from the audioband and up to a higher frequency, where it is inaudible. Although the amount of noise in the signal remains constant, the noise in the audioband is radically reduced. Fig.7-16 shows how noise shaping shifts noise away from the audioband.

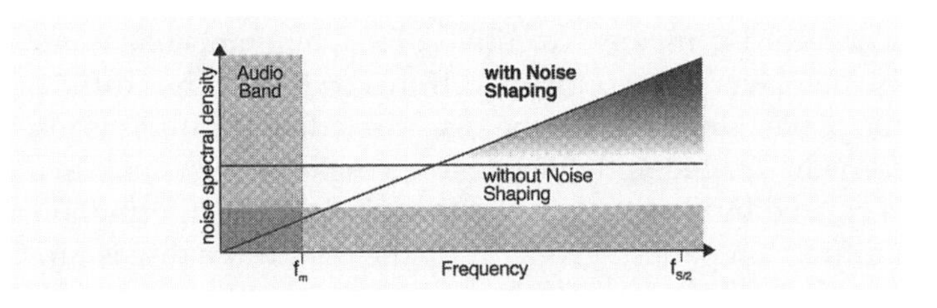

Fig. 7-16 Noise shaping shifts noise out of the audioband. (Courtesy Sony Corporation)

DSD can record frequencies of up to 100kHz, far higher than the 20kHz of conventional 44.1kHz-sampled digital audio. In addition, the signal-to-noise ratio (the difference between the loudest sound that can be recorded and the system's noise level) is a whopping 120dB. This is contrasted with the 98dB signal-to-noise ratio of 16-bit quantization. Note, however, that DSD cannot achieve 120dB signal-to-noise ratio across the full 100kHz bandwidth, only over the audioband of 20kHz. That's because the noise shaping described earlier reduces the signal-to-noise ratio above the audioband. One of the benefits of DSD is the elimination of complex filters from the signal path. Conventional digital audio coding requires so-called decimation and oversampling filters, which introduce noise and errors. DSD is a much more direct method of storing music digitally.

Moreover, the DSD pulse train can be converted to 16-bit/44.1kHz form with fairly simple processing. The same DSD master tape can be the source for both the high-resolution and standard-resolution layers on an SACD. Interestingly, many of the sonic benefits of DSD are apparent in the 16-bit/44.1kHz signal. I've heard comparisons of music recorded on DSD then downconverted to 16-bit/44.1kHz, and the same music originally recorded at 16-bit/44.1kHz. The DSD/downconverted version sounded significantly better than the conventionally coded disc.

Before being recorded on an SACD master disc, the DSD bitstream is compressed so it takes up less space on the disc. Unlike "lossy" compression systems that throw away portions of the signal considered to be "inaudible," the compression system used in SACD ensures perfect bit-for-bit accuracy with the source data by more efficiently coding the audio data. For example, a string of eight consecutive 0s may be coded as 8x0. (In practice, the compression system is vastly more complex.) This compression system is what allows SACD to carry both 2-channel and 6-channel representations of the music on the high-resolution layer.

A drawback of DSD is that all professional recording, mixing, mastering, and signal-processing equipment is based on pulse-code modulation digital audio, not the 1-bit representation of DSD. As a result, many SACD titles have been recorded direct to two-track (or multitrack) DSD, with no mixing or processing after the recording. This approach maintains the greatest sonic purity and is the preferred process for audiophile labels, but isn't practical for the entire music industry. Sony has worked to provide recording and mastering engineers with DSD-based tools, but access to those tools is still limited.

DSD encoding provides the intriguing possibility of converting the pulse-width modulated bitstream directly to analog in a switching amplifier. Rather than convert the DSD bitstream to analog, and then amplify it conventionally, the DSD bitstream can be fed to a switching amplifier, with the pulses turning on and off the amplifier's output transistors. An SACD player's DSD output is fed directly into the amplifier, with the amplifier's output stage functioning as the digital-to-analog converter. (Switching amplifiers are explained in Chapter 5.)

DSD Sound Quality

I've listened to DSD extensively under ideal conditions, both in demonstrations and in my own reference system. For my first demonstration, early in DSD's development, a pair of Wilson Audio Specialties WATT/Puppies, driven by Cello electronics, was installed in the control room of Sony Music Studios in New York. The setup was optimized for a single listening seat. As a four-piece acoustic jazz band (including Randy Brecker on trumpet) played live in the adjacent studio, I could sit in the sweet spot and switch between the live microphone feed, the DSD-processed signal, and a 20-bit/48kHz analog-to-digital and digital-to-analog converter chain. Levels were matched every hour to within 0.1dB. Having a live microphone feed for reference is an extremely powerful tool for assessing the sonic effects of the encoding system.

The sound of DSD was most impressive. The music had a lack of hardness and glare through the midrange and treble that sounded much closer to the live microphone feed. I heard a sense of ease, accompanied by fine resolution of detail. In addition, DSD better preserved the sense of depth, space between instrumental images, and overall soundstage size. I've also heard a number of SACD players in my reference system, and overall am greatly impressed with its sound quality. You can easily compare SACD to CD with a hybrid disc by simply switching between the two layers using the player's remote control. It takes perhaps 30 seconds to switch because the player must switch lasers and read the other layer's table of contents before it will play. DSD is, in my view, a significant advance in sound quality over 16-bit/44.1kHz compact disc.

There's an important caveat to be aware of when evaluating SACD (or DVD-Audio): some titles have been remastered in the new format from 44.1kHz, 16-bit master tapes. This means the SACD or DVD-A release will contain no more information than the CD version; the 44.1kHz, 16-bit master is the limiting factor in the disc's fidelity. The remastered disc can, however, sound better than the original CD because post-processing is performed at a higher sampling frequency (and the errors introduced are distributed over a wider bandwidth), and the filtering performed in the player can be more gentle. Nonetheless, the best way to judge SACD is to choose titles originally recorded in DSD, and to evaluate the DVD-A format using recordings originally made at high sampling frequencies and long word lengths.

DVD-Audio

Although DVD-A is essentially a dead format, I've included an overview here for completeness. The DVD-A specification calls for a disc that can contain 6-channel music along with a 2-channel mix on the same disc, and with sampling rates as high as 192kHz and word lengths of up to 24 bits. You can think of a DVD disc as a "bit bucket" that can hold a wide variety of sampling frequencies, word lengths, playing times, and channels.

Specifically, DVD-A supports sampling rates of 44.1kHz, 48kHz, 88.2kHz, 96kHz, 176.4kHz, and 192kHz, with word lengths between 16 and 24 bits, in one-bit increments. The disc can contain a 2-channel mix, a multichannel mix (any number of channels up to six), or both. The highest sampling frequencies of 176.4kHz and 192kHz are available only for 2-channel discs, not multichannel ones. That's because these high-sampling rates have a very high data rate (number of bits per second) that exceeds the maximum data rate that can be read from the disc. (DVD's maximum data-transfer rate is 9.6 million bits per second.)

Record producers can choose from these sampling frequencies and resolutions (number of bits per sample), keeping in mind that faster sampling and higher resolutions shorten the disc's playing time. To save disc space, producers can elect to include only a multichannel mix, and allow the DVD player to "downmix" a 2-channel version on the fly. Codes inserted in the bitstream (called "Smart Content") tell the player how to perform the downmix conversion.

Even DVD-A's storage capacity and maximum transfer rate are not high enough to deliver multichannel audio with high resolution on all channels. To overcome this limitation, DVD-A employs a lossless coding system developed by Meridian Audio called Meridian Lossless Packing (MLP). Unlike "lossy" compression systems such as MP3 or Dolby Digital that remove information and reduce fidelity, MLP is a perfectly lossless process, producing identical bit-for-bit data on playback with no sonic degradation.

Without MLP, a single-layer, single-sided DVD-A could hold up to 65 minutes of 5-channel audio with a sampling rate of 96kHz and 20-bit resolution. 6-channel audio with 96kHz sampling frequency and 24-bit resolution isn't possible because the data rate exceeds DVD-A's maximum transfer rate of 9.6Mbs.

MLP eliminates this bottleneck by reducing the number of bits needed to represent the audio signal, expanding the possibilities for playing time, sampling frequency, and resolution.

In addition to a multichannel mix and an optional 2-channel mix on the disc, DVD-Audio carries a Dolby Digital-encoded surround track to make the disc compatible with all DVD players, not just those with DVD-A capability. The data rate from the multichannel Dolby Digital track is either 384,000 or 448,000 bits per second—a tiny fraction of even CD's data rate of 705,600 bits per second per channel. Ironically, most listeners' first experience with DVD-A will be via the sonically compromised Dolby Digital track, which is inferior to CD.

Bass Management in SACD and DVD-A Players

The term bass management describes a system in which bass is filtered out of some channels and directed to a subwoofer, if the system employs one. Bass management is necessary in multichannel audio systems, and is found in most SACD and DVD-A players, as well as in multichannel controllers.

To see why we need bass management, consider a recording containing full-bandwidth information (low bass as well as midrange and treble) in every channel. A good example is a recording of popular music in which the bass guitar and kick drum are at least partially positioned in the center channel. Most multichannel loudspeaker arrays employ large, floorstanding loudspeakers in the left and right positions, and smaller speakers in the center and rear positions. Low frequencies from the bass guitar and kick drum would overload the small center speaker, introducing distortion or damaging the speaker. Bass management selectively filters bass from certain channels you specify according to the kind of loudspeakers in your system. That low bass can be directed to a subwoofer, if your system includes one.

If you've set up a home-theater receiver or controller and selected Large or Small from the menu for each of your speakers, you've worked with bass management. The Small setting simply engages a high-pass (low-cut) filter on that channel, preventing low bass from reaching the speaker. A typical cutoff frequency is 80Hz.

Ideally, bass management would be performed in the multichannel controller or preamplifier the player drives, and the connection between them would be digital. But because of copy-protection issues (more on this later), DVD-A and SACD players are connected to a controller or multichannel preamplifier via the analog outputs. The player manufacturers can't assume the multichannel preamplifier has bass management, so they include it in the player.

Bass management in the DVD-A format is performed via on-screen menus. SACD players typically have a selection of common loudspeaker configurations from which to choose; you simple engage one from the player's front panel with no need for a video display. For example, Setting #1 may be ideal for a system with large left and right speakers, small center and surround speakers, and a subwoofer.

The bass-management set-up controls are sometimes accompanied by a variable delay to each channel. You set the amount of delay according to how far you sit from each loudspeaker so that the sound from each speaker reaches you simultaneously. For example, if you sit 12′ from the left and right speakers, but only 6′ from the rear speakers, six milliseconds (6ms) of delay to the rear speakers results in coincident arrival of front sounds with those from the rear.

Blu-ray Disc

The Blu-ray Disc format (BD) was developed as a high-definition replacement for DVD-Video. By using smaller pits and a shorter wavelength playback beam, BD can store much more information on a 120mm optical disc than DVD can. Specifically, a Blu-ray Disc can store 25GB on a single-sided, single-layer disc (contrasted with 4.7GB for single-sided, single-layer DVD). Blu-ray's greater storage capacity provides several intriguing possibilities for high-resolution digital audio.

You might know that the Dolby Digital format used in DVD is limited to 448kbps (kilobits per second) for 5.1-channel audio. This is about one-tenth the data rate of CD-quality audio. By contrast, BD's maximum audio data rate is a whopping 24Mbps, or more than 53 times the data rate possible on DVD. This greatly increased data rate allows BD to deliver high-resolution multichannel digital audio with unprecedented sound quality, either in conjunction with video or as an audio-only format. On concert or music BD titles, the audio stream might be encoded in the Dolby TrueHD or DTS HD Master Audio formats, which provide high-resolution multichannel digital audio with bit-for-bit fidelity to the original high-res master. These two formats, which are described in detail in Chapter 12 ("Audio for Home Theater"), use a lossless compression system that roughly halves the signal's bit rate for storage on the disc.

The Blu-ray specification also calls for an audio-only disc (no video) that contains up to 7.1 channels of uncompressed high-resolution digital audio with a sampling rate up to 192kHz and bit lengths of up to 24 bits. Only a few specialty record labels have introduced titles in this format. The major record companies have so far ignored the massive sonic potential of Blu-ray Disc.

FireWire (IEEE1394) and Digital Transmission Content Protection (DTCP)

To meet the challenge of creating a digital interface that can transmit high-resolution, multichannel audio data, yet protects copyright holders from unauthorized duplication, a consortium of five companies, dubbed "5C," developed a scheme called Digital Transmission Content Protection (DTCP). The term 5C refers to the five companies who contributed to the standard: Hitachi, Intel, Panasonic, Sony, and Toshiba.

DTCP allows content providers to specify the manner in which their content can be recorded by consumers. Specifically, four scenarios are possible:

1. Copy-Never—no copies allowed
2. Copy-One-Generation—one copy allowed
3. Copy No-More—prevents making copies of copies
4. Copy Freely—no restrictions on copying

This control over copying is realized via special codes embedded in the digital audio datastream carried on a FireWire cable. Also known as i.LINK (a Sony tradename) or by its official name IEEE1394 (the Institute of Electrical and Electronics Engineers codified the standard on January 3, 1994, hence the name), FireWire is a bi-directional interface that can carry digital audio, digital video, computer data, and control signals on one cable. In Sony's SACD players equipped with a high-resolution digital output, the interface carries only audio data and the control signals needed for copy protection. FireWire is bi-directional—an essential element of Digital Transmission Content Protection. Specifically, DTCP establishes two-way communication between components with a challenge and response. When a device has been authenticated as authorized to record the digital datastream, the decryption key is downloaded into the receiving device. Content that has been encoded Copy-Never will send the decryption key only to decoding devices such as digital-to-analog converters, not digital recorders. This authentication and key exchange, which occurs in just 30 milliseconds, is invisible (and inaudible) to the user.

In fact, the entire DTCP system works behind the scenes. If you're not trying to record the high-resolution digital bitstream, you won't even know 5C is being used. DTCP simply secures the content of the FireWire digital bus between components. The only time you'll be aware of DTCP is if you attempt to record material restricted by the content provider.

High Definition Compatible Digital (HDCD)

The fundamental problem with CD sound is that the amount of digital data representing the music isn't nearly enough to correctly encode all the musical information we can hear. As explained earlier in this chapter, the CD's sampling frequency of 44.1kHz isn't fast enough, and 16 bits per sample are too few. The challenge, then, is to put more information about the musical signal through the compact disc's 44.1kHz, 16-bit pipeline.

But how can you pack more information into a pipeline that's already filled to capacity? High Definition Compatible Digital (HDCD) uses some ingenious techniques to accomplish just that. The result is musical and technical performance that exceeds the limits of conventional 44.1kHz, 16-bit digital audio. The process was invented by Keith Johnson and Michael Pflaumer, who, with Michael Ritter, formed Pacific Microsonics Inc. to develop and market HDCD. In 2000, Pacific Microsonics was acquired by Microsoft, which has a keen interest in the art of transmitting more information in fewer bits. Despite the increasing availability of music in SACD or DVD-A (technologies which obviate the need for HDCD), many discs are encoded with HDCD, players with HDCD decoding are still available, and HDCD offers, in my view, a significant improvement in CD performance.

Recordings encoded with HDCD can be decoded and played back on any CD player, with some improvement in fidelity. But if your CD player is equipped with an HDCD decoder, HDCD recordings can be reproduced with much better sound quality than is possible with conventional CDs. Here's how it works.

The HDCD encoder takes in an analog audio signal (either from analog tape or a live microphone feed) and converts it to digital with a very fast sampling rate (176.4kHz) and 24-bit resolution. The amount of data generated is far too great to store on CD, so the HDCD encoder analyzes the high-resolution signal to determine what musically significant information will be lost when that signal is converted into a 44.1kHz, 16-bit signal for storage on CD. The encoder then outputs a 44.1kHz, 16-bit digital audio signal, but with a control code hidden in the 16th bit. This hidden and inaudible control code tells the HDCD decoder in the CD player to perform certain processes on the audio signal to restore some of the information present in the original high-resolution digital signal. (Note that you don't give up one bit of resolution with an HDCD recording; the control code is present only 3–5% of the time.)

The small amount of information in the control code can produce large changes in the audio signal when decoded. That's because the control code is a kind of shorthand that the decoder recognizes. Here's an analogy: Say we want to transmit Beethoven's Ninth Symphony from one place to another. Rather than transmit the entire symphony, we instead send just a few bits that tell another device to output Beethoven's Ninth Symphony. That's how HDCD works: instead of transmitting the musical information that would normally be lost in conventional coding, the control code transmits instructions to the decoder telling how to restore that information. Because so much of HDCD's "intelligence" is in the decoder, a few bits in the control code accomplish the work of many.

Although HDCD uses many techniques to improve sound quality, let's look at the way in which HDCD overcomes the limitations of the CD's 20kHz bandwidth.

While it's true that humans can hear pure tones only up to 20kHz (if you're 16 years old; at age 40 you're lucky to hear 15kHz), that doesn't mean there's no perceptible musical information above 20kHz. Very high frequencies imply a steep attack; the ear/brain can reportedly detect the difference in steepness of a signal with a bandwidth of 20kHz and one of 40kHz, according to HDCD's inventors.

The bandwidth of an audio system implies how quickly it can respond to musical signals. Earlier in this chapter I described a subwoofer, a low-bandwidth device, trying to reproduce the attack of a triangle. The woofer simply can't move fast enough to accurately reproduce the steep transient of the triangle's waveform.

That's exactly what happens when we record music with a bandwidth of only 20kHz. The 20kHz-bandwidth digital audio system simply can't encode the parts of the music that quickly change in intensity. The musical result is a closing-in of the sound and a reduced ability to pinpoint the positions of instruments. Experiments have shown that the ear/brain combination uses that steep transient information as directional cues. Moreover, limiting the speed of musical signals changes their timbre. For example, the component of an oboe's sound that includes the tiny "clicks" of the moving reeds disappears. We don't hear this as distortion, but as a dilution of realism. This is one reason that a live oboe has a vitality and presence we don't hear from a standard-resolution digital reproduction of an oboe. Because the CD's bandwidth isn't wide enough, instrumental timbres are changed, along with their directional cues.

Remember that hidden control code buried in the 16th bit of an HDCD-encoded recording? Part of that control code tells the decoder in your CD player that a particular signal had its transient attack slowed; the decoder then restores the signal's original transient steepness. This is how HDCD can reproduce signals whose speed implies a bandwidth greater than 20kHz within the limitations of the CD's 44.1kHz sampling frequency.

This is just one of the many techniques HDCD uses to squeeze more music through the 44.1kHz, 16-bit pipeline. Similar techniques address the dynamic-range limitations of 16-bit coding. The system is extremely sophisticated in practice, and reflects a decade of research into human hearing, recording, electronic design, and digital signal processing.

The first HDCD-compatible products used the Pacific Microsonics PMD100 HDCD decoder, which incorporated an 8x-oversampling digital filter that is integral to HDCD decoding. This filter also operates on conventionally encoded discs, which makes all recordings sound better, not just HDCD-encoded ones. That's because its integral digital filter operates with a higher mathematical precision than previous filters, and employs some tricks to improve sound quality. Today, HDCD decoding is more likely to be incorporated into a digital-signal processing chip that performs many other functions.

HDCD is used primarily at the mastering stage of the recording chain. After an album has been recorded and mixed, the producer and artist take the final tape to a mastering studio to generate the digital master tape from which CDs are made. The mastering engineer, artist, and producer decide if they want to use the HDCD process.

Microsoft sells professional HDCD encoders and licenses HDCD decoding to other companies. For example, Motorola, Analog Devices, and Sanyo manufacture audio-processing chips that include HDCD decoding, and pay Microsoft a royalty on every chip sold. This licensing agreement is similar to that used by Dolby Laboratories in licensing its ubiquitous Dolby B noise-reduction circuit and Dolby Digital decoding technology.

In my view, HDCD is a significant advancement in digital audio sound quality. I've had the opportunity to compare an original analog master tape with an HDCD-encoded/decoded version, along with conventional 44.1kHz, 16-bit coding. I've also heard comparisons of a live microphone feed (and the instrument in the hall) against state-of-the-art analog tape and HDCD. My experience suggests that HDCD improves sound quality in these ways: with a finer resolution of detail, more accurate rendering of timbre, smoother and less synthetic treble, wider dynamic contrast, clearer articulation of loud and complex passages, bigger soundstage, tighter image focus, more front-to-back layering, more bloom around instrumental outlines, and a greater projection of quiet instruments in the presence of louder instruments.

Interestingly, many CDs that are HDCD encoded lack any indication on the disc or packaging that the disc is encoded. The Pacific Microsonics analog-to-digital converter and HDCD encoder are used in a large number of mastering rooms primarily because they're generally considered the best-sounding A/D converters extant. Albums mastered through a Pacific Microsonics A/D are often HDCD encoded by default, and the record company simply doesn't know enough to indicate on the packaging that the disc has been HDCD encoded. It's a pleasant surprise to buy a CD and discover, when you play the disc for the first time and see the HDCD indicator on the CD player illuminate, that the disc has been HDCD encoded.

How a DAC Works

For the technically minded, the rest of this chapter delves into the inner workings of the compact disc, CD transports, the S/PDIF interface, and DACs. You don't need to know this information to enjoy music in your home, but I've included it for those who want to understand what goes on inside those sculpted metal boxes in their equipment racks. If you like, simply skip ahead to the beginning of the next chapter.

Fig.7-17 shows a block diagram of an outboard DAC. The main components are a power supply, input receiver, digital filter, digital-to-analog conversion stage, current-to-voltage converter, and analog output stage.

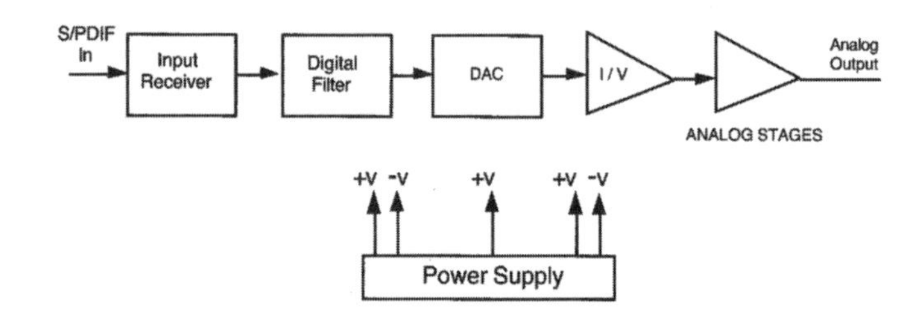

Fig. 7-17 A DAC's main subsections are the input receiver, digital filter, digital-to-analog converter chip, current-to-voltage converter, and analog output stage.

The input receiver takes the S/PDIF signal from a digital source and converts the serial datastream into raw digital audio data. It also generates a clock signal based on the clock embedded in the datastream. A Phase Locked Loop (PLL) compares the input signal frequency (the clock) with a reference frequency (usually generated by a crystal oscillator) and creates a new clock locked to the incoming datastream's clock. This so-called "recovered" clock becomes the master clock for the processor. The input receiver is a significant source of clock jitter and can have a large effect on how the processor sounds. Recent attempts to minimize jitter created by the input receiver include dual PLLs, custom low-jitter input receivers, and external clocks.

Before we move to the next step in digital-to-analog conversion, let's look at what happens when we convert a digital signal to analog. The discrete samples that represent the analog signal don't form a smooth and continuous waveform; they instead create a stairstep waveform. The energy in the stairstep waveform's sharp edges appears as images of the signal centered at multiples of the sampling frequency (Fig.7-18). These images must be removed by an analog low-pass filter.

In the first Sony players, these images were removed and the stairstep waveform smoothed by a brickwall analog filter (the top drawing in Fig.7-18). This filter was extremely steep, and close to the audioband, resulting in phase shift, ripple, and other audible artifacts.

Today's CD players and DACs use an oversampling digital filter that increases the sampling frequency by (typically) eight times by interpolating seven new samples for each audio sample. This raised the sampling frequency to 352.8kHz (44.1kHz x 8), which pushed the first image to 352.8kHz, where it could be filtered by a much gentler and better-sounding analog filter well away from the audioband (lower drawing in Fig.7-18). Again, the spurious images at multiples of the sampling frequency still appear, but are now shifted well away from the audioband. Contrary to some beliefs, oversampling's sonic benefits are not the result of adding new information between the existing samples, nor from "smoothing out" the waveform with more samples. Moreover, oversampling doesn't increase the audio bandwidth; an 8x-oversampling CD player still has a bandwidth of 20kHz. The sonic advantage is achieved by replacing the brickwall filter with a gentler analog output filter.

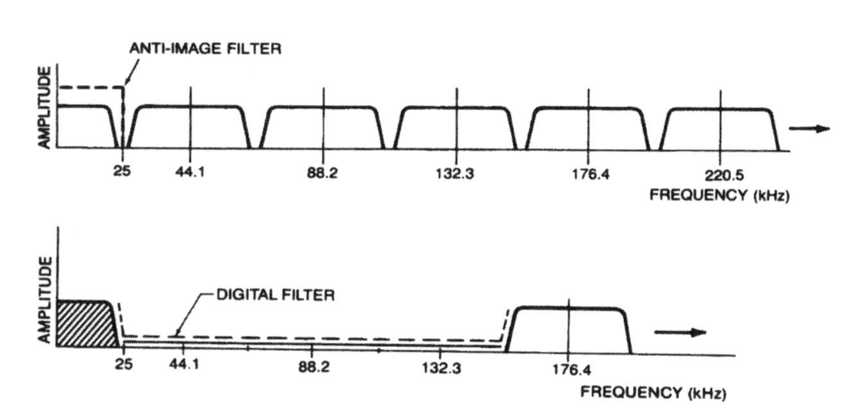

Fig. 7-18 Spurious images are created at multiples of the sampling frequency (top). An oversampling digital filter shifts the first image to a higher frequency that is more easily filtered. (Courtesy Ken Pohlmann, *Principles of Digital Audio*)

In addition to increasing the effective sampling rate, oversampling digital filters increase the output word length from 16 bits to (typically) 24 bits. Note that no new information is generated by the filter when outputting longer words. Instead, the longer words reduce errors resulting from the filter's internal calculations. The audio data being filtered may temporarily have very long word lengths—perhaps 30 bits—that need to be reduced before being output from the filter. Because simply truncating (cutting off) all but the 16 needed bits results in audible distortion, some filters use redithering techniques to reduce round-off errors. These techniques attempt only to compensate for errors introduced in the filter; they cannot improve incoming data.

Custom Digital Filters

Manufacturers of CD players and digital processors have two choices: buy an "off-the-shelf" filter chip or create a custom filter from general-purpose Digital Signal Processing (DSP) chips. The custom filter designer must write the software that controls the DSP chips, an expensive and time-consuming task. Consequently, custom filters are quite a bit more expensive, but allow the DAC designer greater creative control over the product's sound. In addition, custom digital filters can run faster than the typical 8x oversampling of chip filters.

Proponents of this approach believe their custom filtering software is better than that found in conventional integrated circuit chips. Specifically, most custom digital filters are optimized for time-domain performance rather than frequency-domain performance.

Digital-to-Analog Conversion

After digital filtering, the oversampled data are input to the digital-to-analog converter or DAC. The DAC converts digital data at its input to an analog output signal. The most common type of DAC is the "ladder," or "R/2R" DAC. Both names describe the same operating principle. Ladder DACs have "rungs" of resistors, each rung corresponding to one bit in the digital code. A 16-bit DAC will thus have 16 resistor rungs, which allow 65,536 (216) possible input codes. The input data act as switches on the rungs; a binary "1" closes the switch at that particular bit and allows current to flow through the resistor, while a binary "0" opens the switch and no current flows. The DAC's output current is thus determined by the digital input code. (An R/2R resistor-ladder DAC is shown in Fig.7-19.)

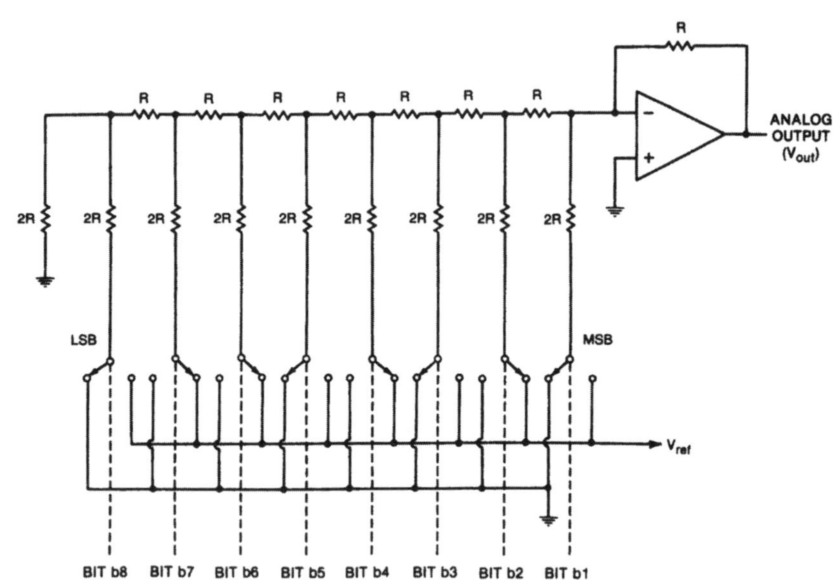

Fig. 7-19 A DAC's main subsections are the input receiver, digital filter, digital-to-analog converter chip, current-to-voltage converter, and analog output stage. (Courtesy Ken Pohlmann, *Principles of Digital Audio*)

Each rung has an effective resistor value half that of the adjacent rung; the resistor determines how much "weight" the bit has by determining how much current flows through the "switch." The LSB (Least Significant Bit) has the lowest weight, the MSB (Most Significant Bit) has the highest. Each bit upward from the LSB should produce an exact doubling of current, in the progression 1, 2, 4, 8, 16, 32, 64, 128, 256, and so on, up to 524,288 in 20-bit DAC. The LSB has a value of 1, the MSB a value of 524,288 (in a 20-bit DAC). When the MSB value of 524,288 is combined with the 19 lower bits, a 20-bit DAC can accommodate up to 1,048,576 input codes.

The precision of these resistor values is crucial: Any deviation from a doubling of resistance causes that bit to have more or less value than it should. These bit-weighting errors result in poor linearity. That is, the analog output level doesn't track the digital input code. A perfectly linear DAC will produce an analog output at exactly the same level as the digital input. For example, if the DAC is driven by the code representing a –90dB dithered sinewave, the analog output should be exactly 90dB below full scale. A non-linear DAC would produce an output level of perhaps –93dB or –86dB. Linearity errors generally occur at very low signal levels.

Delta-Sigma DACs

An alternative to the resistor-ladder DAC (sometimes called a "multibit" DAC) is the delta-sigma DAC, also called a "1-bit" DAC. These converters are sometimes called by their trade names: Bitstream (Philips), MASH (Matsushita's process, developed by Nippon Telephone and Telegraph), and PEM (Pulse Edge Modulation, developed by JVC). All of these converters work on the same principle: instead of using a resistor ladder with different bit weightings to convert a binary code into analog, 1-bit DACs operate in only two states: zero and one. The 1-bit code is a series of varying-length pulses with a single amplitude. The pulse width determines the analog output voltage. This is why 1-bit coding is also called pulse-width modulation. The DSD encoding system used in SACD is Delta-Sigma modulation.

The single-bit code is of a high enough frequency (typically between 64x and 256x oversampling) that the audio signal can be reconstructed from these two logic states with a switched-capacitor network. Consequently, Delta-Sigma DACs don't require the amplitude precision of ladder DACs. Delta-Sigma DACs trade resolution in amplitude for resolution in time. They have inherently good linearity and don't require a current-to-voltage converter. Because Delta-Sigma DACs operate at such high frequencies, noise shaping is used to shift quantization noise from the audio band to just above the audioband.

The British company Data Conversion Systems (dCS) has developed a novel DAC topology that is a hybrid of multi-bit and Delta-Sigma techniques. The so-called Ring DAC is a five-rung resistor-ladder DAC in which each resistor value is identical. A modulator before the DAC converts the PCM (or DSD) code into a five-bit form running at 2.8224MHz (four times the data rate of CD). Using identical resistor values solves the problem of multi-bit DACs in which any variation from a doubling of resistor value between adjacent rungs introduces linearity error. Because even nominally identical resistors have some manufacturing tolerances, the signal is shifted (passed around the "ring") between resistors so that any errors are randomized and converted into a small amount of white noise. Fig.7-20 is a graphic comparison of sampling rates and quantization levels between 44.1kHz/16-bit PCM, high-resolution PCM, the five-bit Ring DAC format, and Delta-Sigma (DSD) conversion.

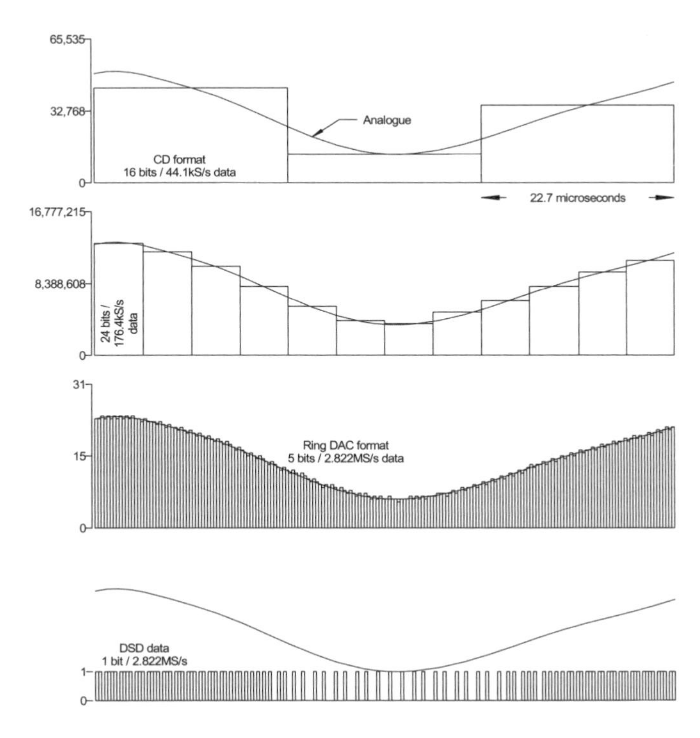

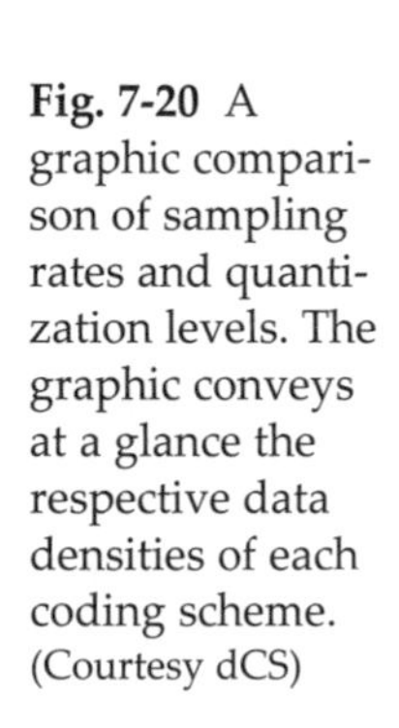

Fig. 7-20 A graphic comparison of sampling rates and quantization levels. The graphic conveys at a glance the respective data densities of each coding scheme. (Courtesy dCS)

Analog Stages

The output of a multi-bit resistor-ladder DAC is a series of currents at varying levels. This current output must be converted into a voltage that can be amplified by the processor's analog output stage. This is the next step after the DAC, called current-to-voltage conversion, or I/V conversion (I is the electronic symbol for current, V the symbol for voltage). The I/V converter takes the DAC's current output and converts it to a voltage. Virtually all digital processors use an op-amp (operational amplifier, an integrated circuit that takes the place of a circuit using separate transistors) for I/V conversion. In addition to being easier and cheaper to implement, an op-amp is generally faster than a discrete (separate transistor) I/V stage. This is because the circuitry is on a single silicon chip rather than spread out in discrete components on a circuit board.

After I/V conversion, the analog voltage is put through an analog output stage. This circuit has several functions. First, it can provide gain to bring the final output level to the standard of 2V RMS (with a full-scale input signal). Second, the output stage incorporates an output filter to remove high-frequency noise above the audioband. This filter is often a gentle first-order type. The output filter can be an elaborate active circuit, or as simple as a capacitor put across the analog output jacks. Finally, the output stage acts as a buffer between the I/V converter and the final analog output. The output stage generally has a low output impedance and is designed to drive cables and a preamplifier. The analog output stage can be based on an operation amplifier (op-amp) or, in more expensive products, a discrete analog stage composed of separate transistors. A direct-coupled output stage has no capacitors in the signal path to block direct current (DC) from getting into the audio signal. Most direct-coupled output stages use a DC servo to prevent DC from appearing on the analog output jacks. A DC servo is a circuit that monitors the DC level at the amplifier's output and makes adjustments to reduce the amount of DC.

How a CD Transport Works

A CD transport has four main sections: the transport mechanism, the servo systems, the decoding electronics, and the output driver stage. The transport mechanism spins the CD, reads the CD's pits with a laser, and outputs a signal to the decoding electronics. The decoding circuits, usually on one integrated circuit chip, demodulate the datastream, perform error correction, and convert the data into a form that can be transmitted from the transport to a digital processor. The servo systems maintain correct disc speed (the rotational servo), keep the laser on the spiral track of pits (tracking servo), and maintain laser focus on the spinning disc (focus servo). The output stage is a digital line driver—sometimes including a small transformer called a pulse transformer—to drive the cable and digital processor input.

Let's look at each of these subsystems in more detail.

The compact disc and laser pickup mechanism constitute a remarkable feat of engineering. The CD has a spiral track of alternating pits and land, recorded from the inside of the disc to the outside. Pits are indentations in the disc surface; land is the disc surface itself (Fig.7-21). The space between adjacent tracks of pits is 1.6μm (micrometer), or 1.6 millionths of a meter. The pits vary in length from 0.8μm to 2.8μm, and are 0.56μm wide. To put these numbers into perspective, a human hair has a diameter of about 75µm. If each pit on a CD were the size of a grain of rice, the CD itself would be the size of a baseball diamond. The more advanced disc formats (DVD, SACD, and Blu-ray Disc) achieve greater data density with smaller pits and a shorter-wavelength laser beam. The pits on a DVD, and on an SACD's high-density layer, are about half the size of CD pits, and the tracks are more closely spaced. Nonetheless, the operating principle is identical to that of CD.

Fig. 7-21 Data on a CD are encoded in a spiral track of "pits" and "land."

The laser beam is reflected from the CD's surface to a photodetector, a device that converts light into an electrical signal. When the beam strikes either land or pit bottom, it is reflected to the photodetector at full strength. This condition represents binary "zero." But when the laser beam strikes a land-to-pit or pit-to-land transition, part of the beam is reflected from the land and part is reflected from the pit bottom. Because the pit depth is one quarter the wavelength of the laser light, the beam portion reflected from the pit bottom is 180° out of phase with the beam reflected from the land, as shown in Fig.7-22. The two beam portions—now out of phase with each other—cancel each other, reducing the output from the photodetector. This reduced output, caused by a pit-to-land or land-to-pit transition, represents binary "1."

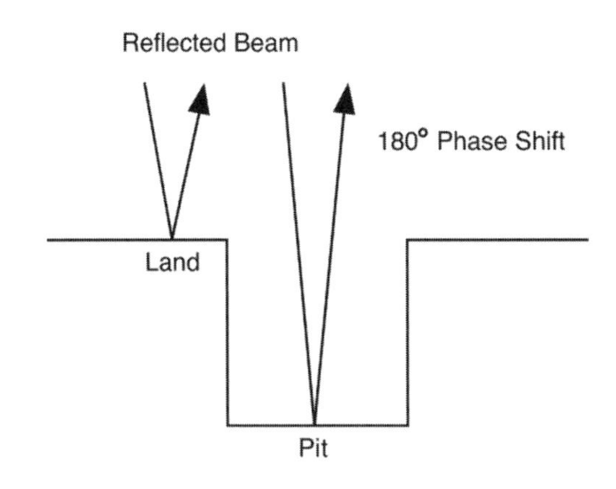

Fig. 7-22 A pit depth of one-quarter the wavelength of the playback beam creates a 180° phase shift in the beam portion reflected from the pit bottom.

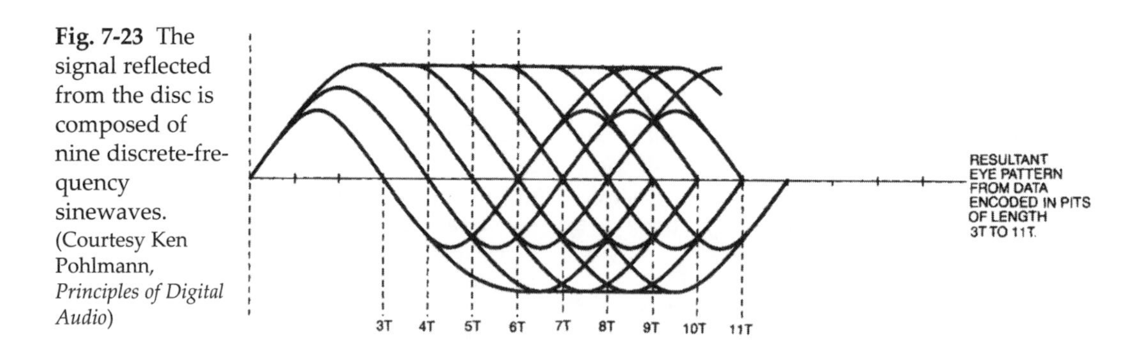

Fig. 7-23 The signal reflected from the disc is composed of nine discrete-frequency sinewaves. (Courtesy Ken Pohlmann, *Principles of Digital Audio*)

The pits thus produce a reflected beam of varying intensity, which produces a varying voltage at the photodetector output. The output signal is a series of sinewaves, the peaks representing land or pit bottom, and the zero-crossing transitions representing the pit edges. Because the pits and lands have nine discrete lengths, the recovered signal is composed of nine discrete-frequency sinewaves. The recovered signal, called an "eye pattern" or "RF" signal, is shown in Fig.7-23. Although the RF signal looks like an analog signal, the digital data are encoded in the zero-crossing transitions. The shortest pit or land length represents the digital data 1001; the ones are the pit edges and the zeros are the pit bottoms or land. The longest pit length represents the digital data 100000000001. The shortest pit produces a frequency (in the RF signal) of 720kHz, and the longest pit produces a frequency of 196kHz.

The data read from the disc don't directly represent digital audio data. Instead, a coding method is used to increase the disc's storage density and provide a clock signal within the data. The coding method is called Eight-to-Fourteen Modulation, or EFM. EFM coding takes a block of 8 bits and assigns it a unique 14-bit word. These 14-bit words are joined together by three "merging" bits to form the data pattern recorded on the CD. EFM coding inserts a minimum of two zeros between successive ones (remember, the shortest pits represent the data 1001), and a maximum of ten zeros between ones. The result is a pit-and-land pattern of nine discrete lengths. Fig.7-24 shows the relationship between the original data, EFM-coded data, and pit structure. SACD and DVD use a modified version of this encoding system that achieves slightly greater packing efficiency.

Although it is counterintuitive to think that increasing the number of bits by EFM coding stores data more efficiently, that's exactly what it does. Moreover, the specific pattern of ones and zeros creates a clock signal at the data rate read from the disc: 4.3218 million bits per second. Of these, only 1.41 million bits are audio data. The rest are error-correction, subcode, and the result of EFM coding. EFM coding also inherently makes the signal self-clocking.

Subcode is non-audio data recorded on the CD. It contains track time, track number, whether the CD has been recorded with pre-emphasis, and other such "housekeeping" information. When the de-emphasis circuit in your CD player or DAC is activated, it has been triggered by a bit in the subcode. The track and time information displayed on your CD player is also recovered from the subcode.

The EFM signal is input to the decoding electronics, which convert the 14-bit words back into the original 8-bit data. These chips also perform error-correction and strip out the subcode. (A general description of error-correction is included in Appendix C.) The Cross Interleaved Reed-Solomon Code (CIRC) error-correcting system used in the CD has two levels, called "C1" and "C2." The first decoder (C1) corrects as many errors as it can before

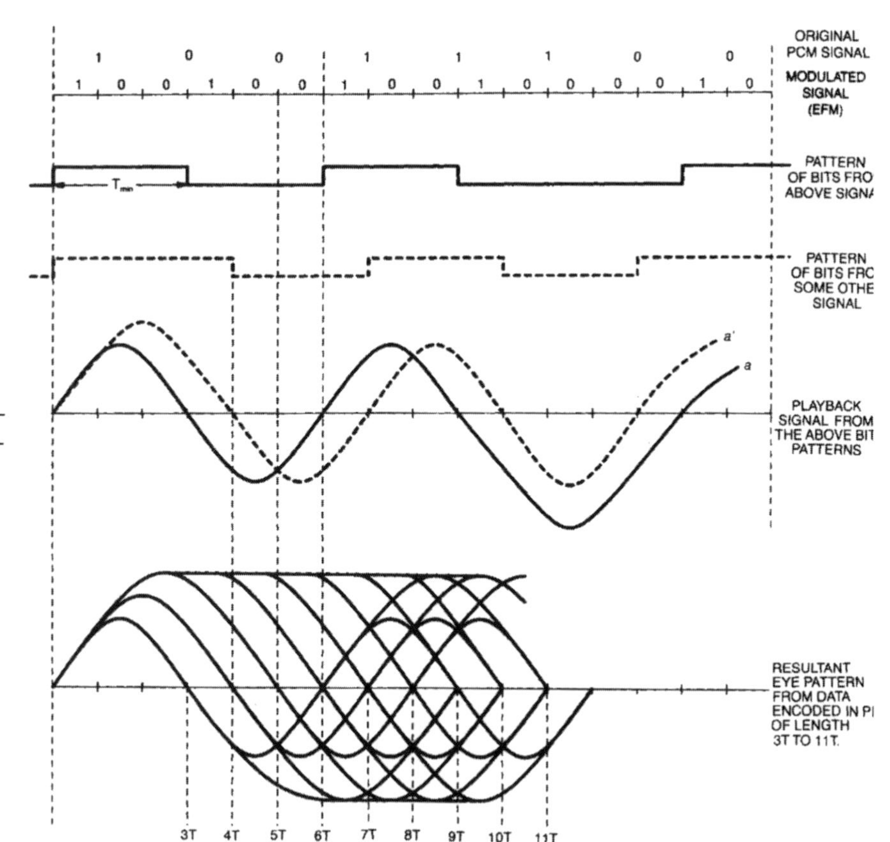

Fig. 7-24 The relationship between original data, EFM-encoded data, pit structure, and the RF signal. (Courtesy Ken Pohlmann, *Principles of Digital Audio*)

sending the data to the C2 decoder. The C2 decoder de-interleaves the data (puts it back in its original order), thereby converting a long-burst error into many shorter errors, which are more easily corrected.

The rotational servo uses an elastic buffer between the disc motor and the decoding electronics. Data are fed into the buffer at one end and clocked out at the other end. When the buffer is more than half full, a signal tells the rotational motor to slow down, decreasing the rate at which data are read from the disc. When the buffer is less than half full, the motor speed is increased to fill up the buffer faster. This servo mechanism always keeps the buffer half full. Data can be read in at an irregular speed, but are clocked out at a precise rate.

The focus servo system maintains the laser's objective lens at a precise distance from the spinning disc. The tracking servo keeps the laser centered on one track of the spinning disc.

A transport's output stage takes the S/PDIF signal from the S/PDIF encoder circuit and presents it to the final output jack. The job of driving the interface cable and digital-processor input falls on the output stage. Some transports have a reclocking circuit at the output stage to reduce jitter in the datastream. The output stage determines the transport's output impedance, which should be 75 ohms. Most transports use a pulse transformer at the output to couple the S/PDIF signal to the outside world.

8 Music Servers and Computer-Based Audio

Imagine sitting in your listening chair and deciding that you want to hear a particular piece of music. You get up, walk over to the racks upon racks of CDs, turn your head sideways to read the labels, and scan the CD spines for the desired disc. You then open the jewel case, take out the CD, put it in a transport mechanism, and start the disc playing.

Imagine instead that there's a touchscreen display next to your listening seat that shows you your entire music collection as album covers. The titles can be presented alphabetically, by genre, by composer, conductor, orchestra, and, in many cases, by what musician plays on the album. You touch the album cover and the music begins playing instantly.

Welcome to the world of music servers.

That's just the tip of the iceberg of how a music server can transform your relationship to your music library. Say in that piece of music just selected you found yourself greatly appreciating, for example, the drumming of Peter Erskine. On what other records in your collection does Erskine play? A couple of finger-taps on the touchscreen display shows you all the titles featuring this virtuoso drummer, ready for instant listening. Or consider this scenario we've all found ourselves in; we know what *kind* of music we're in the mood to hear, but not the specific title. Choose a title on the touchscreen that's in the genre you're interested in hearing and the music server rearranges the album art on the screen to show you titles you *might* want to hear. Or simply choose a genre and allow the server to surprise you by playing music from that genre for as long a time as you specify (a feature that's great for parties).

The server is not just a storehouse for your music collection, but also a portal for accessing new music. Say you're driving home from work one evening and hear a piece of new music on the radio that grabs you. When you get home simply turn on the server and download the album to make it part of your library. In fact, the distinction between the music on your server and music available for download is blurred; you have instant access to a vast music library from the listening seat, all organized in a way that lets you find exactly what you want. Speaking from personal experience, having a 24-hour, seven-day-a-week music store next to the listening seat can get costly very quickly. But it offers unprecedented opportunities to make musical discoveries. Anyone who has lived with a music server quickly finds it indispensable. In fact, the concept of music being stored on individual pieces of plastic that must be found and inserted into a machine will one day seem like a quaint anachronism—and already does to many listeners.

Music Server Sound Quality: Better Than Disc

A music server not only allows you to explore your music collection in new and rewarding ways, it also has the potential of delivering better sound than is available from CD. Music read from a hard-disc drive sounds better than when read from an optical disc. That's correct—music ripped from a CD to a music server can offer greater fidelity on playback from the server than from the original disc itself when decoded through the same digital-to-analog converter.

How can a copy sound better than the original? Consider what goes on during playback of an optical disc: a laser reads the data off a disc spinning at 200-500 RPM and performs error correction on that data on the fly. The optical system has one pass to get it right. Moreover, the crucial clock that controls the digital-to-analog conversion process is tied to the clock that controls the spinning disc.

Now consider what happens when you rip that data to a hard-disc drive on a server. The disc-ripping software can read areas of the disc repeatedly if errors are detected, moving on only when the data are perfectly accurate. Those data are stored on a hard drive and treated like any other file by the computer. Reading those data off a hard drive is a cake-walk compared with the challenge of keeping a laser tracking a spinning, wobbling, eccentric optical disc that has but one pass at getting all the information off the disc correctly. Moreover, there's evidence that the quality of the disc itself (specifically in the pit and land structures that encode the audio data) affects sound quality, a factor obviated by hard-disc storage.

I've compared the sound of state-of-the-art CD transports to music servers playing the same music through the same digital-to-analog converter and digital cable. I first ripped the CD to the server, put the CD in the transport, and compared the two. The sound from the server was smoother in the treble with a more delicate rendering of treble detail. The soundstage was also improved when music was sourced from the server—more spacious, greater image focus, increased depth, and an overall more convincing impression of instruments existing in space within the recorded acoustic. There was also a greater sense of ease and musical involvement. In short, transferring your CDs to a server will improve the sound of your CD library—provided that the server is configured and set up correctly.

High-Resolution Sound Quality From Music Servers

The second way in which a music server can deliver better sound is its ability to play high-resolution music files. High-res files are available via Internet download, or from WAV files stored on a DVD (described in the previous chapter). As I've mentioned before, the 44.1kHz sampling and 16-bit quantization used in the CD format are limiting factors in sound quality. Faster sampling rates (88.2kHz, 96kHz, 176.4kHz, 192kHz) and longer word length (20 or 24 bits) have the potential of delivering significantly better sound quality than CD's 44.1kHz sampling rate and 16-bit word length.

The combination of a music server, high-quality digital-to-analog converter, and true high-resolution source material is revelatory. In fact, high-res digital done right bears little resemblance to most CD sound. I had an epiphany after hearing the Keith Johnson-recorded Reference Recordings HRx titles in high-res (176.4kHz/24-bit) on my server. HRx is

Reference Recordings' name for storing high-res WAV files on a DVD that are then ripped to your music server. Here's an excerpt from my review of HRx, first published in the January, 2009 issue of *The Absolute Sound*.

I've long admired Keith Johnson's orchestral recordings on CD for their dynamics, truth in timbre, spatial detail, and low-level resolution. But hearing these familiar pieces for the first time in high resolution was an absolutely mind-blowing experience. There's so much more information in high-res that it fundamentally changes the musical experience. Yes, the HRx-sourced music has more low-level detail, wider dynamics, more realistic rendering of instrumental timbres, and a more fully developed soundstage. But that description doesn't begin to describe the profound effect these improvements have on the sense of musical involvement. The sound has such a sense of ease that I instantly entered that zone of complete immersion in the performance and expression, and stayed there until the piece was over. Listening to these high-res files wasn't a *quantitative* improvement in the listening experience, but a *qualitative* transformation.

I'll try to describe specific aspects of the presentation that led to this experience. If I had to define the overall sound of HRx with a single word, that word would be "texture." First, instruments and voices have an almost tangible texture, so realistic is the rendering of timbre. The timbral distinctions between brass, woodwinds, strings, and percussion were heightened to the point that other recordings sounded as though all the timbres were overlaid by a common synthetic texture that obscured the instruments' true natures. I heard a vivid realism in timbres that conjured an instant impression of the instrument itself. Even an instrument like the triangle (in the climax of the last movement of Rachmaninoff's *Symphonic Dances,* for example) jumped from the soundstage with a clarity and realism I've never heard from recorded music; the impression of a piece of metal being struck, and then ringing, was startling. Throughout the listening there was the distinct feeling of knowing the mechanism by which an instrument made sound.

This was true for every instrument or section. Basses sounded like big wooden cavities resonating, with the sound of the bows moving across the strings vivid and distinct. The rich darkness of the bass clarinet suggested the instrument's size, composition, and even the reed vibrating. Percussion instruments had such a steep attack that they seemed to jump instantly into existence. Cymbals didn't sound like trash-can lids or spray cans—and even at high levels didn't hurt my ears or make me want to turn down the volume. Moreover, the sound lacked the synthetic character overlaying instrumental timbres so prevalent in standard-resolution digital audio. There was no shiny glare on strings, no grain polluting woodwind textures, and no sense of hardness on brass.

Although the presentation was startling vivid in this way, another single word that describes the overall sound is "gentle." There was just this complete sense of ease and grace. The resolution I heard isn't what we've come to accept as "resolution" in standard-definition digital audio, but something very different. There is absolutely no hype, no etch, and no fireworks. Instead, you melt into the listening chair with an addictive combination of physical relaxation and mental exhilaration.

I'll use another single word to describe HRx: "atmosphere." I heard a nearly tangible sense of the venue's acoustic, of the exact location of each instrument in it, and of the precise spatial relationships between each instrument with the acoustic. The soundstage was exactly sculpted and defined, but not in the way that "edge enhancement" in video attempts to create a sharper picture. Rather, there was a spatial vividness that didn't call attention to itself as stylized or cartoonish. It was instead perfectly organic and natural in a way that one hears in live music.

Many of these qualities I've described are the result of the extraordinary resolution of very fine timbral and spatial details. The information about how an instrument makes sound, for example, is contained in the very finest micro-harmonic and micro-dynamic structures—the very components of the waveform that are simply not captured by standard-resolution digital audio. These waveform components are also the first to be destroyed by less-than-state-of-the-art circuits or signal-processing techniques. I've had the experience many times of listening to a live microphone feed, and then to the recording of that mike feed (analog and digital), and then to the LP or CD made from that recording. Each stage is a disappointing step down from the mike feed's resolution and clarity. Hearing HRx reminded me of listening to live microphone feeds.

Finally, HRx renders music's large-scale dynamics with a visceral realism that's physically thrilling. This ability to instantly and effortlessly swing from resolving the quietest ambient information at the back of the hall to the orchestra at full tilt is alone worth the price of admission. So much of music's expression is in dynamic shadings; to hear them reproduced with no constraints is such a liberating experience. Moreover, the system's complete lack of glare and hardness allowed me to set a higher playback level than usual, further enhancing the visceral thrill of full-scale orchestral climaxes.

High-Resolution Downloads—A Caveat

More and more high-resolution music is available from a variety of audiophile-oriented Web sites. Some of it sounds spectacular, but much of it offers only mediocre sound quality. The reason is that many "high-res" files, although delivered at 96kHz/24-bit, are actually standard-resolution in sound quality. In many cases the "high-res" file is simply a standard-resolution (44.1kHz/16-bit) signal upconverted to 96/24. Moreover, even if the file started life as a true high-res signal, subsequent signal processing can destroy the qualities we seek in high-resolution digital audio. If the file was ever converted to 44.1kHz, the damage has been done and nothing can return it to its former glory. In addition, some recordings from the 1980s through the late 1990s were recorded on professional digital recorders that recorded at a sampling rate of 48kHz. If that's the case, downloading a version at 96kHz offers no sonic advantage. The consumer has the right to know the "provenance" of the high-resolution file offered for sale. The best-sounding high-res files were recorded in high-resolution at the session, kept in their native format through subsequent signal processing, and delivered to the consumer with perfect bit-for-bit accuracy to the high-res file created during the recording session. Look for audiophile-oriented sites with a reputation for sound quality. HDTracks.com and AIXRecords.com are two examples of sites offering true high-resolution downloads.

A somewhat related issue is that high-resolution versions of recordings are usually subjected to the same detrimental signal processing used on the standard-resolution release. It's become standard practice for mastering engineers to severely compress a recording's dynamic range so that it sounds "louder" when played on the radio. Record producers feel the need to make their "product" more "competitive" when played alongside other recordings. The result of the "loudness wars" is, in my view, an unnecessary and tragic destruction of so much of music's expression. Highly compressed recordings sound uninvolving, flat, hard, and lack rhythmic drive.

Unfortunately, virtually all CD titles that have been mercilessly compressed are also compressed in the high-resolution versions. The high-res version will sound better than the CD for a variety of reasons, but won't deliver the full potential of which high-resolution digital audio is capable.

Turnkey Music Servers vs. PC-Based Servers

The music server described at the beginning of this chapter is what I'll call a "turnkey" system. That is, you take it out of the box and start loading your music into it (or have the music loaded into it for you—more on this later). The turnkey server requires absolutely no effort or expertise on the user's part.

The alternative is to create your own server from a personal computer. It's been said, only half-jokingly, that the turnkey music server is for the ultra-rich and the do-it-yourself (DIY) PC-based server is for the ultra-geeky. That's an exaggeration, but nonetheless highlights the fundamental differences between the two approaches—turnkey servers are expensive, and PC-based servers are technically challenging. Let's look at the spectrum of turnkey music servers first, and consider the DIY server later in this chapter.

The Turnkey Music Server

The turnkey music server is at its heart a personal computer, but that fact is transparent to the user. The familiar computer user interface is replaced by an extremely elegant control system that is very skillfully tailored to making the system perfectly suited for organizing a music library. In fact, much of what you're paying for in a turnkey system is the software development behind the system's ease of use. You don't need to worry about installing and configuring sound cards or music management software, going into the PC's operating system and making sure that all the settings are correct, and spending time on the phone with technical support trying to get the system to work correctly.

Most turnkey systems, such as the one shown in Fig.8-1, consist of a touchscreen display along with one other chassis containing the computer hardware hard-disc drives that store the music. The photo is of the Meridian Sooloos system, which we'll look at in depth as an example of a feature-rich music server. I've chosen Meridian Sooloos for this example because it is the state-of-the-art in servers, and because I've lived with one and am familiar with its operation.

The Meridian Sooloos system is based on a 17" touchscreen display that also contains the CD drive for ripping music to the hard drives, which are contained in a separate chassis or are integral to the touchscreen (depending on the model and drive capacity). When you turn on the system your entire CD library is displayed as album cover art alphabetically by artist (you can choose other criteria as the default, such as album release date). You can manually scroll through the covers with back and forth arrows (18 albums per "page"), or jump to a specific artist's name by touching the letters on an alphabetic display. Touching an album cover brings up a screen showing information about the record, including the tracks. You can select specific tracks to add to the playlist, or add the entire album. You can add as many tracks or albums to the playlist as you like.

Fig. 8-1 A turnkey touchscreen music server. (Courtesy Meridian Sooloos)

This method of selecting music to play is far more convenient than finding a CD on a shelf, taking it out of the case, and putting it in a CD player. But that's not Meridian Sooloos' *raison d'être*. Most of us have music libraries that are far too big to scroll through, and we often approach a listening session not knowing specifically what we want to hear. If that's the case, you press the Focus icon on the home page to bring up the Focus page. The Focus page allows you to narrow down the range of displayed albums according to several criteria: genre, credits (who plays on the record), label, release date, and "mood"—or any combination of these. For example, I selected Genre from the Focus page which brought up a list of musical styles. From that list I selected Blues, and then from the 65 (!) sub-categories in Blues, chose Texas Blues. The main screen then displayed from the total of 2536 albums in the system just nine titles—five by Stevie Ray Vaughan, two by Clarence Gatemouth Brown, one by Lightnin' Hopkins, and one by the trio of Albert Collins, Robert Cray, and Johnny Copland. This is a perfect example of how the system allows you to instantly drill down into your music collection and offer you choices you might otherwise not have thought of.

You can Focus in multiple layers if the first Focus is too broad. For example, if you focus on Progressive Rock, you can then narrow down the range of titles by specifying only those albums released in the 1970s. Focus will also winnow your collection by showing you what you've played most recently, and over a selectable time window. Focus narrows down your music collection from several thousand titles to a more manageable dozen or so that you're interested in with just six finger-taps. The entire Texas Blues Focus I just described took about 15 seconds to execute, from the moment I thought about playing Texas Blues to the moment I started listening to Stevie Ray.

A simpler way to Focus is to touch the button "Focus on Albums Like This," which appears every time you've selected an album. If you're not sure exactly what you want to listen to, but have selected an album, this feature suggests records of a similar style.

Another way of finding the music you're looking for is through Sooloos' Search feature. You type in the name of an artist, album, or song, and the system displays all relevant albums. You can also explore your music collection by musician. For example, I was listening to Steps Ahead's first domestic album and was struck by Peter Erskine's terrific drumming. What other albums did he play on? A Search of the library, this time by album credit, turned up the other records featuring Erskine's drumming. Just those albums were displayed as cover art until I reset the Focus.

A feature called Swim plays randomly selected tracks. Swim on its own isn't a compelling feature—the opportunity for jarring juxtapositions is high. But Swim combines synergistically with Focus to create a stream of randomly selected music from within a genre or sub-genre that evokes a sense of surprise, serendipity, and discovery. Swim and Focus together bring out tracks and artists you might not have listened to for a long time. (When importing a CD, you can flag favorite tracks so that they are more likely to be played in Swim.)

Albums culled through Focus can be stored as a Collection. A Collection is a group of albums to which you can add or from which you can remove specific titles. A good use for a Collection is if different family members, each of whom might have diverse musical tastes, access the system. Each member would have a Collection, so that he or she sees only those titles of interest. In other words, your teenage daughter's Britney Spears albums will be invisible to you, and she'll never see your Billy Holiday records. You can, however, introduce your daughter to Billy Holiday by transferring an album from your Collection to hers. Separate Collections can not only be a way of preserving different family members' tastes, but also of sharing and suggesting music.

All this sounds a bit tedious when explained in words, but it's actually incredibly simple, intuitive, easy to learn, and fun. More important, the ability to dig down and instantly find exactly the music you want to hear (or think you *might* want to hear) completely transforms the way you interact with your music library.

The system requires an always-on Internet connection. When you insert a CD into the Control's slot to import to the hard drive, Meridian Sooloos downloads the album art and meta-data about that disc from the Internet (specifically, from All Media Guide). In addition to the usual meta-data (album credits, playing times, track titles, release date, etc.), Meridian Sooloos downloads brief reviews as well. Being an Internet-enabled device allows Meridian Sooloos to download updated software and add features for existing users.

You have several options for getting music into Meridian Sooloos. The first, obviously, is to manually load each CD in your collection into Meridian Sooloos' transport. At about eight minutes per disc, importing your music library soon becomes tedious. Another option is to send your CDs to Meridian Sooloos and they will (for a fee) load your library into the system before they ship it to you. A third option is to specify at the time a purchase a list of CDs you want loaded. Meridian Sooloos will buy the discs, import them, and ship the CDs along with the system. Finally, you can purchase blocks of music already loaded on the server, such as a package of essential rock records or the entire output from a specific record label over a specific time frame (all 1960s Prestige titles, for example). Whatever your solution, getting lots of music into the server is paramount—the more music loaded the better the experience and the greater the benefits of Sooloos' sophisticated search and narrowing features.

Meridian Sooloos supports FLAC, WAV, MP3, AAC and Apple Lossless files at any sampling rate up to 96kHz and any word length up to 24 bits. Music can be streamed to computers throughout your home, or even to a remote device accessed through a Web browser. You can also transfer files from Meridian Sooloos to portable devices such as an iPod, iPhone, or iPad.

Meridian Sooloos is an example of what is possible in music servers. Other turnkey systems may incorporate some of these same features.

Another Type of Turnkey Music Server

In the previous chapter I described the PS Audio Perfect Wave Transport, a new breed of component that bridges the gap between a disc player and a music server. The Perfect Wave Transport is an Internet-connected CD and DVD transport that will output an audio stream just like a conventional CD transport, but that also offers a host of additional features. When a CD is loaded into the Perfect Wave, the album art, song titles, and track times are displayed on a color touchscreen (Fig.8-2). You can play the CD through an outboard DAC, or rip the CD to a network-attached storage (NAS) drive connected to the Perfect Wave. Your music files stored on the NAS can be accessed through the Perfect Wave's front-panel touchscreen, or via an iPhone, iTouch, iPad, or similar device. You can create your own playlists, disc compilations, and other organizational features through the remote-controlling device. A companion piece, called The Bridge, connects the Perfect Wave Transport to your network so that you can stream music to other devices.

Fig. 8-2
Another example of a touchscreen music server
(Courtesy PS Audio)

To recap the hardware aspects of the Perfect Wave, the device incorporates a CD-ROM transport mechanism that will re-read the disc multiple times if necessary until the data on the disc are retrieved with no bit errors. It can accomplish this feat because the data from the disc are read into a 64MB buffer memory that can take in data in chunks, and then clock the data out with high precision. This technique isolates the transport's digital-output signal from the disc playback mechanism for better sound. The transport is compatible with the DVD format, meaning it can play DVD-Audio discs or WAV files (up to 192kHz/24-bit) stored on a DVD.

The transport has a variety of digital connections in addition to S/PDIF, including I^2S (on an HDMI connector), which allows it to connect to the companion Perfect Wave DAC with the potential for lower interface-induced jitter. The device's operating software can be updated via a rear-panel slot for an SD memory card.

Other examples of turnkey servers include the Linn DS Series of products, the Olive line, and the Squeezebox. The Linn DS (Digital Stream) devices incorporate in one chassis a computer interface for downloading music, along with a distribution system for streaming music over a home network. The user interface can be a computer, an ultra-mobile PC, or a PDA, all of which show you album art or text-based lists.

Turnkey Server Considerations

When choosing a server, take into consideration the amount of disc-drive capacity supplied with the machine, along with the cost of adding drives. Some servers allow you to add any relatively inexpensive off-the-shelf NAS (Network Attached Storage) drive; others force you to buy expensive proprietary hardware from the server manufacturer.

You should consider buying an outboard digital-to-analog converter for your turnkey server. The DACs and analog output circuits in many turnkey servers are of lower quality than those in outboard DACs. This is particularly true of turnkey servers that are created as "lifestyle" products that will largely be used for whole-house background music.

As servers have gained in popularity, high-end audio companies, which have traditionally been in the business of making amplifiers, disc players, and DACs, are increasingly offering music servers of their own. But creating a music server is a daunting task, even for a company that's highly computer-savvy. This company might not have the resources to invest in developing a great user interface, instead coming to market with something of marginal utility. There are examples of traditional high-end companies developing great servers, but be aware that bringing to market a music server requires a completely different skill set than creating, for example, a preamplifier.

Conversely, some music servers have been developed as "lifestyle" products aimed at the casual music listener. These products tend to have outstanding user interfaces, but often have poor sound quality. Their manufacturers assume that the sound quality is "good enough" and that their customers will choose a server based on features and interface, not sonic performance.

Using an iPod as a Music Server

The ubiquitous and amazing iPod (I own three, and use them regularly) was built with a feature that allows it to be turned into a good-sounding music server. That feature is a digital-output that taps into the digital datastream. When this datastream is fed to a high-quality DAC, the sound quality is determined by the DAC, not by the iPod, provided that the music is stored on the iPod in a lossless format such as WAV or Apple Lossless.

You'll need an iPod dock that can access the iPod's digital output (Fig.8-3), a digital cable, and an outboard DAC. This solution allows you to access your music through the iPod user interface. Using an iPod as a music server is less convenient than a turnkey or PC-based server, but much cheaper, particularly if you already own an iPod and outboard DAC.

Fig. 8-3 An iPod can function as a music server with the addition of a docking station that taps into the digital bitstream for decoding by a high-quality outboard DAC. (Courtesy Wadia Digital)

Internet Radio and Music Servers

Some music servers (and products such as integrated amplifiers) provide integral Internet-radio capability. Internet radio, also called Webcasting or streaming radio, delivers radio stations from around the world to your computer via the Internet. The audio is coded with a lossy system such as MP3 or WMA and streamed to your computer rather than downloaded as a file. Internet radio gives you access to virtually any radio station in the world.

A related technology is the Pandora service that tailors the music you receive to match your musical tastes. For example, you can type in the name of an artist or musical genre, and Pandora will deliver music with similar characteristics to the artist or genre you selected. Pandora is uncannily accurate, in my experience. This ability is made possible by the Music Genome Project on which Pandora is based. Here's Pandora's description of the Music Genome Project, taken from the Pandora Web site:

"We believe that each individual has a unique relationship with music—no one else has tastes exactly like yours. So delivering a great radio experience to each and every listener requires an incredibly broad and deep understanding of music. That's why Pandora is based on the Music Genome Project, the most sophisticated taxonomy of musical information ever collected. It represents over eight years of analysis by our trained team of musicologists, and spans everything from this past Tuesday's new releases all the way back to the Renaissance and Classical music.

"Each song in the Music Genome Project is analyzed using up to 400 distinct musical characteristics by a trained music analyst. These attributes capture not only the musical identity of a song, but also the many significant qualities that are relevant to understanding the musical preferences of listeners. The typical music analyst working on the Music Genome Project has a four-year degree in music theory, composition, or performance, has passed through a selective screening process and has completed intensive training in the Music Genome's rigorous and precise methodology. To qualify for the work, analysts must have a firm grounding in music theory, including familiarity with a wide range of styles and sounds. All analysis is done on location.

"The Music Genome Project's database is built using a methodology that includes the use of precisely defined terminology, a consistent frame of reference, redundant analysis, and ongoing quality control to ensure that data integrity remains reliably high. Pandora does not use machine-listening or other forms of automated data extraction.

"The Music Genome Project is updated on a continual basis with the latest releases, emerging artists, and an ever-deepening collection of catalogue titles.

"By utilizing the wealth of musicological information stored in the Music Genome Project, Pandora recognizes and responds to each individual's tastes. The result is a much more personalized radio experience—stations that play music you'll love—and nothing else."

Some turnkey music servers offer a built-in interface for Rhapsody, a streaming music service that offers access to a huge library of music (more than 8 million songs at press time) for a monthly fee. Rhapsody also offers downloadable MP3 files for sale.

File Formats

All music-management software for DIY music servers, as well as many of the turnkey systems, offer you the choice of storing your music in one of many file formats. Your choice of file format is important on several levels; it affects sound quality, the amount of disc-storage space required for a given amount of music, and future compatibility and flexibility.

File formats can be divided into three categories: uncompressed, lossless compression, and lossy compression. The first category consists of two commonly used file formats called WAV and AIFF. A WAV file is an unadulterated bitstream that is identical to the source data. That is, a WAV file created from ripping a CD contains the identical PCM data that was recorded on the CD. Consequently, WAV files consume the most disc space for a given playing time—10.5MB per stereo minute for CD-quality audio (44.1kHz/16-bit). The alternative uncompressed file format is AIFF, or Audio Interchange File Format, first developed by Apple

Computer. The advantage to WAV and AIFF files is that they are "universal currency," in that they can be converted to another file format in the future with no loss of quality.

You can cut nearly in half the amount of disc capacity consumed by a music file without sacrificing sound quality simply by using a lossless compression format such as Free Lossless Audio Codec (FLAC). FLAC produces perfect bit-for-bit accuracy to the original data and is sonically transparent. That is, a file converted to FLAC consumes about 50% to 60% of the space of a WAV file, but when "uncompressed" results in a bitstream that is indistinguishable from the original data. Other lossless compression formats are Apple Lossless Audio Codec (ALAC) and Windows Media Audio (WMA) Lossless. FLAC is more likely to be found on Windows-based servers, and ALAC on Apple-based servers. You can rip CDs to a Macintosh in FLAC and play them back, but the process requires additional third-party software.

The third category of file formats, the lossy compression schemes, are anathema to high-end values and should be avoided. These systems reduce the number of bits in the datastream by throwing away musical information. Unlike the uncompressed and lossless compression formats that provide a perfect bit-for-bit-accurate reconstruction of the original datastream, lossy compression formats permanently remove information that cannot be recovered. Examples of lossy compression format are MP3, WMA, and the Dolby Digital format used to encode film soundtracks. These formats can operate at a variety of bit rates, with the higher bit rates delivering better sound quality than the lower bit rates. Nonetheless, with the cost of storage so low, there's no reason to use a lossy compression format. Note that iTunes delivers music using a lossy compression format.

The Do-it-Yourself PC-Based Music Server

A PC-based server can be built on a laptop, notebook, or desktop machine running under the Windows or Macintosh operating systems. Just about any recent-vintage personal computer can be turned into a music server. While a turnkey server can cost five figures, a do-it-yourself system can be had for a few hundred dollars. The disadvantage is that the DIY system lacks the turnkey system's elegant user interface and requires some technical skill (and patience) to get the whole thing to work. Some audiophiles, however, see building a music server as an ongoing hobby.

The main components of a DIY server are the PC, a large-capacity hard drive (internal or external), CD/DVD drive, input/output connections (either integral to the computer or via add-on boards), and music-management software. We'll cover the fundamentals of each of these components conceptually. This section is meant as a general guide to DIY music servers, not as specific instructions on how to build one. There are too many variables that will determine the server's configuration and intended use. Moreover, specific hardware and software recommendations would quickly become dated given the rapid advances in computer technology.

The Importance of Bit-Transparency

You need a fairly high degree of computer knowledge to get a DIY server to work and to sound better than a CD source. That's because the hardware and software must be configured perfectly for the server to be "bit transparent." This term describes the ability of a device to pass a bitstream from input to output with no change in the data. If the data are changed, the sound will change as well, and not for the better. Data corruption can occur in hardware or software, making it difficult to track down. It's fairly common for DIY servers to degrade sound quality in a way that is hard to troubleshoot. One useful tool for determining whether a server is bit transparent is to connect its digital output to a digital-to-analog converter that is HDCD compatible. Rip an HDCD-encoded CD to the server, play it back, and see if the DAC's HDCD indicator LED illuminates. If it does, the server is bit transparent. The least significant bit in an HDCD-encoded bitstream includes a flag that tells the decoder that this is an encoded bitstream. If the data are corrupted, that flag is lost and the HDCD light doesn't illuminate.

The music-management software is frequently a source of data corruption. The software is often loaded with obscure settings that you must check off correctly to avoid having the software corrupt the data. Adding to the confusion, software updates of the music-management system can introduce data corruption *after* you've assured that the system is bit-transparent. Any time you change any hardware or install a newer version of the software, you should re-check that the system is still bit-transparent.

Computer Requirements

Although CPU power is important in many computing applications, it's not a factor in a PC-based server. Today's CPUs provide more than enough horsepower to run a server. The computer must, however, be able to support the music-management software you plan to run. The computer should also be able to support a large amount of RAM—the more the better. A large amount of RAM reduces the frequency at which the CPU accesses the disc drive rather than reading the information from RAM, making the system run more smoothly. (I should point out that there's a fringe view among the server techno-geeks that the best sound quality is realized by having just enough RAM to avoid glitches, and no more.)

If the computer is in the same room as the loudspeakers, fan noise is a major consideration. The constant background sound produced by computer fans is a significant distraction, particularly with classical music recorded in high-resolution. The very low noise floor of high-res digital can be ruined by fan noise. The solutions are to put the computer in another room, or build a "hush box" around the computer to damp the noise, or buy a fan-less PC. The fan-less PC uses cooling structures on the chassis to draw heat away from internal components. My own PC-based server is fan-less.. It is also drive-less, using 64GB of solid-state memory in place of a hard drive, making the server perfectly silent because it has no moving parts. (I use this server for 176.4kHz/24-bit playback, and a Meridian Sooloos server for everyday music listening.) Hard-drives produce their own noise, either from the spinning discs or from their own internal fans (in some external drives).

Operating System and Playback Software

The optimum operating system will depend on your choice of audio-interface hardware, music-playback software, and computer platform. For example, most USB DACs rely solely on the operating system's built-in audio capabilities. With this class of audio hardware, Windows Vista, Windows 7, and Mac OSX will arbitrarily re-sample your music to the highest sample rate supported by the USB DAC. These operating systems deliver bit-transparent playback only if you are willing to manually re-configure their audio settings every time that you play a file with a different sample rate.

This awkward step can be circumvented on Windows-based machines by choosing an audio interface that is supplied with hardware-specific drivers that enable your music playback program to bypass the operating system's limitations. Note, however, that some music-playback programs, most notably Apple's ubiquitous iTunes, lack the essential functionality to access these dedicated hardware drivers. In order to ensure that you can take full advantage of the capabilities of your audio interface hardware, look for playback programs that support either ASIO or Kernel Streaming such as MediaMonkey, J. River Media Center, and Foobar 2000.

The only currently available work-around for the sample-rate-switching problem on the Macintosh platform is to purchase a software program called Amarra, which essentially duplicates the hardware-driver functionality provided for free with the professional-grade soundcards and USB or FireWire audio interfaces used on the PC platform.

These scenarios are likely to change as technology progresses. When selecting the primary components that comprise a DIY music server, it is imperative to consider how these elements ultimately will work together.

Hard-Disc–Drive Storage

The DIY server can use the computer's internal disc drive for storing music, or rely on one or more external drives. How much music storage capacity you need depends on the amount of music you plan to store, the resolution of the files, and the file format selected. We'll assume that you are not willing to sacrifice sound quality for greater storage capacity, and will store the files as WAV, AIFF, FLAC, or Apple Lossless. There's no reason to use inferior-sounding formats such as WMA or MP3 because storage is so inexpensive (and getting cheaper all the time).

The WAV and AIFF formats store the original PCM datastream with no compression or algorithms to reduce the storage space required. WAV files consume 10.5MB per stereo minute (630MB per hour) for CD-quality audio (44.1kHz/16-bit). That means, for example, that a 500GB disc drive can store about 800 hours of music. High-res WAV files encoded at 96kHz/24-bit consume about 35MB per stereo minute (2GB per hour), reducing the disc's capacity to 250 hours. The highest resolution files reduce this capacity further; 192kHz/24-bit files consume 70MB per stereo minute (125 hours of storage on the 500GB disc). In short, the more music you want to store, and the higher the resolution of that music, the greater the disc capacity you'll need. When deciding on disc capacity, allow plenty of space for expanding your music library. It's amazing how fast disc space is consumed. Finally, check to see if the drive's total capacity is provided by a single drive inside the chassis or if that capacity is split between multiple drives. The single-drive option is preferable because there is less chance for drive failure. One factor with a very large drive you should be aware of is that the operating system may take several seconds longer (compared with a smaller drive) to catalog and display your entire music library if the drive has been in "sleep" mode. This could lead to erratic performance.

When allocating your music-server budget to a hard drive, consider that you should have at least one backup drive in case the main drive crashes. It's a fact of life that hard drives crash and can't be resurrected. Your music library on a hard drive represents a considerable amount of work; protect that investment of time by backing up your library on a redundant external drive. It's cheap insurance.

The disc's rotation speed and amount of on-board buffer memory are also important factors. While faster hard drives make sense for general computer operations, the latest "green" hard drives that spin at 5400 or 5900 RPM are a better choice than the faster 7200 RPM discs for a music server; the data transfer rate from a 5400 RPM disc is fast enough for high-resolution stereo audio, and the slower disc speed allows the drive to run quieter and cooler. Slower drives are also less expensive. Drives with smaller platters (2.5" as opposed to those with 3.5" platters) tend to be quieter, all other factors being equal. All disc drives have integral memory that acts as a temporary storage area between the CPU and the drive. A larger data buffer allows the system to operate faster and also perform more thorough error-correction.

A fundamental choice is whether to use a drive inside your computer or one or more external drives. The advantage of an internal drive is one less interface between the computer and drive, resulting in faster operation. On the other hand, if you change computers with an internal drive, you'll be faced with the challenge of moving your music library from the internal drive to the new computer. If you stored your library on an external drive, you simply move the external drive to the new computer. External drives also have the advantage of being isolated from the heat produced inside a PC; exposure to high temperatures increases the potential for drive failure. One practical solution is to use an internal drive for your music library, along with an external USB drive as a back-up. You then have the best of both worlds—the faster access of an internal drive along with the portability of your library with the external drive. The newer eSATA protocol now allows external drives to achieve transfer rates equivalent to internal drives.

A related issue is where to store the computer's operating system. Ideally, the computer's OS will reside on a solid-state drive (SSD), with the music library stored on a high-capacity hard drive. The solid-state memory is lightning fast; my PC-based server boots up in a few seconds from its SSD.

You can use standard off-the-shelf drives or buy drives specially selected and modified for music storage. A number of companies supply server-specific drives that are relatively easy to install and configure. These drives are significantly more expensive than generic hard drives, but include features such as built-in wireless capability (so that you can house the drives remotely), integral software and hardware that allows you to stream music from the drive to a networked music server such as the Logitech Squeezebox or the Sonos system without turning on your computer, a CD drive that allows you to rip CDs directly to the drive (a so-called "ripping NAS"), high-quality mechanisms built to withstand the rigors of a music server, and tech support that's highly attuned to the needs of DIY server users. Fig.8-4 is an example of such a drive.

Fig. 8-4 A ripping NAS offers many more features and capabilities than a standard hard-disc drive. (Courtesy Sound Science)

Two terms related to hard drives you should know are Network Attached Storage (NAS) and Direct Attached Storage (DAS). A NAS drive is more sophisticated than a standard hard drive you would buy for backing up, for example, your word-processing files. The NAS drive is a computer in its own right (although of limited capacity) that incorporates an operating system and software that are tailored to managing and supplying files stored on the drive to other devices connected to the network (your PC server as well as other computers, devices, or drives on the network). The server-specific drive described in the previous paragraph is an example of a NAS drive (although not all NAS drives have the extensive feature-set of the given example).

The DAS drive is much simpler. It lacks the networking capability of the NAS, instead being controlled (data reading and writing) by a single source, the host computer. An off-the shelf drive connected to your computer via USB, ATA, SATA, SCSI, or similar interfaces is a DAS drive. NAS drives are better for music servers because of their ability to be connected through a router to other devices and share data or storage capacity with those devices.

If you opt for a ripping NAS, be aware that most don't offer the ability to verify that the ripped file is identical to the source file. Moreover, some ripping NAS's may introduce bit errors in the ripped file. The best way to assure that you've achieved a bit-accurate rip is to use a program such as Exact Audio Copy, dBpowerAmp (for Windows) or Max (for Mac OSX). Many novice users are unaware that the "use error correction" feature in iTunes merely activates one rudimentary error-checking feature and doesn't guarantee accurate rips. Ripping software is described in more detail later in this chapter.

If you don't need huge capacity, you might consider a storing your library on the solid-state drive (SSD) described earlier. One advantage of an SSD is that it has no moving parts and is thus offers perfectly silent operation. It's also blazingly fast; your computer will boot up virtually instantly from an SSD, and accessing music is instantaneous. The downsides are limited capacity and high cost. However, when used in conjunction with a fan-less computer, the SSD enables your server to have no moving parts whatsoever. Since most SSDs incorporate a standard SATA interface, they can also be installed in inexpensive external drive enclosures that connect to the computer via USB, FireWire, or eSATA.

Loading Your Server with Music

There are generally three ways of getting music into the server: by "ripping" a CD (transferring music from the CD to the computer's hard drive), buying music downloads from an Internet music site, or paying a service to rip your music library to disc.

If you choose to rip your own library, you can use the standard ripping software that comes with your computer's operating system or music-management software, or opt for specialized software such as Exact Audio Copy (EAC) or dB PowerAmp (for Windows), or Max (for Macintosh OSX). These programs "grab" data off the CD and convert the music into a variety of file formats suitable for disc storage. They will also transcode files into virtually any format, such as MP3 or FLAC. Their major advantage over conventional ripping software is their ability to insure verifiable, accurate rips. First, they can detect errors in the datastream and re-read those error-prone sections of the CD multiple times until an error-free bitstream is obtained. They can even deliver a perfect bitstream from damaged CDs, although the process might take several hours. Perhaps even more significantly, they can provide you with confirmation of rip accuracy. This may be accomplished through several methods. The most common is for the program to rip the music twice (automatically deleting the "test" rip when finished), and then compare the two rips; if they are identical, then they are probably an exact representation of the data on the CD. Another technique incorporates a feature called AccurateRip that compares your CD rip to an Internet database of other rips of the same disc to check the accuracy of your rip.

Downloading music from a legitimate Internet music site is a great way of acquiring music and expanding your music library. Be aware, however, that the quality varies considerably between sites. Some offer music encoded with one of the lossy compression formats, resulting in grossly inferior sound quality. Other sites advertise "CD quality" downloads, but you should verify that claim before spending your money. The term "CD quality" has been misused to mean everything from low-bit-rate MP3 files all the way up to true CD quality (44.1kHz/16-bit with no compression). True CD quality is the minimum standard you should accept. Even better are high-resolution downloads that deliver better-than-CD quality for a small incremental price increase. Some sites offer real 96kHz/24-bit downloads sourced from high-resolution master recordings. When played back on a bit-transparent server and decoded through a high-quality DAC, these files can sound spectacular—and far better than CD.

Some Internet music sites advertise "high-resolution" and "96/24" quality but the files they supply are nothing more than standard-resolution files that have been upconverted to 96kHz/24-bit. These files will consume more than three-times the disc space of 44.1kHz/16-bit files and cost more to buy, but offer no improvement in sound quality over standard-resolution files. Once a digital audio signal has been subjected to 44.1kHz/16-bit encoding, the damage has been done. Choose a download site that has high audiophile standards. Some of them wouldn't think of offering anything less than the best sound quality possible.

Finally, you can pay an outside company to rip your music library. Some server companies offer this service for an additional fee when you buy their server. You simply order the server, ship your CDs to the company, and they return your CDs along with the music on those CDs ripped to the server's drive. Independent companies will also rip your CDs to a hard disc, DVDs, or to a portable music player (the latter is not recommended). You must take the CDs out of the cases and load them onto spindles supplied by the company. Be sure that the ripping service offers ripping to an uncompressed or lossless compression format. While you might be inclined to choose uncompressed WAV or AIFF files because they are a universal standard that can be later converted into any other file format, neither the WAV nor AIFF format incorporates a universally compatible tagging structure for storing the music's metadata. For this reason, it may be more prudent to choose one of the lossless formats such as FLAC, WMA Lossless, or Apple Lossless that provide robust, widely-supported, and easily converted metadata tags. Remember, however, that these ripping services are based on fast turnover and may not provide documented evidence of the accuracy of their rips, such as that provided by a program such as Exact Audio Copy or dBpoweramp.

An important issue with any music server, whether turnkey or do-it-yourself, is the accuracy of the "tags" applied to the ripped files. A tag is a package of non-audio data attached to the audio file that identifies the artist, the track, the playing time, and other information. When you rip a CD to iTunes, for example, iTunes automatically downloads these tags (also called "metadata") from a service called Gracenote. Other on-line tagging services are available such as All Media Guide (AMG) and FreeDB. AMG (as used by Meridian Sooloos, Windows Media Audio, and others) is widely regarded as the most accurate tagging service, and also has the highest accuracy and greatest depth of information. These services are generally free to consumers; the companies generate revenue by licensing their services to consumer-electronics manufacturers. When ripping a CD, the service compares the table-of-contents of the CD being ripped to its database of tens of millions of CD titles.

It's impossible to overstate the importance of tagging your library accurately and organizing your music in a logical and consistent way. Your library should be tagged with the same conventions, such as the artist's last name, followed by the first name. The various tagging services follow different conventions, resulting in inconsistent organization of your music. Moreover, the tags applied to a file are sometimes incorrect, or contain small errors. You must therefore monitor the tags and manually edit the tags for consistency and to correct mistakes. The PerfectMeta feature from dBPowerAmp gathers metadata from multiple on-line databases (including AMG) and intelligently compares the results to create a tag that is most likely to be correct.

Accurately tagging classical music is a challenge because the industry-standard tagging structure is designed for popular music with an artist, album, and song rather than the composer, conductor, orchestra or ensemble, and soloist structure. You must decide in advance the convention you would like to use, making sure that the convention selected is transportable between different music-playback programs. Alternately, you can use extended non-standard tags, or use a music-playback program's proprietary database structure to store the additional information. The danger of creating a large library of tags in this way is that if you switch to a different music-playback program in the future, the new program won't be able to read the tags.

Getting Music from the Server to Your Playback System

Now that we've covered getting music into the server, let's look at options for getting it out and to your hi-fi system for listening. The simplest method is to connect a USB (Universal Serial Bus) cable between the computer's USB port and a USB-capable DAC. (USB is covered in detail later in this chapter.) The USB DAC's analog outputs feed one of your preamplifier inputs. If you own a DAC that lacks a USB input, you can buy a USB-to-S/PDIF format converter (Fig.8-5). This small, inexpensive device takes in USB from the computer and converts it to S/PDIF on an RCA jack. Even if you own a USB DAC, you might consider a USB-to-S/PDIF converter because many USB-capable DACs sound better when fed an S/PDIF signal. Some computers offer TosLink optical or FireWire digital outputs.

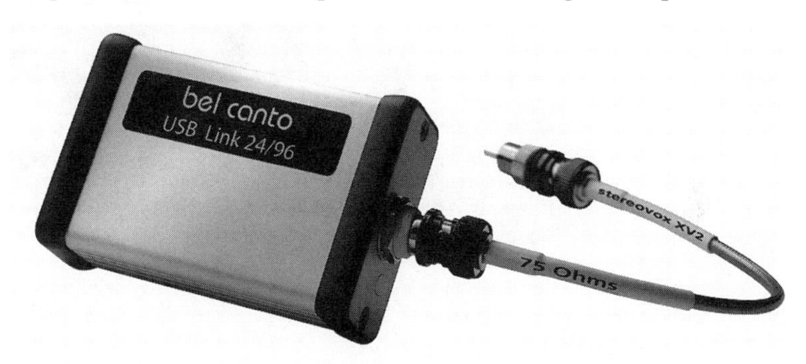

Fig. 8-5 A USB-to-S/PDIF converter. (Courtesy Bel Canto Design)

A more elaborate, but often better-sounding route is to install a sound card inside the PC. The soundcard's S/PDIF output on an RCA jack, or AES/EBU output on an XLR connector, feeds your external DAC. Some soundcards, such as the Lynx AES16, have a "D" connector on the rear panel to which you connect a breakout cable with an XLR connector at the other end. These soundcards replace the digital output that's built into the computer, which is usually USB but sometimes TosLink optical. Replacing the computer's stock digital output with a specialized soundcard usually results in better sound because of the specialized card's superior output circuitry, high-precision clocking, and low-jitter output.

Other soundcards incorporate digital-to-analog converters, outputting analog audio on RCA jacks. These cards don't require an outboard DAC, but are limited in sound quality. The inside of a PC is a hostile environment for analog signals; noise can contaminate the analog output. Moreover, the soundcard offers the designer very little circuit-board space with which to work, as well as limiting power-supply sophistication. Although some soundcards with integral DACs are designed for music playback (rather than just gaming or general-purpose computing), they do not provide the ultimate in sonic performance.

The third option is to stream music wirelessly from your PC to your DAC. You'll need WiFi capabilities in your PC, along with a WiFi base station to convert the WiFi signal to S/PDIF (either TosLink optical or coaxial). WiFi is subject to signal dropouts, is limited in distance (25 feet is generally considered the maximum), and is affected by your home's construction.

The USB Interface

The Universal Serial Bus (USB) was originally designed to connect computer peripherals to a PC, replacing the serial and parallel ports on earlier-generation machines. The popularity of PC-based music servers has caused the high-end audio industry to adopt the interface for digital audio, specifically to connect the output of a PC-based music server to an outboard DAC. Most DACs today offer USB compatibility. An audio-grade USB cable is shown in Fig.8-6.

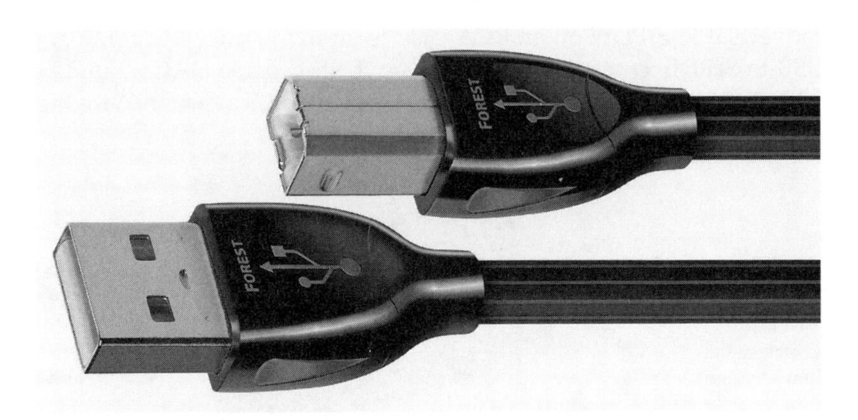

Fig. 8-6 An audio-grade USB cable. (Courtesy AudioQuest)

Connecting a PC to a USB DAC is "Plug and Play"; you simply connect the USB cable and the PC (the "host" in USB parlance) recognizes the DAC (the "device"). The alternative to the USB interface for PC-based music servers is to install a card inside the PC that presents the digital audio bitstream on an AES/EBU cable. Obviously, USB is a much simpler and less expensive proposition.

The DACs with USB inputs available as of this writing conform to the USB 2.0 protocol. Although USB 2.0 has a bandwidth of 480Mbps (far faster than any high-resolution digital audio signal), most of today's USB DACs cannot transmit digital audio with a sampling rate greater than 96kHz. That's not a limitation of USB but rather of the chipsets available to high-end DAC manufacturers. This situation is changing, however, as newer USB devices are now being produced that can accommodate all common sample rates up to 192kHz

In most digital interfaces, including S/PDIF and its variant AES/EBU, the source component (the CD transport or PC music server, for examples) is the master clock to which the receiving device must lock. Virtually all USB DACs operate in this way, which is known as "Adaptive Mode." Several high-end manufacturers have developed proprietary USB interfaces in which the DAC's critical clock runs independently of the PC's clock. This technique, called "Asynchronous Mode," greatly reduces interface-induced jitter and results in better sound. "Asynchronous Mode" transforms the USB interface into a high-quality interface. Only a few manufacturers offer Asynchronous Mode USB because it requires writing custom software for the USB chipset along with considerable expertise.

A good example of Asynchronous Mode USB is the U-Clock from England's Data Conversion Systems (dCS). The U-Clock interfaces to a PC via Asynchronous USB, converts USB to S/PDIF, and presents that S/PDIF signal on an RCA jack for connection to a S/PDIF DAC. Even if your DAC has both S/PDIF and USB, the U-Clock's Asynchronous Mode operation, conversion to S/PDIF, and ultra-precise clocking will provide better sound via S/PDIF than if the DAC were connected directly to the PC via USB.

You can also buy very simple USB-to-S/PDIF converters for a fraction of the U-Clock's price, but many don't offer Asynchronous Mode or high-precision clocking. Nonetheless, connecting a DAC to a PC through such a devices generally results in better sound than simply plugging the USB DAC into the PC, unless the USB DAC already incorporates an advanced (usually Asynchronous) USB interface. Many professional audio companies manufacture USB and FireWire external audio interfaces that circumvent the hardware limitations through the use of custom hardware-specific drivers, as described earlier in this chapter.

Music-Management Software and User Interface

The final element of a music server is the music-management software. This software organizes your music files, provides the interface for importing and playing music, assembles playlists, converts file formats, and performs other useful functions.

The most popular and familiar music-management software is Apple's iTunes, which is free, easy to learn, and reliable. Moreover, if you own an iPod and already use iTunes, you won't have to go through the learning curve required by different software. The downside is that iTunes is structured around artists and songs rather than composers, orchestras, and conductors, making it less than ideal if your music collection is largely classical. ITunes has another drawback for the serious music enthusiast; it is not built for handling high-resolution files. It will often down-convert high-res to standard-res, or corrupt the high-res datastream.

An alternative to iTunes (and to Windows Media Audio that is part of Windows) is third-party music-management software. Popular choices for Windows-based servers are MediaMonkey and Foobar 2000, both of which are free (retailed versions with upgraded feature sets are also available). Macintosh users might opt for Amarra, a sophisticated software package that uses the iTunes user interface and music-management functions, but provides its own audio processing. Amarra also automatically makes the correct settings in iTunes for the resolution of a particular file, so that you don't have to. Amarra allows iTunes to handle high-res files without you having to set the sampling rate every time you play files at different rates.

All these applications rely on text-based lists to find and select the music you want to hear. You can, however, add an Apple iPhone, iTouch, iPad, or similar device to control your server.

Transferring an LP Collection to a Server

Once you have a music server up and running, it's inevitable that you'll want to digitize your LP collection (if you have one) so that you can access that music via the server. Although the prospect of subjecting analog sources to digital conversion is anathema to many audiophiles, it's nonetheless handy to have your entire music collection—LP and digital sources—on the server. I recommend, however, that you keep your LP collection and turntable so that you can continue to enjoy the sound of analog for those focused listening sessions.

Transferring your LPs to digital can be a fairly quick and simple process, or an involved and exacting procedure. As with everything else, how much effort you put into the transcription process will determine the quality of the result. Big-box retailers sell "USB turntables" for $99 that plug into your computer; your only contribution is a few mouse clicks. At the other end of the spectrum is an elaborate process of cleaning the record, fine-tuning the turntable and cartridge, buying a high-quality analog-to-digital converter, installing high-end LP-transcription software, and recording your LPs in high resolution. We'll cover the various methods and technologies, but keep in mind that you'll probably only transfer your LP collection once. A listener's LP library represents a lifetime of musical exploration; preserving as much of that legacy as possible should be a high priority.

I can't recommend highly enough the step of cleaning your LPs before converting them to digital files. If you don't have an LP cleaning machine, this is a good excuse to buy one. Not only will your LPs sound better for normal listening, but they will produce much cleaner-sounding versions on your server. Chapter 9 covers LP cleaning machines in detail.

The quality of the turntable, tonearm, cartridge, and phono preamplifier will be a significant factor in the final sound quality. If you have an inexpensive LP front-end and many treasured records, considering renting or borrowing a better 'table. Fine-tune the tracking force and vertical tracking angle (VTA) before beginning the transfer, and adjust turntable speed with a strobe (this is explained in Chapter 9). Phono cartridges have widely varying frequency responses, transient fidelity, soundstaging, and other characteristics. Keep in mind that these qualities will be imprinted on every LP you transfer. Similarly, the phono preamp's sonic signature will be overlaid on your digitized LP library. Be sure to set the phono preamp's cartridge loading optimally. If it is a tubed unit, consider replacing the tubes with specially selected low-noise versions before a transcription session. The phono preamp can be a stand-alone unit or part of a full-function preamplifier or integrated amplifier.

Some PC soundcards allow you to connect the output from your turntable directly to the soundcard. The soundcard amplifies the very low-level signal from the cartridge, and also performs RIAA equalization (RIAA equalization is explained in Chapters 4 and 9) via the software supplied with the soundcard. This obviates the need for a phono preamplifier, but is sonically inferior to a phono preamp. If you opt for the soundcard path, be sure that the soundcard has enough gain for your cartridge's output voltage. If the soundcard lacks adequate gain, the resulting digital files will be noisy and lacking in dynamics.

The phono preamp will output a line-level signal that's ready for conversion to digital. This analog-to-digital conversion can take place in a stand-alone analog-to-digital converter (ADC), the ADC that's built into digital recorders, or the ADC in a soundcard. Before choosing an ADC, you should decide on the resolution with which you want to encode your LPs. The lowest resolution to consider is a 44.1kHz sampling frequency with 16-bit word length. This is "CD quality" resolution. But with the ready availability of high-resolution ADCs and cheap storage, there's no reason not to digitize your LPs with the best possible quality. Keep in mind, however, that files made with 96kHz/24-bit resolution consume more than three times the disc space of 44.1kHz/16-bit files. Be sure the ADC you choose can support the maximum resolution you would like to use.

An important consideration in the ADC is the ability to adjust and monitor the signal level. If the signal level feeding the ADC is too high, musical peaks will sound distorted. If the signal level feeding the ADC is too low, you are throwing away resolution. For example, if the highest signal level is 12dB below the maximum allowable level without distortion, you are losing two bits of information. If you started with 16 bits, it's like having a 14-bit converter. Setting the level so that the highest peak on an LP uses nearly all the bits without overload is vital to making a great-sounding transcription.

You'll need software to manage the transfer from the ADC to the computer's disc drive. Again, be sure that the software is compatible with your computer and operating system, and will support high-sample rates if you choose. Virtually all software packages include equalization, hiss filtering, pop removal, and other signal processing. Although some of these can improve the sound of the transfer, these "improvements" often come at the price of reduced sonic transparency.

After you've transferred an LP side to a digital file, you'll need to divide that file into the individual songs and identify them. Some software provides a graphic interface that looks like an LP side, complete with visual bands between songs that allows you to easily find the song starting points. Other transfer software packages recognize the gaps between songs and automatically separate the songs into separate files.

Undoubtedly the most tedious—but essential—step in converting LPs to digital files is "tagging" those files with the names of the artist, album, and song. When creating your own digital music files from LP, you can create this metadata yourself by typing it out on a keyboard or by buying an additional piece of software that allows Gracenote (described earlier in this chapter) to identify the music and attach the tags. After the file has been tagged, your files transferred from LP will appear as just another title on your music server that's ready to play with a couple of fingertaps on a touchscreen or mouse clicks. The same problems noted earlier regarding the accuracy of metadata tags, and whether the library is organized in a logical and consistent manner, apply to tagging files transferred from LP.

9 Turntables, Tonearms, and Cartridges: The LP Playback System

Introduction

In 2009, sales of LP records more than doubled the figures for 2008. The vinyl record was the only packaged music format to report increased sales in 2009. Vinyl records outsold SACD and DVD-Audio combined in that year. Although records still account for only a small percentage of all music sold, the fact that this archaic technology is thriving in a world of MP3 players and Internet downloads is remarkable.

The resurging popularity of the vinyl record is a testament to its fundamental sound quality. Not only do LPs deliver a musical realism unmatched by standard-resolution digital audio, they also represent a completely different approach to music listening. At one extreme is the person who downloads MP3 files at the lowest possible bit rate, and listens to them in the background on a cheap portable player and $5 ear buds, while engaging in some other primary activity. Listening to an LP couldn't be more different. Placing an LP on a turntable signifies your intention to set aside all other activities and turn your full attention to the music.

The high-end audio industry has responded to vinyl's spectacular comeback with a flood of interesting new turntables, tonearms, cartridges, and phono preamplifiers, not to mention accessories for getting the most out of a record collection. Specialty audiophile labels have licensed classic recordings of the past and reissued them in magnificent sound. Today's LPs are mastered and manufactured with a level of care unimaginable 20 years ago. The LP was once a cheap commodity; now it's an art. The result is that we're now hearing, for the first time, the full measure of the LP format's greatness.

The renewed interest in vinyl playback is largely driven by two demographic groups. The first is young adults who value the LP's sound quality, tactile nature (including the cover art and liner notes), and "retro-cool" factor—qualities missing from MP3 downloads. The second group driving vinyl sales is the music lover who grew up listening to records, but abandoned them in favor of CD. After 25 years of listening to digital this kind of listener is now dusting off his records and discovering anew the musical pleasures and sonic quality possible with today's LP playback systems.

If you haven't heard a good LP playback system, I encourage you to visit a specialty retailer who can demonstrate the potential of vinyl, or attend a consumer-audio show where turntables are prevalent (more than a third of the 120 exhibit rooms at the 2009 Rocky Mountain Audio Fest featured LP playback).

The general public's perception that CD is vastly superior to LP (remember the "Perfect Sound Forever" marketing campaign for CD?) is perpetuated because very few listeners have heard high-quality LP playback. When done right, LP playback has an openness, transparency, dynamic expression, and musicality that cannot be matched by CD. There's just a fundamental musical rightness to a pure analog source (one that has never been digitized) that seems to better convey musical expression. There's a saying in high-end audio circles: CDs sound great—if you never listen to vinyl. That is, you don't know what you're missing until you hear good LP playback.

This isn't to say that LP playback is perfect. It suffers from a variety of distortions such as mistracking, ticks and pops, speed instability, surface noise, cartridge frequency-response variations, inner-groove distortion, wear, and susceptibility to damage. But for many listeners, these problems are less musically objectionable than the distortions imposed by digitally encoding and decoding an audio signal. Moreover, the best turntables, tonearms, and cartridges substantially reduce these annoyances while maximizing the strengths of LP playback. In fact, today's best LP playback systems are remarkable in their ability to make you forget that you are listening to anything but music.

Converting the binary ones and zeros of digital audio back into music has been likened to trying to turn hamburger into steak. Some listeners can hear past the LP's flaws and enjoy the medium's overall musicality. Other listeners can't stand the ritual of handling and cleaning records—not to mention keeping the turntable properly adjusted—and think CD is just fine. I think of it this way: LP's distortions are periodically apparent, but separate from the music; digital's distortions are woven into the music's fabric. Consequently, analog's distortions are easier to ignore. If you're inclined to think CD is without fault, and you've never heard a properly played LP, give yourself a treat and listen to what vinyl can do before you write off the possibility of owning a high-quality turntable.

LP Playback Hardware Overview

The long-playing (LP) record playback system is a combination of a turntable, tonearm, and phono cartridge that converts the mechanical information encoded on vinyl records into an electrical signal that can be amplified by the rest of your playback system. The turntable spins the record, the tonearm holds the cartridge in place, and the cartridge converts the wiggles in the record's groove into an electrical signal. Each of these elements, and how they interact with each other, plays a pivotal role in getting good sound from your system.

System Hierarchy: Why the LP Front End Is So Important

My suggestion later in this chapter that some audiophiles spend as much as 40% of their total system budget on an LP front end may strike some as excessive. After all, how much of an effect can the cartridge, tonearm, and especially the turntable have on the reproduced sound compared to the contributions of amplifiers and loudspeakers?

The answer is that the components at the front of the playback chain should be of at least as high quality as those at the end of the chain (loudspeakers). The LP playback system's job is to extract as much musical information from the LP grooves as possible, and with the greatest fidelity to the signal cut into the LP. *Any musical information not recovered at the front end of the playback system cannot be restored later in the chain.* It doesn't do any good to have superlative electronics and loudspeakers if you're feeding them a poor-quality, low-resolution signal from an inadequate LP front end. If the music isn't there at the start, it won't be there at the finish.

The importance of the turntable was first brought to the world's attention by Ivor Tiefenbrun of Linn in the early 1970s. He sold his turntable by walking into audio dealerships unannounced with the turntable under his arm, and asking the store owner to listen with him to the best system in the store with the store's turntable. Then they listened to the store's least expensive system hooked up to the Linn LP12 turntable. When the store owners heard their most modest system outperform their most expensive system, many became convinced of the turntable's importance. (When asked by the store owner "But how can I sell this to my customers?" Tiefenbrun responded "The same way I just sold it to *you*.") It took years of these kinds of demonstrations to convince the audio community that the turntable was a significant variable in a system's sound quality. Today, no debate exists over the turntable's importance in achieving good sound. (To date, more than 150,000 Linn LP12s have been sold worldwide.)

How to Choose an LP Playback System

Because there are many more variables to account for in LP playback, how much of your audio budget you should spend on this part of your system is a more complicated decision than setting a budget for, say, a power amplifier. Let's look at two hypothetical audiophiles, one of whom should spend much more on a turntable, tonearm, and cartridge than the other.

Our first audiophile has a huge record collection that represents a lifetime of collecting music. Her record collection is a treasure trove of intimately known music that she plays daily. Conversely, she has very few CDs, buying them only when her favorite music isn't available on vinyl. She much prefers the sound of LPs, and doesn't mind the greater effort required by LP playback: record and stylus cleaning, turning over the record, lack of random access.

The second audiophile's record assortment represents a small percentage of his music collection—most of his favorite music is on CD. His LP listening time is a fraction of the time he spends listening to CD. He likes the convenience of CD, and can happily live with the sound of his excellent disc player and music server.

The first audiophile should commit a significant portion of her overall system budget to a top-notch LP front end—perhaps 40%. The second will want to spend much less—say, 10 to 15%—and put the savings into the components he spends more time listening to.

In the previous edition of this book (2004), I recommended that readers invest immediately in a high-quality LP playback system on the assumption that fewer manufacturers would offer turntables, tonearms, and cartridges in the future. How wrong I was. Since that edition, we've seen an explosion of new LP playback gear at all price levels. Moreover, today's products deliver much better sound at a given price point because of economy-of-scale manufacturing and increased competition for the turntable buyer's dollar.

A decent entry-level turntable with integral tonearm runs about $300, with an appropriate cartridge adding from $60 to $150 to the price. A mid-level turntable and arm costs $800 to $1500; the cartridge price range for this level of turntable is from $200 to $700. Many systems in these price ranges are supplied with a tonearm and cartridge. These complete packages are ideal for the first-time turntable buyer; they take the guesswork out of matching the component parts, as well as the work involved in mounting the cartridge and aligning it in the tonearm. Fig.9-1 shows a "plug and play" entry-level turntable.

Fig. 9-1 Today's entry-level turntables are supplied with a tonearm and cartridge, and come calibrated out of the box. (Courtesy Pro-Ject and Sumiko)

There are roughly two quality and price levels above the $2000 mark. The first is occupied by a wide selection of turntables and arms costing between $3000 and $6000. At this price, you can achieve outstanding performance. Plan to spend at least $1000 for a phono cartridge appropriate for these turntables. A turntable/arm combination costing from $10,000 to $20,000 will buy you very nearly state-of-the-art performance (if you choose the right model). The sound quality of such a system (Fig.9-2) will be significantly better than that of a turntable/arm combination in the lower tiers, and comes surprisingly close to the ultra-exotic models that sometimes rely on "bling" for their appeal. Later in this chapter we'll consider the sonic differences between turntables, tonearms, and cartridges.

Fig. 9-2 A high-quality turntable and tonearm. (Courtesy Basis Audio)

The very highest price level is established by turntables costing between $25,000 and $150,000. A topflight phono cartridge can add as much as $10,000 to the price. These systems are characterized by extraordinary and elaborate construction techniques such as vacuum hold-down systems to keep the record in intimate contact with the platter, massive vibration-resistant bases, exceptional vibration isolation systems, lots of high-precision machining, and a high-gloss finish (Fig.9-3). Turntables at this price level often provide the option of using more than one tonearm. The second tonearm is often fitted with a cartridge of different sonic characteristics that is best suited to certain kinds of music. Or the second arm can hold a monophonic cartridge for those listeners who want to hear mono recordings without compromise. (Although you can play a mono record with a stereo cartridge, the best sound quality is achieved with a mono cartridge.) Many of these mega-buck turntables are sonically superlative, some turntables in the $30,000 to $50,000 range are equal (or even superior) to the costliest units in construction quality, finish elegance, reliability, and sound quality.

Once you've decided on an LP front-end budget, allocate a percentage of that budget to the turntable, tonearm, and cartridge. As I mentioned, most lower-priced turntables come with an arm and cartridge already fitted and included in the price. At the other end of the scale, the megabuck turntables also include a tonearm—the turntable/arm system is engineered as one piece. In the middle range, you should expect to spend about 50% of your budget on the turntable, 25% on a tonearm, and 25% on a cartridge. These aren't hard figures, but an approximation of how your LP front-end budget should be allocated. As usual, a local audio retailer with whom you've established a relationship is your best source of advice on assembling the best LP playback system for your budget.

An item of utmost importance in achieving good sound from your records is a good turntable stand or equipment rack, particularly if the turntable has a less than adequate suspension. (Turntable suspensions are described later in this chapter.) I cannot overstate how

Fig. 9-3 An upper-end turntable with a tangential-tracking air-beari-ng tonearm. (Courtesy Walker Audio)

vital a solid, vibration-resistant stand is in getting the most from your analog front end. Save some of your budget for a solid equipment rack or you're wasting your money on a good turntable, arm, and cartridge. Note, however, that the very best turntables have extraordinary mechanical systems to isolate the turntable from vibration, and thus don't benefit as much from a solid stand.

The importance of a good rack was underscored for me while I was reviewing a turntable isolation system. I moved the turntable (a model that lacked a suspension) from its usual place on a 350-pound sand-filled and spiked turntable stand (the Merrill Stable Table) to a generic wooden "stereo rack" to test the effectiveness of the turntable isolation system. The difference was like night and day—and far greater than you'd expect. I'll talk more about turntable stands later in this chapter, under "LP System Setup," and also in Chapter 15.

I'll also go into some detail about how turntables, tonearms, and cartridges are built, and what to look for when choosing each element of the LP front end. You can use this information to assemble your own turntable system, or just brush up on the subject so you can talk confidently with the dealer who puts together your LP playback system. Whichever route you choose, you'll still want to evaluate the turntable system's sonic merits by listening to it—at length.

What to Listen For

Judging the sonic and musical performance of an LP playback system is more difficult than evaluating any other component. If you want to audition a phono cartridge, for example, you cannot do so without also hearing the turntable and tonearm, as well as how the cartridge interacts with the rest of the LP playback system. The same situation applies to each of the elements that make up an analog front end; you can never hear them in isolation. Further, how those three components are set up greatly affects the overall performance. Other variables include the turntable stand, where it's located in the room, the phono preamp, and the load the preamp presents to the cartridge. Nonetheless, each turntable, tonearm, and cartridge has its own sonic signature. The better products have less of a sonic signature than lower-quality ones; i.e., they more closely approach sonic neutrality.

A high-quality LP front end is characterized by a lack of rumble (or low-frequency noise), greatly reduced record-surface noise, and the impression that the music emerges from a black background. Low-quality LP front ends tend to add a layer of sonic grunge below the music that imposes a grayish opacity on the sound. When you switch to a high-quality front end, it's like washing a grimy film off of a picture window—the view suddenly becomes more transparent, vivid, and alive.

Even if the LP front end doesn't have any obvious rumble, it can still add this layer of low-level noise to the music. The noise not only adds murkiness to the sound, it also obscures low-level musical detail. A better turntable and tonearm strip away this film and let you hear much deeper into the music. It can be difficult to identify this layer of noise unless you've heard the same music reproduced on a top-notch front end. Once you've heard a good LP playback system, however, the difference is startling. To use a visual analogy, hearing your records on a good analog front end for the first time is like looking at the stars on a cloudless, moonless night in the country after living in the city all your life. A wealth of subtle musical detail is revealed, and with it comes a much greater involvement in the music.

Another important characteristic you'll hear on a superlative system is the impression that the music is made up of individual instruments existing in space. Each instrument will occupy a specific point in space, and be surrounded by a halo of air that keeps it separate from the other instruments. The music sounds as if it is made up of individual elements rather than sounding homogenized, blurred, congested, and confused. There's a special realness, life, and immediacy to records played on a high-quality turntable system.

A related aspect is soundstage transparency—the impression that the musical presentation is crystal-clear rather than slightly opaque. A transparent soundstage lets you hear deep into the concert hall, with instruments toward the rear maintaining their clarity and immediacy, yet still sounding far back on the stage. The ability to "see" deep into the soundstage provides a feeling of a vast expanse of space before you in which the instruments exist (if, of course, the recording engineers have captured these qualities in the first place). Reverberation decay hangs in space longer, further conveying the impression of space and depth.

Conversely, a lower-quality LP front end will tend to obscure sounds emanating from the rear of the stage, making them seem undifferentiated and lacking in life. The presentation is clouded by an opaque haze that dulls the music's sense of immediacy, prevents you from hearing low-level detail, and tends to shrink your sense of the hall's size. The differences between superb and mediocre LP reproduction are much like those described Chapter 1's Grand Canyon analogy.

Other important musical qualities greatly affected by the LP front end are dynamic contrast and transient speed. A top-notch LP replay system has much wider dynamic expression than a mediocre one; the difference between loud and soft is greater. In addition to having wider dynamic range, musical transients have an increased sense of suddenness, zip, and attack. The attack of an acoustic guitar string, for example, is quick, sharp, and vivid. Many mediocre turntables and arms slow these transient signals, making them sound synthetic and lifeless. A good LP front end will also have a coherence that makes the transients sound as if they are all lined up in time with each other. The result is more powerful rhythmic expression.

Better LP front ends are also characterized by deeper bass extension and a more solid bottom end. With a great turntable, arm, and cartridge, the bass takes on a whole new level of impact, power, and dynamic agility. Moreover, it's easier to hear subtleties of bass pitch and dynamics with a good turntable system. Finally, a great LP playback system will reveal a richness and density of tone color with instruments such as acoustic bass, electric bass, cello, bass clarinet, piano, and organ.

Just as we want the LP front end to portray the steep attack of a note, we want the note to decay with the actual decay characteristics of the original musical event. In some cases with some instruments played a certain way, there will be a sudden end to the note. In other cases, such as a struck cymbal, a piano played without using the damper pedal, and any bowed or plucked string instrument, the decay might last for several seconds. In some cases the decay will still be very audible even after the next notes of the same instrument have been struck. This might not seem like a significant factor in the enjoyment of music, but correctly reproducing these decays results in denser and richer tone colors as the decay of one note changes the character of the next note. The composer or performer intended for the music to be heard this way, not with a premature truncation of decays. The ability of the best LP front ends to portray and unravel this amazingly delicate and intricate low-level information can be astonishing. A first-rate LP front end is characterized by its ability to clearly articulate each note with a sense of silence between the notes, rather than blurring them together, while retaining decay information within the inter-note silence. A good test for an LP front end's transient characteristics is intricate percussion music. Any blurring of the music's dynamic structure—attack and decay—will be immediately obvious as a smearing of the sound, lack of immediacy, and the impression that you're hearing a replica of the instruments rather than the instruments themselves. Hearing live, unamplified musical instruments periodically really sharpens your hearing acuity for judging reproduced sound.

All turntables, tonearms, and cartridges influence the sound's overall perspective and tonal balance. Even high-quality components can have distinctive sonic signatures, although the very best analog components can be remarkably transparent. Careful matching is therefore required between the turntable, arm, and cartridge to achieve a musical result. Matching a bright, forward cartridge to an arm with the same characteristics, and mounting both on a somewhat aggressive-sounding turntable, could be a recipe for unmusical sound. Those same individual components may, however, be eminently musical in a mix of components that tend to complement each other.

The very best turntables, tonearms, and cartridges are more sonically neutral than lesser products. System matching is therefore less critical as you go up in quality. State-of-the-art turntables and tonearms tend to be so neutral that you can put together nearly any combination and get superlative sound. Even upper-end cartridges, however, exhibit significant sonic differences in frequency response, detail retrieval, spatial perspective, dynamics, and richness of tone color.

You should listen for two other aspects of turntable performance: speed accuracy and speed stability. Speed accuracy is how close the turntable's speed is compared to 33-1/3 rpm. You need to worry about speed accuracy only if you have absolute pitch (the ability to identify a pitch in isolation). Speed instability, however, is easily audible by anyone, and is particularly annoying. Speed stability is how smoothly the platter rotates. Poor speed stability causes *wow* and *flutter*. Wow is a very slow speed variation that shifts the pitch slowly up and down and is most audible on solo piano with sustained notes. Flutter is a rapid speed fluctuation that almost sounds like tremolo. Together, wow and flutter make the sound unstable and blurred rather than solidly anchored. A good turntable with no obvious flutter can still suffer from speed instability. Instead of hearing flutter overtly, you may hear a reduction in timbral accuracy—an oboe, for example, will sound less like an oboe and more like an undifferentiated tone. Speed variations also seem to erase low-level details such as the finest information in an instrument's harmonic structure, or the tail end of reverberation decay. In addition, tiny speed variations affect the quality described earlier of the impression of the presentation being composed of individual instruments occupying specific points in space, surrounded by an enveloping acoustic.

Technical Aspects of Choosing an LP Front End

Matching a turntable, tonearm, and cartridge involves some technical decisions, not just aesthetic choices about which combinations sound the most musical.

First, the tonearm must be able to fit the turntable's arm-mounting area. Many turntables have an *arm-mounting board* on which the tonearm is fastened. The arm-mounting board must be at least as big as the arm's base, and be able to securely hold the arm. Any looseness here will seriously degrade the sound. When mounted to the armboard, the tonearm's cartridge end should be positioned within a range that allows the cartridge to be positioned at exactly the correct distance from the tonearm's pivot point, a parameter called *overhang*. Overhang (described later in this chapter) can be set using the turntable manufacturer's template, or a third-party alignment protractor.

The turntable's suspension should be stiff enough to support the tonearm's weight. If the tonearm is too heavy for the turntable's suspension, the turntable won't be level. Turntable manufacturers will specify a range of tonearm weights appropriate for their turntables.

Next, the tonearm's *effective mass* must be matched to the cartridge's *compliance*. Let's define these terms before examining how they interact.

An arm's effective mass isn't the tonearm's weight, but rather the amount of mass in combination with where along the arm that mass is located. For example, one gram of additional mass located at the pivot point represents almost no contribution to the arm's effective mass. But one gram located at the stylus position contributes exactly one gram to the effective mass. The effective mass specification conveys the amount of inertia the cartridge "feels" as it tries to move the arm when the stylus is modulated by the record groove. (Technically, the arm's effective mass is the amount of mass that, if positioned at the stylus, would equal the total rotational inertia of the arm's moving parts relative to the distance from the pivot. This rotational inertia is called the moment of inertia.) Less than 10 grams of effective mass is considered low mass, 11 to 20 grams is considered medium mass, and more than 20 grams is high mass.

A cartridge's compliance describes how stiffly or loosely the suspension holds the *cantilever*. The cantilever is the thin tube that emerges from the cartridge *body* and holds the *stylus*. If the cantilever is easily moved, the cartridge is high compliance. If the cantilever is stiffly mounted, the cartridge is said to have low compliance.

Compliance is expressed as a number indicating how far the cantilever moves when a force is applied. Specifically, a force of 10^{-6} dynes is applied, and the cantilever's movement in millionths of a centimeter is the cartridge's compliance. For example, a low-compliance cartridge (a stiff suspension) may move only ten millionths of a centimeter; we say the cartridge has a compliance of 10. Because this method of expressing compliance is standardized, the reference to millionths of a centimeter is dropped, leaving only the value 10. Moderately compliant cartridges have compliances of 12 to 20, and high-compliance cartridges are any value above 20.

A cartridge's compliance and the tonearm's effective mass form a resonant system, called a "mass-spring system" in engineering jargon. That is, the combination will vibrate much more easily at a particular frequency than at other frequencies. Think of a weight attached to the bottom of a Slinky, a popular children's toy. Imparting motion to the weight will cause it to begin cycling up and down at its resonant frequency, which is determined by the spring rate (analogous to the cartridge compliance), the amount of mass (analogous to the tonearm's effective mass), and whether the weight was bounced with a large or small motion. Similarly, a tonearm and cartridge will resonate at its own resonant frequency when put into motion. Energy is imparted to the tonearm and cartridge by record warp, turntable rumble, the turntable's resonance, record eccentricity (the center hole isn't exactly centered), and footfalls (the vibrations of someone's footsteps transmitted from the floor to the turntable). These energy sources are all of very low frequencies, perhaps below 8Hz, and can set the tonearm/cartridge system into resonance at the tonearm/cartridge resonant frequency. The tonearm/cartridge resonant frequency must be lower than any musical signal, and higher than the resonant frequency of the turntable suspension, or great distortion can result. Although we can't avoid resonance in the tonearm and cartridge, we can adjust it so that their resonant frequency falls above the very low frequency of rumble and record warp, but below the lowest musical pitch recorded in the record grooves. By matching the arm's effective mass to the cartridge's compliance, we can tune the resonant frequency to fall between the sources of vibration. Fig.9-4 shows the resonant frequencies of different values of arm mass and cartridge compliance.

Another, but distinctly different, type of tonearm resonance is the arm itself resonating independently of the cartridge, which we'll call structural ringing. Just as a bell being struck will ring at its own natural frequency, the tonearm will vibrate for a finite period of time after energy is imparted to it. Different parts of the tonearm, headshell, and armtube can ring at their own structural resonant frequencies. Good tonearm design isolates these resonances and damps them or changes the structure or materials to reduce ringing.

Preventing the tonearm and cartridge from ringing is of utmost importance. The audio signal is generated in the phono cartridge by the motion of the cantilever relative to the cartridge. If the arm and cartridge are vibrating even slightly, that vibration is converted into an electrical signal by the cartridge. Because the cartridge can't distinguish between groove modulation (the musical information) and tonearm resonance, distortion is mixed in with the music. Tonearm resonance distorts the music's tonal balance, colors instrumental timbre, changes the music's dynamic structure (the way notes start and stop), and destroys the sense of space and imaging on the recording.

There's an important caveat to this discussion of matching a tonearm's effective mass with a cartridge's compliance: Very often the stated specifications are wrong. It has

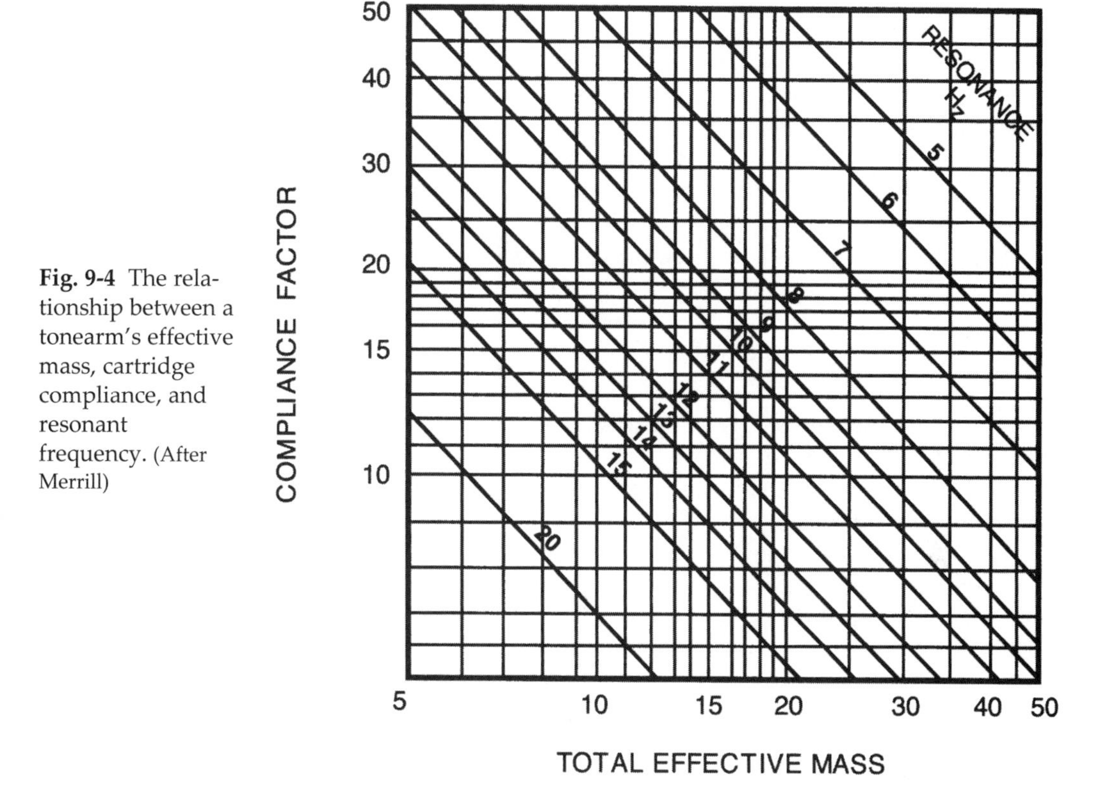

Fig. 9-4 The relationship between a tonearm's effective mass, cartridge compliance, and resonant frequency. (After Merrill)

been brought to my attention that the effective mass specification of most tonearms is grossly incorrect. Similarly, the compliance spec of most cartridges doesn't describe the cartridge's true compliance. It's therefore difficult to match a cartridge and tonearm by looking at specs alone. This is where your retailer's skill and knowledge come into play; he will have had experience setting up hundreds of different arm/cartridge combinations over the years, and knows which pairings deliver the best performance.

The Turntable

It's easy to think of the turntable as having a minor role in a playback system's sound quality. After all, the turntable only spins the record and holds the tonearm; how much sonic influence could it have?

The answer is surprising: A high-quality turntable is absolutely essential to getting the best performance from the rest of your system. A good turntable presents a solid, vibration-free platform for the record and tonearm, allowing the cartridge to recover the maximum amount of information in the grooves while minimizing interference with the audio signal.

The turntable is composed of a base, platter, platter bearing, plinth, drive system, and often a sub-chassis. The *base* is the turntable's main structure; it holds all the components, and is usually finished in black or natural wood. The *platter* is the heavy disc that supports the record; it rests on the *bearing assembly*. The *plinth* is the top of the turntable beneath the platter. The drive system conveys the motor's rotation to the platter. Some turntables have a *sub-chassis* suspended within the base on which the platter and tonearm are mounted. Every turntable will also have an armboard for mounting a tonearm, or a slot into which the tonearm fits. Many turntables have no base or plinth, instead suspending the sub-chassis in open air.

Let's look at how each of these components is assembled into the modern turntable.

The Base and Plinth

A turntable's base and plinth play important roles in sound quality. The base must be a rigid, vibration-resistant structure on which the other turntable components can be mounted. If the base is flimsy, it will vibrate and transmit that vibration to the platter and tonearm, thus degrading the sound.

A turntable system can be set vibrating by four forces: 1) acoustic energy impinging on the turntable (called *feedback*); 2) structure-borne vibration traveling through the turntable stand (primarily when the stand is located on a suspended floor); 3) the turntable's mechanical systems, such as the platter bearing and motor vibration; and 4) the vibration imparted to the tonearm by groove modulations.

This vibration creates relative motion between the stylus and the cartridge. Because the cartridge can't distinguish between groove modulation and turntable resonance, this vibration is converted to an electrical signal and amplified by your system. This is why turntable designers go to elaborate measures to prevent vibration.

Let's first take the case of acoustic energy impinging on the turntable. If the base and plinth aren't rigid, they're more likely to be set in motion by sound striking the turntable. In extreme cases, the loudspeakers and turntable create an *acoustic feedback loop* in which sound from the loudspeakers is converted into an electrical signal in the cartridge through vibration, which is amplified by the loudspeakers, which causes even more feedback to be produced by the cartridge, and so forth. This acoustic feedback can muddy the music, or even make it impossible to play records at a moderately high playback level. You can hear this phenomenon by putting the stylus on the lead-in groove without the record spinning, then gradually turning up the volume. You'll start to hear a howling sound as the acoustic feedback loop grows strong enough to feed on itself, and a "runaway" condition develops in which the sound keeps getting louder. If you try this, keep your hand on the volume control and be ready to turn down the volume as soon as you hear the howl—if you don't turn down the volume *immediately*, the system could be damaged. The more vibration-resistant the turntable, the less severe this phenomenon. You might try this with and without the turntable's dustcover in place. A turntable's dustcover can catch acoustic energy and feed it into the base and plinth, degrading the sound. Other dustcovers protect the turntable from vibration.

Turntable bases and plinths are designed to resist vibration by making them very massive (so it's hard to put them in motion), and by damping any vibrations that do exist. Materials such as Medium Density Fiberboard (MDF) and Fountainhead are often used in turntable bases. High-quality turntables are often made from machined acrylic. Vibration damping is achieved with layers of material applied to or within the base so that the vibration stops rather than being allowed to ring. An example of vibration damping is putting a wet rag on a steel plate before striking the plate. Instead of hearing the plate ring, you hear a dull thud—the vibrational energy is dissipated more quickly. In practice, damping materials are often exotic compounds (one turntable uses the same sound-deadening material that lines a submarine's hull). One technique, called constrained-layer damping, puts layers of soft damping material between layers of harder materials to more effectively dissipate vibrational energy. The rationale is when vibration encounters a discontinuity in the material, shear forces between layers produce shear strain, which causes frictional losses. The mechanical energy is converted into a minute amount of heat. In another technique lead inserts are distributed throughout the base and platter. Some designers believe that this discontinuity in materials is in fact a mistake, and rather than absorbing vibration, simply reflects it. These designers hold that solid blocks of ideally chosen material, coupled with a sophisticated isolation system, deliver optimum performance.

The armboard should have vibration resistance and damping properties similar to those of the base. Some armboards of acrylic and lead weigh up to six pounds.

Another theory of turntable design suggests that the less massive the base and plinth, the better. This school argues that a very rigid, low-mass base is ideal because there's less mass to resonate. The theory, used in only a few turntable models, has been largely discredited by proponents of high-mass turntables.

So much for the problem of acoustic energy putting the turntable in motion. Now let's look at how turntable design addresses the problems of structure-borne vibration. Vibration entering the turntable through the stand or rack can be greatly reduced by mechanically isolating the turntable's key components (platter, armboard, and tonearm/cartridge) with a suspension system—the sprung turntable.

Sprung and Unsprung Turntables

Most turntables are *sprung*, meaning that the platter and armboard are mounted on a sub-chassis that floats within the base on springs. The terms *suspended* and *floating* describe the same construction.

Sprung turntables can be one of three designs. In one method, the sub-chassis sits on springs attached to the base's bottom. In the second method, the sub-chassis hangs down from the plinth on springs. The third technique, shown in Fig.9-5, dispenses with the base entirely and hangs the sub-chassis in open air on pillars. The turntable is suspended between the four pillars at each corner of the turntable.

Whichever technique is used, the goal of all sprung designs is to isolate the platter and tonearm from external vibration. Any vibration picked up by the supports on which the turntable rests won't be transferred as effectively to the platter and tonearm. The primary sources of structure-borne vibration are passing trucks, footfalls, air conditioners, and motors attached to the building. Structure-borne vibration is much less of a problem in a single-family home with a concrete floor than in an apartment building or frame house with suspended floors.

The suspension on which the sub-chassis is mounted is carefully tuned to a very low frequency. The vertical resonant frequency of the suspension can also be tuned to a

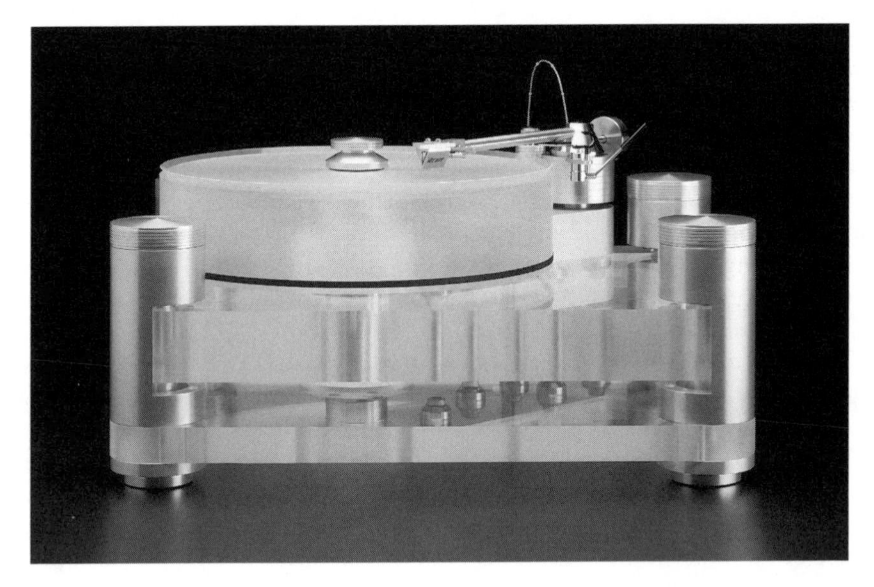

Fig. 9-5 The sub-chassis can also be mounted in open air, with no base. (Courtesy Basis Audio)

slightly different frequency than the horizontal resonant frequency. In addition, the suspension can be immersed in a viscous fluid to damp any remaining vibration. An advantage of viscous damping over mechanical damping such as foam or Sorbothane (a rubberlike material) is that the viscous fluid doesn't compress, retaining its damping properties under all conditions.

You can judge a suspension's effectiveness by putting the stylus on a stationary record and tapping on the stand on which the turntable is mounted. The volume should be turned up to what would be a moderate listening level if the record were spinning. You should start by gently tapping the stand, then gradually increasing the force. The less sound you hear from the loudspeakers, the more effective the turntable's suspension. With the best models, you can even hit the stand as hard as you like with a hammer and hear no sound from the loudspeakers; the suspension completely isolates the platter and tonearm from the hammer blow. In less effective designs, a small tap on the stand can be heard through the loudspeakers.

A minority of turntable designers believe, however, that the springs used in suspended turntables can resonate and actually introduce vibration into the platter and arm. They believe the answer is the unsprung turntable, in which the platter and tonearm are mounted directly to the base. If the base is made rigid and non-resonant enough, they argue, the platter and arm won't vibrate—and you don't have to worry about springs resonating. Unsprung turntables sometimes use constrained-layer damping in the base in an attempt to dissipate mechanical energy before it reaches the platter and tonearm.

Whatever the design, you can judge its isolation properties by putting a stethoscope on the turntable and scratching or tapping the equipment rack beneath the turntable. The greater the isolation, the less sound you will hear.

The Platter and Bearing Assembly

The platter not only provides a support for the record, it also plays two other important roles: as a flywheel, to smooth the rotation; and as a "sink," to draw vibration from the record. Many platters are very heavy (up to 80 pounds), with most of their mass concentrated toward the outside edge to increase their moment of inertia. This high mass also counters bearing friction and stylus drag. Irregular (non-linear) bearing friction can create rapid irregularities in the platter's speed, which frequency-modulate the recovered audio signal. Massive platters are not a cure-all for poor-quality bearings or a high degree of runout (the platter isn't perfectly round). A heavy platter cannot make up for the additional noise created by high bearing friction. Precision tolerances in centering and roundness, and low friction, are much more important than weight. This high-precision machining is part of what you're paying for in a more expensive turntable.

Most platters are made of a single substance such as acrylic, stamped metal (in the cheaper turntables), cast and machined aluminum (in mid-level turntables), or exotic materials such as ceramic compounds. The platter sometimes has a hollowed-out ring around the outer edge that is filled with a heavy material to increase the platter's mass, or is loaded with a damping substance to make the platter more inert and resistant to vibration.

These techniques also attempt to make the platter act as a "sink" for record vibration. When the record is clamped to the platter, some degree of record vibration will be transferred to the platter. How much of that energy is transferred, and over what frequency range, is determined by the integrity of the contact between record and platter, as well as the properties of the platter material. Vacuum hold-down systems, described later in this chapter, intimately couple the record to the platter for minimum record vibration.

Because the platter spins on a stationary object (the rest of the turntable), there must be a bearing surface between the two. With one technique, the bearing is mounted at the end of a shaft to which the platter is attached. This shaft—often made of stainless steel—extends down a hollow column in the base. The shaft has some form of bearing on the end, either a chrome-hardened steel ball, tungsten-carbide, Zirconium, ceramic, or even a very hard jewel such as sapphire. The bearing often sits in a well of lubricant. A second technique puts the bearing surface on top of a stationary shaft, with the bearing surface between the platter and shaft.

Whichever technique is used, the bearing must provide smooth and quiet rotation of the platter. Any noise or vibration created by the bearing will be transmitted directly to the platter. Turntable bearings are machined to very close tolerances, and are often highly polished to achieve a smooth surface.

An alternative bearing that offers extremely low friction is the air bearing. The platter rides on a cushion of air rather than on a mechanical bearing. A pump forces compressed air into a very tiny gap between the platter and an adjacent surface. This air pressure pushes the platter up slightly so that the platter literally floats on air. Air-bearing turntables are generally expensive and more fussy about set up.

Another method of lifting a platter so that it doesn't spin on a bearing is magnetic suspension. The platter and plinth contain powerful magnets (usually neodymium) whose north-south orientations repel each other. This lifts the platter off its mechanical bearing a fraction of an inch, resulting in the platter spinning on a cushion of air.

Platter Mats, Record Clamps, and Vacuum Hold-Down Systems

Platter mats are designed to minimize record vibration as well as to absorb what vibration remains. Designers of soft mats suggest that an absorbent felt mat works better in drawing vibration away from the record. Designers of hard mats contend that a stiffer mat material better couples the record to the platter. Finally, some turntable manufacturers discourage using any mat at all, believing that their platter design provides the best sink for record vibration. Putting marketing claims aside, physics dictate that the greatest energy transfer occurs when the platter's key physical properties match as closely as possible the record's properties. This phenomenon of a mechanical wave being transferred from one solid to another is well researched in other engineering fields. Consequently, one could make a good argument that correct choice of platter material is a better solution than a mat. Moreover, some soft mats attract and hold dust that can be transferred to the record.

Record clamps couple the record to the platter so that the record isn't allowed to vibrate freely. Ideally, the platter acts as a vibration sink, draining vibration from the record. By more intimately coupling the record to the platter, the record clamp improves the sound.

Record clamps come in three varieties. First, the clamp can simply be a heavy weight put over the spindle. The clamp's weight squeezes the record between clamp and platter. Other clamps have a screw mechanism that threads down onto the spindle. The third type is the "reflex" clamp, in which a locking mechanism pushes the clamp down onto the record. Which type works best should be decided by your listening, the turntable manufacturer's recommendations, or your local dealer's suggestion. Note that very heavy clamps can put a strain on some sprung turntables, compressing the springs (or expanding them if the sub-chassis is hung from the plinth). Another type of device designed to more closely couple the record to the platter is a heavy 12"-diameter ring placed over the record's outer edge. This device couples the record's outer diameter to the platter, but is cumbersome to use in practice (it's difficult to center the ring on the record). Whatever route you chose, some form of record clamping is a must for any high-end turntable.

Perhaps the best technique for coupling the record to the platter is the *vacuum hold-down system*. An external pump creates a vacuum that is transferred by tubes to the platter. The platter contains small channels that distribute the vacuum, and the platter's edge is surrounded by a flexible lip that couples to the record and forms the vacuum seal. The vacuum holds the record tightly to the platter surface so that vibration is drained away from the record. A vacuum hold-down also makes the record more flat, even if it is mildly warped. Incidentally, LP mastering lathes use vacuum hold-down to secure the master lacquer to the lathe's turntable. One can easily hear the effect of a vacuum hold-down system simply by turning it off and comparing the sound. In my experience, vacuum hold-down offers a significant sonic improvement in clarity, instrumental timbre, and image focus. Vacuum pumps can be noisy, however, and some benefit from acoustic isolation from the listening room. Bright Star Audio makes a small acoustic isolation chamber, called "The Padded Cell" (their motto: "Every audiophile needs a padded cell"), for isolating vacuum hold-down and air-bearing pumps. Detractors of vacuum hold-down suggest that the vacuum system causes dust on the platter to become embedded in the underside of the record, contributing to surface noise. I've used a turntable with vacuum hold-down for several years and have not encountered this phenomenon.

The Drive System

A turntable's drive system transfers the motor's rotation to the platter. Virtually all high-end turntables currently made are belt-driven; the platter is spun by a rubber belt or silk thread stretched around the motor pulley and outer rim of the platter.

Mass-market mid-fi turntables are usually *direct-drive*; i.e., the motor is connected directly to the platter. The motor's spindle is often the spindle over which you place the record. Direct drive was sold to the public as superior to belt drive—there are no belts to stretch and wear, and a direct-drive motor can be electronically controlled to maintain precise speed and exhibit low wow and flutter. The wow and flutter specifications of a direct-drive turntable can be better than all but the best belt-drive turntables. Direct-drives often have an electronic speed control and sometimes a *Phase Lock Loop* (PLL) servo mechanism that are said to ensure good speed stability. In a servo system, the platter's speed is sensed and a signal is fed back to the motor, which responds to keep the platter turning at a constant speed. As with all servo systems, the correction signal is generated only after the system has detected a speed error, which introduces some degree of pulsation in the rotational speed. Servo systems are also prone to long-term drift, a drawback for listeners with perfect pitch.

The vast majority of high-end turntables are belt-driven. Rather than directly coupling the motor's vibration to the platter as in direct-drive, the drive belt acts as a buffer to decouple the platter from the motor. Motor noise is isolated from the platter, resulting in quieter operation than is possible from a direct-drive turntable. Belt drive also makes it easier to suspend the platter and drive system on a sub-chassis.

No elaborate speed controls are used on many belt-drive turntables; the motor just sits there spinning at a fairly high speed (as fast as 1000 rpm). This high-speed rotation is coupled to the large platter with a small pulley, resulting in 33-1/3 rpm rotation of the platter. Belt-drive motors are usually *AC-synchronous*, meaning that their rotational speed is determined by the frequency of the AC voltage driving them. But because the AC line is subject to noise and fluctuation, the platter's speed can be very slightly affected by dirty AC power. Although this speed variation is minute, it doesn't take much to destroy the music's pitch definition, timbre, and image stability. Consequently, some manufacturers synthesize their own 50Hz or 60Hz AC sinewave with an outboard drive unit. This device creates a pure, stable sinewave with a precision oscillator, then amplifies it to 120V to drive the turntable's AC synchronous motor. In one such drive system, the stock power supply produces a small but noticeable vibration of the motor. After switching to the upgraded supply, the motor runs so smoothly that one can hold it and not be able to determine if the motor is running or not.

The drive motor can be a source of turntable vibration. As the motor spins, it produces vibration that can be transferred to the other components in the turntable, producing a low-frequency rumble. Even if you don't hear rumble directly, motor vibration can still degrade the sound. The motor assemblies of some turntables are completely separate from the base and encased in damping material. Other designs mount the motor to the sub-chassis, and isolate its vibration from the other turntable components.

The Tonearm

The tonearm's job is to hold the cartridge over the record and keep the stylus in the groove. We want the tonearm to be an immovable support for the cartridge, yet also be light enough to follow the inward path of the groove, track the up-and-down motions of record warps, and follow any record eccentricity caused by an offset center hole—all without causing undue wear on the delicate grooves themselves. As we'll see, this is a challenging job.

Tonearms come in two varieties: *pivoted* and *tangential-tracking*. A pivoted tonearm allows the cartridge end of the arm to traverse the record in an arc while maintaining a fixed pivot point. A tangential tonearm (also called a *linear-tracking, radial-tracking,* or *parallel-tracking* tonearm) moves the entire tonearm and bearing in relation to the record.

Let's first look at the pivoted tonearm, by far the most popular type of arm. Its major components are, from back to front, the counterweight, bearing, armtube, and headshell. (These elements are shown from left to right in the photograph of Fig.9-6.) The counterweight counteracts the weight of the armtube and cartridge; its weight and position determine the downward force of the stylus in the groove. The bearing provides a pivot point for the arm, in both the vertical and horizontal planes. The armtube extends the cartridge position away from the pivot point to an optimum position over the record. The headshell is attached to the end of the armtube and provides a platform for mounting the cartridge to the armtube. The small, flat disc near the bearing in Fig.9-6 sets the anti-skating compensation.

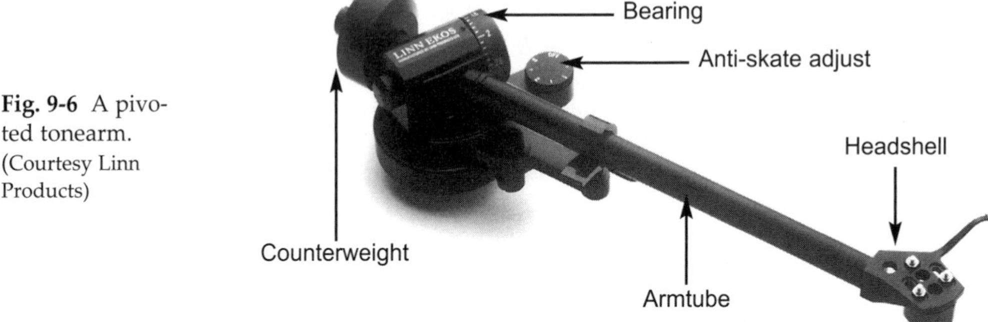

Fig. 9-6 A pivoted tonearm. (Courtesy Linn Products)

A tonearm's bearing is an important aspect of its design. Bearings are required to allow the tonearm to move up and down as well as side-to-side. The bearing should provide very low friction and not impede the arm's movement. If the bearing is sticky, the stylus will be forced against the groove wall, causing distortion and record wear. Loosening the bearings reduces friction, but can cause the bearings to "chatter" as the tonearm is rattled by the motion of the stylus in the groove, or by other sources of tonearm vibration. Remember, any movement of the tonearm in relation to the stylus in the groove is interpreted by the cartridge as a groove modulation, and is converted into an electrical signal that appears at the cartridge output along with the musical signal. Tightening the bearings decreases chatter but also increases friction. Tonearm designers must balance these tradeoffs.

The most common type of pivoted tonearm bearing is the *gimbal,* in which a set of rings attached to bearings allows the arm to move in any direction. Gyroscopes are sometimes mounted in gimbal bearings. Another bearing type, called the *unipivot*, has recently come back into favor, partially from its exemplary use in the Graham tonearm and in an innovative variation in the Basis Vector arm. A unipivot bearing is similar to a ball within a cup, allowing motion in any direction. The unipivot is the simplest design, and offers the lowest friction of any bearing (except an air bearing). Gimbal and unipivot bearings are lubricated internally, and some are immersed in a silicon fluid to damp resonances.

An interesting approach to tonearm bearings is the Well Tempered Tonearm. The tonearm is attached to a vertical post and horizontal paddle suspended in a cup of silicon by monofilament line. In essence, the WTA has no bearing; the arm moves in a fluid with no hard connection between the arm and the rest of the turntable.

The Basis Audio Vector tonearm provides a twist on the unipivot arm. The Vector's innovation is to asymmetrically weight the arm (with a cutout in the counterweight) so that it "leans" over onto a second bearing. This technique eliminates the possibility of dynamic azimuth error despite the fact that the vertical load rests on a point-loaded pivot. (*Azimuth* is the perpendicular relationship between the stylus and the groove.) Record warps can cause conventional pivoted arms to "roll" or constantly change their azimuth alignment. The Vector maintains the same azimuth because it rests on the second bearing.

A tonearm's armtube is designed to be rigid, low in mass, resistant to vibration, and have the ability to damp any resonances that do occur. Armtubes are often made of exotic materials to combine rigidity with low mass (carbon fiber is a popular armtube material). Armtubes can be filled with damping material to kill resonances, and one particular arm (the Graham) is even made of concentric tubes separated by damping material to help damp resonances. The goal is to make the arm inert and less likely to ring and transmit vibration to the cartridge. A well-damped armtube correlates with an improved retrieval of low-level musical detail.

A tonearm's effective mass (defined earlier in this chapter) is carefully chosen to strike a balance between the ability to hold the cartridge as its stylus is moved from side to side (where high mass is desired), and the ability to follow record warps (where low mass is desired). When a massive tonearm encounters a record warp, it will increase the downward force of the stylus on the "uphill" side of the warp, then tend to reduce the downward force of the stylus on the "downhill" side. In some cases, the stylus can even fly out of the groove. Although this phenomenon occurs with all tonearms, it is magnified by high effective tonearm mass.

A tonearm's *effective length* is the distance from the pivot point to the stylus tip. Most tonearms measure between 210 and 230 millimeters (nearly 9") of effective length. (The effective length is shorter than the actual length due to the tonearm's offset.) The tonearm's length affects the amount of *overhang*. The technical definition of overhang is the distance the stylus extends past the spindle. In general parlance, "overhang" means where the stylus falls in relation to the pivot point. Overhang can be adjusted by moving the cartridge forward or backward in the headshell. Some tonearms have adjustable-length armtubes for setting overhang. These alignment procedures are described later in this chapter.

Let's consider a fundamental limitation of pivoted tonearms, which will explain the reason for the tangential tonearm's existence.

An LP is cut with a tangential-tracking mechanism—the cutting head moves in a straight line across the lacquer master. But because the pivoted tonearm traverses an arc across the record, the playback stylus "rotates" slightly in relation to the groove as it swings in this arc. This rotation causes one side of the stylus to contact the groove wall slightly ahead of its nominal contact point, and the other side of the stylus to contact the groove wall slightly behind its nominal contact point. This difference in the angular relationship of the playback stylus to the groove is called *lateral tracking error*, or simply tracking error.

More technically, tangential cutting results in groove modulation that is always at a right angle to the groove's tangent line. But the back and forth motion of the playback stylus in response to this groove modulation deviates from a right angle to the groove's tangent line over most of the record. Specifically, the angle is typically a few degrees less than 90° at the outermost grooves. As the stylus swings in an arc toward the record center, the stylus is rotated, reducing this angular error until it reaches a point at which there's no error. At this point the playback stylus has the same geometric relationship to the groove as the cutting stylus had to the groove. As the stylus continues in its arc and rotates further, the angular error increases in the opposite direction. The error reaches a peak and then, counter-intuitively, begins decreasing until the error reaches zero at a second point on the record. The two points of zero tracking error are typically at 2.4″ and 5″ from the record's outer edge.

Lateral tracking error introduces harmonic distortion, primarily second-harmonic, which is fairly benign sonically. Other harmonic products are present, but are lower in amplitude than the second-harmonic. The amount of tracking-error-induced distortion is not just a function of the tracking error, but also of the groove speed, which varies from the inside to outside grooves. The distortion introduced by tracking error is more than double in the innermost grooves compared with the outermost grooves. That's because one revolution of the LP beneath the stylus at the outer radius represents about 2.5-times the distance as one revolution at the inner radius.

Tracking error is minimized by putting a bend in the tonearm, called the *offset*. Offset can be a sudden angle at the end of the tonearm, or a gradual bend that gives the armtube a J or S shape. The offset angle in modern 9″ tonearms is 22°–25°, which provides the least tracking error. Note that adding an offset to the tonearm doesn't eliminate tracking error. Instead, the offset produces the least *overall* tracking error, with two points along the record at which the error is zero.

Various cartridge alignments have been developed to minimize this lateral tracking error. The two factors that affect tracking error are the amount of offset and the overhang, or the distance between the stylus and the pivot point. The optimum alignments were worked out in the 1930s and 1940s; E. Löfgren and H. G. Baerwald independently developed slightly different calculations of optimum tonearm/cartridge alignment geometry in 1938 and 1941, respectively. Today, they are referred to as Löfgren or Baerwald alignments, and you'll see cartridge-alignment protractors that give you a choice of Löfgren or Baerwald settings.

Another approach to reducing lateral tracking error is the 12″ tonearm, which was once popular and has recently made a limited comeback. In theory, an arm of infinite length would have zero tracking error. The 12″ arm's additional length (compared with a 9″ arm) reduces lateral tracking error because the arc's curvature is smaller. The 12″ arm, however, pays a penalty of greater mass, greater counterweight inertia (which increases record wear), and greater potential for misalignment. Small alignment errors are magnified in a 12″ arm compared with a 9″ arm—the 12″ arm's offset and overhang are reduced, magnifying any alignment errors. Moreover, a 12″ arm is more prone to resonances than a 9″ arm, all other factors being equal.

The tonearm offset needed to reduce tracking error introduces a problem in pivoted arms called *skating*. Skating is a force that pulls the tonearm toward the center of the record. The result is a force acting on the stylus that must be compensated for by applying an equal but opposite force on the cartridge. This compensation, called *anti-skating*, counteracts the skating force caused by tonearm offset. Anti-skating allows the stylus to maintain equal contact pressure with both sides of the groove, and prevents the cartridge's cantilever from being displaced from its center position in the cartridge. Anti-skating can be generated by springs, weights with pulleys, or mechanical linkages. (I'll talk about anti-skating adjustment later in this chapter under "LP Playback System Setup.")

On a more technical level, skating is caused solely by the friction between the groove and stylus. Frictional force always acts in the direction of the attempted motion; the frictional force of the record pulling the stylus is in the tangential direction of the record groove. This force vector doesn't pass through the tonearm pivot. As with any vector quantity, this force can be resolved into two component forces at right angles to each other. Because one of the component forces is in-line with the pivot, the second (and much smaller) force is at a right angle to the primary force, causing the arm to skate inward.

As you can see, the problems of pivoted tonearms—lateral tracking error and skating—don't exist with tangential-tracking arms. Since the entire arm moves across the record, the angular relationship between the playback stylus and record groove is identical to that of the LP cutting lathe at any record radius. And because the arm has no offset, no skating force is generated. Tangential-tracking arms have another advantage: Because they're generally very short, tangential-tracking arms can have lower vertical mass than pivoted arms. Tangential-tracking arms, however, have higher horizontal mass. Because they don't pivot, the effective horizontal mass of a tangential arm is the arm's actual mass, which is higher than the effective horizontal mass of a pivoted arm. Tangential arm design is a compromise between too-low vertical mass and too-high horizontal mass.

Tangential tracking is also much more complex, and consequently more expensive to execute well. The entire arm must be moved along a track tangential to the record—a difficult engineering challenge. In mid-fi tangential-tracking turntables, the arm is driven by a servo motor whose motion may not exactly match the record's pitch (spacing between record grooves), forcing the stylus first against one side of the groove, then the other. Although the servo system attempts to correct for this error, servo systems by their very nature work only after an error has occurred.

The solution in high-end tangential-tracking arms is the *air bearing*, in which the arm rides on a cushion of air surrounding a tube. The only force moving the tonearm is the gentle force of the stylus. Owing to the air bearing, the friction between the arm and the tube on which it is mounted is practically zero. (The turntable in Fig.9-3 is fitted with a tangential-tracking, air-bearing tonearm.)

Tangential-tracking arms are more difficult to correctly set up, require periodic adjustment, and have air pumps that can be audible if not properly isolated from the listening area. The air rushing through the air bearing can also be audible.

Proponents of pivoted tonearms argue that the inherent geometric advantages of tangential arms are far outweighed by concerns over the tangential arm's rigidity and resonance damping. They suggest that the distortion produced by tracking error isn't a significant source of sonic degradation. Indeed, when listening to a record side all the way through, you don't feel a sense of relief at the two points of perfect tangency, nor do you notice the distortion gradually increase as the stylus leaves each of the tangency points.

The Phono Cartridge

The phono cartridge has the job of converting the modulations of the record groove into an electrical signal. Because the cartridge changes one form of energy into another (mechanical into electrical), the cartridge is called a transducer. There's one other transducer in your system—the loudspeakers at the other end of the playback chain.

A phono cartridge consists of the cartridge body, stylus, cantilever, and generator system. The body is the housing that surrounds the cartridge, and comprises the entire surface area. The stylus is a diamond point attached to the cantilever (the tiny shaft that extends from the bottom of the cartridge body). The stylus is moved back and forth and up and down by modulations in the record groove. This modulation is transferred by the cantilever to the generator system, the part of the cartridge where motion is converted into an electrical signal.

Moving-Magnet and Moving-Coil Cartridges

Cartridges are classified by their principle of operation: *moving-magnet* or . In a moving-magnet cartridge, tiny magnets attached to the cantilever move in relation to stationary coils in the cartridge body. The movement of the magnetic field through the coils induces a voltage (the audio signal) across the coils. (This phenomenon, called electromagnetic induction, is explained in detail in Appendix B.) Less common variations on the moving-magnet cartridge are the induced magnet and moving-iron types. (Fig.9-7 shows the essential elements of a moving-magnet cartridge.)

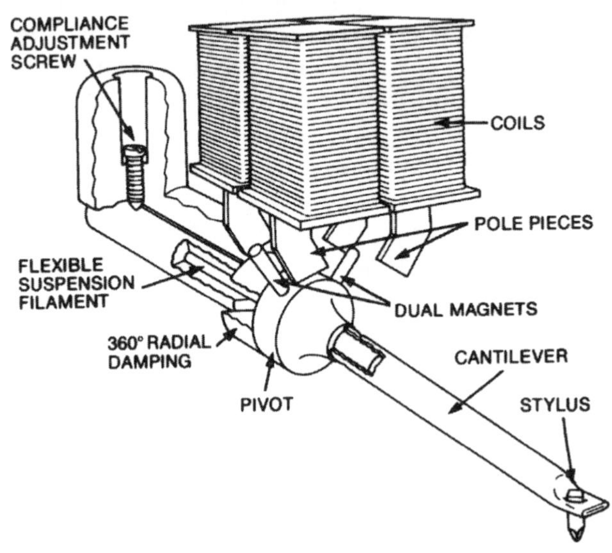

Fig. 9-7 Moving-magnet cartridge construction. (Courtesy Audio-Technica U.S., Inc.)

A moving-coil cartridge works on exactly the same physical principles, but the magnets are stationary and the coils move. A moving-coil cartridge generally has much less moving mass than a moving-magnet cartridge. Consequently, a moving-coil cartridge can generally track better than a moving-magnet type, and also have better transient response.

With less mass to put into motion (and less mass to continue moving after the motivating force has stopped), moving-coil cartridges can better follow transient signals in the record. Because of their construction, moving-coil cartridges generally don't have user-replaceable styli; you must return the cartridge to the manufacturer.

Cartridge output voltage varies greatly between moving-magnet and moving-coil operation. A moving-magnet's output ranges from 2mV (two thousandths of a volt) to about 8mV; a moving-coil cartridge's output is typically between 0.15mV and 2.5mV. Although moving-coil cartridges generally have lower output voltage than moving-magnet types, some so-called "high-output" moving-coils have higher output voltage than some moving-magnet cartridges. A cartridge's output voltage is specified when the stylus is subjected to a recorded lateral groove velocity of 5cm per second RMS at 1kHz.

This wide range of cartridge output voltage requires that the phono preamplifier's gain be matched to the cartridge's output voltage. The lower the cartridge output voltage, the higher the gain needed to bring the phono signal to line level. (A full discussion of matching cartridge output voltage to phonostage gain is included in Chapter 4.)

A moving-coil's output voltage is determined largely by the number of turns of wire on the coil: The greater the number of turns, the higher the output voltage. More turns means less gain is needed in the phono preamp, but most designers use the minimum number of turns possible to keep the moving mass low.

The method of generating the audio signal, whether moving-magnet or moving-coil, affects how sensitive the cartridge is to the effects of preamplifier "loading." The cartridge output "sees" an electrical load of the preamplifier's input impedance and capacitance, and the tonearm cable's capacitance. The load impedance is in parallel with the capacitance.

The frequency response of a moving-magnet cartridge is greatly influenced by the amount of capacitance in the tonearm cable and preamplifier input. A moving-magnet cartridge terminated with the correct impedance (47k ohms), but with too much capacitance, can have frequency-response errors in the upper midrange and treble of 5dB. Most cartridge manufacturers specify the ideal load impedance and capacitance for their cartridges. A typical load for a moving-magnet cartridge is 47k ohms in parallel with 200-400pF (picofarads) of capacitance. Moving-coil cartridges are nearly immune to capacitance loading effects.

The Strain Gauge Cartridge

Not all phono cartridges work on the principle of electromagnetic induction. The *strain gauge cartridge,* first introduced in the 1960s and making a very limited comeback today, translates an LP's groove modulations into an electrical signal using a completely different principle. In the strain gauge cartridge, the cantilever is mounted between two silicon semiconductor elements that change their electrical resistance in response to mechanical pressure. A special preamplifier designed specifically for the cartridge supplies continuous current flow through the semiconductors. The LP's groove modulation causes the cantilever to press against the semiconductor crystals, varying the current flow through them. It is this variation in current flow that represents the audio signal. Note that all strain gauge cartridges are sold with a preamplifier; the two components are integral to the system's operation. Moreover, the strain gauge cartridge doesn't require RIAA equalization.

Stylus Shapes and Cantilever Materials

Styli (the plural of stylus) come in a variety of shapes, the simplest and least expensive of which is the conical or spherical tip. The conical stylus is a tiny piece of diamond polished into a cone shape. Because the stylus tip must be smaller than the record groove, the radius of a conical stylus tip is about 15μm, or about 0.0006".

The conical stylus, however, cannot perfectly track the groove's modulations because its shape is different from that of the cutting stylus used to make the record. As shown in Fig.9-8, the conical stylus contacts different points on the groove wall compared to the cutting stylus, causing *tracing distortion*. In a related problem, called *pinch effect*, the stylus is pushed up out of the groove. Although the cutting stylus created a groove of equal width, the conical playback stylus may see a narrower groove because it touches different places on the left and right groove walls in relation to the cutting stylus. Pinch effect is most likely to occur where there are large high-frequency groove modulations.

The solution to these two related problems is the elliptical stylus. Instead of using a rounded tip to track the groove modulations, an elliptical stylus has an oval cross section, with two flattish faces. Because this shape more closely approximates the shape of the cutting stylus, it results in lower tracing distortion and eliminates the pinch effect. An elliptical stylus can also track high-frequency modulations better, and distributes its force over a wider area to reduce record wear. (Fig.9-8 compares the stylus/groove relationships of cutting, conical, and elliptical styli.)

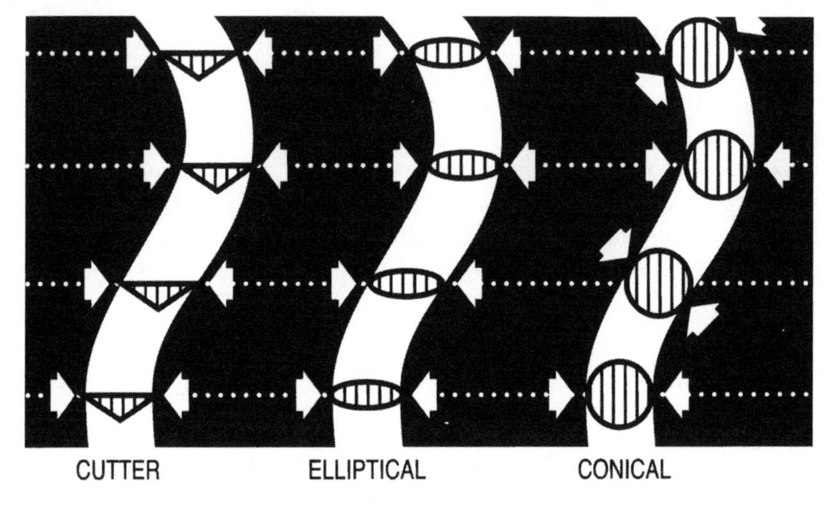

Fig. 9-8 Comparison of conical and elliptical styli in relation to the cutting stylus. (After Shure)

A third type of stylus takes the elliptical shape one step further by making the tip even narrower. These stylus shapes are called Shibata, line-contact, fine-line, van den Hul, and hyperelliptical. A cartridge with any of these stylus shapes requires more precise alignment and setup than conical or standard elliptical types.

A variation of the elliptical stylus is the microridge, which has a tiny groove cut into each edge. As the stylus wears, the microridge stylus maintains the same shape, and thus its relationship to the record groove.

As described later in this chapter, keeping your stylus clean is of paramount importance. A speck of dust or dirt is like a boulder attached to the stylus, grinding away at the groove walls. An appreciation of the enormous pressure a stylus imposes on the groove further highlights the need for a clean stylus. For example, a tracking force of 1.4 grams applied to a typical stylus contact area (0.2×10^{-6}") results in a pressure of nearly four tons per square inch. This pressure is enough to momentarily melt the outer layer of the groove wall. It's easy to see how stylus motion through the groove is much smoother with a clean stylus, and produces much less record wear. A clean stylus sounds better, too.

With good maintenance, a stylus should last for about 1000 hours of use. It's a good idea to have the stylus examined microscopically after about 500 hours, then again at 800 hours to check for irregular wear that could damage records.

Because the cantilever transfers stylus motion to the generator, its construction is extremely important. Cantilevers are designed to be very light, rigid, and non-resonant. The lower the cantilever mass, the better the cartridge's *trackability*, all other factors being equal. Trackability refers to the ability of the stylus to maintain contact with the groove walls, particularly during passages of high groove velocity. When the groove modulation is so severe that the stylus cannot accelerate quickly enough to follow the groove wall, the stylus momentarily loses contact with both groove walls, resulting in tracking distortion. A stiff, low-mass cantilever contributes to good trackability. To obtain stiffness with low mass, exotic materials are often used in cantilever design, including boron, diamond, beryllium, titanium, ceramic, ruby, and sapphire. Cantilevers are often hollow to reduce their mass, and are sometimes filled with a resonance-damping material. Most test records include bands of increasing groove velocity that allow you to assess a cartridge's trackability.

The cantilever is mounted in a ring or block made of a compliant rubber, called the damper, inside the cartridge body at the end opposite to that which bears the stylus. The damper allows the cantilever to move, yet keeps it in position.

Some audiophiles remove the cartridge body to run the cartridge "naked." The body can be a source of resonance, and adds a fair amount of effective mass to the tonearm. Removing a cartridge body should be attempted only if you are very skilled in working with tiny precision devices, and then only if you're willing to risk destroying your cartridge.

LP Playback System Setup

Correctly playing back a record is a delicate art. It takes patience, skill, a keen ear, a delicate touch, an appreciation for the forces involved—and a high-quality LP front end. Doug Sax, co-founder of Sheffield Lab and the pioneer of the modern direct-to-disc recording, has said, "Probably the easiest thing in the world is to play an LP record incorrectly."

What prompted Sax's observation is the great variability in setup possible with a turntable, tonearm, and cartridge. When you add to the equation how the phono preamp loads the cartridge, the turntable stand, and its placement in the room, you have the potential for not getting all the music out of the record grooves.

But this situation also offers the promise of improving the sound of your LP front end without spending any more money. By just setting up the system correctly, you can realize far better sound from a modest turntable than from a poorly set-up front end costing thousands more. Great musical rewards await those willing to tweak their LP front ends.

The best sources of advice for LP system setup are the manufacturers of the components in the LP playback system, and your dealer. The dealer will have undoubtedly set up many turntable/tonearm/ cartridge combinations like yours, and will have the techniques down pat. A good dealer will set up your system when you buy a product from him.

Nonetheless, you should know the procedure for getting the best sound from your LP front end. This will let you converse with the dealer from a position of knowledge, and also allow you to set up your system if you just change cartridges or move your turntable from one location to another. You might also consider the excellent DVD *21st Century Vinyl: Michael Fremer's Practical Guide to Turntable Set-Up*. The DVD walks you through all the steps in optimizing an analog front-end. Another excellent visual demonstration of cartridge and tonearm setup is the DVD that's supplied with the Basis Vector tonearm.

The first rule of LP system setup is to put the turntable on a good platform. I can't emphasize this enough; a flimsy rack, or one prone to resonating, will seriously degrade the sound of your system. The less good the turntable's suspension, the more important the rack. The very best turntables (those having superlative isolation from structure-borne vibration) are affected very little by the stand. With all but the very best turntables, you won't hear your system at its best without a solid, non-resonant support structure for your turntable. Many stands and racks are made specifically for turntables. Invest in one.

The most massive racks can, however, put too much stress on a suspended wood floor. In addition, footfalls and other vibrations can be transmitted right through the rack and to the turntable if the turntable has inadequate suspension. If you live in a building with a shaky floor, you have several options. The first is to use a turntable shelf that mounts to the wall instead of to the floor. The second option is to locate the rack directly above the point in your room where the floor joists meet the foundation. The third alternative, which can be used in conjunction with the second option, is to support the joists with house jacks. Put a concrete block on the ground below the floor joists, put the house jack between the concrete block and the joist, and raise the jack to meet the joist. Continue raising the jack about half a turn past where contact is made between the jack and the joist. Don't try to move the joist; all you want to do is provide a stable support. Finally, you can further increase the turntable's isolation from vibration by using a commercially available product such as the Marigo Turntable Isolation System. Such products decouple the turntable from the stand. Finally, the best solution is to buy a turntable with a superior suspension system.

Once you've got a good rack, where you place it in the room can have a large effect on the sound. Here's why: When the air in a room is excited by sound from the loudspeakers, the pressure isn't evenly distributed throughout the room. Instead, there are stationary pockets of high and low pressure, called standing waves. You can demonstrate this by playing an organ recording and walking around the room; you'll hear the sound get louder and softer. Moreover, the pressure tends to build up along walls and in corners. If the turntable is in an area of high pressure, it will be more prone to resonate from the acoustic energy impinging on it. Conversely, a turntable located in a room "null" won't be as affected by sound from the loudspeakers striking it. You can avoid the worst turntable placements by following a few simple rules. First, don't put the turntable in a corner where bass tends to build up. Second, keep the turntable away from the wall, if possible. An extra 8" or 12" can make a difference.

With the turntable on your stand, make sure that the turntable is level by using a pair of small round bubble levels and adjusting the turntable's feet. If you have a sprung turntable, make sure the tonearm cable doesn't impede the motion of the sub-chassis.

I'll assume you've followed the manufacturer's directions for installing the tonearm and are ready to mount the cartridge. After you've attached the four tiny tonearm cable clips to the cartridge pins, mount the cartridge to the headshell with minimal torque on the bolts—we'll be adjusting the cartridge position in the headshell later. Just before attaching the tonearm cable clips to the cartridge, consider applying a contact enhancer on the cartridge pins such as Tweak from Sumiko or a similar product from Caig.

Move the counterweight forward or backward until the arm floats, evenly balanced, then move it slightly forward to apply some tracking force. Use a stylus pressure gauge (also called a vertical tracking force gauge) to set the tracking force (Fig.9-9) to the cartridge manufacturer's tracking-force specification. (Some tonearms include built-in calibrated tracking-force gauges.)

Fig. 9-9 A vertical tracking force gauge. (Courtesy AcousTech)

The next step roughly sets the *vertical tracking angle* (VTA), also called the stylus rake angle. Adjusting the VTA changes the angle at which the stylus enters the groove. Ideally, the VTA of the playback stylus should match the VTA of the cutting stylus. Most cartridges work best when the cartridge body is exactly parallel with the record surface (Fig.9-10a). VTA is adjusted by moving the tonearm rear up or down at the pivot point. Negative VTA is when the cartridge rear is below the level of the cartridge front (Fig.9-10b). Positive VTA is when the cartridge rear is above the level of the cartridge front (Fig.9-10c). Follow the cartridge manufacturer's suggestions for rough setting of VTA. (We'll fine-tune the VTA later.)

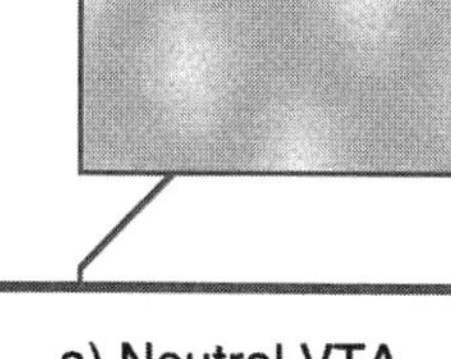

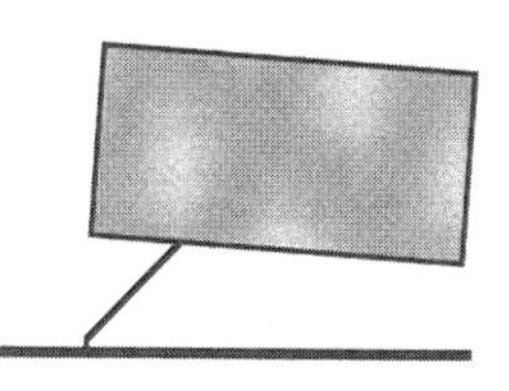

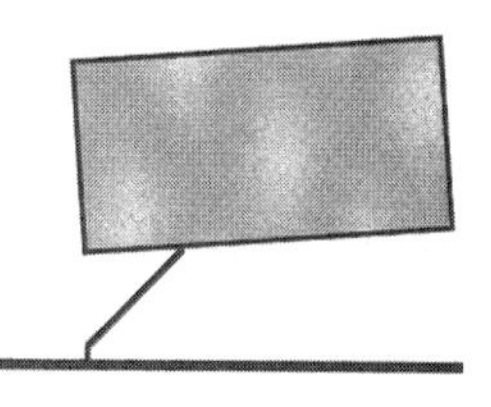

Fig. 9-10 Vertical Tracking Angle (VTA) adjustments

A precision protractor is essential to correctly setting the overhang and the offset. As we saw earlier, setting the overhang and offset correctly minimizes lateral tracking error. Overhang is adjusted by moving the cartridge forward and backward in the headshell, and offset by twisting the cartridge body in the headshell. The single-point templates sometimes provided with tonearms are accurate only if the templates are designed for the specific tonearm, and include an outline of the tonearm printed on the template.

Fig.9-11 shows a high-quality protractor that offers both Löfgren and Baerwald alignments. A quality protractor is considerably more expensive than a paper template, but much more accurate. Note that all alignment devices (see above comment) assume that the cartridge's cantilever is parallel with the cartridge body, which isn't always the case.

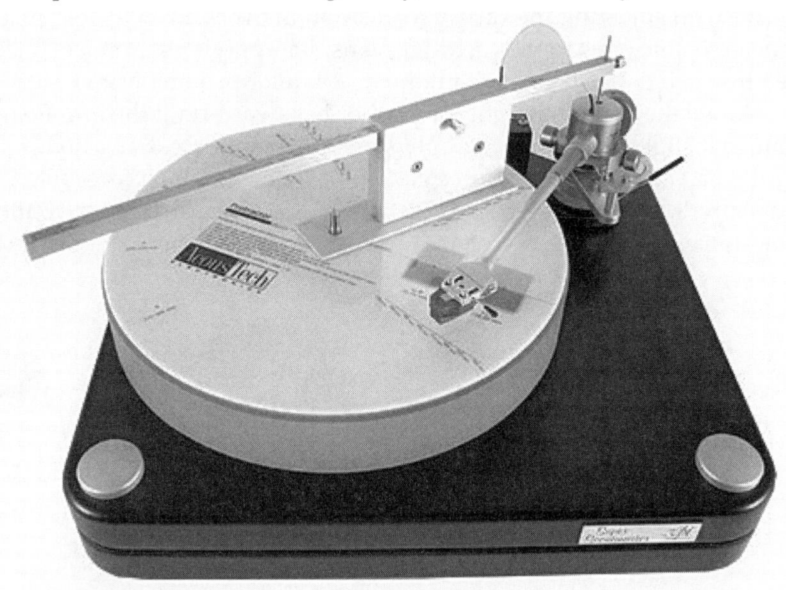

Fig. 9-11 A high-quality protractor allows you to optimally set up your tonearm and cartridge. (Courtesy Feickert and AcousTech)

Azimuth describes the cantilever's perpendicular relationship to the record groove. A cantilever with perfectly aligned azimuth will form a right angle with the record (Fig.9-12b). Notice that I said the cantilever's—not the cartridge's—relationship to the groove. Many cantilevers are slightly skewed in their mountings, making it impossible to achieve perfect azimuth alignment by looking at the cartridge body alone. Instead, you can put the stylus tip on a small mirror and look for a perfectly straight line through the cantilever and its reflection. If there's a bend in the reflection at the mirror, the azimuth is wrong. Some tonearms are designed to allow for ease of azimuth adjustment; those that aren't require you to insert tiny shims between headshell and cartridge. Azimuth alignment is vital to correct left-right channel balance, and consequently to soundstaging and imaging. Azimuth can also be checked with a dual-trace oscilloscope and a test record. With each channel feeding one input of the 'scope, play a test record with a pure tone. Perfect azimuth alignment will produce identical amplitudes from both 'scope channels. A simpler method for assuring perfect azimuth alignment without an oscilloscope is the Fosgate "Fozgometer," a $250 device that, when used with a test record, allows you to quickly dial-in perfect azimuth alignment.

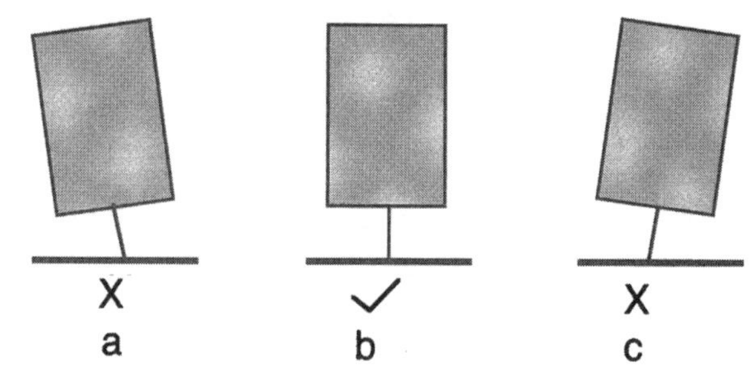

Fig. 9-12 Azimuth is the perpendicular relationship of the cantilever to the record.

Recheck the cartridge alignment and overhang with the protractor; when these are correct, gently tighten the cartridge mounting bolts. The tightness of the bolts should strike a balance between making the cartridge tight in the headshell and distorting the cartridge body. Don't overtighten them.

Following the tonearm manufacturer's instructions, set the anti-skating force. If you don't have the instructions, and the tonearm doesn't have a calibrated anti-skating scale, you can set the anti-skating with a grooveless record (test pressings generally have one grooveless side), or a sheet of Mylar cut into the shape of a record with a spindle hole. Put the stylus about halfway between the spindle and the outer edge, and spin the grooveless record or Mylar sheet. Adjust the anti-skating so that the arm moves very slowly toward the outside of the record. If the anti-skating is set incorrectly, mistracking on loud passages will occur more in one channel than in the other. Too much anti-skating will cause the left channel to distort first; too little will make the right channel distort first.

Another trick for setting the anti-skating force is to sum the left and right channels into mono, and invert the polarity of one channel. When playing a record, the anti-skate will be properly set when the least sound is heard from the loudspeakers. Correct anti-skating adjustment will produce equal output from both channels; when one channel is inverted, the greatest cancellation will occur (the least sound heard) when the output levels are most closely matched. To make an inverting cable, buy a Y-adapter with two female jacks and one male. Cut off one lead and reverse the shield and center conductor to invert the polarity. The turntable output drives the two female jacks, and the single output feeds one phono input on the preamplifier. This method works very well, and costs about $3 for the Y-adapter.

The next step is to fine-tune the VTA by listening. Mark on the tonearm rear the position that produces a parallel line between the cartridge bottom and the record. Raise the tonearm rear a few millimeters, then listen. Lower the tonearm rear a few millimeters below the parallel line and listen again. Find the position that provides the smoothest tonal balance and best imaging. Use several records for setting VTA, not just one or two.

There are many more ways to improve the sound of your LP front end. You can replace the stock tonearm cable with a high-end one; place isolation feet under the turntable; put contact enhancer on the cartridge pins; or put damping material on the tonearm tube, cartridge, finger lift, and counterweights. Damping material should be applied with care; it may upset the careful balance and resonance-tuning designed into the turntable and tonearm.

You can check the turntable's speed with a strobe disc and light. You hold the light above the disc as it spins on your turntable. When spinning at exactly the right speed, the disc's dots appear stationary. Some outboard motor drives have fine-tuning adjustments for turntable speed.

There's another way of getting the best performance from your analog front-end: hire a local turntable set-up expert. Such experts often work in high-end retail stores; others are independent but can be recommended by your dealer. Watching him fine-tune your 'table can also be a great learning experience.

Record Care and Cleaning

How your records and stylus are cared for will greatly affect the sound quality of your analog front-end. Many of the sonic defects the general public associates with LPs (ticks and pops, skipping) are the result of mishandled records. I have records bought in the early 1970s that have been played countless times that are still in good condition. Picking up a few simple habits will allow your record collection to sound its best as well as deliver decades of musical pleasure. For a thorough visual presentation on record care, I recommend *It's a Vinyl World, After All* by Michael Fremer. The DVD also takes you behind the scenes at an LP mastering session as well as showing you how LPs are made.

Records should be stored vertically and packed tightly enough so that the records remain at a right angle to the floor (not leaning), but not so tightly that they are pressed together. If you've seen an LP jacket with a white ring in the center (an outline of the label), that record has been stored incorrectly. The pressure from too-tight storage creates a raised area on the jacket around the label that wears quickly. Records should also be stored away from heaters and direct sunlight. A record left in even a slight leaning position at a moderately high temperature will loose its flatness. Incidentally, you can buy an LP flattener that gently heats the records and applies even pressure across the entire surface. It looks like a waffle iron (but with flat interior surfaces).

The most important rule of record care is to never touch the disc surface, even on the outside edge. You should instead remove the record and sleeve together from the jacket, and then slide your hand into the sleeve to the label without touching the record. Remove the record with your index finger on the label and your thumb on the record's edge. Once the record is clear of the sleeve, hold the record by its edges with both hands. I've seen people who should know better (manufacturers at trade shows, for example) pull a record out of its sleeve by grasping the outer edge of the record. This approach smears body oils on the outside grooves of the record, which are then distributed by the stylus through the grooves. Even worse, the oil acts to hold dirt and debris on the stylus.

Speaking of record sleeves, it's a good idea to replace paper and plastic sleeves with those made from rice-paper. Paper sleeves leave tiny fibers on the disc surface, and plastic sleeves leech chemicals on your records.

Before playing an LP, remove dust from the turntable with a micro-fiber cloth. After putting the LP on the turntable, a quick clean of the record with a carbon-fiber brush removes surface dust. Hold the brush gently above the spinning record for a few revolutions.

Ideally, the stylus should be cleaned before playing each record side. The stylus picks up dirt as it is dragged through the record groove, degrading the sound quality and accelerating groove wear. Cleaning the stylus with a commercially available brush and liquid solution will remove this dirt. A stylus should be cleaned with a back-to-front motion so that the brush follows the record's motion. Some manufacturers recommend that no cleaning fluid be used; others suggest that a fluid is essential to removing accumulated dirt. There is also debate over the best type of brush. Some have short, stiff bristles, while other cleaners resemble nail-polish brushes. Your best bet is to follow the cartridge manufacturer's cleaning instructions.

Dirt isn't the only stylus contaminant; the friction between the stylus and groove heats the stylus enough to slightly melt the vinyl record. Some of this vinyl sticks to the stylus, with obvious detrimental effects. This melted vinyl cannot be removed with most commercial stylus cleaners, but there's a simple and free technique that removes even the most stubborn gunk from your stylus. Hold a matchbook from which all the matches have been removed underneath your stylus and scrape the striking area over the stylus in a back-to-front motion (the same direction as the stylus moves through the record). This abrasive action removes the accumulated coating of vinyl on the stylus.

The coils of moving-coil cartridges become slightly magnetized over time by their proximity to the cartridge's fixed magnets. This magnetization degrades fidelity. The solution is to demagnetize the cartridge with a small battery-operated device that connects to the tonearm's RCA cables. Some phonostages have a built-in demagnetization circuit; you simply press a front-panel button and your cartridge is demagnetized. Note that moving-magnet cartridges don't need demagnetizing, and can be ruined by trying to demagnetize them.

This routine for playing each LP should be augmented with occasional deep record cleaning with a heavier-duty brush and cleaning solution, or with a record-cleaning machine. If you don't want to invest in a full-bore vacuum record cleaner, you can use a brush such as the one shown in Fig.9-13 and cleaning solution. You apply the fluid, scrub the record, and then wipe off the record and let it dry (a dish rack can hold records while they dry). A better approach is to use a cleaning machine with vacuum action (Fig.9-14). After placing the LP on the machine's turntable, apply fluid and allow the machine's integral brush to scrub dirt out of the grooves as the record spins. It's a good idea to spin the record in both directions during this cycle. You then move a vacuum arm over the record and allow the vacuum system to suck up the fluid in which the dirt is suspended. Incidentally, you should periodically clean a machine's brushes with another brush and record-cleaning fluid. You must also keep the machine's turntable clean. If you clean one side of a record, and then turn that record over to clean the other side, the cleaned side will be in contact with the turntable, potentially picking up any dirt on the turntable. Avoid felt mats on cleaning machines; they are dust magnets and are hard to clean.

Fig. 9-13 A brush, cleaning solution, and replacement sleeves are essential parts of a high-quality LP-playback system. (Courtesy Music Direct)

Cleaning a record with a good vacuum system renders amazing results. The surfaces get much quieter, with a blacker background. Even new records benefit from cleaning. Not only do new records have dust on their surfaces that was picked up at the pressing plant, they are coated with mold release. This is a chemical in the raw vinyl that rises to the surface as the record is pressed, allowing the record to be more easily removed from the press.

Records become electrostatically charged, which causes them to attract and hold dust. You can reduce this phenomenon with a device that removes the static charge. One such anti-static device looks like a gun; squeezing the trigger puts pressure on a piezoelectric crystal, causing it to generate an electric field.

It might sound like a lot of work to properly care for records and your stylus, but it seems like second-nature once these processes become a habit. You'll be rewarded with better sounding LPs that can last a lifetime.

Fig. 9-14 A vacuum-based record-cleaning machine. (Courtesy VPI Industries and Music Direct)

Vinyl as Art: Half-Speed Mastering, 45rpm Pressings, 180-Gram Vinyl, and Direct-to-Disc LPs

The resurgence of vinyl playback has fueled a demand for the highest quality records. Several specialty companies have risen to the challenge, creating LPs of exceptional sound. These endeavors are often labors of love, with the proprietors painstakingly researching the mastertape provenance to find the true original master, digging up the original photography from classic recording sessions, creating the LP with no-holds-barred technology, and presenting an album in better-than-original sound and packaging. We'll look at several of the special techniques used in making today's audiophile LPs.

The first technique is *half-speed mastering* in which the tape is played back at half its normal speed, and the mastering lathe's turntable spins at half the normal speed. If the original tape was recorded at 30 ips (inches per second), the tape is played back at 15 ips to cut the record. Concomitantly, the turntable spinning the master lacquer runs at 16 2/3 rpm rather than 33 1/3 rpm. The net result is that the record plays back the proper pitch when played at 33 1/3 rpm. One advantage to half-speed mastering is that the signal is lowered in frequency by half; 20kHz becomes 10kHz, for example. The cutting stylus has more time to precisely cut the signal into the master lacquer (technically, the groove velocity is reduced by half). In addition, the amplifier driving the cutting head need produce only half the power because amplifier power is proportionate to stylus velocity. The cutting amplifier isn't stressed, and less heat builds up in the cutting head's coils. Of course, it takes twice as long to cut a half-speed lacquer, but the effort is worth it; half-speed mastered LPs sound better than conventional LPs, all other factors being equal. (Fig.9-15 shows an LP mastering lathe.)

Fig. 9-15 A pair of LP mastering lathes. (Courtesy The Mastering Lab; photo by Neil Gader)

Some records are reissued on 12" LPs cut at 45 rpm. Playback requires a turntable that will spin at 45 rpm (virtually all do). Because records cut at 45 rpm have shorter playing time, a regular 40-minute record is often released on two discs. The advantage of cutting and playing back a disc at 45 rpm records is significant. With the record moving faster under the cutting stylus, wavelengths become longer and are easier to cut and to play back. The cutting stylus has more time to cut the signal, and the playback stylus more time to trace the groove modulation.

You've probably noticed that the inner LP grooves don't sound as clean, open, and extended as the outer grooves. The reason is that the recorded wavelengths become shorter toward the center of the record. Because the record spins at the same speed regardless of where the stylus is (a format called *constant angular velocity*), each revolution of the record with the stylus in the outermost groove travels a much greater distance than the stylus tracing the innermost groove (specifically, the distance around the outermost modulated groove is 2.5 times the distance around the innermost modulated groove). The signal gets "scrunched up" toward the center of the record. High frequencies, in particular, are affected by this phenomenon because they have the shortest wavelengths to start with. A 45 rpm record ameliorates this phenomenon by making all wavelengths longer.

Incidentally, the CD is a *constant linear velocity* format in that the rotational speed varies according to the playback radius. The CD spins at about 500 rpm near the innermost radius and 200 rpm at the outermost radius, maintaining the same speed (constant linear velocity) as seen by the laser.

Most audiophile LPs today are pressed on 180-gram or even 200-gram vinyl. This thicker vinyl makes for a more solid record that is less prone to vibration. Today's premium LPs are also manufactured to a much higher standard than LPs made when records were a mass-market item. For example, the record is held in the press longer, which assures that the molten vinyl completely fills the void between the stampers. You might have noticed that if an LP is noisy, the noise usually occurs at the outermost grooves. This is because the small vinyl "biscuit" that is pressed into a record flows outward toward the stamper edges; a too-short press cycle doesn't allow the vinyl to fully fill in the outer edges of the record, resulting in surface noise. With LPs selling at premium prices, record manufacturers can lavish on these records longer press-cycle times as well as 180-gram premium-quality vinyl.

The highest quality LP is the *direct-to-disc,* in which the master lacquer is cut in real-time as the musicians perform. Let's consider a little history before getting to modern direct-to-disc LPs.

The first sound recording, made 133 years ago, was created by cutting into a piece of tinfoil a physical representation of the acoustic waveform. This direct-to-disc process (actually, direct-to-cylinder in Edison's case) would be the only available recording method for many decades until the invention of the wire recorder, and later, the tape recorder. The tape recorder freed musicians from the burden of performing the entire side of a record live in the studio. Each track could be recorded individually, edited into an album, and later cut into a record. The tape machine also divided the recording and disc-mastering processes into separate events. Before the widespread use of tape that began in 1948, a musical group would go into the studio in the morning and have a finished record later that day.

With all these advantages of tape, recording to a transcription disc was happily abandoned by musicians and engineers despite its sonic superiority. The world forgot about disc recording until 1968 when Doug Sax and Lincoln Mayorga created Sheffield Lab. Their stupendously-great sounding LPs demonstrated beyond any doubt that the purest recording format extant was the direct-to-disc LP. Sax is known today as the father of modern direct-to-disc recording (and of *stereo* direct-to-disc recording; all disc recording before the advent of tape was mono).

A few other purist labels followed Sheffield's lead and made direct-to-disc recordings, but the process was largely abandoned because of the immense difficulties involved. Making a direct-to-disc recording isn't for the faint-of-heart. In conventional LP mastering, the physical cutting of the record—the last creative step and the first manufacturing step—takes place in an environment of perfect calm. The mastering engineer and the record's producer work with a finished mastertape, transferring an electrical signal from the tape machine to the physical motion of the lathe's cutting head. If something goes wrong, or the results aren't optimal, they simply put on a new lacquer and start over. There are no musicians, no microphones, no mixing, and no performances to worry about.

Contrast that scene with a direct-to-disc session. The band plays together in the studio—no overdubbing less-than perfect parts or "fixing it in the mix." The group must not only play together, but perform the entire LP side with no mistakes. Once the cutting head starts at the outside of the record, there's no turning back.

The challenge for the recording engineer is to mix all the microphones to stereo on the fly as the band plays. If he misses a cue (bringing up the microphone on a soloist, for example), the band has to start over and play the entire record side again. For the mastering engineer cutting the LP in real-time as the band plays, the challenge is to cut that signal into a groove with the highest possible sound quality but with no technical errors. In cutting a disc from tape, an additional playback head (called the "preview" head) picks up the signal before the playback head. This preview signal is used solely to control the groove spacing, called the "pitch." When the preview head picks up a high-level signal, it causes the groove spacing to become wider to accommodate the greater excursion of the cutting head. When the signal is low in level, the preview system packs the grooves more tightly together. The invention of variable pitch is what made the long-playing record possible. The direct-to-disc mastering engineer enjoys no such safety net. Because the signal is coming right off the recording console, there's no preview, which means the mastering engineer must constantly adjust the pitch by hand during the entire LP side based on his anticipation of the music's dynamics and bass content. If the average pitch is too wide, he'll run out of room on the lacquer. If he fails to anticipate a high-level signal, the grooves will crash into each other and ruin the take.

Getting every element of the musical performance, engineering, and disc-cutting absolutely right—all at the same time—is a daunting challenge. But consider the sonic advantages of removing from the signal path a tape machine and all its electronics (not to mention the second pass of the signal through a recording console). Direct-to-disc recording not only greatly shortens the signal path for better sound, it has enormous implications for the entire recording process and consequently, the musical result. Musicians in a typical multitrack recording session listen to the rhythm tracks through headphones and play their parts—sometimes dozens of times—until they're right. Records are made in which the musicians aren't even in the studio at the same time. It can be a soul-sapping experience.

A direct-to-disc session couldn't be more different. The band not only plays together, but shares the adrenaline rush of knowing that *this is the take that's going on the record.* It's a transformative experience. The result is a musical energy that just doesn't exist in a multitrack session.

For the music lover and audiophile, there's no closer experience to a live musical event than a direct-to-disc LP played on a first-rate analog front-end.

10 FM Tuners, Satellite Radio, HD Radio, and Internet Radio

Introduction

A tuner is a source component that receives radio transmissions from the air and converts them into a line-level audio signal that feeds an input on a preamplifier. The tuner has an antenna input (or inputs) and usually one pair of unbalanced line-level outputs.

Compared with other source components, the tuner plays a different role in a high-end system. On one hand, FM is a lower-quality source than LP and CD, and doesn't let your system live up to its full sonic potential. On the other hand, a tuner is an unending source of free music, and an invaluable way of being exposed to new musical forms and artists you might not have otherwise heard. How important a tuner is in your listening life is greatly influenced by the musical programming of stations in your area, how far away you live from transmitters, the care (or lack of care) taken by the station in achieving good sound, and the quality of your tuner.

The digital revolution offers listeners a wide range of new technologies that augment, or even supplant, FM radio. These include satellite radio (familiar from its inclusion in automobile audio systems), HD Radio, and Internet radio. We'll look at each of these in detail, but we'll first consider FM radio, the radio format of choice for many music lovers who are fortunate to live within reception range of some of the country's great FM stations.

The limiting factor in broadcast sound quality is often the radio station, not the FM transmission format or even the tuner. FM broadcasts of live concerts by radio stations that care about sound, and received by a good tuner, can be spectacular. I haven't any firsthand experience, but the live broadcasts by WGBH in Boston of the Boston Symphony Orchestra reportedly sounded superb.

Further, it's possible to set up a closed-circuit FM broadcast in a listening room using the highest-quality source components, then receive that broadcast signal with a tuner under evaluation. The source signal can thus be compared directly with the closed-circuit

FM transmission. Listening tests and demonstrations of this technique reveal that FM is capable of extremely high sound quality.

How to Choose a Tuner

Choosing a tuner for your system is a little different from choosing other components. When auditioning other components, we're primarily concerned with their sound quality, not their technical performance. If, for example, a preamplifier under audition sounds good, we don't need to worry much about its technical performance—if it sounds good, it's probably working well.

Tuners, on the other hand, exhibit great variability in their technical performance. We're interested not only in aspects of a tuner's sound—tonal balance, soundstaging, portrayal of timbre, etc.—but also in more basic characteristics such as the ability to pick up weak or distant stations, reject adjacent stations, provide a noise-free audio signal, and stay tuned to a station without drifting. A tuner's performance in these areas can be accurately characterized by measurement; this makes tuner specifications much more meaningful than those of other audio components. There is a direct correlation between a tuner's specifications and its sonic performance. You'll still need to listen to the tuner before you buy, but you can often separate poorer-performing models from better units by looking at the specification sheets. And unlike most audio products, in which the highest-performance units have the fewest features, the best high-end tuners have more features, front-panel controls, and displays than the lower-end products.

How much of your audio budget you should allocate to a tuner varies greatly with the individual. Some listeners just want a tuner for background music and National Public Radio, not for serious music listening. If this describes you, or if you listen to FM only occasionally, you may consider a tuner from a mid-fi company. The performance will be acceptable (if you choose wisely), and you can put more of your budget into components that matter more to you musically and sonically.

Those listeners who want the best possible technical and musical performance from FM will opt for a tuner designed and made by a high-end company. Several companies have dedicated themselves exclusively to the design and manufacture of tuners: their products are often superb, but expensive. Without the economies of scale provided by a factory making an entire line of audio products, the dedicated tuner company must charge more for its products. It can, however, focus its talents and engineering expertise on making the best tuner possible. If you want the highest performance and are willing to pay for it, seek out companies that have established reputations for making superlative tuners.

The price range for a good tuner from a mass-market manufacturer is between $400 and $1000. Some of the higher-end models from mass-market companies offer excellent performance. The price range from $750 to $1200 is very competitive, with many superb units to choose from. The very best tuners cost as much as $12,000.

Although we'll discuss tuner specifications in detail later in this chapter, I'll briefly describe here the primary differences between mediocre and excellent tuners.

Good tuners are characterized by their *sensitivity*, or ability to pull in weak stations. The greater a tuner's sensitivity, the better it can pick up weak or distant stations. This aspect of a tuner's performance is more important in suburban or rural areas that are far from radio transmitters.

A tuner characteristic of greater importance to the city dweller is *adjacent-channel selectivity*—the ability to pick up one station without interference from the station next to it on the dial. The *alternate-channel selectivity* specification defines a tuner's ability to reject a strong station two channels away from the desired channel. When stations are packed closely together, as they are in cities, adjacent-channel and alternate-channel selectivity are more important than sensitivity.

Equally important to all listeners is the tuner's signal-to-noise ratio, a measure of the difference in dB between background noise and the maximum signal strength. A tuner with a poor signal-to-noise ratio will overlay the music with an annoying background hiss. In short, a poor tuner will have trouble receiving weak stations, may lack the ability to select one station when that station is adjacent to another station, have high background noise, and be overloaded by nearby FM transmitters or other radio signal sources (taxi dispatchers, for example).

Tuners can be roughly divided into two categories: analog tuning and digital tuning (the latter is more correctly called *frequency-synthesized* tuning). If the tuner has a dial and pointer moved with a flywheel knob, it uses analog tuning. Frequency-synthesized tuners move in discrete jumps; from, say, 88.1MHz to 88.3MHz. A frequency-synthesized tuner will lock on to the station and can't be mistuned (unless the tuner is misaligned). (Note that the presence of a digital station readout doesn't mean that the tuner is synthesized, only that the display is digital.) Synthesized tuners provide features such as *seek*—to move to the next station—and the ability to store many stations in a preset memory, recalled at the touch of a button. Another common feature of synthesized tuners is *scan*, which stops briefly at each station until you find a station you want and tell it to stop. *Memory scan* samples only the stations preset in the tuner's station.

Although it would appear that synthesized tuners have a big advantage over analog tuners, the very best tuners are all analog. Analog tuners have lower noise, and also allow fine-tuning to find the center of a station. Synthesized tuners jump in discrete steps of at least 25kHz, precluding the precise degree of fine-tuning possible with an infinitely variable analog tuner.

Better tuners have a feature called *selectable IF bandwidth* that adjusts the bandwidth in the tuner's intermediate frequency (IF) stage for best sound quality and minimum interference from adjacent stations. When in the "wide" mode, the audio quality improves at the expense of lower selectivity (less adjacent-channel rejection) and lower sensitivity. Specifically, a wide IF bandwidth provides the lowest distortion and the best high-frequency audio response. The high-frequency audio response of some tuners isn't affected by IF bandwidth, but all tuners have better imaging with wider IF bandwidths: phase shift in the IF filter may compromise stereo separation (and add distortion) at high frequencies. Wide bandwidth is used when no strong adjacent stations are present; narrow bandwidth is selected when the dial is crowded near the desired station. The best tuners have three selectable IF bandwidth settings.

Nearly every tuner has a *signal-strength meter* to indicate the strength of the received signal. The meter, usually a row of Light Emitting Diodes (LEDs), is helpful in positioning an antenna for best reception. The most sophisticated tuners have an oscilloscope display that provides a means of perfectly locating the center of a station. This expensive feature is found only on analog tuners.

Another useful tuner feature, used in conjunction with the signal-strength meter, is a *multipath indicator*. Multipath occurs when the broadcast signal reaches the tuner's antenna directly, in addition to reflections of that same signal off objects such as buildings. The slight delay between the two signals causes loss of audio-channel separation and increased distortion. In a car stereo, multipath distortion is sometimes called "picket-fencing": the signal swells and fades as the car travels quickly in and out of multipath reflections. A tuner's multipath indicator allows you to rotate your antenna for least multipath. A signal-strength meter alone doesn't discriminate between a clean signal and a signal with multipath, making a multipath indicator a very useful feature.

The high-end tuner shown in Fig.10-1 has a multipath indicator (the meter on the left side of the front panel). The middle meter shows center tuning (how closely to the station's center frequency the tuner is set). The right-hand meter indicates signal strength.

Fig. 10-1 A high-end analog tuner (Courtesy Magnum Dynalab)

Nearly all tuners have a *local/distant* switch. When in the Local position, the tuner simply attenuates the antenna's signal level before it gets to the tuner's input circuit, and prevents the tuner from being overloaded by strong stations.

A tuner's *muting* function mutes the audio output when the signal received by the tuner drops below some defined level. Muting prevents a blast of noise through your system when the tuner is between stations, or on a very weak station. A tuner's muting can be turned off (if the tuner has a muting on/off switch) to receive very weak stations.

The *mono/stereo* button found on most tuners allows you to reduce noise from weak stations by making the signal mono instead of stereo. The complete loss of left/right stereo separation is more than offset by the significant reduction in noise.

Many tuners have a *high-blend* circuit that automatically switches the signal to mono when the signal strength falls below a certain level. The difference between high-blend and the mono/stereo switch just described is that the high-blend circuit puts only the treble into mono, leaving the rest of the spectrum in stereo. This gets rid of most of the noise, but maintains stereo separation through most of the midrange and bass.

Some tuners have a switch marked "MPX" that invokes a 19kHz filter to remove the 19kHz pilot tone from the broadcast signal. This feature is necessary when recording on an analog tape machine: the 19kHz pilot tone can interfere with the tape machine's Dolby noise-reduction tracking. Some analog recorders have a switchable 19kHz filter (also marked "MPX") in case your tuner lacks this feature.

Finally, all good tuners have a 75 ohm coaxial antenna input as well as the more commonly used 300 ohm flat lead input. The coaxial input should be used for best signal transmission between the antenna and tuner.

What to Listen For

If the tuner has high technical performance standards—particularly sensitivity and selectivity—we can then characterize the tuner's musical performance using the listening techniques described in Chapter 3.

In addition to the usual sonic checklist, tuners should be auditioned with an ear to certain qualities unique to tuners. Many tuners overlay the music with a whitish haze that makes listening fatiguing. This sound isn't perceived as noise, but as a fuzz that rides over the music. It sounds as though the channel isn't perfectly tuned, even though it's tuned in as well as the tuner allows.

Better tuners have a more extended and open treble. Lower-quality units tend to sound closed-in, lacking air, and even rolled-off in the treble. This characteristic can obscure low-level detail and make the music bland.

The music's dynamic contrast is also greatly affected by the tuner's audio quality. Some tuners squash dynamics to a point where the sound is lifeless and flat. Others have a much wider variation between loud and soft. Note, however, that most radio stations compress the music's dynamic range before the transmitter to achieve a higher average signal level, making the station seem louder and "stick out" when someone is scanning the dial. The best stations—primarily classical music stations—leave the dynamic range intact. When evaluating a tuner's dynamic range, be sure to use a variety of stations, particularly high-quality classical ones.

A mediocre tuner can compress soundstage depth to the point where the music is a flat, sterile canvas, not a huge, spacious panorama. The music becomes congested, thick, cold, and uninviting. The best tuners can present a deep, spacious soundstage, with real depth and precise imaging.

Many of the negative sonic qualities I've described aren't inherent in the FM format; an excellent radio station broadcasting to a high-end tuner is capable of outstanding musical performance.

When evaluating tuners, keep in mind that a tuner is only as good as the signal supplied to it from the antenna. A high-quality antenna is essential to getting the best technical and musical performance from your tuner. If you're auditioning a tuner at a dealer's showroom or a friend's house, remember that the antenna's quality is a significant variable. Depending on your antenna, you may get better or worse performance than you expect. Further, the very tiny voltages transmitted down the antenna lead, through the connections, and to the tuner input require tight, clean connections. A loose-fitting jack, or some corrosion on the terminals, can greatly degrade a tuner's performance.

Tuner Specifications and Measurements

The earlier part of this chapter described some of a tuner's fundamental specifications and how they relate to choosing a tuner. In this next section, we'll take a closer look at what constitutes good and poor tuner performance.

Sensitivity, or the ability to receive weak or distant stations, is defined technically as the radio frequency (RF) signal strength required to produce an audio signal output with a specified signal-to-noise ratio: the lower the sensitivity specification, the better (less signal required for good reception). Sensitivity is expressed as a voltage across the antenna in microvolts (μV), or as signal power in dBf (decibels referenced to one femtowatt, 10^{-15}W, or one trillionth of a watt). Two methods of expressing the signal strength across the antenna exist because of the differences between the two antenna impedances (300 ohms and 75 ohms). For example, 30dBf represents 18μV across 75 ohms, but only 9μV across 75 ohms. If you compare dBf specifications, you don't need to worry about this difference. But when looking at microvolts, be certain that the impedance is specified. If one tuner's sensitivity is specified across the 75 ohm input, and the other across a 300 ohm input, simply double the 75 ohm figure and the two specifications will be directly comparable.

Tuner sensitivity is more precisely defined by specifying the voltage across the antenna required to produce an audio signal with a specified signal-to-noise ratio, usually 50dB. This specification is called the 50dB *quieting sensitivity*. A less stringent specification, called *usable sensitivity*, is the voltage required to achieve a 30dB signal-to-noise ratio—which is barely listenable. When comparing tuner specifications, be sure that the manufacturer specifies the 50dB quieting-sensitivity figure.

Sensitivity is different for mono and stereo reception: Mono requires less RF signal strength to achieve a specified quieting sensitivity. In fact, a tuner may require more than double the signal strength to achieve the same quieting sensitivity in stereo than in mono. The very best tuners have a sensitivity (for 50dB quieting) of 30dBf (stereo) and 10dBf (mono). Usable sensitivity (for 30dB quieting) figures for an excellent tuner are 10dBf (stereo) and 8dBf (mono). Lower-sensitivity tuners may have a 50dB quieting specification of 40dBf (stereo) and 20dBf (mono). The lower the number, the better the tuner.

With alternate-channel selectivity, however, the higher the number, the better. This specification indicates how well the tuner can reject strong stations near the desired station. Technically, alternate selectivity is the ratio (in dB) of the signal strength needed to produce a reference output level on the wanted channel to the signal strength needed to produce an audio output level 30dB below the reference level from a station two channels away. The higher the selectivity, the greater the tuner's ability to reject unwanted adjacent or alternate stations. Selectivity may be the most important tuner specification, particularly in areas with a crowded FM dial.

Selectivity is often specified with both narrow and wide IF bandwidths. The narrower bandwidth improves the selectivity figures, but decreases audio quality. The best tuners have an alternate-channel selectivity of 100dB in narrow mode, and an adjacent-channel selectivity of 40dB (also in narrow mode). In wide mode, a tuner's alternate-channel selectivity may be reduced to 30dB. Lower-quality tuners may have an alternate-channel selectivity specification of 40dB. These tuners generally don't have selectable IF bandwidth.

When a tuner receives two stations at the same frequency, the tuner must suppress the weaker one and capture the stronger one. *Capture ratio* is the difference in dB between the strengths of the two stations needed before the tuner can lock to the stronger station and reject the weaker one. The lower the capture ratio, the better; the stronger station doesn't have to be that much stronger for the tuner to reject the weaker one. Although it is relatively uncommon to receive two stations at the same frequency, a good capture ratio helps in preventing multipath distortion; the weaker reflected signal is more easily rejected. A capture ratio of 1dB is excellent, 1.5dB is average, and 2dB is poor. This specification, along with adjacent-channel selectivity, is of paramount importance to the city dweller who receives strong multipath reflections from buildings. Capture ratio has a high degree of correlation with sound quality, particularly for those listeners who can't put up a directional antenna to minimize multipath. Multipath interference often produces severe amplitude modulation of the FM carrier, an aberration that multiplex stereo decoders are particularly sensitive to.

A tuner's signal-to-noise ratio specifies the level of background noise, usually in both mono and stereo. Signal-to-noise must be specified with a given input signal, usually a very high 65dBf. Some manufacturers specify signal-to-noise ratio at 85dBf, which yields a better number than 65dBf. With a roof antenna located ten miles from a transmitter, many signals may be in the 65dBf range, and the strongest signals may exceed 85dBf. When comparing specifications, be certain that the signal-to-noise ratio is specified with the same signal strength. A good signal-to-noise specification is 90dB (mono) and 80dB (stereo) with a 65dBf signal. The higher the signal-to-noise ratio, the quieter the tuner (all other factors being equal).

Stereo separation is the degree of isolation, measured in dB, between the left and right audio channels. Greater separation correlates with increased spaciousness to the sound. Stereo separation is measured at 1kHz with a 50dB quieting input, and the IF bandwidth adjustment in the wide position (if the tuner has selectable IF bandwidth). A tuner designer can hype a separation specification by tailoring the tuner for wide separation at 1kHz at the expense of the rest of the audio band. It is more difficult to achieve wide separation consistently over the band. This requires excellent phase linearity in the IF stage, or careful phase-compensation of the feed to the multiplex decoder. The IHF/EIA standards for tuner measurements include separation figures up to a frequency of 6kHz. The separation at 6kHz is a more reliable indicator of a tuner's stereo separation than the 1kHz specification. Unfortunately, not all tuner manufacturers adhere to the IHF/EIA measurement standards. Tuners range from about 40dB of stereo separation, to 70dB in the best models (both are specified at 1kHz).

Satellite Radio

Satellite radio is an extremely popular technology that delivers hundreds of commercial-free stations to your car, home, or portable stereo. Unlike AM or FM radio, satellite radio is not limited to a narrow geographic range; you can pick up the same stations anywhere the service is available (nearly all of North America). Moreover, satellite radio can offer many more station choices than terrestrial radio, including niche or specialty programming of interest to a relatively small group of listeners. The downside is that satellite radio is a subscription-based service for which you pay a monthly fee.

Satellite radio established itself in the car with the competing Sirius and XM services. Virtually all car manufacturers included either Sirius or XM as original equipment. This incompatibility was resolved when Sirius acquired XM in July, 2008. Although satellite radio is thought of as primarily for the automobile, you can buy satellite receivers for the home. Some tabletop audio systems such as the Polk iSonic include satellite-receiver compatability. The Sirius system is also delivered to many homes via cable or satellite television. Some personal portable satellite receivers have the ability to record the live stream. In addition, a free iPhone app lets you stream Sirius to your iPhone, iTouch, and iPad. Of course, you much purchase a subscription to access the service.

Perceptual coding reduces the number of bits in the transmitted signal. The number of bits available varies with the channel, resulting in varying sound quality between channels. Some sound fairly good, but none is as good as uncompressed linear PCM audio at 44.1kHz/16-bit as used on CD. The major benefit over FM is the lack of multipath distortion and signal "fade," particularly in mobile applications.

HD Radio

HD Radio is a digital replacement for terrestrial AM and FM broadcasts. The digital HD Radio signal is broadcast in parallel with the AM or FM signal by your local radio station. If you have an HD Radio receiver, you can listen to the digital version of the radio program rather than the analog AM or FM version. Unlike satellite radio, HD Radio is subscription-free. You must, however, buy an HD Radio tuner to receive the digital broadcasts. Don't be confused by the "HD" moniker; HD Radio is not the radio equivalent of HDTV. Rather, HD originally stood for "Hybrid Digital" but now, according to iBiquity, the company who developed the technology, HD is not an acronym for anything. Also unlike HDTV, the FCC has not mandated a date at which AM and FM broadcasts must cease. HD Radio was chosen by the FCC in 2002 for digital audio broadcasting in the U.S.

Technically, HD Radio uses a proprietary codec called COFDM that is based on MPEG-4. the data rate varies with the station, from a low of 100kbps (kilobits per second) to a high of 300kbps. At this highest mode, 5.1-channel surround-sound is possible, although the sound quality is poor. For comparison, the Dolby Digital format on DVD typically runs at a data rate of 384kbps for 5.1-channels. If the digital datastream is lost or corrupted, the HD Radio receiver automatically switches to the analog signal, resulting in a seamless signal. Metadata in the digital stream can identify the artist, song, and other information that is displayed on the HD Radio receiver.

Internet Radio

Some music servers (and products such as integrated amplifiers) provide integral internet-radio capability. Internet radio, also called webcasting or streaming radio, delivers radio stations from around the world to your computer via the internet. The audio is coded with a lossy system such as MP3 or WMA and streamed to your computer rather than downloaded as files. Internet radio gives you access to virtually any radio station in the world. Chapter 8 ("Music Servers") covers Internet Radio in more detail.

11 Cables and Interconnects

Introduction

Loudspeaker cables and line-level interconnects are an important but often overlooked link in the music playback chain. The right choice of loudspeaker cables and interconnects can bring out the best performance from your system. Conversely, poor cables and interconnects—or those not suited to your system—will never let your system achieve its full musical potential. Knowing how to buy cables will provide the best possible performance at the least cost.

In this chapter we'll look at all aspects of loudspeaker cables and interconnects. We'll cover balanced and unbalanced lines, bi-wiring, matching cables to your system, and how to get the most cable for your money. Moreover, we'll see why the most expensive cables and interconnects aren't always the best.

But first, let's start with an overview of cable and interconnect terms.

Cable: Often used to describe any wire in an audio system, "cable" more properly refers to the conductors between a power amplifier and a loudspeaker. Loudspeaker cables carry a high-current signal from the power amplifier to the loudspeaker.

Interconnect: Interconnects are the conductors that connect line-level signals in an audio system. The connection between source components (turntable, CD player, tuner, tape deck) and the preamplifier, and between the preamplifier and power amplifier, are made by interconnects.

Unbalanced Interconnect: An unbalanced interconnect has two conductors and is usually terminated with RCA plugs. Also called a *single-ended* interconnect.

Balanced Interconnect: A balanced interconnect has three conductors instead of two, and is terminated with 3-pin *XLR* connectors. Balanced interconnects are used only between components having balanced inputs and outputs.

***Digital Interconnect*:** A single interconnect that carries a stereo digital audio signal in the S/PDIF format, usually from a CD transport, music server, or other digital source to a DAC.

USB Cable: A single cable that carries digital audio from a computer-based music server to a USB-capable DAC.

FireWire Cable: Also called IEEE1394, FireWire is a bi-directional interface that can carry high-resolution digital audio.

***Bi-wiring*:** Bi-wiring is a method of connecting a power amplifier to a loudspeaker with two runs of cable instead of one.

***RCA Plug and Jack*:** RCA plugs and jacks are the most common connection termination for unbalanced signals. Virtually all audio equipment has RCA jacks to accept the RCA plugs on unbalanced interconnects. RCA jacks are mounted on the audio component's chassis; RCA plugs are the termination of unbalanced interconnects.

***XLR Plug and Jack*:** XLR plugs are three-pin connectors terminating a balanced interconnect. XLR jacks are chassis-mounted connectors that accept XLR plugs.

***Binding Post*:** Binding posts are terminations on power amplifiers and loudspeakers that provide connection points for loudspeaker cables.

***Five-way Binding Post*:** A type of binding post that can accept bare wire, spade lugs, or banana plugs. Five-way posts are found on most power amplifiers and loudspeakers.

***Spade Lug*:** A flat, pronged termination for loudspeaker cables. Spade lugs fit around power-amplifier and loudspeaker binding posts. The most popular kind of loudspeaker cable termination.

***Banana Plug and Jack*:** Banana plugs are sometimes found on loudspeaker cables in place of spade lugs. Banana plugs will fit into five-way binding posts or banana jacks. Many European products use banana jacks on power amplifiers for loudspeaker connection.

***AWG*:** American Wire Gauge: a measure of conductor thickness, usually in loudspeaker cables. The lower the AWG number, the thicker the wire. Lamp cord has an AWG of 18, usually referred to as "18 gauge."

HDMI: An acronym for High-Definition Digital Multimedia Interface, HDMI was developed for home theater to carry high-definition video along with high-resolution digital audio in the same cable. Chapter 12 ("Audio for Home Theater") includes a detailed description of HDMI.

Note: AC power cords are covered in Chapter 15 along with AC power conditioners.

Some of these cables, interconnects, and terminations are shown in Fig.11-1.

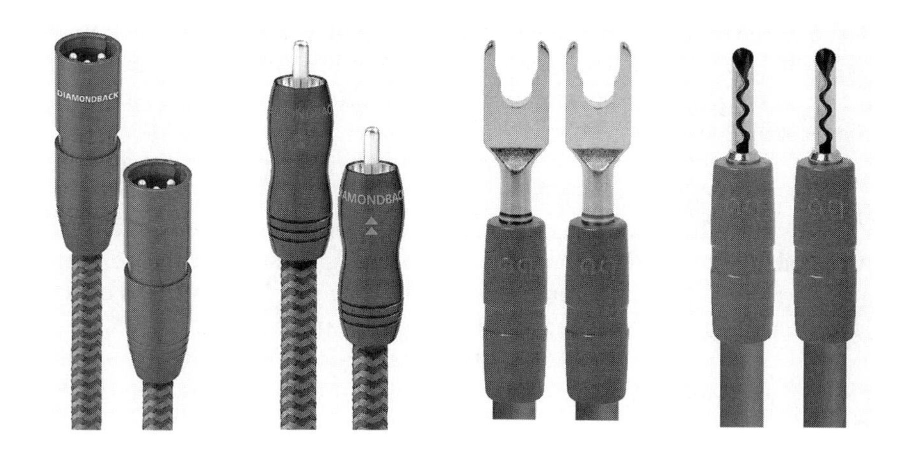

Fig. 11-1 From left to right: balanced interconnects terminated with XLR connectors, unbalanced interconnects terminated with RCA plugs, a spade lug, and a banana plug (Courtesy AudioQuest)

How to Choose Cables and Interconnects

Ideally, every component in the system—including cables and interconnects—should be absolutely neutral and impose no sonic signature on the music. As this is never the case, we are forced to select cables and interconnects with colorations that counteract the rest of the system's colorations.

For example, if your system is a little on the bright and analytical side, mellow-sounding interconnects and cables can take the edge off the treble and let you enjoy the music more. If the bass is overpowering and fat, lean- and tight-sounding interconnects and cables can firm up and lean out the bass. A system lacking palpability and presence in the midrange can benefit from a forward-sounding cable.

Selecting cables and interconnects for their musical compatibility should be viewed as the final touch to your system. A furniture maker who has been using saws, planers, and rasps will finish his work with steel wool or very fine sandpaper. Treat cables and interconnects the same way—as the last tweak to nudge your system in the right direction, not as a Band-Aid for poorly chosen components.

Cables and interconnects won't correct fundamental musical or electrical incompatibilities. For example, if you have a high-output-impedance power amplifier driving current-hungry loudspeakers, the bass will probably be soft and the dynamics constricted. Loudspeaker cables won't fix this problem. You might be able to ameliorate the soft bass with the right cable, but it's far better to fix the problem at the source—a better amplifier/loudspeaker match.

Good cables merely allow the system's components to perform at their highest level; they won't make a poor system or bad component match sound good. Start with a high-quality, well-chosen system and select cables and interconnects that allow that system to achieve its highest musical performance. Remember, a cable or interconnect can't actually effect an absolute improvement in the sound; the good ones merely do less harm.

A typical hi-fi system will need one pair of loudspeaker cables (two pairs for bi-wiring), one long pair of interconnects between the preamplifier and power amplifier, and several short interconnect pairs for connections between source components (such as a turntable or CD player) and the preamplifier.

If the power amplifier is located near the loudspeakers, the loudspeaker cables will be short and the interconnects between the preamplifier and power amplifier will be long. Conversely, if the power amplifier is near the source components and preamplifier, the interconnects will be short and the loudspeaker cables long. There is no consensus among the experts about which method is preferable, but I use long interconnects and short loudspeaker cables. Ideally, interconnects *and* loud-speaker cables should be short, but that often isn't practical.

Once you've got a feel for how your system is—or will be—configured, make a list of the interconnects and cables you'll need, and their lengths. Keep all lengths as short as possible, but allow some flexibility for moving loudspeakers, putting your preamp in a different space in the rack, or other possible changes. Although we want to keep the cables and interconnects short for the best sound, there's nothing worse than having interconnects 6" too short. After you've found the minimum length, add half a meter for flexibility.

Interconnects are often made in standard lengths of 1, 1.5, and 2 meters. These are long enough for source-to-preamplifier connections, but too short for many preamplifier-to-power-amplifier runs. These long runs are usually custom-made to a specific length. Similarly, loudspeaker cables are typically supplied in 8' or 10' pairs, but custom lengths are readily available. It's best to have the cable manufacturer terminate the cables (put spade lugs or banana plugs on loudspeaker cables, and RCA or XLR plugs on interconnects) rather than trying to do it yourself.

Concentrate your cable budget on the cables that matter most. The priority should be given to the sources you listen to most. For example, you may not care as much about the sound of your tuner as you do your DAC. Consequently, you should spend more on interconnects between the DAC and preamplifier than between the tuner and preamp. And because all your sources are connected to the power amplifier through the interconnect between the preamplifier and power amplifier, this link must be given a high priority. But any component—even an iPod's analog output—will benefit from good interconnects.

Should all your interconnects and loudspeaker cables be made by the same manufacturer? Or is it better to mix and match brands? There are two schools of thought on this issue. The first holds that an entire system with one brand of cable and interconnect is the best route. If one interconnect works well in your system, use it throughout. This argument also suggests that the cable designer made his interconnects and loudspeaker cables to work together to achieve the best possible sound.

The second school of thought holds that different brands are best. Because each cable or interconnect affects the sound in a certain way, using the same interconnect and cable throughout the system will only reinforce the cable's sonic signature. By using cables and interconnects from different manufacturers, the characteristic sonic signature of a cable won't be superimposed on the music by every interconnect.

This second theory has an analog in the recording world. Engineers will record through one brand of recording console, then mix the record through a different brand of console. They don't want to hear the console's sound in the final product, so they don't subject the signal to the same sonic signature twice.

My experience suggests that the only way to determine the best cable or interconnect for your system is to experiment and listen. In some cases, the best results will be achieved with all the interconnects and cables made by the same manufacturer. In others, a mix of different interconnects will work best. It's impossible to predict which cables will sound best in your system.

Most dealers will let you take home several cables at once to try in your system. Take advantage of these offers. Some mail-order companies will send you many cables to try: you keep the ones you want to buy—if any—and return the others. Compare inexpensive cables with expensive ones—sometimes manufacturers have superb cables that sell for a fraction of the price of their top-of-the-line products.

If you're starting a system from scratch, selecting cables is more difficult than replacing one length in your system. Because different combinations of cables will produce different results, the possibilities are greatly increased. Moreover, you don't have a baseline reference against which to judge how good or bad a cable is. In this situation, the best way of getting the ideal cables for your system is your dealer's advice. Try the cables and interconnects he suggests, along with two other brands or models for comparison.

How Much Should You Spend on Cables and Interconnects?

At the top end of the scale, cable and interconnect pricing sometimes bears little relationship to the cost of designing and manufacturing the product. Unlike other audio products, whose retail prices are largely determined by the parts cost (the retail price is typically four to six times the cost of raw parts), cables and interconnects are sometimes priced according to what the market will bear. This trend began when one company set its prices vastly higher than everyone else's—and saw its sales skyrocket as a result. Other manufacturers then raised *their* prices so they wouldn't be perceived as being of lower quality. Although some very expensive cables and interconnects are worth the money, many cables are ridiculously overpriced.

The budget-conscious audiophile can, however, take advantage of this phenomenon. Very often, a cable manufacturer's lower-priced products are very nearly as good as its most expensive models. The company prices their top-line products to foster the impression of being "high-end," yet relies on its lower-priced models for the bulk of its sales. When shopping for loudspeaker cables and interconnects, listen to a manufacturer's lower line in your system—even if you have a large cable budget. You may be pleasantly surprised.

Because every system is different, it's impossible to be specific about what percentage of your overall system investment you should spend on cables and interconnects. Spending 5% of your system's cost on cables and interconnects would be an absolute minimum, with about 15% a maximum figure. If you choose the right cables and interconnects, they can be an excellent value. But poor cables on good components will give you poor sound and thus constitute false economy.

Again, I must stress that high cost doesn't guarantee that the cable is good or that it will work well in your system. Don't automatically assume that an expensive cable is better than a low-priced one. Listen to a wide variety of price levels and brands. Your efforts will often be rewarded with exactly the right cable for your system at a reasonable price.

What to Listen For

Cables must be evaluated in the playback system in which they will be used. Not only is the sound of a cable partially system-dependent, but the sonic characteristics of a specific cable will work better musically in some systems than in others. Personal auditioning is the *only* way to evaluate cables and interconnects. Never be swayed by technical jargon about why one cable is better than another. Much of this is pure marketing hype, with little or no relevance to how the cable will perform musically in your system. Trust your ears.

Fortunately, evaluating cables and interconnects is relatively simple; the levels are automatically matched between cables, and you don't have to be concerned about absolute-polarity reversal. One pitfall, however, is that cables and interconnects need time to break in before they sound their best. Before break-in, a cable often sounds bright, hard, fatiguing, congested, and lacking in soundstage depth. These characteristics often disappear after several hours' use, with days or weeks of use required for full break-in. You can't be sure, however, if the cable is inherently bright- and hard-sounding, or if it just needs breaking-in. Note that break-in wears off over time. Even if a cable has had significant use, after a long period of not being used it may not sound its best until you've put music through it for a few days.

With those cautions in mind, you're ready to evaluate cables and interconnects. Listen to the first interconnect for 15 minutes to half an hour, then replace it with the next candidate. One way of choosing between them is merely to ask yourself which interconnect allows you to enjoy the music more. You don't need to analyze what you're hearing; just pick the interconnect that makes you feel better.

The other method is to scrutinize what you're hearing from each interconnect and catalog the strengths and weaknesses. You'll often hear trade-offs between interconnects: one may have smoother treble and finer resolution than another, but less soundstage focus and transparency. Another common trade-off is between smoothness and resolution of detail: The smooth cable may lose some musical information, but the high-resolution cable can sound analytical and bright. Again, careful auditioning in your own system is the only way to select the right cables and interconnects. Keep in mind, however, that a better cable can sometimes reveal flaws in the rest of your system. You should also know that cables and interconnects sound better after they have "settled in" for a few days.

Cables and interconnects can add some annoying distortions to the music. I've listed the most common sonic problems of cables and interconnects. (A full description of these terms is included in Chapter 3.)

***Grainy and hashy treble*:** Many cables overlay the treble with a coarse texture. The sound is rough rather than smooth and liquid.

***Bright and metallic treble*:** Cymbals sound like bursts of white noise rather than a brass-like shimmer. They also tend to splash across the soundstage rather than sounding like compact images. Sibilants (s and sh sounds on vocals) are emphasized, making the treble sound spitty. It's a bad sign if you suddenly notice more sibilance. The opposite condition is a dark and closed-in treble. The cable should sound open, airy, and extended in the treble without sounding overly bright, etched, or analytical.

***Hard textures and lack of liquidity*:** Listen for a glassy glare on solo piano in the upper registers. Similarly, massed voices can sound glazed and hard rather than liquid and richly textured.

***Listening fatigue*:** A poor cable will quickly cause listening fatigue. The symptoms of listening fatigue are headache, a feeling of relief when the music is turned down or stopped, the need to do something other than listen to music, and the feeling that your ears are tightening up. This last condition is absolutely the worst thing any audio component can do. Good cables (in a good system) will let you listen at higher levels for longer periods of time. If a cable or interconnect causes listening fatigue, avoid it no matter what its other attributes.

***Lack of space and depth*:** Using a recording with lots of natural depth and ambiance, listen for how the cable affects soundstage depth and the sense of instruments hanging in three-dimensional space. Cables also influence on the sense of image focus. Poor cables can also make the soundstage less transparent.

***Low resolution*:** Some cables and interconnects sound smooth, but they obscure the music's fine detail. Listen for low-level information and an instrument's inner detail. The opposite of smoothness is a cable that's "ruthlessly revealing" of every detail in the music, but in an unnatural way. Musical detail should be audible, but not hyped or exaggerated. The cable or interconnect should strike a balance between resolution of information and a sense of ease and smoothness.

***Mushy bass or poor pitch definition*:** A poor-quality cable or interconnect can make the bass slow, mushy, and lacking in pitch definition. With such a cable, the bottom end is soggy and fat rather than taut and articulate. Low-frequency pitches are obscured, making the bass sound like a roar instead of being composed of individual notes.

***Constricted dynamics*:** Listen for the cable or interconnect's ability to portray the music's dynamic structure, on both small and large scales. For example, a guitar string's transient attack should be quick, with a dynamic edge. On a larger scale, orchestral climaxes should be powerful and have a sense of physical impact (if the rest of your system can portray this aspect of music).

I must reiterate that putting a highly colored cable or interconnect in your system to correct a problem in another component (a dark-sounding cable on a bright loudspeaker) isn't the best solution. Instead, use the money you would have spent on new cables toward better loudspeakers—*then* go cable shopping. Cables and interconnects shouldn't be Band-Aids; instead, cables should be the finishing touch to let the rest of your components perform at their highest level.

Binding Posts and Cable Terminations

Binding posts vary hugely in quality, from the tiny spring-loaded, push-in terminal strips on cheap loudspeakers to massive, custom-made, machined brass posts plated with exotic metals. Poor binding posts not only degrade the sound, they also break easily. When shopping for power amplifiers and loudspeakers, take a close look at binding-post quality.

The most popular type is the five-way binding post. It accepts spade lugs, banana plugs, or bare wire. Some five-ways are nickel-plated; higher-quality ones are plated with gold and won't tarnish. Five-way binding posts should be tightened with a 1/2" nut driver or tool specifically designed for binding posts, not a socket and ratchet or wrench that could easily overtighten the nut. The connection should be tight, but not to the point of stripping the post or causing it to turn in the chassis. When tightening a five-way binding post, watch the inside ring or collar next to the chassis; if it begins to turn, you've overtightened the post and are in danger of damaging the power amplifier or loudspeaker.

Custom posts of heavy-duty machined metal are more robust than five-way posts—they can accept more torque without stripping or coming loose in the chassis. Custom posts are often found on the most expensive equipment. Some binding posts have such a large center post that spade lugs won't fit around them. These posts have large holes in the center for accepting large bare wires or banana jacks. Although these posts are expensive and appear to be of high quality, they're inconvenient to use. If your equipment has this sort of post, the best solution is to terminate your loudspeaker cables with oversized spade lugs. Most spade lugs have a distance between the prongs of 1/4" to 3/16"; oversized spades have prongs 5/16" apart—enough of a difference to fit the large-holed posts.

If you have a choice of bare wire, banana plug, or spade lug on loudspeaker cable terminations, go with the spade lug. It forms the best contact with a binding post and is the most standard form of connection. Many European products provide only banana jacks, forcing you to use sonically inferior banana-plug terminations.

You should be aware that any termination slightly degrades the sound of your system. Consequently, some audiophiles have gone to the trouble of removing all plugs, jacks, spade lugs, and binding posts from their systems and hard-wiring everything together. This is an extreme measure and makes switching equipment difficult or impossible. Hard-wiring is an option, but not one that should be undertaken without considerable deliberation and technical expertise.

Bi-Wired Loudspeaker Cables

Bi-wiring is running two lengths of cable between the power amplifier and loudspeaker. This technique usually produces much better sound quality than conventional single-wiring. Most high-end loudspeakers have two pairs of binding posts for bi-wiring, with one pair connected to the crossover's tweeter circuit and the other pair connected to the woofer circuit. The jumpers connecting the two pairs of binding posts fitted at the factory must be removed for bi-wiring.

In a bi-wired system, the power amplifier "sees" a higher impedance on the tweeter cable at low frequencies, and a lower impedance at high frequencies. The opposite is true in the woofer-half of the bi-wired pair. This causes the signal to be split up, with high frequencies traveling mostly in the pair driving the loudspeaker's tweeter circuit and low frequencies conducted by the pair connected to the loudspeaker's woofer circuit. This frequency splitting reportedly reduces magnetic interactions in the cable, resulting in better sound. The large magnetic fields set up around the conductors by low-frequency energy can't affect the transfer of treble energy. No one knows exactly how or why bi-wiring works, but on nearly all loudspeakers with bi-wiring provision, it makes a big improvement in the sound. Whatever your cable budget, you should bi-wire if your loudspeaker has bi-wired inputs, even if it means buying two runs of less expensive cables.

You can bi-wire your loudspeakers with two identical single-wire runs, or with a specially prepared bi-wire set. A bi-wire set has one pair (positive and negative) of terminations at the amplifier end of the cable, and two pairs at the loudspeaker end of the cable. This makes it easier to hook up, and probably offers slightly better sound quality.

Loudspeakers can also be connected with a *single bi-wire* set in which a single cable with multiple internal conductors has two pairs of terminations on one end and a single pair of terminations at the other end. Although this approach is much less expensive than two runs of cable, you lose the benefit of magnetically isolating the low- and high-frequency conductors from each other. Two bi-wiring options are shown in Fig.11-2.

Most bi-wired sets use identical cables for the high- and low-frequency legs. Mixing cables, however, can have several advantages. By using a cable with good bass on the low-frequency pair, and a more expensive but sweeter-sounding cable on the high-frequency pair, you can get better performance for a lower cost. Use a less expensive cable on the bass and put more money into the high-frequency cable. If you've already got two pairs of cable the same length, the higher-quality cable usually sounds better on the high-frequency side of the bi-wired pair. If you use different cables for bi-wiring, they should be made by the same manufacturer and have similar physical construction. If the cables in a bi-wired set have different capacitances or inductances, those capacitances and inductances change the loudspeaker's crossover characteristics.

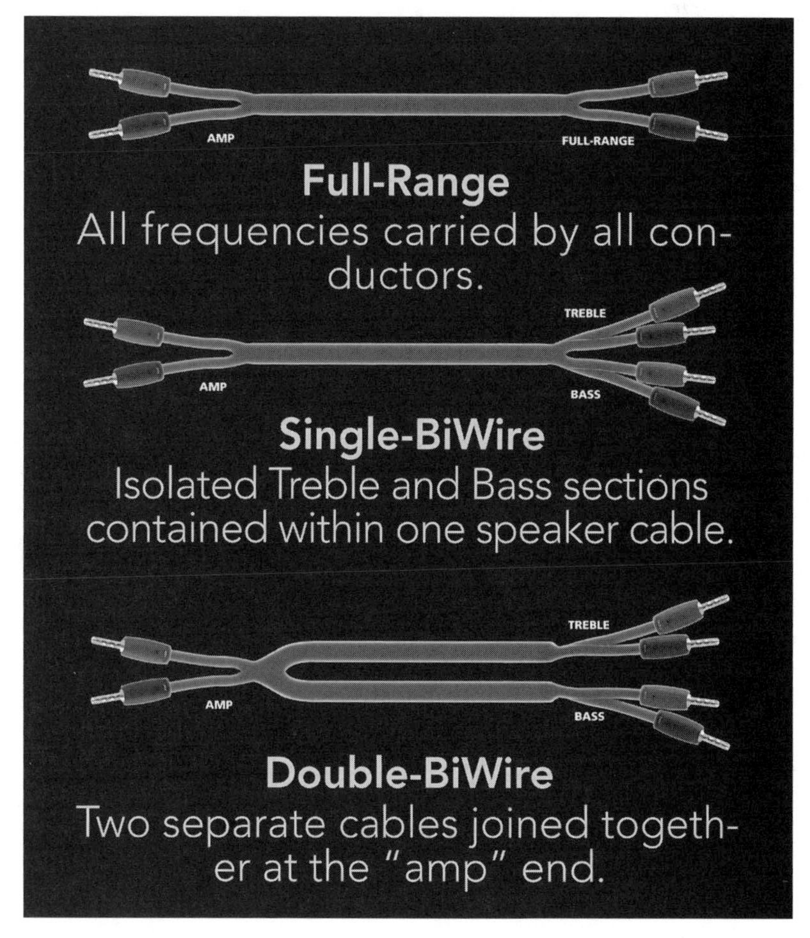

Fig. 11-2 Loudspeakers can be connected with one cable (top), a single bi-wire connection (middle), with a double-bi-wire cable (bottom), or with two completely separate runs of cable (not shown). (Courtesy AudioQuest)

Balanced and Unbalanced Lines

Line-level interconnects come in two varieties: balanced and unbalanced. A balanced interconnect is recognizable by its three-pin XLR connector. An unbalanced interconnect is usually terminated with an RCA plug. Balanced and unbalanced lines are shown in the photograph at the beginning of this chapter.

Why do we use two incompatible systems for connecting equipment? At one time, all consumer audio hardware had unbalanced inputs and outputs, and all professional gear was balanced. In fact, balanced inputs are often called "professional inputs" to differentiate them from "consumer" unbalanced jacks. Balanced connection was considered both unnecessary and too expensive for home playback systems.

The emergence of high-end audio changed that thinking. Instead of using the least expensive connection method, high-end product designers started using higher-quality balanced lines and terminations in consumer gear. The better the equipment, the more likely that it has at least some balanced connections in addition to unbalanced jacks. Moreover, more and more manufacturers are offering balanced connections on their equipment. This is why we have two standards—balanced and unbalanced. The technical and sonic advantages of balanced connection, once the exclusive domain of audio professionals, are becoming increasingly available for home playback systems.

But what exactly is a balanced line, and how is it different from a standard RCA cable and jack?

In an unbalanced line, the audio signal appears across the center pin of the RCA jack and the shield, or ground wire. Some unbalanced interconnects have two signal conductors and a shield, with the shield not used as a signal conductor. If this unbalanced interconnect happens to be close to fluctuating magnetic fields—an AC power cord, for example—the magnetic field will induce a noise signal in the interconnect that is heard as hum and noise reproduced through the playback system's loudspeakers.

In a professional application, this hum, buzz, and noise is unacceptable, leading to the development of an interconnection method that is immune to noise interference: the balanced line. A balanced line has three conductors: two carrying signal, and one ground. The two signals in a balanced line are identical, but 180° out of phase with each other. When the signal in one of the conductors is at peak positive, the signal in the other conductor is at peak negative (see Fig.11-3). The third conductor is signal ground. Some balanced interconnects use three conductors plus a shield.

When the two identical but opposite polarity signals carried on the balanced line are input to a *differential amplifier* in the component receiving the signal, noise picked up by the interconnect is rejected. Here's why: a differential amplifier amplifies only the *difference* between the two signals (see Fig.11-4). If noise is introduced into the line, the noise will be common to both conductors and the differential amplifier will reject the noise. This phenomenon of rejecting noise signals common to both conductors in a balanced line is called *common-mode rejection*. Differential inputs are specified according to how well they reject signals common to both conductors, a measurement called *Common-Mode Rejection Ratio*, or *CMRR*. Note that a balanced line won't make a noisy signal clean; it just prevents additional noise from being introduced in the interconnect. If the noise is common to both halves of the balanced line, however, common-mode rejection will eliminate the noise.

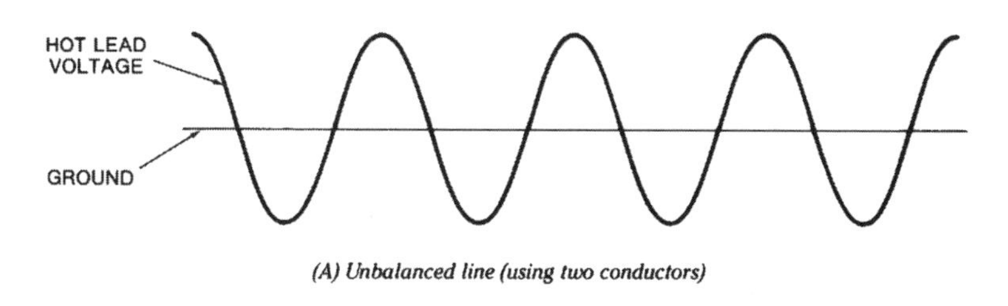

(A) Unbalanced line (using two conductors)

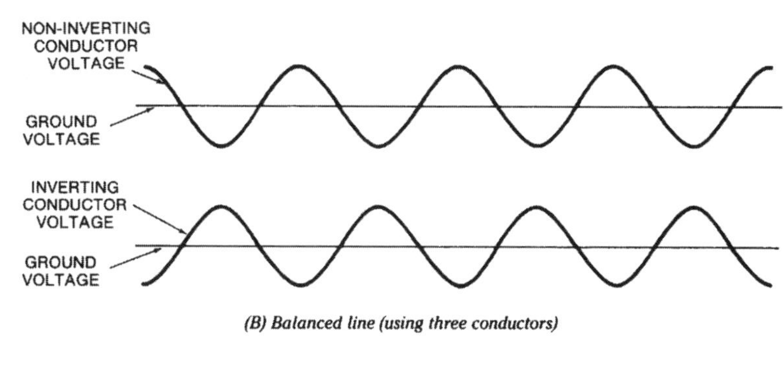

(B) Balanced line (using three conductors)

Fig. 11-3 Signal voltages applied to an unbalanced and a balanced line (Copied with permission from *Audio Technology Fundamentals* by Alan A. Cohen)

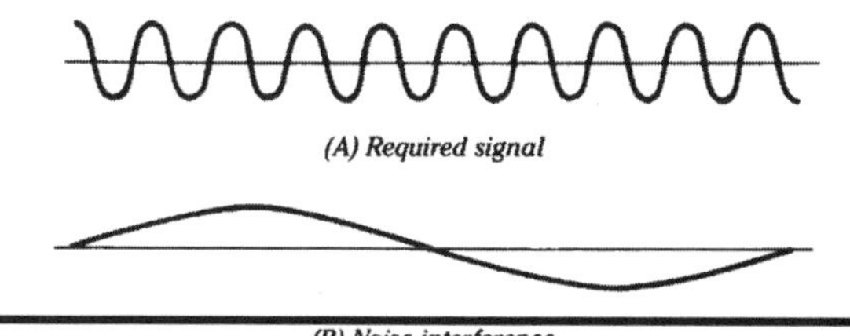

(A) Required signal

(B) Noise interference

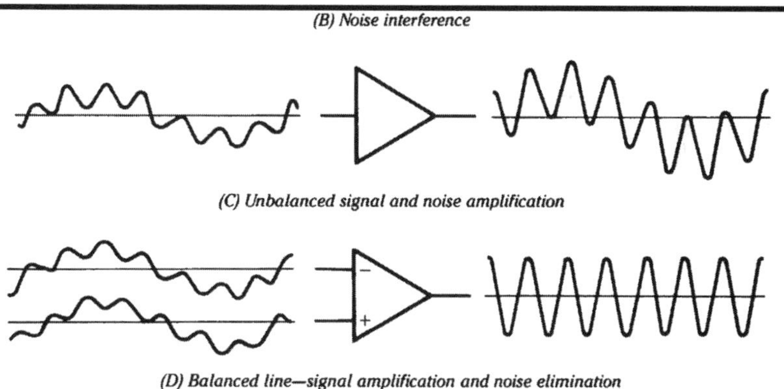

(C) Unbalanced signal and noise amplification

(D) Balanced line—signal amplification and noise elimination

Fig. 11-4 When the two signals carried by the balanced line are input to a differential amplifier, noise common to both conductors is rejected (bottom). (Copied with permission from *Audio Technology Fundamentals* by Alan A. Cohen)

In professional applications, a transformer sometimes serves the same function as the differential amplifier, passing only the difference between the two signal conductors and rejecting signals common to both conductors. Consumer audio equipment uses differential amplifiers, not transformers, but the concept is identical.

In a balanced line terminated with XLR connectors, pin 1 is always signal ground. There is, however, no universal convention for which of the two signal conductors carries the non-inverted signal and which carries the inverted signal. The non-inverted conductor is often called the "hot" conductor, with the inverted conductor designated "cold." After decades of no clear standard, the Audio Engineering Society recently adopted the North

American tradition of having pin 2 carry the non-inverted ("hot") signal, pin 3 the inverted ("cold") signal.

How the balanced line is wired (pin 2 or pin 3 "hot") can determine if your playback system is inverting or non-inverting of absolute polarity (see the section on absolute polarity in Appendix A). If your system is non-inverting—that is, a positive-going signal on the LP or CD produces a positive-going signal at the loudspeakers—substituting a pin 3 "hot" balanced input power amplifier for a pin 2 "hot" power amplifier will make your system inverting. When switching balanced components—DACs, preamps, or power amplifiers—you should know if the new component's XLR jacks are wired the same—either pin 2 or pin 3 "hot"—as the existing component. You can also change your system's absolute polarity by rewiring the balanced interconnect to swap the wires going to pins 2 and 3. It's far better, however, to simply reverse absolute polarity by switching red for black and black for red on both your loudspeaker cables if you want to change your system's absolute polarity. The freak interconnect with pins 2 and 3 switched at one end may wind up in another system or application where you don't want the polarity switched.

Quite apart from the advantage of noise cancellation in a balanced line, a balanced connection often sounds better than an unbalanced line. A system connected by balanced interconnects can, however, also often sound less good than one connected with unbalanced lines. Say you have a disc player or DAC that takes an unbalanced signal from the digital-to-analog converter chip and converts it to a balanced signal so that the DAC manufacturer can tout the product as having "balanced outputs" (see the section on balanced DACs in Chapter 7). Inside the DAC, the unbalanced signal is converted to a balanced signal by a *phase splitter*, a circuit that takes a signal of one polarity and turns it into two signals of opposite polarity. Phase splitting subjects the unbalanced signal to an additional active (transistor- or op-amp– based) stage and puts more circuitry in the signal path.

The balanced DAC's output is then input to a balanced-input preamplifier. Because all but the very best balanced preamplifiers convert a balanced input signal to an unbalanced signal for the preamplifier's internal gain stages, the preamplifier's input converts this balanced signal to an unbalanced signal—adding yet another active stage to the signal path. After the unbalanced signal is amplified within the preamplifier, it is converted back to balanced with another phase splitter.

The preamplifier's balanced output is then sent from the preamplifier output to the power amplifier's balanced input where it's—that's right—converted to unbalanced with yet another active stage. The result of these unbalanced/balanced/unbalanced/balanced/unbalanced conversions is additional electronics in the signal path—just what we don't want. This is why you can't assume that balanced components sound inherently better than unbalanced ones. Magazine reviews of audio components should include musical and technical comparisons of the product's balanced and unbalanced modes.

Some products, however, are truly balanced and don't rely on phase splitters and unbalancing amplifiers. For example, a DAC may create a balanced signal in the digital domain (at no sonic penalty but, indeed, a sonic gain) and convert that balanced signal to analog with four digital-to-analog converters and analog output stages (left channel + and –, right channel + and –). Similarly, some preamplifiers are truly balanced and have double the circuitry to operate on the non-inverting and inverting signals separately. You can tell a truly balanced preamplifier by the number of elements in the volume control. A preamplifier that operates on an unbalanced signal internally will have two volume-control elements: one for the left channel, one for the right. A fully balanced preamplifier will have four elements: ± left channel and ± right channel. The signal thus stays balanced from before the DACs inside the digital processor all the way to the final stages in the power amplifier.

As in all things audio, the proof is in the listening. When shopping for a component, listen to it in both balanced and unbalanced modes. Let your ears decide if the component works best in your system when connected via the balanced or unbalanced lines.

Cable and Interconnect Construction

Cables and interconnects are composed of three main elements: the signal conductors, the dielectric, and the terminations. The *conductors* carry the audio signal; the *dielectric* is an insulating material between and around the conductors; and the *terminations* provide connection to audio equipment. These elements are formed into a physical structure called the cable's *geometry*. Each of these elements—particularly geometry—can affect the cable's sonic characteristics.

Conductors

Conductors are usually made of copper or silver wire. In high-end cables, the copper's purity is important. Copper is sometimes specified as containing some percentage of "pure" copper, with the rest impurities. For example, a certain copper may be 99.997% pure, meaning it has three-thousandths of one percent impurities. These impurities are usually iron, sulfur, antimony, aluminum, and arsenic. Higher-purity copper—99.99997% pure—is called "six nines" copper. Many believe that the purer the copper, the better the sound. Some copper is referred to as *OFC*, or *Oxygen-Free Copper*. This is copper from which the oxygen molecules have been removed. It is more proper to call this "oxygen-reduced" copper because it is impossible to remove all the oxygen. In practice, OFC has about 50ppm (parts per million) of oxygen compared to 250ppm of oxygen for normal copper. Reducing the oxygen content retards the formation of copper oxides in the conductor, which can interrupt the copper's physical structure and degrade sound quality.

Another term associated with copper is *LC*, or *Linear Crystal*, which describes the copper's structure. Drawn copper has a grain structure that can be thought of as tiny discontinuities in the copper. The signal can be adversely affected by traversing these grains; the grain boundary can act as a tiny circuit, with capacitance, inductance, and a diode effect. Standard copper has about 1500 grains per foot; LC copper has about 70 grains per foot. Fig.11-5 shows the grain structure in copper having 400 grains per foot. Note that the copper isn't isotropic; it looks decidedly different in one direction than the other. All copper made into thin wires exhibits a chevron structure, shown in the photograph of Fig.11-5. This chevron structure may explain why some cables sound different when reversed.

Conductors are made by casting a thick rod, then drawing the copper into a smaller gauge. Another technique—which is rare and expensive—is called "as-cast." This method casts the copper into the final size without the need for drawing.

The highest-quality technique for drawing copper is called "Ohno Continuous Casting" or *OCC*. OCC copper has one grain in about 700 feet—far less than even LC copper. The audio signal travels through a continuous conductor instead of traversing grain boundaries. Because OCC is a process that can be performed on any purity of copper, not all OCC copper is equal.

Fig. 11-5 Copper in cables and interconnects has a grain structure. (Courtesy AudioQuest)

The other primary—but less prevalent—conductor material is silver. Silver cables and interconnects are obviously much more expensive to manufacture than copper ones, but silver has some advantages. Although silver's conductivity is only slightly higher than that of copper, silver oxides are less of a problem for audio signals than are copper oxides. Silver conductors are made using the same drawing techniques used in making copper conductors.

The Dielectric

The dielectric is the material surrounding the conductors, and is what gives cables and interconnects some of their bulk. The dielectric material has a large effect on the cable's sound; comparisons of identical conductors and geometry, but with different dielectric materials, demonstrate the dielectric's importance.

Dielectric materials absorb energy, a phenomenon called *dielectric absorption*. A capacitor works in the same way: a dielectric material between two charged plates stores energy. But in a cable, dielectric absorption can degrade the signal. The energy absorbed by the dielectric is released back into the cable slightly delayed in time—an undesirable condition.

Dielectric materials are chosen to minimize dielectric absorption. Less expensive cables and interconnects use plastic or PVC for the dielectric. Better cables use polyethylene; the best cables are made with polypropylene or even Teflon dielectric. One manufacturer has developed a fibrous material that is mostly air (the best dielectric of all, except for a vacuum) to insulate the conductors within a cable. Other manufacturers inject air in the dielectric to create a foam with high air content. Just as different dielectric materials in capacitors sound different, so too do dielectrics in cables and interconnects.

Terminations

The terminations at the ends of cables and interconnects are part of the transmission path. High-quality terminations are essential to a good-sounding cable. We want a large surface contact between the cable's plug and the component's jack, and high contact pressure between them. RCA plugs will sometimes have a slit in the center pin to improve contact with the jack. This slit is effective only if the slit end of the plug is large enough to be compressed by insertion in the jack. Most high-quality RCA plugs are copper with some brass mixed in to add rigidity. This alloy is plated with nickel, then flashed with gold to prevent oxidation. On some plugs, gold is plated directly to the brass. Other materials for RCA plugs and plating include silver and rhodium.

RCA plugs and loudspeaker cable terminations are soldered or welded to the conductors. Most manufacturers use solder with some silver content. Although solder is a poor conductor, the spade lugs are often crimped to the cable first, forming a "cold" weld that makes a gas-tight seal. In the best welding technique, *resistance welding,* a large current is pulsed through the point where the conductor meets the plug. The resistance causes a small spot to heat, melting the two metals. The melted metals merge into an alloy at the contact point, ensuring good signal transfer. With both welding and soldering, a *strain relief* inside the plug isolates the electrical contact from physical stress.

Geometry

How all of these elements are arranged constitutes the cable's *geometry*. Some designers maintain that geometry is the most important factor in cable design—even more important than the conductor material and type.

An example of how a cable's physical structure can affect its performance: simply twisting a pair of conductors around each other instead of running them side by side. Twisting the conductors greatly reduces capacitance and inductance in the cable. Think of the physical structure of two conductors running in parallel, and compare that to the schematic symbol for a capacitor, which is two parallel lines.

This is the grossest example; there are many fine points to cable design. I'll describe some of them here, with the understanding that I'm presenting certain opinions on cable construction, not endorsing a particular method.

Most designers agree that *skin effect,* and interaction between strands, are the greatest sources of sonic degradation in cables. In a cable with high skin effect, more high-frequency signal flows along the conductor's surface, less through the conductor's center. This occurs in both solid-core and stranded conductors (Fig.11-6). Skin effect changes the cable's characteristics at different depths, causing different frequency ranges of the audio signal to

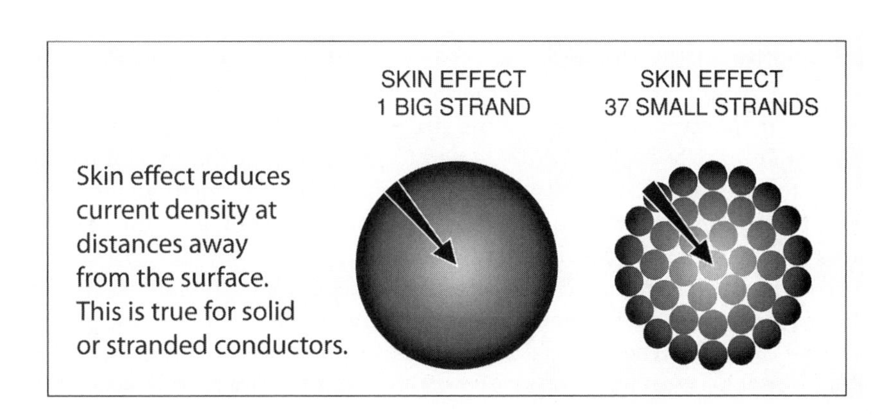

Fig. 11-6 Skin effect causes more of the audio signal to travel on the outside of the conductor. (Courtesy AudioQuest)

be affected by the cable differently. The musical consequences of skin effect include loss of detail, reduced top-octave air, and truncated soundstage depth.

A technique for battling skin effect is *litz* construction, which simply means that each strand in a bundle is coated with an insulating material to prevent it from electrically contacting the strands around it. Each small strand within a litz arrangement will have virtually identical electrical properties. Litz strands push skin-effect problems out of the audible range. Because litz strands are so small, many of them bundled together in a random arrangement are required to achieve a sufficient gauge to keep the resistance low.

A problem with stranded cable (if it isn't of litz construction) is a tendency for the signal to jump from strand to strand if the cable is twisted. One strand may be at the outside at a point in the cable, then be on the inside farther down the cable. Because of skin effect, the signal tends to stay toward the outside of the conductor, causing it to traverse strands. Each strand interface acts like a small circuit, with capacitance and a diode effect, much like the grain structure within copper.

Individual strands within a conductor bundle can also interact magnetically. Whenever current flows down a conductor, a magnetic field is set up around that conductor. If the current is an alternating-current audio signal, the magnetic field will fluctuate identically. This alternating magnetic field can induce a signal in adjacent conductors (see Appendix B), and degrade the sound. Some cable geometries reduce magnetic interaction between strands by arranging them around a center dielectric, which keeps them farther apart. These are just a few of the techniques used by cable designers to make better-sounding cables.

Terminated Cables and Interconnects

Some cables and interconnects are not merely pieces of wire, but contain electronic components. These cables are identified by "boxes" on one or both cable ends housing the resistors, inductors, and capacitors that form an electrical "network." These products are often called terminated cables. To my knowledge, two companies offer terminated cables and interconnects: Music Interface Technologies (MIT) and Transparent Audio (Figs.11-7 and 11-8).

According to MIT, some of the audio signal's voltage is stored in the cable's capacitance, and some of the signal's current is stored in the cable's inductance. The amount of energy stored in the cable varies with frequency, which is manifested as tonal and dynamic emphases at those frequencies. Moreover, the energy stored in the cable is released over time, rather than being delivered to the loudspeaker along with the rest of the signal. This frequency-dependent, non-linear energy storage distorts the size and shape of the soundstage, according to MIT.

Terminated cables are designed to prevent such energy storage, and to deliver all the signal to the loudspeaker with the correct phase (timing) relationship. MIT claims that terminated cables are essential to realizing correct bass weight, smooth tonal balance, and a full-sized soundstage with precise image focus.

Because terminated cables form a low-pass filter (i.e., they don't start filtering until about 1MHz, well above the audioband), they are useful for connecting very-wide-bandwidth electronics. The Spectral brand of electronics, for example, has bandwidths as high as 3MHz—not because the designer wants to pass those frequencies, but to make the circuits behave better in the audioband. Terminated cables filter those very high frequencies, ensur-

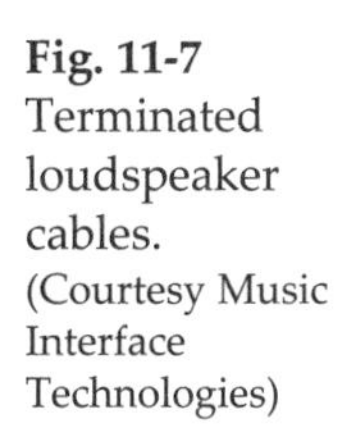
Fig. 11-7
Terminated loudspeaker cables.
(Courtesy Music Interface Technologies)

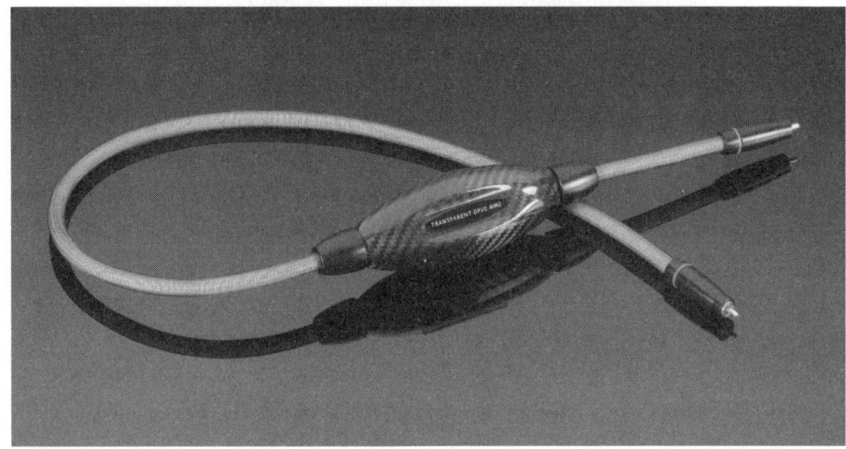
Fig. 11-8
Terminated interconects.
(Courtesy Transparent Audio)

ing that loudspeakers are never fed energy in the megahertz range, and making the system more stable. Cable manufacturer Transparent Cables also makes the case for terminated cables and interconnects, and offers a complete line of products.

Battery Bias in Cables and Interconnects

Anyone who has listened to loudspeaker cables and line-level interconnects will have discovered that cables and interconnects start to sound better (smoother treble, greater ease, less grain, more dimensional soundstage) after they've been in the system for a few days. This phenomenon is often called "break-in." It's easy to understand "break-in" of loudspeakers and phono cartridges (the rubber surround in loudspeaker drivers and the rubber suspension in cartridges relaxes with use), but why would putting audio signals down a conductor cause that conductor's sound to change?

The answer is that the cable's dielectric "forms" to the charged state created by current flowing through the conductor. The same phenomenon occurs in capacitors, and is one reason why electronics sound better when left on for a few days. After a cable or capacitor "forms," it returns to its uncharged state slowly. This means that cables that have been "broken-in" will slowly return to their original state when the system is not in use. The process is not fully understood, but extensive listening experience confirms this cycle.

Rather than rely on the audio signal to "form" the cable's dielectric, the cable designer can apply a relatively high DC voltage across the dielectric when the cable is made, in essence "breaking-in" the cable more effectively than could an audio signal, and keeping the cable in that state indefinitely.

That is the theory behind a new breed of cable that is supplied with a battery attached. AudioQuest, which has applied for a patent on the idea, calls the technique Dielectric-Bias System (DBS). A wire connected to a 24V battery's positive terminal runs down the cable's length, and is not connected to the cable or to the battery's negative terminal. The battery's negative terminal is connected to a shield around the conductors. Because there is no path for current between the battery's positive and negative terminals, the battery experiences no current drain and will last for years. Keep in mind that the battery system is completely outside the audio signal path, and doesn't interact with the audio signal. An interconnect employing this technology is shown in Fig.11-9.

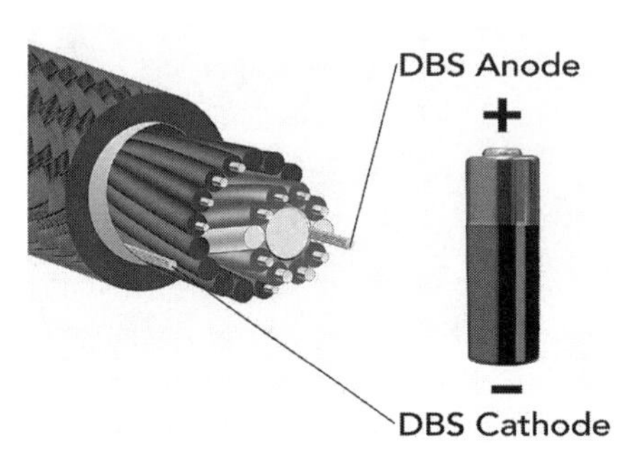

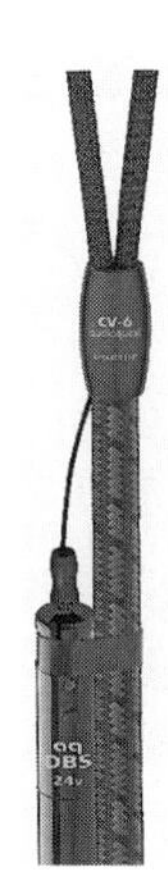

Fig. 11-9 A loudspeaker cable employing the Dielectric-Bias System (Courtesy AudioQuest)

A related technique is employed by Richard Vandersteen in his Model 5 and Model 7 loudspeakers. The two loudspeakers use an integral powered woofer and an external crossover to roll off bass driving the loudspeaker. Vandersteen included a battery-bias system across the capacitors in this crossover which kept the capacitors biased at a relatively high voltage. This DC shifts the audio signal away from the zero-crossing point, improving sound quality. This technique is different from AudioQuest's Dielectric-Bias System in that the audio signal is combined with the battery's DC voltage.

Cable and Interconnect Specifications

There's a lot of hype and just plain misinformation about cables and interconnects. Manufacturers sometimes feel the need to invent technical reasons for why their cables sound better than the competition's. In reality, cable design is largely a black art, with good designs emerging from trial and error (and careful listening). Although certain conductors, dielectrics, and geometries have specific sonic signatures, successful cable designs just can't be described in technical terms. This is why cables should never be chosen on the basis of technical descriptions and specifications.

Nonetheless, some cable and interconnect specifications should be considered in some circumstances. The three relevant specifications are *resistance, inductance,* and *capacitance.* (These terms are explained in detail in Appendix B.)

A cable or interconnect's resistance, more properly called *DC series resistance,* is a measure of how much it opposes the flow of current through it. The unit of resistance is the *Ohm.* The lower the number of ohms, the lower the cable or interconnect's resistance to current flow. In practice, cable resistance is measured in tenths of ohms. Resistance isn't usually a factor in interconnect performance (except in some of the new non-metallic types), but can affect some loudspeaker cables—particularly thin ones—because of their higher current-carrying requirements.

The sounds of interconnects and loudspeaker cables can be affected by inductance. It is generally thought that the lower the inductance, the better, particularly in loudspeaker cables. Some power amplifiers, however, need to see some inductance to keep them stable; many have an output inductor connected to the loudspeaker binding post (inside the chassis). When considering how much inductance the power amplifier sees, you must add the cable inductance to the loudspeaker's inductance.

Capacitance is an important characteristic of interconnects, particularly when long runs are used, or if the source component has a high output impedance. Interconnect capacitance is specified in picofarads (pF) per foot. What's important isn't the interconnect's intrinsic capacitance, but the total capacitance attached to the source component. For example, 5′ of 500pF-per-foot interconnect has the same capacitance as 50′ of 50pF-per-foot interconnect. High interconnect capacitance can cause treble rolloff and restricted dynamics. (A full technical discussion of interconnect capacitance is included in Appendix B.)

Cables in the Power Amplifier/Loudspeaker Interface

The interface between a power amplifier and a loudspeaker through a cable is a critical point in a playback system. Unlike interconnects, which carry low-level signals, loudspeaker cables carry much higher voltages and currents. Loudspeaker cables thus react more with the components they are connected to.

Damping factor is an amplifier's ability to control the woofer's motion after the drive signal has ceased. For example, if you drive a loudspeaker with a bass-drum whack, the woofer's inertia and resonance in the enclosure will cause it to continue moving after the signal has died away. This is a form of distortion that alters the music signal's dynamic envelope.

Fortunately, the power amplifier can control the motion; the degree of this control, or damping factor, is expressed as a simple number.

Damping factor is related to the amplifier's output impedance. The lower the output impedance, the higher the damping factor. When you connect a power amplifier and loudspeaker with cable, the cable's resistance decreases the amplifier's effective damping factor. For example, an amplifier's damping factor of 100 may be reduced to 40 by 20′ of moderately resistive loudspeaker cable. The result is reduced tightness and control in the bass. Loudspeaker cables should therefore have low resistance and be as short as possible.

12 Audio for Home Theater

Introduction

The massive growth of home theater over the past 15 years has had a profound effect on high-end audio. On one hand, home theater—the combination of high-quality audio and video presentation in the home—has opened new markets to high-end companies and new possibilities for the music enthusiast. On the other hand, combining a music-reproduction system with a video display and multichannel surround-sound all too often compromises the system's musical performance. As audiophiles, we go to great lengths to achieve the highest possible sound quality and get closer to the music. Any compromise of that pursuit is anathema to the goals and spirit of high-performance audio.

Yet the application of high-end products to movie watching can create an enormously compelling experience. Film is as important an art form as music; just as good technical quality in music reproduction brings us closer to the musician's expression, home theater done right conveys to us more of the filmmaker's art. Moreover, the availability of a significant number of multichannel titles in the SACD format has fueled a push toward more than two loudspeakers in the listening room—loudspeakers that can reproduce film soundtracks as well as multichannel music programs.

I explored this conflict between high-end audio and home theater in the following short essay, first published as "From the Editor" in Issue 140 of *The Absolute Sound.*

Now I know why I hate home theater," said a colleague to me after a series of home-theater demonstrations at a Consumer Electronics Show. The fellow journalist, a dedicated 2-channel music enthusiast, voiced a familiar refrain in high-end audio circles—namely, that music is a pure art form and home theater is merely glorified television at best, and an abomination at worst. At the time (1992), I had to agree; the demonstration consisted of subjecting a group of audio writers to a series of the loudest and most violent five-minute clips from the loudest and most violent movies at ear-splitting levels. The manufacturer was trying to impress us, but succeeded only in perpetuating the myth that home theater and high-quality music reproduction are antithetically opposed.

I say "myth" because since that demonstration ten years ago, I've come to believe that the antipathy between music and home theater is a false dichotomy. High-end audio equipment is about getting closer to the artistic expression of composers and musicians. Home theater is (or should be) about getting closer to the artistic expression of filmmakers. The greater the technical quality of the presentation, the more deeply we can experience the artists' intent—whether musical or cinematic. So what went wrong with this picture?

The home-theater industry has, unfortunately, given those of us in the high end many reasons to abhor its intrusion into music reproduction: gimmicky surround-sound modes; amusical, pant-rattling subwoofer bass; excessive volume levels; and an aesthetic that values impact over nuance. But an entire concept—the pursuit of high-quality film reproduction in the home—cannot be dismissed solely because of the excesses of those who appeal to the lowest common denominator. Rather than viewing home theater as a bastardization of audio, I've come to the belief that home theater, when done right, is an expansion of the high-end ethic, and one that forges an entirely new path for creating an immensely rewarding connection between artist and audience.

This expansion has been made possible by the entry of true high-end audio companies into the home-theater arena. These companies, driven by high-end ideals, create products that are faithful to the source, eschew gimmicks, and don't sacrifice musical performance for explosions and sound effects.

Moreover, home theater has been a boon to high-end audio in several ways. First, the exploding market for home-theater equipment has kept many traditional 2-channel companies prosperous. This allows them to stay in business, and continue to offer 2-channel and well as multichannel products. Second, home theater causes many who once had a "stereo" with loudspeakers in opposite corners of a room to set up the loudspeakers correctly and, for the first time, consider the spatial aspects of reproduced music. Third, a decent-quality home theater that's well setup can inadvertently become a vehicle for rediscovering the joys of music. I once recommended the Paradigm Atom multichannel loudspeaker system to an "average joe" friend ($1106 complete at the time). I ran into him a few weeks later, and the first thing he said to me was that he and his wife now spend much more time listening to music, and that they are happily rediscovering their music collection. Multiply this scenario by millions, factor in upward mobility, and you have a recipe not only for a robust high-end audio industry, but more importantly, the return of music to a prominent place in daily life.

§

With that introduction, let's look at the nuts and bolts of high-quality audio for home theater. Note that this chapter cannot cover the field in the same detail as other topics throughout this book, and I've avoided describing the video aspects of home theater.

Overview of Home-Theater Systems

Home theater is made possible by multichannel surround sound. Instead of two speakers in front of you as in a stereo system, home theater surrounds you with five, six, or even seven. Three speakers are placed in the front of the room, with two, three, or four surround speakers located to the rear or sides of the listening/viewing position. Surround sound is what provides the sense of envelopment, of being in the action, that makes the home-theater experience so engrossing.

Multichannel surround sound is encoded on DVDs, Blu-ray Discs, some satellite transmissions, and over-the-air high-definition television (HDTV) broadcasts in the Dolby Digital format. (Blu-ray Discs can include more advanced surround-sound formats, covered later in this chapter.) The Dolby Digital format, introduced in 1997 to replace Dolby Surround (and its decoding variant, Dolby Pro Logic), encodes six discrete audio channels in a single digital bitstream. The six channels are called left, center, right, left surround, right surround, and low-frequency effects (LFE), and correspond to loudspeakers arrayed in the pattern shown in Fig.12-1. The LFE channel carries only high-impact, low-bass information (below 100Hz), and can be reproduced by any of the loudspeakers, or by a separate subwoofer. Typically, the LFE channel drives a subwoofer. With five full-bandwidth channels plus the 100Hz LFE channel, we call Dolby Digital a 5.1-channel format. The ".1" in 5.1-channel sound is the LFE channel. The competing Digital Theater Systems (DTS) surround-sound format, found on some DVDs and Blu-ray Discs as an option, also delivers 5.1 discrete audio channels to your home-theater system.

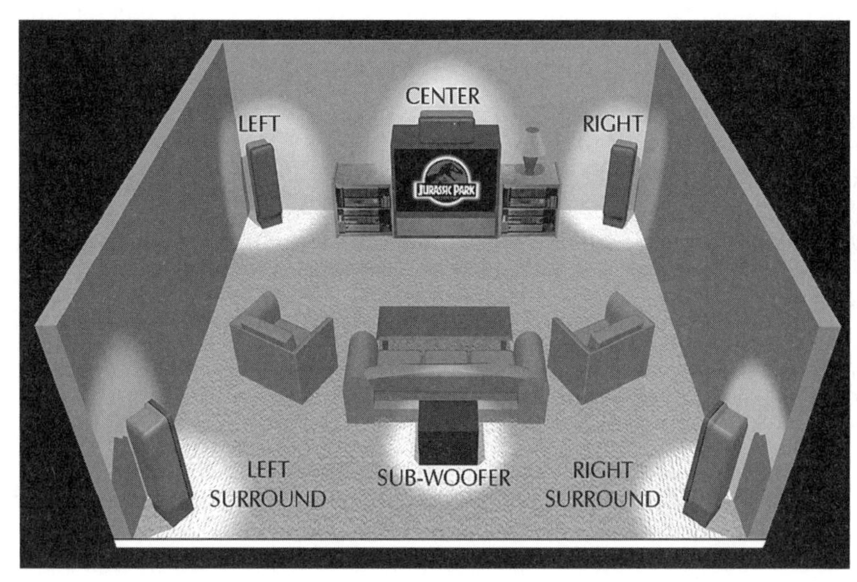

Fig. 12-1 A home-theater loudspeaker array uses three front loudspeakers, two surround loudspeakers, and a subwoofer. (Courtesy DTS)

The newer, more sophisticated surround-sound formats available on Blu-ray Disc (Dolby Digital Plus, DTS HD, Dolby TrueHD, and DTS HD Master Audio) provide for seven discrete channels plus a subwoofer (7.1-channel sound). We'll look at these formats in detail later in this chapter.

The job of decoding the Dolby or DTS bitstreams (whatever the format) into separate audio channels, and then converting those digital data to analog signals, falls to the home-theater controller, also known as a surround-sound processor. In addition to performing surround-sound decoding, the controller receives audio and video signals from the source components (satellite, disc player, tuner, music server) and selects which one is decoded and amplified by your home-theater audio system and sent to your video monitor for display. The controller also performs digital signal processing, adjusts the overall volume, and fine-tunes the levels for individual channels.

The controller's output is eight line-level signals: the left, center, right, left surround, right surround, left back, right back, and subwoofer channels. Note that a home-theater system can be based on a 5.1-channel or a 7.1-channel speaker array. A 7.1-channel array uses four surround speakers, two on the sidewalls and two on the rear wall. The newer surround formats, Dolby Digital Plus, Dolby TrueHD, DTS HD, and DTS HD Master Audio are all 7.1-channel formats, although they can be reproduced on a 5.1-channel system. All modern controllers have eight audio channels so that they are compatible with the 7.1-channel surround-sound formats.

The controller's eight separate outputs feed a 7-channel power amplifier (plus an optional subwoofer), where they are amplified to a level sufficient to drive the home theater's loudspeaker system. A home-theater power amplifier can have five, six, or seven amplifier channels in a single chassis. Alternately, some power amplifiers have three channels, others have two, and some are available as monoblocks: one amplifier channel per chassis. A 3-channel power amplifier allows you to use your existing 2-channel stereo power amplifier on the surround channels and the 3-channel amplifier on the front three loudspeakers, for example. Most subwoofers have built-in power amplifiers to drive their woofer cones.

An audio/video receiver (AVR) incorporates, in one chassis, an A/V controller and multichannel power amplifier. AVRs are less expensive than separate controllers and power amplifiers. If you are installing a second system dedicated to home theater, the AVR is a good option. If you are adding home-theater capability to an existing 2-channel music system, or building a dual-purpose system from scratch, a separate controller and power amplifier is the better route.

A home-theater system's left and right loudspeakers reproduce mostly music and sound effects. The center loudspeaker's main job is reproducing dialogue, and anchoring onscreen sound effects on the television screen. Having three loudspeakers across the front of the room helps the sound's location to more closely match the location of the sound source in the picture. For example, in a properly set-up home-theater system, if a car crosses from the left side of the picture to the right, you hear the sound of the car move from the left loudspeaker, through the center loudspeaker, and then to the right loudspeaker. The sound source appears to follow the image on the screen.

The surround loudspeakers have a different job. They're generally smaller than the front loudspeakers, and handle much less energy. Consequently, they can be mounted unobtrusively on or inside a wall. Surround speakers mostly reproduce "atmospheric" or ambient sounds, creating a diffuse aural atmosphere around the listener. In a jungle scene, for example, the surround loudspeakers would re-create sounds such as chirping birds, falling raindrops, and blowing wind; in a city scene, the viewer would be surrounded by traffic sounds. The surround loudspeakers' contribution is subtle, but vitally important to the overall experience. Correctly set-up surround loudspeakers should not be able to be heard directly, but should instead envelop the viewer/listener in a diffuse soundfield.

Should You Choose a 5.1-Channel or a 7.1-Channel System?

A fundamental question facing the home-theater shopper is whether to buy a 5.1- or 7.1-channel audio system. The vast majority of today's film soundtracks are encoded in 5.1 channels, but newer films are usually mixed in 7.1 channels for theatrical exhibition as well as for home theater. The newer surround-sound formats of Dolby Digital Plus, Dolby TrueHD, DTS HD, and DTS HD Master audio are all based on 7.1 channels (although they can be played on a 5.1-channel speaker array). The future might be 7.1, but the present is rooted in 5.1 channels.

There are several arguments for choosing a 5.1-channel system. First, most films are mixed in 5.1 channels. Second, mounting and running cables to the additional surround-back speakers isn't worth the small improvement in performance (except in very large rooms). Third, many extremely satisfying home-theater systems are based on 5.1-channel playback. Fourth, a 7.1-channel loudspeaker package distributes your speaker expenditure over seven speakers plus a subwoofer rather than over five speakers and the sub, which could compromise the quality—two good surround speakers are better than four mediocre ones. Finally, if your living room dictates that your couch be positioned against the wall opposite the video display, there's little point in adding surround-back speakers on that wall; it's not likely that they will perform as intended.

The argument for purchasing a 7.1-channel system goes like this: Even though most films are mixed in 5.1 channels, today's A/V receivers and controllers have processing circuits to synthesize the surround-back channels from 5.1-channel sources. Although this trick doesn't provide true 7.1-channel playback (the surround-back signals are simply derived from the left- and right-surround channels), it nonetheless increases the feeling of immersion in the soundfield. When playing true 7.1-channel sources, the film-sound mixers are able to position sounds directly behind you, an impossible feat with just 5.1 channels. Moreover, we are moving into an increasingly 7.1-channel world with the new audio formats on Blu-ray Disc; it's better to get a 7.1-channel system now rather than have to upgrade later.

Here's my view. If you have a moderately sized room and you're adding surround speakers for the first time, go with a 5.1-channel system. If your room is large, or if you're building a theater room as part of new construction, select a 7.1-channel package. Larger rooms benefit from the surround-back speakers because they help to fill the "hole" between widely spaced left- and right-surround speakers. If you're building a home with surround-speaker wiring in the walls, there's very little cost and effort required to run wires for surround-back speakers.

Let's take a closer look at the components of a home-theater system and how to choose the features best suited to your system.

Home-Theater Controllers

No product better exemplifies the fundamental shift in home-entertainment technology than the controller. The controller combines many diverse functions and sophisticated technologies in a single chassis. A modern controller, shown in Fig.12-2, performs the source-switching functions of a preamplifier, decodes all the surround-sound formats, provides eight channels of digital-to-analog conversion, and incorporates an electronic crossover to split up the frequency spectrum. While power amplifiers and loudspeakers change relatively little over time, the controller has undergone a radical transformation in recent years to keep up with new surround-sound formats, digital signal processing, and connectivity.

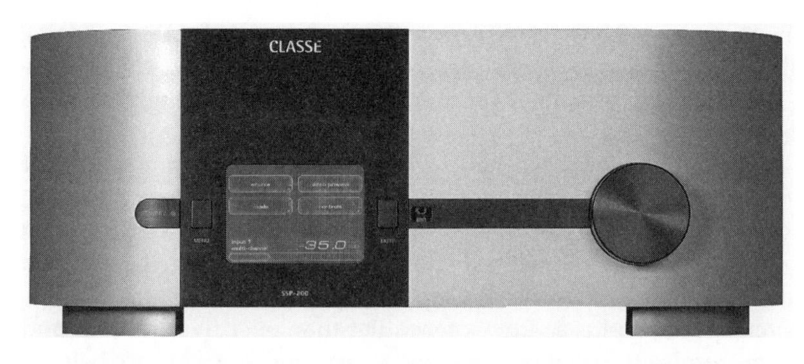

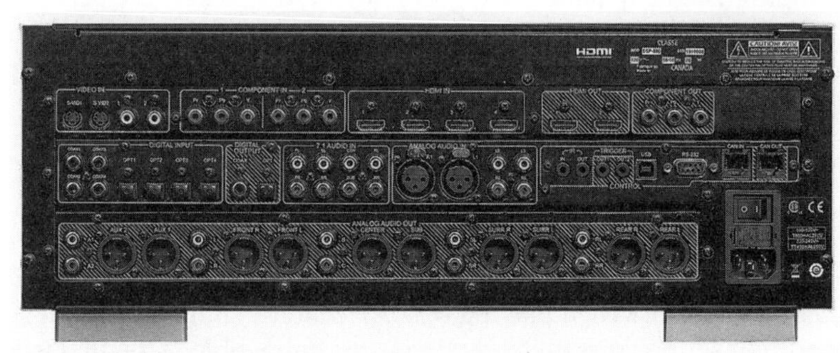

Fig. 12-2 A modern high-end digital controller. (Courtesy Classé Audio)

Specifically, a controller performs these major functions:

1) receives video and audio signals from various source components (DVR, dish, disc player, etc.) and selects which are sent to the video monitor and home-theater audio system (this function is called "source switching");

2) performs surround decoding;

3) controls playback volume;

4) makes adjustments in system setup, such as the individual channel levels;

5) directs bass to the appropriate loudspeakers or subwoofers (called bass manage ment); and

6) performs THX processing (if so equipped).

A controller has multiple audio and video inputs to accept signals from your source components. It then performs surround decoding to create a multichannel signal for output to the separate power amplifiers. As mentioned earlier, one of the controller's outputs is marked LFE (Low-Frequency Effects), the ".1" channel in 5.1-channel and 7.1-channel sound that carries additional bass signals for high-impact, very low bass sounds, such as explosions. The LFE output is connected to a powered subwoofer. Bass from other channels can be selectively added to form the subwoofer drive signal. For example, if you have small front speakers, you can direct the front-channel bass to the subwoofer along with the LFE channel. The bass that would have driven the front speakers is instead mixed with the LFE signal and sent to the subwoofer, relieving the front speakers from the burden of reproducing low bass.

The controller is the master control center of your home-theater system. When you first set up a controller, its onscreen display will guide you through selecting such things as whether or not you're using a subwoofer, if your surround loudspeakers are large or small, and calibrating individual channel levels and delay times. You will also control the system on a day-to-day basis through the controller's front panel and remote control. This includes source selection, volume adjustment, and the setting of individual channel levels (described later in this chapter).

How to Choose a Controller

Controllers range from entry-level units costing about $1500 to state-of-the-art models that will set you back as much as $20,000. This wide price range reflects the significant differences between controllers in their features, connectivity, future expandability, and sonic performance.

Let's look at some controller features so that you'll know what to look for when shopping for a controller.

Inputs, Outputs, and Source Switching

Let's start with the controller's most basic function: selecting the source you listen to or watch. The controller accepts audio or A/V (audio and video) signals from all your source components and lets you select which source signal is sent to the power amplifiers and video monitor. A basic controller will offer two analog-audio inputs (for a tuner and disc player, for example), three digital audio inputs, and perhaps four audio/video (A/V) inputs, with the latter on HDMI connectors. In addition to the main outputs that drive your TV and power amplifiers, a record output is often provided to drive a recording device.

When choosing a controller, make sure its array of inputs matches or exceeds the number of source components in your system. Your system is likely to expand in the future, so look for a controller with at least one more input than you need right now.

All controllers have inputs for digital audio signals as well as for analog. These inputs receive the digital-audio output of a DVD player, dish, digital video recorder (DVR), or music server. The signals carried on these digital connections include Dolby Digital, DTS, Dolby Surround, and 2-channel PCM (Pulse Code Modulation) signals.. These inputs are in addition to the HDMI inputs that have essentially replaced a wide range of video and audio connections found on previous-generation home-theater equipment. HDMI carries high-definition video along with high-resolution multichannel audio in a single cable. Most controllers have at least four High-Definition Digital Multimedia Interface (HDMI) inputs along with one or two HDMI outputs, with the latter feeding your video display. Note that a standard coaxial or TosLink optical connection cannot carry the newer surround-sound formats (Dolby Digital Plus, Dolby TrueHD, DTS HD, DTS HD Master Audio). These formats can be carried only on HDMI. In fact, component video, composite video, and S-video jacks are quickly becoming a thing of the past as they are supplanted by HDMI. Virtually all televisions, Blu-ray Disc players, AV receivers and controllers, and set-top boxes use the HDMI interface. (An HDMI cable is shown in Fig.12-3.)

Fig. 12-3 An HDMI connector (Courtesy Monster Cable Products)

Unlike AV receivers, with their RCA jacks for inputs and outputs, higher-end controllers sometimes have balanced inputs and outputs. A balanced cable, described in Chapter 11, is found only on higher-end controllers and power amplifiers. An RCA cable, also called an unbalanced or single-ended cable, carries the signal on one conductor plus a ground conductor. A balanced cable carries the signal on two conductors, plus a ground. Balanced cables are used in professional audio, and have some advantages over unbalanced cables. Controllers with balanced inputs and outputs will usually offer both balanced and unbalanced connection jacks to ensure compatibility with a wide range of power amplifiers and source components. To use a controller's balanced inputs, your source components must also have balanced output jacks. And a controller's balanced output jacks work only with a power amplifier with balanced input jacks. If your source components or power amplifier lack balanced jacks, you can still connect them through the unbalanced (RCA) connections. The controller shown in Fig.12-2 has balanced outputs on all channels.

Automatic Calibration

A controller must be calibrated for the particular loudspeakers and room with which it is used. Specifically, you must adjust the levels of each channel so that they are the same loudness, as well as tell the controller how far you are sitting from each speaker so that the controller can impose a delay on signals feeding the speakers closest to the listener. This last provision insures that the sound from each speaker reaches your ears at the same time. The calibration process is described later in this chapter.

Some controllers, however, provide automatic calibration that adjusts the level and delay of each channel with the push of one or two buttons. You position the supplied calibration microphone at the listening seat, press a button or two, and the controller does the rest. The controller emits a series of test signals that are reproduced by your speakers, picked up by the microphone, and analyzed by the controller. The controller is now precisely calibrated for your loudspeakers, their positions, and your listening position.

DSP Speaker and Room Correction

Many controllers have the ability to measure the frequency response of your loudspeakers and room, and then apply a correction in the digital domain to deliver perfectly flat frequency response at the listening position. As with automatic calibration, you position a microphone at the listening position and wait for the system to reproduce and analyze a series of test signals. If the system detects, for example, a peak in the frequency response of +4dB at 125Hz, the correction system creates a filter that attenuates the signal by 4dB at 125Hz. In essence, the DSP correction system "pre-distorts" the signal so that when it is reproduced by your loudspeakers in a room, the result is flat frequency response. In practice, DSP correction systems are vastly more complex and sophisticated than suggested by this simple example. The more sophisticated correction systems such as Audyssey can have a profound effect on the quality of the bass, making it smoother and less colored. DSP room-correction systems generally operate over the entire frequency range, and can significantly change your system's tonal balance, sometimes for the worse.

Analog Bypass Modes

For the music lover shopping for a controller that will also serve as a 2-channel preamplifier for his system, one of the most significant considerations is its performance with 2-channel analog sources (especially if you have an extensive vinyl collection) and signals from a high-resolution disc player. In that case, you'll want a controller that has an analog bypass mode. Without an analog bypass mode, the analog signal will be converted to digital and back to analog as it passes through the controller. Digital conversion is far from sonically transparent, so the sound will suffer.

There are two catches to look out for regarding the bypass mode. First, the controller must have an analog volume control. Most modern controllers adjust the volume digitally in their DSP (digital signal processing) chips, which usually don't sound as good as an old-fashioned volume knob. Second, whenever you engage bass management, even a controller with a bypass mode will convert the analog signal to digital because bass management is performed by the DSP chips. If you have a subwoofer with satellite speakers and use the controller's crossover to divide the frequency spectrum, the bypass mode won't remove the A/D and D/A conversions from the signal path. This is a serious limitation for music lovers who demand the ultimate in sound quality.

One novel approach used by a few manufacturers is to split the incoming analog signal and filter bass from the left and right outputs with an analog filter. This signal, which has never been digitized, drives the left and right channels. The other half of the split signal is digitized and processed through the controller's bass management, which derives a subwoofer signal. It's not pure analog bypass because the signal driving the subwoofer has been digitized, but bass frequencies are less prone to sonic degradation by digital conversions.

If you've found a controller that meets your home-theater needs but lacks an analog bypass mode, you can still enjoy uncompromised music performance by adding an analog preamplifier to your system. Your analog source components connect to the analog preamplifier, as do the left and right outputs from the controller. In essence, the preamplifier is inserted in the signal path between the controller and left- and right-channel power amplifiers. When you play movies through the controller, the analog preamp acts as if it's not there, and has no effect on the signal. Some preamps have a feature (called something like "theater pass-through") that sets their gain (volume) at a predetermined level when you play movies, so that you don't have to adjust the preamp's volume control to achieve correct volume from the left and right loudspeakers in relation to the other loudspeakers. This set-up is explained in detail later in this chapter under "Adding Home Theater Without Compromising Music Performance."

Bass-Management Flexibility

An important controller function is bass management, the subsystem that lets you selectively direct bass information in the soundtrack to the main loudspeakers or to the subwoofer. Bass management allows a controller to work correctly with a wide variety of speaker systems. For example, if you have five small loudspeakers and a subwoofer, you tell the controller to filter bass from each of the five channels and to direct it, in sum, to the subwoofer. With a Dolby Digital or DTS movie, the bass from the left, center, right, and surround channels is mixed with the LFE channel to drive the subwoofer. The controller's "subwoofer" output is the LFE channel selectively mixed with bass from any of the other channels. The bass management in most controllers lets you direct the full frequency range to the left and right channels (including the LFE channel), but filter bass from the center and surround channels.

A feature of the most advanced controllers is their ability to specify the crossover frequency and slopes between the subwoofer and main speakers. The crossover is implemented in the digital domain with DSP. That is, the audio spectrum is divided by performing mathematical computations on the ones and zeros representing the audio signal. Splitting the frequency spectrum into bass and treble in the controller is a vastly better approach than subjecting the analog audio signal to the capacitors, resistors, and inductors found in the crossovers built into subwoofers.

The most sophisticated controllers let you specify the crossover frequency (40Hz, 80Hz, 120Hz, for examples), as well as the crossover slope or phase characteristics. The greater the flexibility in specifying crossover frequency, slope, and phase, the greater the likelihood that you can seamlessly integrate the subwoofer with the main speakers. (Crossover slope and phase are described in Chapter 6.)

Keeping low bass out of smaller loudspeakers confers large advantages in the speakers' power handling, dynamic range, midrange clarity, and sense of ease. When the woofer doesn't have to move back and forth a long distance trying to reproduce low bass, the midrange sounds cleaner and the speaker can reproduce loud peaks without distortion.

8-Channel Analog Input

If you plan to use a multichannel SACD player with a controller, you'll need a feature called "discrete multichannel analog input." This is an input consisting of six or eight RCA jacks that accept the six outputs from a multichannel SACD player. This input bypasses all digital processing and sends the multichannel signal to the controller's multichannel outputs. Note that SACD players provide bass-management within the player, obviating the need to engage the controller's DSP-based bass management.

Digital-to-Analog Conversion

Built into every 7.1-channel controller are eight digital-to-analog converters (DACs) and eight analog output stages. The DACs convert the digital data for each channel into analog signals. The quality of these DACs and the subsequent analog output stage (which drives the power amplifier through interconnects) is crucial to realizing good sound quality. DACs vary greatly in their sound, and a poor-sounding DAC (or a poor implementation of a good one) can ruin an otherwise excellent controller. More expensive controllers use higher-quality parts and design techniques, including metal-film resistors, polystyrene capacitors, four-layer circuit boards, and exotic circuit-board material. Also look for analog stages made from discrete transistors instead of inexpensive operational-amplifier (or op-amp) chips. Virtually all of today's controllers are compatible with high-resolution signals, up to 192kHz sampling rate and 24-bit word length. Some high-end companies now have considerable expertise in designing cutting-edge digital converters, expertise they can apply to building multichannel digital controllers. Much of what you're paying for in a high-end controller is the quality of the DACs and analog output stages.

7.1-Channel Playback from 5.1-Channel Sources

Some home-theater receivers and controllers provide 7.1-channel playback from 5.1-channel sources such as Dolby Digital and DTS. These products have eight channels (seven channels plus a subwoofer) rather than the six channels of most Dolby Digital- and DTS-equipped products. The two additional channels drive two extra speakers placed behind the listener, augmenting the two surround speakers at the sides of the listening location.

Making effective use of the additional surround speakers requires sophisticated signal processing in the controller. Controllers with THX Ultra2 processing incorporate a circuit that creates a 7.1-channel signal from any source, allowing you to use your entire 7.1-channel loudspeaker system on all source material, not just on those movies that have been mixed in 7.1-channel sound. Several Dolby and DTS processes (described at the end of this chapter) are also designed to convert 5.1-channel sources into 7.1-channel signals for reproduction on a 7.1-channel loudspeaker array.

THX Certification

Some controllers are THX-certified, meaning that they incorporate certain types of signal processing to better translate film sound into a home environment and meet a set of specific technical criteria established by Lucasfilm to better translate the theater experience into the home. THX processing in a controller includes re-equalization, surround decorrelation, and timbre matching. Re-equalization removes excessive brightness from film soundtracks for home-theater playback. Surround decorrelation improves the sense of spaciousness from the surround channels by introducing slight differences between the two surround signals. Timbre matching compensates for the fact that the ear doesn't hear the same tonal balance of sounds in all directions. The THX timbre-matching circuit ensures that sounds from the front and rear have the same timbre (tonal balance). To keep their products' prices low, some manufacturers of budget controllers license only the THX re-equalization circuit from Lucasfilm. Such controllers cannot, however, be billed as "THX-certified."

A THX-certified controller will also include a crossover circuit to separate the bass from the rest of the audio spectrum when a subwoofer is connected to the system. The crossover keeps bass out of the front-channel signals, and filters mid and high frequencies from the subwoofer output. A THX crossover has a specified cutoff frequency of 80Hz. (The cutoff frequency is the frequency below which bass is sent to the subwoofer output and above which higher frequencies are sent to the front-channel outputs.)

Advanced Features: 3D Capability, Network Connection, Multi-Zone

If you think you will want 3D video in your home, choose a controller that can accommodate 3D signals, which requires HDMI 1.4a or later connectivity. In addition, the Audio Return Channel feature of HDMI 1.4a simplifies your system connection by allowing an HDMI cable to send the audio signal back from the video display to the controller. This feature obviates the need for connecting an audio cable from your video display back to your controller. If you are watching TV from a tuner source, for example, the audio signal from the TV will return to the controller via the HDMI cable. All HDMI cables support Audio Return Channel functionality.

Some controllers provide an Ethernet jack for networking to other devices in the home. This feature allows you to access content (photos and audio files, for examples) stored on a PC. The Ethernet port can be used for software updates, as well as for accessing streaming media such as Pandora and Rhapsody.

Most controllers have multi-zone capability, which allows you to listen/watch one source in a room, and another source in a second room. More elaborate versions of this feature provide greater control over the "Zone 2" signal. Keep in mind that you'll need to run wires to the second zone; it might be more convenient and cost-effective to install separate systems.

Controllers will sometimes offer 9.1-channel capability, with the additional two channels (beyond left, center, right, surround left, surround right, surround back left, surround back right) for reproducing the additional height or width channels offered by Audyssey DSX or the height channels in Dolby Pro Logic IIz.

A front-panel USB port allows you to access content on a USB stick, such as music files, photos, and album art from a connected iPod or iPhone.

Multichannel Power Amplifiers

The power amplifier is the workhorse of your home-theater system. It takes in line-level signals from the output of your controller and amplifies them to powerful signals that will drive your home-theater loudspeaker system. A power amplifier has line-level inputs to receive signals from the controller, and terminals for connecting loudspeaker cables. The power amplifier is the last component in the signal path before the loudspeakers.

A separate multichannel power amplifier can offer higher output power than the power amplifiers found in A/V receivers. The most powerful receivers can have outputs of 140 watts per channel; separate power amplifiers offer as much as 350Wpc.

A power amplifier is described by the number of amplifier channels in its chassis. The most common home-theater power amplifiers have five channels for powering the left, center, right, surround left, and surround right loudspeakers. The next most common number of channels is three. A 3-channel amplifier can power the front three loudspeakers, leaving a 2-channel amplifier to handle the two surround loudspeakers. If you already have a stereo (2-channel) amplifier, you can update your system for home theater by adding a 3-channel home-theater amplifier. Alternatively, a 3-channel amplifier can power the center and two surround channels, leaving your stereo amplifier to drive the left and right loudspeakers. As the Dolby Digital Plus and DTS HD formats become more popular, power amplifiers with seven channels are becoming increasingly available. Note that if you upgrade your controller to a 7.1-channel unit, you can augment your 5-channel power amplifier with a stereo amplifier to power the surround-back speakers.

Some home-theater power amplifiers, called monoblocks, drive only one amplifier channel per chassis. The monoblock approach lets you upgrade more easily by buying power for only as many channels as you need. Monoblocks have the potential of better sonic performance, but also cost more than the same number of amplifier channels in a multichannel power amplifier.

How to Choose a Home-Theater Power Amplifier

High-quality multichannel power amplifiers are big, heavy, and can be expensive. The more output power provided by the amplifier, the bigger, heavier, and more expensive it will be. Most home-theater enthusiasts listen at moderate levels to loudspeakers of average sensitivity (85–88dB) in rooms of average size (3000–4500 cubic feet) with average furnishings (carpet, some drapes), and so will need a minimum of about 70W for each of the five, six, or seven channels.

Some multichannel power amplifiers feature a single large power transformer to supply all five, six, or seven channels. Others provide smaller, separate transformers for each channel. There is no consensus as to which approach is better. The argument for a single large transformer goes like this: For 2-channel music sources, the amplifier benefits from the extra power-supply capacity provided by a huge transformer. Moreover, even when multichannel sources are played, the rear channels don't work as hard as the front, allowing some of their power-supply capacity to be delivered where it's needed. The opposing point of view holds that separate smaller transformers completely isolate the channels from each other and eliminate interaction between channels. For example, a loud explosion in the film soundtrack reproduced by the center channel can pull power from the transformer needed for the musical score that is being reproduced by the left and right channels.

Note that driving five, six, or even seven loudspeakers places extraordinary current demands on the amplifier. There's a way to reduce the strain placed on multichannel amplifiers: use high-sensitivity loudspeakers. Remember that every 3dB increase in sensitivity is equivalent to doubling amplifier power.

More specific information on power amplifiers may be found in Chapter 5. The discussion of sound quality, build quality, and matching to loudspeaker sensitivity applies equally well to multichannel amplifiers as it does to 2-channel amplifiers.

Loudspeakers for Home Theater

A home-theater loudspeaker system will perform double-duty, reproducing 2-channel music sources through the front left and right speakers, and film soundtracks through all five, six, or seven loudspeakers (plus an optional subwoofer). Home-theater loudspeakers can be evaluated using the criteria described in Chapters 3 and 6, with one difference: film soundtracks present somewhat different challenges to a loudspeaker system than music. Specifically, film soundtracks are generally more dynamic than most music, and have much more low-frequency energy. Consequently, great home-theater sound requires loudspeakers that not only meet our criteria for good music reproduction, but also have a wide dynamic range and the ability to handle loud and complex signals without sounding smeared or confused.

The Center-Channel Speaker

Though the center-channel loudspeaker carries a large part of the film soundtrack—nearly all the dialogue, many effects, and some of the music—only recently have center-channel speakers been elevated from afterthoughts to recognition that they are the anchor of the entire home-theater loudspeaker system.

The center speaker is usually mounted horizontally on top of the video monitor. It can also be placed beneath the video monitor or mounted inside a wall above the video display, or, if you're using a front-projection system, behind an "acoustically transparent" screen.

Because a stereo system uses only two loudspeakers—left and right—across the front, you may be wondering why you need a third loudspeaker between them for home theater. The center-channel loudspeaker provides many advantages in a home-theater system. First, it anchors dialogue and other sounds directly associated with action on the screen in the center of the sonic presentation. When we see characters speaking, we want the sound to appear to come from their visual images. Similarly, when sounds are panned (moved from one location to another) across the front, we want the sound to move seamlessly from one side to another. For example, if the image of a car travels from the left side of the screen to the right, the sounds of the car's engine and tires should travel with it, precisely tracking the car's movement. Without a center speaker, we may hear a gap in the middle as the car sounds jump from the left loudspeaker to the right. The center speaker makes sure on-screen sounds come from the screen.

A 2-channel stereo system is capable of producing a sonic image between the two speakers. This so-called "phantom" center image is created by the brain when the same signal is present in both ears. A sound source in front of us in real life produces soundwaves that strike both ears simultaneously. The brain interprets these cues to determine that the sound source is directly in front of us. Similarly, two speakers reproducing the same sound send the same signal to both ears, fooling the brain into thinking the sound is in front of us.

For two loudspeakers to create this phantom center image, they must be precisely set up, and the listener must sit exactly the same distance from each of them—if you sit off to the right, the center image will pull to the right. In addition, creating a phantom center image from two loudspeakers may be a learned skill; some listeners may never hear a phantom center image. There's also evidence that the work the brain must perform to conjure up this phantom center image is distracting and fatiguing.

These problems are overcome by putting a center-channel speaker between the left and right speakers. Dialogue and onscreen sounds are firmly anchored on the screen for all listeners, not just those sitting in the middle. With three speakers across the front, someone sitting way off at the left end of the couch can still hear dialogue coming from the area of the screen—not just from the left loudspeaker. The center speaker also prevents the entire front soundfield from collapsing into the speaker closest to where you're sitting. Moreover, the center speaker provides a tangible sound source directly in front of you; your brain doesn't have to work to create a phantom image between the left and right speakers. Finally, the center speaker reduces the burden on the left and right loudspeakers. With three speakers reproducing sound, each can be driven at a lower level for cleaner sound.

Adding a Center Speaker to Your System

Although you can add a different brand of center speaker to your existing left and right speakers, you're better off with three matched speakers across the front of your home theater. Speakers all sound different from one another. No matter how good the quality of a speaker, it will have some coloration, or variations from accuracy, in its sound. I mentioned earlier that you want to hear a seamless movement of sounds across the front soundfield. If the center speaker has a different sound from the left and right speakers, you'll never achieve a smooth and continuous soundfield across the front—when sound sources move from one side to another through the center speaker, the sound's character will abruptly change.

The solution is to buy three matched front-channel speakers. Their identical tonal characteristics will not only provide smooth panning of sounds, but also produce a more stable and coherent soundfield across the front of the room. If you already have high-quality left and right loudspeakers that you want to keep, buy a center-channel speaker made by the manufacturer of the left and right loudspeakers. They probably won't be as well-matched as a three-piece system, but the added center speaker is much more likely to sound similar to your existing left and right speakers. Separate center speakers sometimes use the same drivers (the raw speaker cones themselves) and other parts as the stereo speakers from the same manufacturer.

For state-of-the-art home theater, the three front loudspeakers should be identical. Although most home-theater enthusiasts will use a smaller center speaker that will fit on top of a television set and doesn't reproduce much bass, more ambitious systems use large, full-range center speakers. These speakers are hard to position, and are generally used only with front-projector systems; the center speaker can be placed behind a perforated projection screen, just as in a movie theater.

Left and Right Speakers

The left and right loudspeakers carry the majority of the film's musical score and many of the effects. And if you aren't using a subwoofer, nearly all the bass will be reproduced by the left and right speakers. Consequently, left and right speakers are often the largest speakers in a home-theater system.

Left and right speakers can also be small units that sit on speaker stands or on an entertainment cabinet. If that's the case, the small speakers must be used with a subwoofer to reproduce bass. The small left and right speakers reproduce the midrange and treble frequencies, and the subwoofer handles all the bass. Such a system, called a subwoofer/satellite system, is ideal if your available space is limited, or if you want the left and right loudspeakers to better blend into your decor. Satellite speakers can be small and unobtrusive, and the subwoofer can be tucked out of the way. If you build the three front speakers into a wall in a custom installation, they will most likely be satellites. All THX-certified loudspeaker systems use a subwoofer.

Surround Speakers

The surround loudspeakers are completely different in design and function from left, center, and right speakers. Their job is to re-create a diffuse atmosphere of sound effects to envelop us in a subtle sonic environment that puts us in the action happening on the screen. Unlike front and center speakers that anchor the sound onscreen, surround speakers should "disappear" into a diffuse "wash" of sound all around us .

A good example of how surround speakers create an atmosphere comes from the film *Round Midnight*. Toward the beginning of the movie the character François is outside a Paris jazz club in a driving rainstorm, listening to Dexter Gordon's character playing inside. The scene cuts between the intimate sound of the jazz club and the rainy Paris street. Inside the club, the sound is direct and immediate to reflect the camera's perspective of just a few feet from the musical group. When the scene cuts to the street, we are surrounded by the expansive sound of rain, cars driving by, people talking as they walk past—in other words, all the ambiance and atmosphere of a rainy night in Paris. This envelopment is largely created by the surround speakers. We don't want to hear the rain and street sounds coming from two locations behind us, but to be surrounded by the sounds, as we'd hear them in real life. The surround speakers perform this subtle, yet vital role in home theater.

Surround speakers envelop us by their design and placement in the home-theater room. They are best located to the side or rear of the listening position, and several feet above ear level. Because they don't have to reproduce bass, surround speakers can be small and unobtrusive, and are often mounted on or inside a wall.

Dipolar and Bipolar Surround Speakers

Most of the surround speakers' ability to wrap us in sound comes from their dipolar design. Dipolar simply means that they produce sound to the front and rear equally. While front speakers have one set of drivers that project sound forward, dipolar surround speakers have two sets of drivers, mounted front and back. This arrangement produces a directional pattern that fires to the front and back of the room (see Chapter 6). Because the surround speakers are positioned to the sides and fire to the front and back of the room, we hear no direct sound from them. The listener sits in the surround speakers' "null"—the point where they don't directly project sound. Instead of hearing direct sound from the surround speakers, we hear their sound after it's been reflected, or bounced off the room's walls and furnishings. The surround speakers' dipolar directional pattern makes their sound diffuse and harder to localize. You shouldn't be able to tell where a properly set-up surround speaker is just by listening.

You may see surround speakers called either dipolar or bipolar. Although both produce sound equally from the front and rear, the bipolar speaker produces sound from the rear that is in-phase with the front sound. A dipolar produces sound from the rear that is out-of-phase with the front sound. In-phase means that the front and rear waves have the same polarity: when the front-firing woofer moves forward to create a positive pressure wave, the rear-firing woofer also moves forward. The opposite is true in a dipole: when the front-firing woofer moves forward, the rear-firing woofer moves backward. In a dipole, the front and back waves are identical, but have the opposite polarity. Both bipolar and dipolar achieve the objective of enveloping the viewer/listener in the film's aural ambiance, but the dipolar type is preferred because it creates a greater "null," or reduction in sound, at the side of the speaker that faces the listener.

A true dipolar or bipolar surround speaker can't be flush-mounted inside a wall. Instead, it must be mounted on the wall so it can radiate sound toward the front and rear of the room. Some surround speakers approximate a dipolar radiation pattern by mounting small drivers on an angled baffle. Although not true dipoles, these flush-mounted surround speakers accomplish the goal of creating a diffuse ambiance and fit unobtrusively in your living room.

The amount of sound produced by the surround speakers is adjustable through the receiver's or controller's remote control. Their volume level should be set so that you never hear them directly; but if you turned them off, the soundfield would collapse toward the front. All too often, consumers set surround speaker levels too high; their thinking seems to be that if they bought speakers, they should be able to hear them.

Surround-Back Speakers

As described earlier, the 7.1-channel formats add two additional surround speakers behind the listener. According to Lucasfilm, the four surround channels are best reproduced by four identical dipolar speakers. The two surround speakers remain at their normal position (normal for 5.1-channel reproduction), which is 110°, or slightly behind the listener on the side walls. The surround back speakers are located at 150° as shown in Fig.12-4 later in this chapter..

The two surround back speakers should be arranged so that their facing sides are in-phase with each other. That is, the side of the dipole that would face the back of the room if mounted on a sidewall should face toward the room's centerline when the speaker is mounted on the rear wall. This arrangement keeps the drivers of two surround back speakers that face each other in-phase, contributing to their ability to create an image between them directly behind the listening position.

Subwoofers

A subwoofer is a speaker that reproduces only low bass. Subwoofers are usually squarish or rectangular cabinets that can be positioned nearly anywhere in the home-theater room. A good subwoofer adds quite a bit of impact and visceral thrill to the home-theater experience.

Most subwoofers for home-theater use are powered or active subwoofers. These terms describe a subwoofer with a built-in power amplifier that drives its large woofer cone. The powered subwoofer takes a line-level output from the jack marked subwoofer out on your A/V receiver or controller, and converts that electrical signal into sound. Because powered subwoofers have built-in amplifiers, they must be plugged into an AC wall outlet. (See Chapter 6 for more on subwoofers.)

The subwoofer is typically driven by the LFE Output jack on an AVR or controller. The LFE output is a monophonic mix of the LFE channel along with the bass from any channel designated as "Small" during the controller's initial set up.

Setting up a Home Theater

How a home-theater system is setup and calibrated has an enormous influence on the quality of the sound you hear. Just as with 2-channel music system, ideal loudspeaker placement and careful attention to detail will reward you with superior performance. The acoustic principles described in Chapter 14 apply equally well to a multichannel audio system.

Basic Setup

The first step in setting up your system is deciding where to install it. More than likely, the room where you watch television will become the home-theater room. But this room may have been arranged for casual TV watching, not for experiencing the full impact of what a home-theater system can deliver. Now is the time to rethink the room's layout and consider making the home-theater system the room's central focus.

The home-theater room should be arranged with the video monitor and front loudspeakers at one end, the couch or viewing chairs at the other. Ideally, the couch will be positioned partway into the room and not against the back wall. This location is better acoustically (excessive bass build-up occurs near the back wall), and allows the surround speakers to better do their job of enveloping you in the film's soundtrack. Of course, the setup should provide room for the left and right speakers on either side of the video monitor.

The optimum viewing distance between you and the video monitor is determined by the video monitor's size. The smaller the monitor, the closer you should sit to maintain the maximum image area in your field of view. Sitting too close to a large monitor, particularly a rear-projection set, exposes the line structure inherent in a video presentation. If you are watching high-definition programming on an HD set, you can sit much closer without seeing the line structure. The best viewing distance offers the largest possible picture without making that picture look grainy. A good rule of thumb is to sit three-to-four picture heights away from the screen. You should also try to sit no more than 15° off to the side of the video monitor. Plasma panels, with their wider field of view, maintain a good picture even when they are watched from an extreme angle.

Acoustical Treatment

Before we get to the specifics of loudspeaker placement, here are a few guidelines on how to make your home-theater room sound better. First, if you have a bare tile or wood floor between you and the front three speakers, covering the floor with a rug is probably the single most important thing you can do to improve your room's acoustics. The rug absorbs unwanted acoustic reflections from the floor, which improves dialogue intelligibility and adds to a sense of clarity. Similarly, absorbing the acoustic reflections from the side walls between you and the left and right speakers also helps you get better sound. Bookcases and hanging rugs are both effective in absorbing or diffusing side-wall reflections. Generally, the front of the room should be acoustically absorptive, with drapes, carpets, and hanging rugs. Conversely, the back of the room should be more acoustically reflective. Bare walls behind the listening/viewing position tend to reflect sound and add to the feeling of spaciousness generated by the surround speakers. The sounds produced by a movie theater's array of surround speakers all reach your ears at different times, increasing the feeling of envelopment. But because a home-theater system uses just two surround speakers, we must help them achieve their goal by providing reflective surfaces near them.

Ideally, the lower portion of the room's front end should be acoustically absorptive, the upper portion acoustically reflective. This situation absorbs unwanted wall reflections from the three front speakers, but lets the dipole surround speakers positioned on the sidewall just below the ceiling create a bigger sense of envelopment by reflecting sound off the front walls as well as the rear walls.

Speaker Placement

Where speakers are located in your room has a huge influence on the quality of sound you will achieve. Correct speaker placement can make the difference between good and spectacular performance from the same equipment. If you follow a few simple speaker-placement techniques, you'll be well on your way toward getting the best sound from your system. Correct speaker placement is the single most important factor in achieving good sound.

Surround Speaker Placement

Let's start with the most difficult of a home-theater system's six speakers: the two surround speakers. As I said earlier, the surround speakers are usually dipolar. That is, they produce sound equally to the front and the rear of the speaker cabinet. Dipoles produce a "null," or area of no sound, at the cabinet sides. Here's a simple rule for positioning dipolar surround speakers no matter what the installation: point the surround speakers' null at the listening position. Never point the surround speakers directly at the listening position.

Ideally, the surround speakers would be located on the side walls directly to the side of the listening position and at least two feet above the listeners' heads when seated. Dipolar surround speakers should never be located in front of the listening position, where they could be heard directly; the surround speakers' sound should be reflected around the room before reaching the listener. This helps to mimic the sound of a movie theater's surround array with just two speakers, and to create the envelopment that's so important to the home-theater experience.

Keep in mind that reflections from the front speakers are bad; reflections from the rear speakers are good. Well-positioned surround speakers make a tremendous difference in the quality of the home-theater experience by re-creating the sense of ambience and space we hear in a movie theater.

The surround-back speakers should be positioned on the rear wall behind the listener at the locations shown in Fig.12-4. This placement provides the greatest sense of envelopment, along with the most continuous soundfield behind the listener. The two surround-back speakers should be positioned so that the speakers facing the room's centerline are in-phase with each other. This produces a central image between the two surround-back speakers.

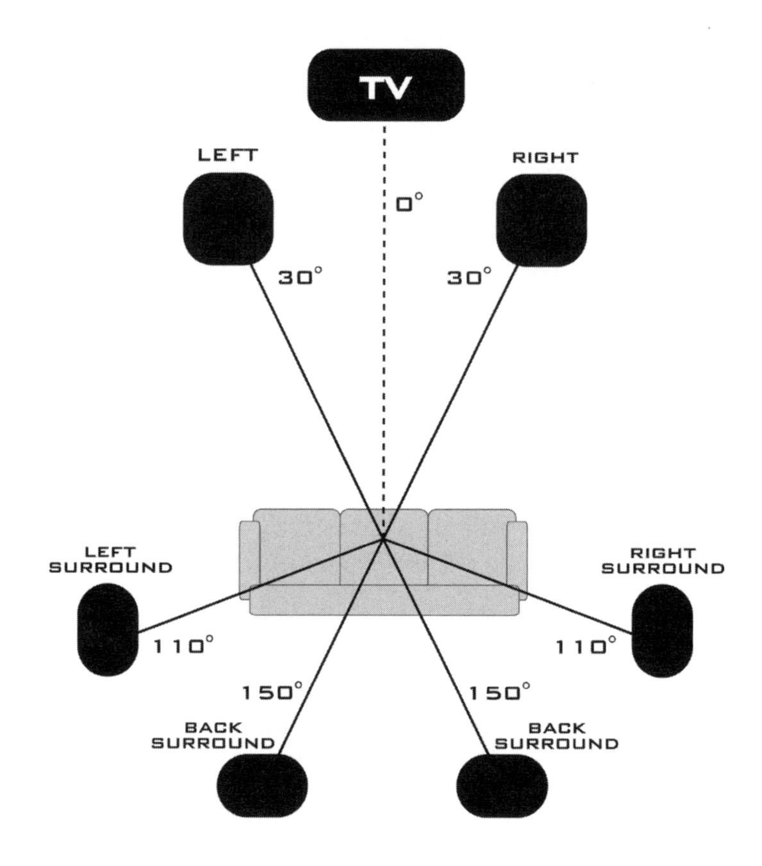

Fig. 12-4 The surround loudspeakers should be positioned at 110° and the surround back loudspeakers at 150°.

Center Speaker Placement

Correct positioning of the center speaker results in better dialogue intelligibility, smoother movement of onscreen sounds, and a more spacious soundstage. As just mentioned, reflections from the front speakers that reach the listener work against good sound. This is particularly true of the center speaker, which carries the all-important dialogue. Because most center-channel speakers are mounted on top of a direct-view TV or rear projector—both of which are highly reflective—there's the potential for unwanted reflections from the video monitor.

To reduce reflections from the center speaker, align its front edge absolutely flush with the front of the video monitor so that they form as contiguous a surface as possible. This positioning reduces a phenomenon called diffraction, which is the re-radiation of sound when the soundwaves encounter a discontinuity. (Specifically, diffraction introduces variations in the speaker's frequency response.) Think of waves rippling in a pond: If you put a stick in the water, the waves will strike the stick and be re-radiated around it. The same thing happens when the sound from the center speaker encounters the "stick" of the video monitor: the monitor re-radiates that sound toward the listener, making the overall sound less accurate. That's why making the center speaker flush with the front of the video monitor is so important.

High-end loudspeaker manufacturers carefully design their enclosures to reduce diffraction. You'll notice that high-quality loudspeakers have smooth surfaces around the speaker cones. Some have rounded fronts and even the mounting bolts securing the individual drivers within the speaker are recessed to reduce diffraction from the speaker cabinet. You should also tilt the center speaker down toward the listening position if it is mounted on a large rear-projection set, or on a direct-view TV positioned on a tall stand. If the center speaker is located below the video monitor, point it up toward the listeners. Direct the center speaker's sound toward the height of the listeners' ears when seated. Finally, be sure the center speaker is less than 2' different in height from the left and right speakers, measured from each speaker's tweeters.

Left and Right Speaker Placement

The left and right speakers should be pulled slightly in front of the video monitor, and not be in a straight line with it (as some owner's manuals erroneously advise). Positioning the left and right speakers so that they form a gentle arc with the center speaker has two advantages. First, acoustic reflections from the video monitor are reduced. Second, the left, right, and center speakers are now all the same distance from the listener. The sounds from each of the three front speakers will reach the listener at the same time. If the front three speakers were lined up, we would hear the center speaker slightly ahead of the left and right speakers, which tends to make sounds bunch together in the center and reduces the soundstage's left-to-right width.

If you're using large, full-range left and right speakers, putting the speakers near the front and side walls will increase the amount of bass you hear. The walls reinforce bass to produce a weightier sound. This can be a good or a bad thing, depending on the speakers and your room. Speakers that have lots of bass already will probably sound boomy and unnatural if put too close to the front wall. Pulling them out into the room often produces a huge improvement in bass definition (the bass no longer sounds like a continuous blur) and midrange clarity. The unnatural boominess of a speaker too close to a wall makes the midrange sound thick, and lacking in openness and clarity. If you're using small left and right speakers with a separate subwoofer, the left and right speaker placement in relation to the front wall is less important.

How far apart the left and right speakers are positioned is also important to achieving a natural and spacious sound. If the left and right speakers are too close to the video monitor, you'll never get a sense of large space at the front of the room. If too far apart, onscreen sounds are reproduced too far away from their apparent sources on the screen. It can be subconsciously confusing and distracting to hear a sound come from a location different from the visual location of the sound source. A good rule of thumb is to position the left and right speakers so that they form a 60° angle when seen from the listening/viewing location.

Left and right speakers can be positioned so that they point straight ahead, or are angled in (called toe-in) toward the listening position. No toe-in (speakers pointing straight ahead) produces a wider and more spacious soundstage, but at the expense of precise localization of sounds. Too much toe-in creates a more tightly focused sound but a smaller sense of overall soundstage size. Toe-in has the advantage of directing more sound at the listener and less sound toward the side walls. Reflection of sound from the side walls is the single worst problem in getting good sound from your speakers.

Some speakers are meant to be toed-in; that position provides their flattest (most accurate) frequency response. Others achieve their best performance with no toe-in. If you toe-in speakers designed for no toe-in, the sound will be too bright. Consult your owner's manual or your dealer for guidance. You can also experiment by playing the same section of a movie soundtrack or a piece of music with varying degrees of speaker toe-in to determine which position sounds best.

Calibrating a Home Theater

Once your system is connected, you'll need to configure it for your particular room and speaker system. This is done through a series of menus or graphical icons generated by the A/V receiver or A/V controller and appearing on your video monitor. You use the receiver's remote control to make selections from the onscreen display. Every A/V receiver or controller includes instructions for setting up your system; the following section explains why you make certain selections, and in what circumstances.

When you turn on your system for the first time, don't be alarmed if you don't hear any sound when you expect to. (This usually happens when I first set up a product for review.) Don't turn up the volume to a high level and then start pushing buttons—when you push the right button, the sound will blast you out, possibly damaging your speakers, your amplifier or receiver, and/or your hearing. Turn the volume to a low level and make sure the correct source is selected. Check to make sure the mute and tape monitor buttons aren't pressed. The lack of sound is probably due to one of these three incorrect settings.

Bass Management

First, you must tell the A/V receiver or controller what kind of loudspeakers you have so it can direct bass appropriately. Most of today's controllers ask you whether each of the speakers is large or small. By selecting small for the left and right speakers, you're telling the receiver to keep low-bass signals out of those speakers. This configuration is used when you have a subwoofer connected to the system to reproduce low bass instead of the left and right speakers. You'll also need to answer yes when asked if the system includes a subwoofer. The bass-management option labeled "THX" in THX-certified products automatically sets all speakers to small and engages the subwoofer.

If you use full-range left and right speakers and no subwoofer, answer large in the setup menu when asked if the left and right speakers are large or small. The large setting directs bass to the left and right speakers. The setup menu will also ask if the surround speakers are large or small. Nearly every installation will use the small setting. Only if you have full-range, floorstanding speakers should you answer large.

Setting Individual Channel Levels

Next you'll need to individually set the loudness of each of the six or eight channels (assuming you are using a subwoofer. In addition to providing an overall volume control, all A/V receivers and controllers provide adjustment of the individual channel volumes. This process begins by turning on the "test signal," a noise-like sound generated by the A/V receiver or controller. The noise is produced by each speaker in turn. Ideally, the noise signal's volume should be the same for each speaker when you're sitting at the listening/viewing position. If it isn't, you can adjust the volume of each speaker independently using the receiver's remote control. Individually adjusting the channels lets you compensate for different loudspeaker sensitivities, listening-room acoustics, and loudspeaker placements.

Although setting the individual channel levels by ear will get your system in the ballpark, a more precise calibration can be achieved by using a Sound Pressure Level (SPL) meter. Available from RadioShack for about $30, an SPL meter lets you accurately calibrate your home-theater system. RadioShack offers digital and analog SPL meters; buy the easier-to-read analog type (catalog #33-2050). Switch the meter to the c-weighted position, slow response, and set the knob to 70. Turn on the test noise on your A/V receiver and set each channel's volume until the meter's display reads 5. This indicates that the noise is being reproduced at a level of 75dB. Repeat this procedure for each channel.

A powered subwoofer usually has a knob on its back panel for setting loudness. When you set the subwoofer's level, the meter will be hard to read because the indicator will be jumping around (it's reading a low frequency). Stare at the meter for a few minutes as it jumps to get an idea of where the average level is.

These settings will get you very close to the optimum volume for each channel, but you should use your ear and some well-recorded film soundtracks in making the final adjustment. If you find that dialogue is hard to hear, a 2dB boost in the center-channel level will help bring it out (although the soundstage will be focused more in the center than spread out across the front of the room). If the bass is thumpy and boomy, turn down the subwoofer. Don't be afraid to adjust the volume of the subwoofer, center channel, or surround speakers—each movie soundtrack is mixed differently. You may find yourself slightly adjusting the surround- and center-channel levels at the beginning of each movie.

Although your ears should be the final judge when setting the individual channel levels, calibrating your system first with the SPL meter (or by ear with the test noise) will at least get you started from the right place. The most common mistakes in setting channel levels are a subwoofer level set too high and too much volume from the surround speakers. The bass shouldn't dominate the overall sound, but instead serve as the foundation for music and effects. Many listeners think the more bass you hear, the more "impressive" the sound. In reality, a constantly droning bass is fatiguing, and robs the soundtrack of impact and surprise when the filmmakers want you to hear bass. Low frequencies are used as punctuation in a film soundtrack; by keeping the subwoofer level appropriate, you'll achieve a more accurate and satisfying sound than if the subwoofer is constantly droning away. Though it's understandable that you paid for this big box in your living room and you want to hear what it does, try to avoid the temptation to set the subwoofer level too high.

Similarly, listeners who have never had surround speakers in their homes think they should be aware of the surround speakers at all times. In truth, film soundtracks don't always contain signals in the surround channels; long stretches of the movie may have nothing in the surrounds. But even when a signal is driving the surround speakers, you should barely notice their presence. Remember, the surround channels provide a subtle ambience and envelopment. If you're consciously aware of the surround speakers because they're set too loud, the illusion they're supposed to be creating is diminished. Just as it's a temptation to set a subwoofer's level too high, don't turn up the surround speakers to the point where you hear them. Note that you won't hear any sound from the surround speakers unless the source program has been Dolby- or DTS-encoded.

Adding Home Theater without Compromising Music Performance

As Editor-in-Chief of *The Absolute Sound* my job involves evaluating cutting-edge, state-of-the-art 2-channel audio products. But my dedicated listening room for the previous 13 years also served as a theater and multichannel music system. Clearly, the performance of my 2-channel system could not be compromised by the presence of home-theater products. Here's what I did to integrate home theater into my music-playback system, along with some tips for adding home theater to your music system. (I now have separate stereo and theater systems.)

First, my video display is a front-projection system with a retractable, motorized screen. The projector is at the back of the room, and when I listen to music, the screen is rolled up into a small enclosure. With the screen in the lowered position, soundstage depth is compromised, as is the precision of image placement. If you use a front projector, a motorized screen's ability to retract for music listening is a big benefit. If you must use a fixed screen, drapes that can be drawn across the screen when you play music are effective at preventing the screen from reflecting sound.

Those who use a direct-view television or rear-projector big-screen are faced with the challenge of having a large, acoustically reflective object near the loudspeakers. As described in Chapter 14, absorbing acoustic reflections at the loudspeaker-end of the room is of paramount importance. To minimize the television's degradation of the sound, move the television back as far as possible, and bring the left and right loudspeakers forward. You can get an idea of how much the television is degrading stereo imaging by throwing a blanket over the TV to absorb its reflections. If you hear a big improvement with the blanket, consider keeping the blanket on hand for critical musical listening.

As mentioned earlier, an analog-bypass mode on a controller is absolutely essential if you want to listen to analog sources with the highest possible fidelity. This feature passes analog input signals through the controller without converting the signal to digital and then back to analog. Keep in mind, however, that even the very best multichannel controllers fall short of the performance standards set by high-end 2-channel preamplifiers, even in analog-bypass mode.

If you want uncompromised musical performance, you'll need a controller and a separate preamplifier. The preamplifier should have a "theater pass-through" mode that sets one of the inputs at unity gain (the output signal's amplitude is the same as the input signal). The left and right outputs from your controller drive this unity-gain input, and the preamplifier is connected to the power amplifier in the usual way. When watching movies, it's as though your preamp isn't there. When listening to music, it's as though your controller isn't there. With this technique, shown in Fig.12-5, your analog sources never go through the controller. If you choose this option, you don't need a controller with analog bypass, except on the multichannel input.

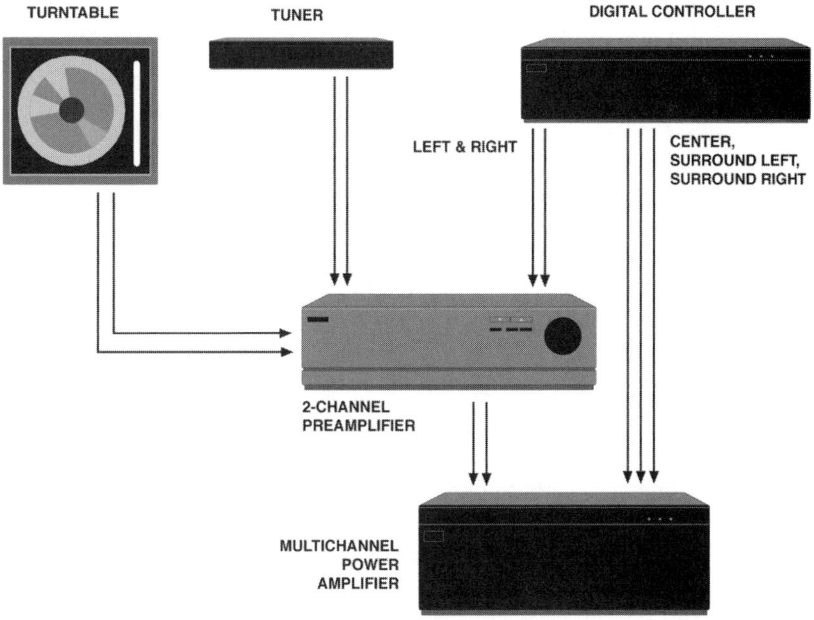

Fig. 12-5 A preamplifier with a "theater pass-through" mode avoids passing music signals through a digital controller.

If you're adding home theater to your music system and are happy with your power amplifier, you should choose a 3-channel amplifier to drive the center and two surround loudspeakers, or a 5-channel amplifier if you want a 7.1-channel system employing two surround back. If you're starting from scratch, or have been thinking about upgrading your amplifier, consider a multichannel amplifier; some of today's multichannel amplifiers are fully the equal of high-end 2-channel models.

Some subwoofers have separate processor and line inputs. The processor input (a single jack) accepts the LFE output from a controller; the line input is stereo, and is connected to a second output from your preamplifier. When watching movies, set the subwoofer's input switch to processor. When listening to music, set the sub's input switch to line. This assumes, of course, that you prefer listening to music with the subwoofer engaged. Unless you have small loudspeakers, you'll probably leave the subwoofer off for music listening by leaving the switch set to processor.

Addendum: Surround-Sound Formats Explained

I've broken out a full explanation of the various surround-sound formats for those readers interested in more technical detail. Even if you're not technically inclined, this section includes useful information about choosing among the vast array of decoding formats in today's AVRs and controllers.

We've already covered Dolby Pro Logic, Dolby Digital, and DTS in the body of this chapter. Now let's look at variations on those formats.

In mid-2001, Dolby Laboratories made available a more sophisticated version of Pro Logic decoding, called Pro Logic II. The idea behind Pro Logic II was to create from 2-channel sources a listening experience similar to that of a discrete 5.1-channel digital format. And with more consumers having 5.1-channel playback available to them for reproducing music sources, Pro Logic II attempts to deliver multichannel sound from 2-channel recordings, even those that have not been surround-encoded.

Found on most A/V receivers and controllers made after early 2002, Pro Logic II offers improved performance over its predecessor in several areas. First, Pro Logic II delivers full-bandwidth stereo surround channels rather than the bandwidth-limited monaural surround channel of conventional Pro Logic decoding. This attribute provides a more enveloping soundfield, greater precision in the placement and pans (movements) of sounds behind the listener, and more natural timbre of sounds reproduced by the surround channels. In this respect, Pro Logic II emulates the experience of listening to a 5.1-channel discrete digital source.

Pro Logic II also uses more sophisticated "steering" circuits that monitor the level in each channel and selectively apply attenuation (reduction in level) to prevent sounds in one channel from leaking into another channel. Pro Logic IIx, announced in late 2003, creates 7.1-channel playback from 2-channel and 5.1-channel sources. In 2009, Dolby introduced Pro Logic IIz, which calls for two "height" speakers to be located in the front of the room above the left and right speakers. Pro Logic IIz extracts this height information from 5.1- or 7.1-channel sources and directs it to the height speakers, which can be small and unobtrusive. The idea is to create a more immersive soundfield by expanding the soundfield vertically.

Not to be outdone, DTS has developed a suite of surround-decoding formats that either enhance the DTS experience or provide decoding of non-DTS sources such as conventional stereo.

DTS Neo:6 Music and Neo:6 Cinema are decoding algorithms that convert stereo or Dolby Surround encoded 2-channel sources into multichannel surround sound. Neo:6 Music leaves the front left and right channel signals unprocessed for the purest reproduction, and extracts center and surround-channel signals from the 2-channel source. DTS recommends this mode for all 2-channel sources, such as CD and FM broadcasts.

Neo:6 Cinema is similar to Dolby Pro Logic II decoding, and can be used with Dolby Surround-encoded sources. Neo:6 Cinema has a much larger effect on the signal, rearranging the signal distribution among the front three channels. Both Neo:6 Music and Neo:6 Cinema create a 7.1-channel signal from 2-channel sources.

These decoding algorithms are very useful to those, like me, who greatly enjoy musical performances on DVD. When a DVD gives me the choice of listening to the Dolby Digital or 2-channel Surround-encoded linear pulse-code modulation (LPCM) mix, I always opt for the 2-channel mix. Although I lose all the advantages of Dolby Digital (complete channel separation, for example), I hear smoother treble, less grainy instrumental and vocal textures, and a greater sense of space. That's because Dolby Digital uses 384,000 bits per second (or 448,000 bits per second) to encode all 5.1 channels; the linear PCM track uses 1.536 million bits per second to encode just two channels. When using Pro Logic II and DTS Neo:6 to decode these two channels, however, I still hear excellent spatial resolution, but with the more natural timbres made possible by the linear PCM track's much higher bit rate.

In 2004, Dolby Labs and Lucasfilm jointly developed the Dolby Digital EX format. Dolby Digital EX encodes a third surround channel in the existing left and right surround channels in the Dolby Digital signal. This additional surround channel, called surround back, is decoded on playback, and the signal drives a loudspeaker (or two loudspeakers) located directly behind the listener. Dolby Digital EX allows filmmakers to more precisely position and pan (move) sounds around the room. For example, the sound of an object moving directly overhead from the front of the room to the back tends to become smeared; the sounds starts in the front speakers, then splits along the sidewalls as it moves toward the rear of the room. By adding a third surround channel directly behind the listener, these "fly-overs" can be made more realistic. Note that an EX-encoded soundtrack is compatible with all Dolby Digital playback equipment because the addition of the surround-back channel is encoded into a conventional 5.1-channel Dolby Digital signal. Dolby Digital EX is not a 6.1-channel format; it is still 5.1-channels, but with the additional channel encoded within the 5.1-channel datastream.

DTS' equivalent format is called DTS-ES. You'll also see it called DTS-ES Matrix, because the additional surround channel is matrix-encoded into the existing left and right surround channels. DTS also offers DTS-ES Discrete, in which the surround-back channel is a discrete channel in the ES-Discrete bitstream. DTS-ES Discrete is a true 6.1-channel format. The advantage of a discrete format is complete separation between channels. Matrix surround systems have limited channel separation.

Note that Dolby Digital and DTS are "lossy" formats, meaning that some information is lost in the encoding and decoding process. This loss of information is intentional so that the number of bits required to represent the signal can be dramatically reduced. Consequently, Dolby Digital and DTS have reduced fidelity to the source.

The massive storage capacity of the Blu-ray Disc format as allowed Dolby Laboratories and DTS to develop better-sounding surround-sound formats with higher bit rates than Dolby Digital and DTS. From Dolby Laboratories, we have two new formats called Dolby Digital Plus and Dolby TrueHD. Dolby Digital Plus is an extension of the conventional Dolby Digital we've been listening to for years on DVD and HD television shows. Dolby Digital Plus offers more channels and a higher bit rate for better sound quality. Where Dolby Digital was limited to a maximum bit rate of 640kbps (kilobits per second), Dolby Digital Plus allows scalable bit rates up to 6Mbps on Blu-ray. In addition to more bits per second, Dolby improved the encoding algorithms in Dolby Digital Plus for better sound.

For the ultimate in sound quality, Dolby has introduced Dolby TrueHD, which delivers high-resolution multichannel audio with perfect bit-for-bit accuracy to the source. With TrueHD, you will hear in your home sound quality identical to what the engineers heard in the studio. Specifically, TrueHD is capable of eight channels of uncompressed audio at 192kHz/24-bit. This is made possible by TrueHD's maximum bit rate (on Blu-ray) of 27Mbps. TrueHD decoding is an option, rather than a requirement, in Blu-ray players. The availability of a consumer release format that delivers uncompressed high-resolution multichannel audio is nothing short of a revolution in home entertainment. Concert videos and musical performances on Blu-ray with uncompressed high-resolution multichannel audio is spectacularly great.

With more than 40 million Dolby Digital decoders in use throughout the world, Dolby made sure the new formats were backward-compatible with your receiver or controller. Blu-ray Disc players output a conventional Dolby Digital or DTS bitstream on the familiar coaxial or optical digital outputs for connection to your receiver. This output will likely sound a little better than Dolby Digital from DVD because it always operates at the highest possible data rate of 640kbps (Dolby Digital on DVD is limited by the DVD format to 448kbps, but is typically 384kbps).

DTS has introduced its own high-resolution surround-sound audio format for Blu-ray called DTS-HD. This new format is an extension of conventional DTS, offering scalable bit rates from conventional DTS at 754kbps, all the way to 6Mbps on Blu-ray. A variant of DTS-HD, called DTS-HD Master Audio, can operate at higher bit rates (up to 24Mbps on Blu-ray) for true lossless encoding of high-resolution multichannel audio. The "Master Audio" tag signifies lossless encoding. As with Dolby TrueHD, HD DVD players are not required to implement DTS-HD Master Audio; however, if a player encounters that format on a disc, it must at least deliver the DTS core audio stream (that is, good old standard DTS).

To hear the full capability of Dolby TrueHD and DTS HD Master Audio, you must have a controller or receiver that can decode the new formats, and connect the Blu-ray player with an HDMI cable.

13 Multichannel Audio

Introduction

Since the commercialization of the stereo LP in 1958, music has been delivered to consumers through two channels: left and right. Storing an acoustic event in two information channels was considered a compromise by early audio researchers (the literature suggests some considered three channels a minimum), but no consumer format could hold more than two channels. The LP groove has only two sides, and multichannel tape machines were impractical for home playback.

The industry attempted to bring multichannel playback to the masses in the 1970s with Quadraphonic. As we all know, Quad was a commercial flop. The idea of a multichannel consumer-audio format died with Quad until the late 1990s, when a new generation of high-capacity 120mm optical discs (SACD and DVD-Audio) was developed that could easily accommodate multichannel audio. SACD and DVD-Audio revived the idea of delivering to consumers more than two audio channels, but DVD-Audio is essentially dead and SACD has become a niche format (although a sizeable niche). The Blu-ray Disc format has a provision in the specification for an audio-only version that can store up to eight channels of uncompressed PCM audio at a sampling rate of up to 192kHz and with 24-bit resolution. A few specialty labels have introduced Blu-ray music titles, but so far the major labels have shown no interest in the format and audiophile labels have not rushed to reissue classic albums on Blu-ray. It's ironic that just as a high-resolution multichannel format has become a reality, the major record companies, who control most of the world's music catalog, are concentrating on low-resolution downloads.

In the past, quality conscious consumers were stuck with compromised formats developed for the mass market.. Creating a new physical format is a massive undertaking, and one that the major technology and music companies embark on only if the financial rewards are great. But the advent of downloadable music renders moot the idea of physical formats (packaged media). We can instead bypass packaged media in favor of high-resolution multichannel downloads. The availability of high-speed Internet connections, coupled with ridiculously cheap hard-disk storage, makes it feasable to download high-res multichannel music. Nonetheless, if you are considering going multichannel, there are a number of factors to consider before making the leap.

Before we look at these new multichannel formats, discuss their merits, and talk about how to get multichannel audio in your home, let's take a brief look back on the history of multichannel audio.

A Short History of Multichannel Audio

Stereo audio recording and reproduction was invented by a young British engineer named Alan Dower Blumlein. At the age of 26, Blumlein was hired by the Columbia Gramophone company to develop a record-cutting system that didn't infringe on the patents held by Bell Telephone Laboratories. Although the cutting system he devised was monophonic, Blumlein was nonetheless thinking about a new technique that would capture audio signals with greater spatial fidelity. Blumlein was aware of the mechanisms by which humans localize sound sources (phase and amplitude differences—see Appendix A), and was working in the background on a 2-channel recording and reproduction chain that would allow spatial positioning of sounds.

By early 1931, at the age of 29, Blumlein had conceptualized and built a system he called "Binaural Sound"—what we today call stereo. In that year he filed his landmark patent "Improvements in and relating to Sound-transmission, Sound-recording and Sound-reproducing Systems" (patent number 394,325, filed December 14, 1931) which described in great detail a new microphone technique, a "shuffler" circuit, and record-cutting system that made stereophonic sound possible. Blumlein described it this way:

The invention relates to the transmission, recording and reproduction of sound, being particularly directed to systems for recording and reproducing speech, music and other sound effects especially when associated with picture effects as in talking motion pictures. The fundamental object of the invention is to provide a sound recording and reproduction system whereby a true directional impression may be conveyed to a listener, thus improving the illusion that the sound is coming from the artist or other sound source as presented to the eyes.

Blumlein employed two bi-directional (also called figure-of-eight microphones because their directional-sensitivity pattern looks like a figure-of-eight) crossed at 90°, with the two positive-lobe axes pointed at the center of the musical group. When their signals are reproduced through two loudspeakers, the listener can localize sound between the two loudspeakers. (This explanation is simplified to remain within the scope of this book. For a detailed analysis, see John Eargle's *The Microphone Handbook*.)

Blumlein also invented other so-called "coincident" microphone techniques (M-S, X-Y) that are still used today, although only the crossed figure-of-eight technique is named after him. (Incidentally, you can hear "Blumlein" miking on many titles released by Water Lily Acoustics, notably the gorgeous *A Meeting by the River*.)

Although Blumlein and his colleagues made many convincing recordings and films (Blumlein himself walking across a stage while counting, a train leaving a station, for examples), none of his contemporaries realized the significance of this work. The world would have to wait 27 years for the first commercial realization of stereo—the long-play record in 1958—long after Blumlein's patent had expired. (Alan Blumlein was killed in a Royal Air Force plane crash in 1942 while working on radar research.)

About the same time, the Bell Telephone Laboratory was furiously researching all aspects of sound recording and transmission. In 1932 Harvey Fletcher of Bell Labs made a 2-channel recording with conductor Leopold Stokowski. Despite the obvious advantages of two channels over one, stereo recording and reproduction lay dormant until the late 1950s—on both sides of the Atlantic. Soon thereafter, of course, it became the standard for all recorded music.

The stereo record-cutting system developed by Blumlein, in which each channel is cut into one side of a 90° groove (the "45/45" system), appeared to limit the number of playback channels to two. In the 1970s, however, engineers developed several different techniques for encoding four audio channels within the two walls of a record groove. Although competing systems were developed, the technique was generally called Quadraphonic. The two major Quad formats are SQ (developed by CBS, later bought by Sony) and CD-4 (developed by JVC). SQ is a matrix system in which the rear channels were matrix-encoded in the front channels. SQ suffered from extremely limited channel separation (as low as 3dB before logic steering became prevalent). CD-4 (Compatible Discrete 4) combined the front and rear channels in one signal, and recorded a difference signal between 30kHz and 50kHz. This high-frequency carrier was demodulated to obtain four discrete channels from a two-walled LP groove. CD-4 required a special phono cartridge and precise turntable setup. (Other proposed Quad systems such as QS and UMX were never fully realized in the marketplace.)

Quad was a musical and commercial disaster, primarily because the market was presented with competing systems and the demonstrations were laughably bad. In an attempt to show off Quad's 4-channel sound, Quad recordings often put the listener in the middle of a musical group, with instruments coming from all directions. Moreover, the theory behind Quad was fundamentally flawed, as was repeatedly pointed out at the time by British audio genius Michael Gerzon. For one thing, Quad called for two loudspeaker pairs spaced at 90 ° (roughly one speaker in each room corner), which is too far apart to produce phantom images between the loudspeakers. The general public never cottoned to this gimmickry and Quad quickly died. (There are, however, Quad aficionados who still debate the relative merits of SQ and CD-4.)

Although CD is a 2-channel medium (it has a provision for four discrete, lower-quality channels, but that capability was never used), it is possible to encode multichannel music on CD. Digital Theater Systems (DTS) has released a small number of multichannel titles encoded with its "perceptual coding" technique that reduces the number of bits required to represent the audio signal. The result is some loss of fidelity (CD's bit-rate of 1.41 million bits per second for two channels must be spread over six channels), but a gain in the number of channels. Such discs must be played in a DTS-compatible player. Titles were also released in the Dolby Digital format, with the best examples coming from the classical label Delos. These formats were, however, a stopgap measure until a true multichannel music carrier became a reality.

The availability of higher-capacity digital audio discs (SACD and Blu-ray) along with multichannel downloads again offer the possibility of playing back music recorded with more than two channels of information. These formats obviate the serious technical compromises inherent in both SQ and CD-4 Quadraphonic, and don't require a reduction in sound quality and lower bit-rates to achieve multichannel capability. Consequently, we're entering another age in which the general public will again be offered multichannel audio for music. The general public has, of course, embraced multichannel audio for film-soundtrack reproduction. The popularity of home theater has greatly increased the number of homes with a multichannel loudspeaker array (more than 60 million as of this writing).

Do We Want Multichannel Music Playback?

But just because we *can* play back multiple channels in the home, should we? Is multichannel audio a significant advance in sound quality, or a marketing gimmick foisted upon us? This issue is perhaps the most polarizing of any current controversy among audiophiles. The majority of audiophiles consider 2-channel music reproduction sacrosanct, and not something to be tampered with. Virtually all the recordings that we consider cultural treasures were recorded in stereo, and playing them back in any way other than exactly as they were intended to be heard is anathema. Moreover, surround-sound for music is associated with surround-sound for film, in which explosions and special effects are presented all around the listener—fine for movies but justifiably abhorrent to the purist music lover. Moreover, 2-channel listening can be immensely satisfying, so why bother with surround sound? Multichannel audio is viewed by many audiophiles as a fad, not a legitimate means of advancing the goal of recreating the original musical event in our homes with the greatest possible fidelity. On the other hand, some of the most ardent and dedicated audiophiles have championed multichannel as the most significant advance in audio since stereo.

In my view, multichannel audio is a quantum leap forward in advancing the music-listening experience—*with a properly set-up system playing recordings made with musical sensitivity*. There's no question in my mind that multichannel audio can greatly increase the spatial realism of reproduced music and deliver a more involving experience. The catch is that the system must be properly configured, and that the recording use the technology in a musically appropriate way rather than as an "effects" gimmick.

Multichannel overcomes a fundamental limitation of stereo, which is that it tries to create a 360° soundfield from two loudspeakers spaced 60° apart. Let's explore this idea with a thought experiment. Suppose we are in a concert hall enclosed in an acoustically isolated shell. If we cut two holes in the front of the shell facing the stage (analogous to listening to the performance through two loudspeakers), we'll hear mostly direct sound from the instruments. If we cut additional holes in the back of the shell, reflections from the back of the concert hall can enter the shell and create a soundfield in the shell that is closer to that in the concert hall. How could anyone argue that putting more than two holes in the shell is a step backward? Reverberation and reflections arrive at our ears in the concert hall from behind us; why shouldn't they in the home?

I've attended many demonstrations of multichannel sound; I'll tell you about two of them that exemplify the problems and promise of multichannel music reproduction. In one demonstration, the sound from the speakers behind the listening position was played as loudly as the sound from the front speakers. I heard a blasting trumpet just behind my left ear, and an electric guitar behind my right ear. Most of the rest of the band was reproduced by the front channels, giving the impression not of hearing a musical group performing in front of me, but of being surrounded by the musicians. Not only was this unnatural and musically distracting, but hearing loud sounds suddenly blare out from behind me caused me to turn my head toward the sound source. Human beings have a survival instinct that makes them turn around when they hear a sharp sound behind them. Putting the listener in an instinctive state of defensive readiness is not conducive to musical involvement. Moreover, hundreds of years of Western musical tradition call for the performers on a stage and an audience in front of those performers. (Some compositions call for instruments behind the audience, but those are the rare exception.) To top it off, the overall sound was far too loud. All of this added up to a most unpleasant experience.

Now consider a second demonstration of surround sound I attended. For the original recording, five microphones had been placed in a large concert hall, three across the front and two in the rear. The five microphone signals were recorded in Sony's DSD format (the encoding format of SACD), and then reproduced in the demonstration through five loudspeakers arranged similarly to the microphone placement. The re-creation of the original acoustic space was breathtaking; instead of hearing reverberation come from the front channels along with the instrumental images, the warmth of the hall was reproduced from behind me. More precisely, I wasn't aware of sound sources behind me, only of being inside a large acoustic space. Only ambience and hall reflections were reproduced by the rear channels, and at a *very* low level. Moreover, the front of the soundstage was more spacious, and the impression of air between instrumental images was greater than I've heard from any 2-channel playback. The soundstage didn't stop at the loudspeaker boundaries; it instead gently enveloped me, just as in live music. This microphone array, along with the multichannel delivery path, captured the spatial aspects of musicians in a hall with far greater realism than is possible from two channels.

The promoters of the first demonstration wanted to hit people over the head with the fact that they could place instruments behind the listener. They had absolutely no regard for whether the technology was being used in a musically appropriate way. In the second demonstration, surround-sound technology increased the realism and provided a more satisfying experience than was possible with two channels. Unfortunately, the first example is the rule, the second the exception. Those who conducted the first demonstration apparently believe that most listeners are not sophisticated enough to discern and appreciate subtlety. It's demonstrations like the first one that will further polarize the audio community over the validity of multichannel audio.

This comparison makes the point that surround sound is neither a miracle technology nor the evil that many 2-channel "purists" consider it to be. Provided that it is used tastefully, multichannel audio has the potential of elevating the music-listening experience to a new level. If you have an antipathy toward multichannel audio, be sure your opposition is based on your own listening experience rather than on dogma or prejudice.

I'd like to tell you about another experience I've had with multichannel audio. Peter McGrath, one of the world's finest classical-music recording engineers, visited my listening room with his 4-channel Nagra D digital recorder and some 4-channel master tapes he'd made. Listening in multichannel mode first (for more than two hours), I was never conscious of sound originating from behind me. Rather, I felt enveloped in a large acoustic space. Halfway through a piece, we switched to stereo, and the magic disappeared. The soundstage collapsed into the front channels, depth was reduced, and I suddenly had the feeling of hearing the music through a window in front of me rather than the illusion of being inside the concert hall. Had I heard the 2-channel recording first, I would have thought it excellent. But after hearing reverberation and hall ambience reproduced from behind me—the same place it originates in live orchestral music—there was no going back.

During this listening session, I had another epiphany. Peter suggested that I walk around the listening room while the music was playing. Standing between the front speakers and facing the listening room's rear, I had the distinct impression of standing on a stage in a concert hall. The direct sound of instruments was to my left and right, and I heard reflections and reverberation in front of me coming from the rear speakers, just as one would when standing on a concert-hall stage. Standing between the rear speakers and facing forward, I heard more sound from the rear channels, which created the impression of being in the back of the concert hall. I could move forward a few steps and hear the spatial perspective change, exactly as one would hear in a hall, as the ratio of direct to reverberant sound changed. The movements were, of course, exaggerated; a few steps in my listening room were equivalent to moving several dozen feet in a hall.

I then stood on the right sidewall facing out into the room, with the front speakers to my right and the rear speakers to my left. Again, the impression of standing in that location in the concert hall—performers on the right, hall reverberation on the left—was eerily palpable.

This loudspeaker array created four separate but connected soundstages: the conventional left/right; between the front and rear left loudspeakers; between the front and rear right loudspeakers; and between the two rear loudspeakers. These soundstages are much easier to hear when moving around the room as I did; when facing forward, precise imaging between front and rear loudspeakers was diminished because the sounds arrive from the side rather than from the front, which the brain processes differently from two front-arriving sounds.

This experience suggests that multichannel audio, done right, creates a more accurate, realistic, and compelling document of the original acoustic event. That the illusion survived my walking around the room says to me that the process is fundamentally valid. By contrast, the imaging we hear from 2-channel stereo is fragile; move your head sideways by a foot—or even a few inches—and the illusion can collapse.

Unfortunately, very few recording engineers currently view multichannel audio as an opportunity to better recreate the spatial aspects of live music. Rather, they see it as a new tool for fundamentally changing the listening experience. For example, AIX Records founder Mark Waldrep, a highly respected recording engineer and musicologist (he holds a Ph.D. in the subject), believes that multichannel audio is a means of escaping from the musical straightjacket that is 2-channel stereo. His recordings are often made with the musicians in a circle around the microphones. The rear channels contain just as much information as the front channels—even with a string quartet recording. In my interview with him in Issue 134 of *The Absolute Sound,* Waldrep articulated his approach: "For me, it's an active choice to *not* create an archival document of a space where musicians are performing, and instead try to enhance or energize that performance with the technology we can bring to the equation." Waldrep's approach may be musically valid at some level—he knows far more about music than I ever will—yet I find such recordings disconcerting. (All AIX titles, which are superbly recorded at 96kHz with 24-bit resolution, contain both multichannel and 2-channel mixes, as well as "stage" and "audience" spatial perspectives—enough choices to satisfy any listener's sensibilities.)

At an Audio Engineering Society convention, I attended a workshop that explored how multichannel music should be recorded and mixed. The panel was a "who's who" of contemporary popular-music engineers. Each panelist took a turn presenting his views on how to use the additional channels. There was absolutely no consensus among the panelists; everyone had a different approach. One engineer believed in placing the lead vocal *only* in the center channel. Someone else stated that the vocal should be mainly reproduced by the left and right channels, with slight augmentation by the center channel. Yet another advocated putting the vocal in the center channel with slight augmentation in the left and right channels. And the discussion hadn't yet considered the rear channels.

(You can judge for yourself the appropriateness of "hard" placement of the lead vocal in the center channel with James Taylor's *Hourglass* SACD. Simply switch between the multichannel and 2-channel mixes. The multichannel mix puts Taylor's voice only in the center channel. The 2-channel mix puts his vocal equally in the left and right channels, which we hear as a phantom image between the two loudspeakers. In my experience with this recording, the hard placement of the vocal in the center channel kills the sense of depth and air around Taylor's voice.)

At any rate when it's used judiciously—however much experimentation that may require—multichannel audio can be a legitimate vehicle for musical expression, freeing the artist from the constraint of positioning sounds solely in front of the listener. In addition to this (potential) expansion of a musical aesthetic, multichannel audio can increase the listener's ability to hear deeper into the music. Psychoacoustic research has shown that humans can absorb more aural information when that information arrives from different locations. That is, a sound arriving from behind the listener is more intelligible than the same sound arriving from the front along with other sounds. We can better differentiate individual sounds, and thus more easily explore their musical meanings, when those sounds arrive from different locations around us.

This phenomenon was graphically demonstrated at a Consumer Electronics Show a few years ago. Tomlinson Holman, inventor of THX (THX stands for "Tomlinson Holman's eXperiment"), set up and conducted the most unusual demonstration I've ever attended. Ushers with flashlights guided groups of showgoers to seats in a huge and nearly pitch-black room. The lack of light deprived us of any clue as to what we were about to hear—and prevented any prejudicial conclusions. Without any announcement describing what the demo was all about, the music began. The piece, an instrumental performed on synthesizer, was densely layered, yet perfectly intelligible on first listening. Each of the many parts seemed to originate from a different point in space surrounding me. There were no discrete sound sources, only a seemingly continuous soundfield. I could easily shift my attention from one musical line to another. In fact, the unique sonic presentation encouraged—perhaps "compelled" is a more appropriate word—such an exploration of the music. It was an unforgettable experience.

After the demonstration, Holman turned on the lights and explained what we had just heard—a 10.4-channel audio system playing a piece composed and performed by Herbie Hancock that had been recorded in 10.4 discrete channels (ten full-range channels and four subwoofer channels). Hancock used this wide sonic palette to expand his musical expression. My enthusiasm for what I'd just heard was tempered by the fact that a 10.4-channel audio system is never likely to become a commercial reality. Still, Holman proved his points: multichannel audio opens up new musical horizons, and more channels are better than fewer channels. (Holman's company, TMH Audio, develops and builds innovative products for the recording, film sound, and music industries.)

An unfortunate side effect of the home-theater boom is that multichannel music formats and playback systems have been dictated by the movie industry rather than developed from first-principles for the highest possible musical fidelity. Today's multichannel loudspeaker layout, channel configurations, and use of a single subwoofer all trace their origins to the film-sound industry. For example, a center loudspeaker is required in movie theaters because of the very wide placement of the left and right speakers. Dialogue must seem to originate on-screen; in fact, the center channel in movie sound is called the "dialogue channel." For music reproduction, we may want a center loudspeaker, but not a center channel driving it. (This is explained later in the discussion of Trifield processing.)

A fundamental problem of judging multichannel audio is the scarcity of good set-ups and demonstrations. Most multichannel loudspeaker arrays at high-end audio dealers are optimized for home-theater playback, not music listening. After all, home-theater is a booming industry, while multichannel audio is not.

Ironically, the best demonstrations of multichannel audio's capabilities may be in the car. The big advantage of a car stereo is that the loudspeaker placement, listener position, and acoustic environment are all known. There's no chance for a botched demonstration because of less-than-ideal loudspeaker placement, or other variables. Of course, a car audio system, no matter how good, won't approach the quality of a true high-end system set up in a room. Nonetheless, the TL's discrete multichannel audio system and overall high quality offer the general public a good representation of multichannel audio's potential.

How to Get Multichannel Audio in Your Home

If you decide to make the multichannel plunge, you can follow one of four paths:

1) Build a system optimized for 2-channel music playback and home-theater reproduction, and accept compromised multichannel music performance.

2) Build a system optimized for 2-channel and multichannel music playback, and accept compromised home-theater sound.

3) Build a system optimized for 2-channel and multichannel audio without any capability for reproducing film soundtracks.

4) Build a system that does it all—that's equally adept at both 2-channel and multichannel music as well as home-theater soundtrack reproduction.

This first approach is the most common, which is why it's difficult to hear state-of-the-art multichannel audio demonstrations. Such a system employs sidewall-mounted dipolar loudspeakers, which are ideal for film-soundtrack reproduction but not for multichannel music. As described in Chapter 12, dipolar surround speakers mounted on the sidewalls better create a sense of envelopment and diffusion in the surround soundfield. They "smear" the rear-channel information by their directional pattern; the listener hears their sound only after the sound has been reflected around the room (this phenomenon is described in more detail in Chapter 12). Although this smearing may be appropriate for the atmospheric sounds in a film soundtrack (a rainstorm, jungle scene, or cityscape, for examples), when applied to a multichannel music mix this smearing is a distortion of the musical presentation created by the recording engineer. Multichannel music is recorded and mixed with uni-polar loudspeakers, and should be played back with an identical loudspeaker array. You will never hear imaging between dipolar surround loudspeakers.

The second option—optimize your system for multichannel music at the expense of compromised film-soundtrack performance—simply involves using uni-polar, rather than dipolar, surround loudspeakers. These need to be positioned behind the listener rather than on the sidewalls, as would dipolar models. When listening to multichannel music, you'll hear a better representation of what the recording engineer intended, but film soundtracks won't sound quite as spacious or enveloping. The atmospheric sounds in the surround channels will appear to originate from two points in space (the left and right surround speakers) rather than enveloping you in a diffuse wash of sound. Nonetheless, many listeners find uni-polar surround loudspeakers perfectly acceptable—even preferable—for film-soundtrack reproduction.

Both these options require a home-theater controller (described in Chapter 12), which will function as a surround-sound decoder for film soundtracks and be the source selector and volume control for multichannel music. The six analog outputs from a universal-disc player feed the controller's multichannel analog input. Alternately, the controller can accept digital multichannel audio from a Blu-ray Disc player (or universal-disc player) on an HDMI cable. The digital outputs from your other sources (satellite, DVD player) feed the controller's digital inputs.

Which brings us to the third option of dismissing film-soundtrack reproduction entirely. If you never intend to use your system for movie watching, you'll obviously choose uni-polar surround loudspeakers. Although you won't necessarily need a home-theater controller, you will need a way of selecting sources (one of which will be multichannel) and controlling the volume. The solution for this system is the multichannel preamp. This new breed of product includes no provisions for surround-sound decoding, bass management, or other home-theater requirements. Rather, it is simply a preamplifier with six channels rather than two. Fig.13-1 shows an example of a multichannel preamplifier. Note the two sets of balanced and unbalanced multichannel inputs on the lower left.

Finally, if you want the ultimate performance for both multichannel music and movies, you'll install two sets of surround speakers, one dipolar on the sidewalls and one uni-polar toward the rear. You'll need two separate 2-channel power amplifiers, one to drive the dipoles and one to drive the uni-poles. A pair of "Y" adapters on the controller's surround-channel outputs splits the signals so that the surround signal can drive two amplifiers. You select which surround loudspeakers are engaged by which power amplifier you turn on. This method is a bit cumbersome, but does solve the music/movie dilemma.

All these approaches require a multichannel source such as a universal-disc or Bluray player, or a music server with multichannel capability. These machines are described in detail in Chapters 7 and 8.

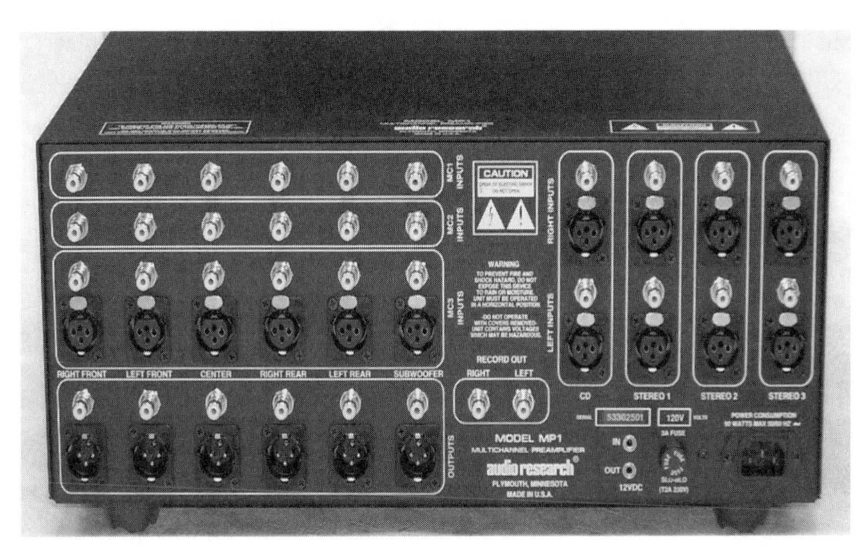

Fig. 13-1 A multichannel preamplifier—rather than a digital controller—provides the highest audio quality in multichannel music reproduction. (Courtesy Audio Research Corporation)

Loudspeaker Types and Placement

The home-theater paradigm calls for five (or seven) loudspeakers and a single subwoofer. The subwoofer is driven by the Low-Frequency Effects (LFE) channel in a Dolby Digital soundtrack and selectively, bass from any of the other channels. In fact, many home-theater systems employ small satellite loudspeakers and a single subwoofer that reproduces all the bass in the system. Even if your home-theater loudspeaker system has full-range left and right loudspeakers, chances are that the center loudspeaker will be the small, horizontally-mounted variety.

This arrangement is, however, not ideal for multichannel music reproduction. The best loudspeaker configuration for multichannel music is five identical full-range loudspeakers in the locations shown in Fig.13-2. No subwoofer is required. Of course, if your system also functions as a home theater, you can have a subwoofer that is engaged only for film-soundtrack reproduction, not music playback. The loudspeakers should be equidistant from the listener. This array creates four separate but connected soundstages: between the left and right loudspeakers; between the surround loudspeakers, between the front left and rear left loudspeakers; and between the front right and rear right loudspeakers.

Chesky Records objected to the film-sound industry dictating multichannel music playback standards, so the company developed its own channel configuration and loudspeaker array and made some recordings for playback on what it considered its ideal system. That system, called Chesky 6.0, uses the loudspeaker array shown in Fig.13-3.

Demonstrations of this technique have sounded exceptionally good. Unfortunately, it's unlikely that this excellent approach will become a standard.

Bass Management

Bass management is a subsystem within a multichannel source machine (Blu-ray Disc player, SACD player) or digital controller that allows you to selectively distribute bass information

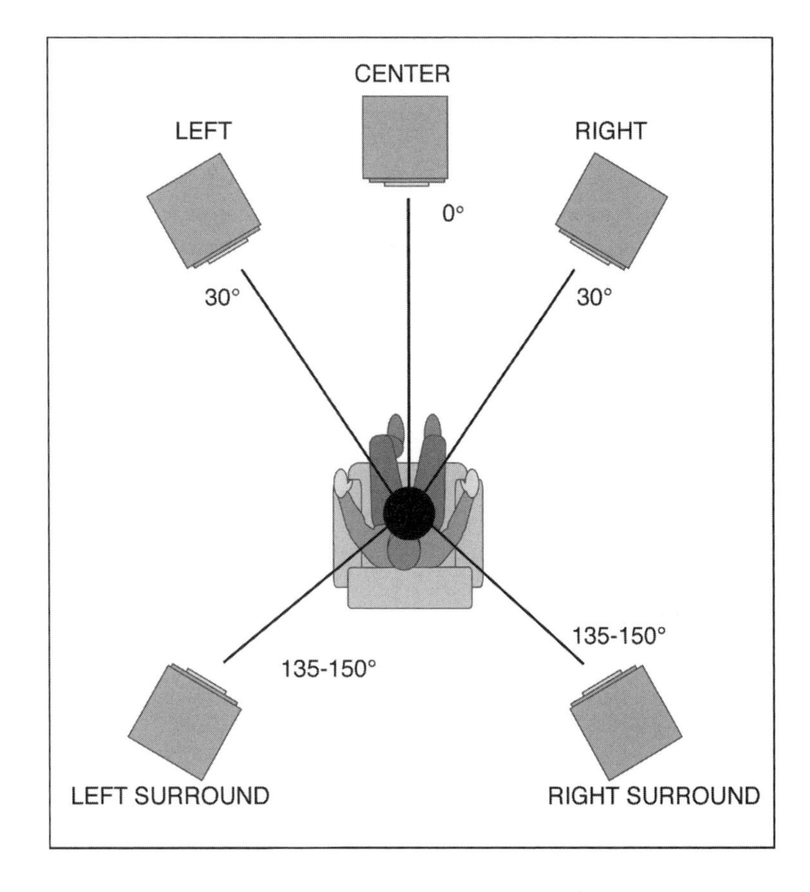

Fig. 13-2 Loudspeaker placement for multichannel music reproduction

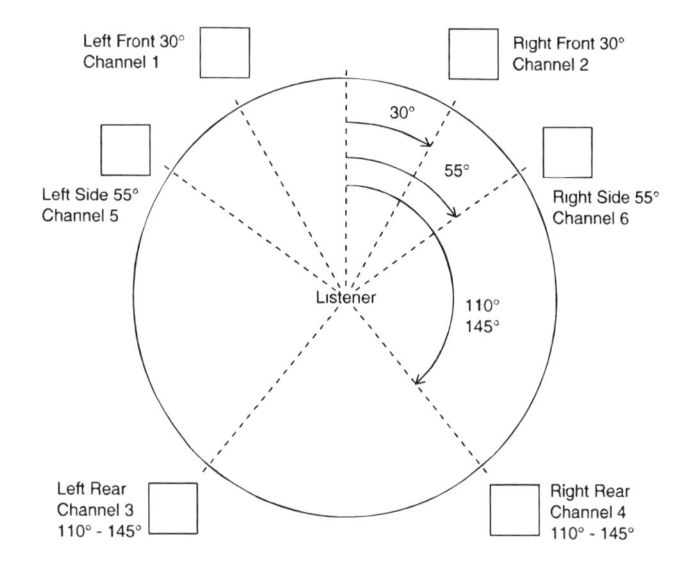

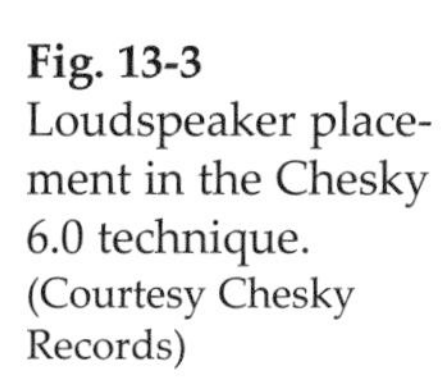

Fig. 13-3 Loudspeaker placement in the Chesky 6.0 technique. (Courtesy Chesky Records)

among your loudspeaker array. For example, if you have full-range left and right loudspeakers, but small center and surround loudspeakers, bass management keeps bass out of the small center and surround speakers and directs that bass to a subwoofer (if the system uses one). Bass management is essential in a multichannel loudspeaker system because not all systems use full-range loudspeakers.

Although bass management is provided in both multichannel players and controllers, you need set up only one device to perform bass management. Note that if you choose a system with a multichannel preamp rather than a digital controller, you'll need to set up bass management in the player.

Calibration

Just as it's vital to 2-channel playback that the left and right loudspeakers produce identical sound-pressure levels, the loudspeakers in a multichannel audio system need to be precisely calibrated so they produce the same level at the listening position. Adjusting the individual channel levels compensates for differences in loudspeaker sensitivity, placement, distance from the listener, and the effects of room acoustics.

All digital controllers are equipped with a pink-noise generator that can be programmed to cycle around all the channels. Pink noise is a hissing-like sound. You'll need to use this test signal in conjunction with a sound-pressure level (SPL) meter; setting the levels by ear is not accurate enough. Available from RadioShack for about $30, an SPL meter lets you accurately calibrate your multichannel audio system. RadioShack offers digital and analog SPL meters; buy the easier-to-read analog type (catalog #33-2050). Switch the meter to the C-WEIGHTED position, SLOW RESPONSE, and set the knob to 70. Position the SPL meter where your head would be in the listening seat, and point the meter toward the ceiling. (This ensures that the meter doesn't favor any one channel.) Turn on your controller's test noise and set each channel's volume until the meter's display reads 5. This indicates that the noise is being reproduced at a level of 75dB. Repeat this procedure for each channel. You may need to repeat this cycle over all the channels a few times to get the levels precisely matched.

After calibration, the controller's volume display will accurately show the level relative to 0dB. A typical listening level may be read as "–10dB" in the display. This may be confusing; why is 0dB the reference level? As explained in Appendix A, decibels express a difference, or ratio, between two values rather than an absolute value. At the 0dB volume setting, the system should produce a peak sound-pressure level of 105dB. This calibration standard comes from film sound, where the maximum sound-pressure level of the loudest peaks is 105dB. All other volume settings are referenced to this maximum value. With the system properly calibrated and the controller set to 0dB, the sound-pressure level in your listening room should be the same as on the film-sound mixing stage.

If the surround channels are too high in level, you'll hear too much ambience and reverberation and not enough direct sound. Set the rear channels too low in level and you won't experience the full surround effect. It's a little like changing rows in a concert hall; the farther from the stage you sit, the lower the ratio of direct to reflected sound. Unfortunately, multichannel music discs in which the surround channels contain reverberant information vary greatly in the amount of signal in the surround channels. This means that one calibration is not ideal for all discs. It's best to start with the precise meter-derived calibration, and adjust slightly from there to suit the recording.

You'll also need to set the controller's individual channel delays so that the sound from each loudspeaker arrives at the listening position at the same time. The controller's set-up menu will ask you for the distance between each loudspeaker and the listening seat. The controller then calculates the amount of delay to apply to each channel. If the center channel has too little delay, soundstage width will be constricted, with instruments tending to bunch up in the middle. The middle of the soundstage will seem more forward than the sides. Conversely, too much center-channel delay will increase the sense of depth, but at the expense of presence and tonal fidelity.

Multichannel Playback from 2-Channel Sources

The proliferation of multichannel loudspeaker arrays, combined with the scarcity of multi-channel recordings, has led to the creation of decoding algorithms that generate 5- or 7-channel playback from unencoded stereo program material. Consumers who paid for those center and surround loudspeakers want to hear them working all the time, not just when watching movies. Aside from that cynical view, unencoded stereo programs often contain ambience information that is sufficiently decorrelated with the direct component of the signal that this ambience can be extracted and reproduced by rear loudspeakers to "unlock" the recording's spaciousness—sometimes with stunning effect.

Among these processes are Dolby Pro Logic II, Pro Logic IIx, DTS Neo:6, Logic 7, and Music Logic. All create signals to drive the center and surround loudspeakers from 2-channel sources. These decoding circuits are standard in virtually all digital controllers, except Logic 7 which is available only in products made by Harman International (Harman Kardon, Lexicon, JBL, for examples). Another example is EARS, developed by NAD Electronics. EARS is relatively gentle, feeding in-phase information common to the left and right channel to the center loudspeaker, and extracting ambience that feeds the rear loudspeakers. The effectiveness of these systems varies greatly with the recording; some sound worse with processing, others sound the same (if there is very little uncorrelated ambient information), and a few sound spectacular—almost like a discrete multichannel recording.

These decoding algorithms all attempt to extract ambience *present in the recording* and deliver that ambience to the rear loudspeakers. These techniques are not to be confused with the digital-signal-processing (DSP) modes on home-theater receivers and controllers with labels such as "Concert Hall," "Jazz Club," or "Stadium." Rather than extract ambience from the original program, these DSP modes synthesize reverberation and add it to the music. Attempts have been made to measure the acoustic properties of concert halls and other venues and encode that acoustic signature in DSP algorithms that can be overlaid on the musical signal. All these schemes are nothing more than marketing gimmicks; many consumers choose audio/video receivers based on the number and variety of these surround modes. (One audio/video receiver I know of offers 1024 possible surround-mode variations.) Needless to say, such manipulation is antithetical to high-end values, and should be avoided. (Some would argue that even ambience recovery techniques that only extract information already in the recording are antithetical to high-end values.)

I've found the most effective and musically appropriate algorithm for playing back 2-channel sources over more than two loudspeakers to be Trifield, a process found in products made by Meridian Audio. According to Meridian, "Trifield extracts the mono and surround components of the original recording. It then calculates the signals for the front left, center, and right speakers, using the phase and amplitude differences between the three front channels, to redistribute the sounds on a frequency-dependent basis. This gives a significant improvement over traditional stereo, which converts the differences between the microphone signals into amplitude differences in the speaker signals. Trifield is recommended for well-made recordings and stereo television broadcasts that are not Dolby Surround encoded."

Trifield, which is derived from Ambisonics (described later in this chapter), is actually two algorithms. The listener has the choice of frontal stereo only or the addition of extracted ambience in the rear channels. Meridian controllers allow you to select side or rear surround loudspeakers, or both. The extracted ambience signal is extremely subtle; the listener is never consciously aware of sound sources behind or to the side. Rather, the extracted ambience seems to make the front soundstage more spacious, open, and detached from the loudspeakers. A "width" control allows you to widen the soundstage. A wider soundstage is possible without the "hole in the middle" imaging because the center loudspeaker anchors the soundstage center.

In my experience, listening in Trifield fills in the center of the soundstage, presents a wider sweet spot, and creates more stable images between the loudspeakers. Moreover, the imaging doesn't shift with small head movements, as it often does when listening to two loudspeakers (the so-called "head-in-the-vice" syndrome).

Here's an interesting story about Trifield: During a visit to my listening room by an electronics designer who was a "2-channel purist" (someone who believes music should be reproduced by just two loudspeakers), I performed an experiment. We listened to music for about an hour, but unbeknownst to him, I had Trifield engaged. Part of the way through a piece of music, I switched to 2-channel stereo and he was instantly and visibly let down by this sudden change. He heard a reduction in sound quality, but had no idea what could have caused it—until I gave away the secret. Trifield is subtle in that it's not an obvious "effect," but profound in its ability to solidify the soundstage.

It's important to keep in mind that there's nothing sacrosanct about using two loudspeakers to reproduce music. Early researchers considered two channels a compromise dictated by practicality and the technology of the day. Further, the term "stereo" doesn't mean 2-channel reproduction. Rather, it comes from the Greek word for "solid" because two channels can create apparently solid images between the loudspeakers.

Ambisonics

The solution to how best to capture a soundfield, and to reproduce that soundfield through more than two loudspeakers, has been solved for more than 35 years. It's called Ambisonics. Invented by Michael Gerzon, Ambisonics is a simple expansion of Alan Blumlein's idea, adding an omnidirectional microphone to the crossed figure-of-eight microphones that constitute a Blumlein pair. This setup captures all the information (in the horizontal plane) present in the original soundfield. When reproduced by four or more loudspeakers (the more loudspeakers the better), Ambisonics reconstructs the original acoustic soundfield. This isn't merely a marketing claim; there's a solid body of mathematics that shows that Ambisonics indeed encodes and reconstructs every aspect of the soundfield.

An additional figure-of-eight microphone can be added to capture the height information as well. In Ambisonic recording, the monaural sound pressure is called the "W" component, the left-right velocity component is called "Y," the front-back component is called "X," and the optional up-down (height) component is called "Z."

The W, X, and Y signals are called the "B-Format" and are obviously not compatible with 2-channel distribution media. The signals can, however, be combined to produce stereo (and mono) compatibility. This encoding method, called UHJ, matrix encodes the three (or four with the Z channel) signals for mono, stereo, and multichannel compatibility. UHJ-encoded material is decoded on playback by an Ambisonic decoder in a digital controller. Ambisonic recordings are made without regard for the number of loudspeakers through which the program will be reproduced. The special microphone samples the soundfield and encodes all the necessary information about that soundfield. On playback, the Ambisonic decoder generates the appropriate output signals based on the number of loudspeakers in the system. Because the multichannel audio is hierarchically represented, it can be played back through as few or as many loudspeakers as your room can accommodate. The B-Format representation is completely independent of the loudspeaker array; the Ambisonic decoder will correctly map the soundfield to the number of loudspeakers.

Many hundreds of Ambisonic recordings have been made in the past 30 years, notably by the classical label Nimbus, which uses the technique exclusively. Nimbus is releasing an extensive series of Ambisonic recordings on DVD-Audio during 2004. Even if the signal will never be decoded, Ambisonics has many advantages for 2-channel recording and reproduction. (Nimbus at one time owned the Ambisonics trademarks, but sold them to Meridian Audio.)

The advent of high-capacity, multichannel digital-audio discs such as Blu-ray provides an opportunity to deliver to the listener the full Ambisonics experience. It is possible to encode the hierarchical Ambisonics information in a multichannel signal and flag that data so it can be reproduced by a 5.1-channel loudspeaker system, even if the playback system doesn't include an Ambisonic decoder. A hierarchically-encoded multichannel signal is not tied to any particular number of loudspeaker feeds; rather it is a method of storing and transmitting multichannel audio to fit into a wide variety of loudspeaker configurations. The hierarchical signal is "scalable" to the number of loudspeakers through which the signal will be reproduced, thus making a multichannel signal perfectly compatible with any number of playback channels (including mono).

Ambisonics was developed long before any standards were in place that specified the number of loudspeakers in homes. However, today's ubiquity of 5.1-channel playback systems presents an opportunity to deliver Ambisonics without a decoder in the playback chain. With the number of loudspeakers known in advance, the Ambisonic signal can be "pre-decoded" for playback on a 5.1-channel loudspeaker array. No decoder is required, although it isn't possible to use more than five loudspeakers with this so-called "G-Format." In addition, height information in the G-Format can't be reconstructed without a decoder. If the system does employ an Ambisonic decoder, that decoder can be programmed to generate the appropriate number of signals to feed the loudspeaker system. This decoding is automatically engaged by the datastream flag, making the process completely transparent to the user.

The ideal loudspeaker placement for Ambisonic playback is four loudspeakers in a rectangle; five loudspeakers in a regular pentagon; or six loudspeakers in a regular hexagon. As mentioned earlier, the greater the number of loudspeakers, the more faithful the reconstruction of the original soundfield that was sampled by the microphone array.
There's no question that Ambisonics is by far the most musically appropriate multichannel recording technology. It reconstructs the soundfield in a way that the ubiquitous 5.1-channel systems can't approach, producing a much more believable sense of space, more precise image placement, and a much larger listening area than the typically small, sharply limited "sweet spot."

So why didn't Ambisonics catch on? First, Ambisonics came to market just as Quad was dying. Ambisonics was lumped in with Quad's failures, and record companies and electronics manufacturers wanted to put that failure behind them. Ambisonics was also seen, mistakenly, as just another competing Quad format. Moreover, no major record labels had any interest in Ambisonics because all had already associated themselves with one of the existing Quad formats. Ambisonics was also seen as an audiophile technique suitable only for capturing classical music in a concert hall with natural ambience, and not appropriate for studio records (although Alan Parsons' Ambisonic-encoded *Stereotomy* destroyed that argument). Finally, Ambisonics was doomed by a complete marketing failure, the result of the patents being turned over to a British government agency whose job was to promote and market British inventions. The agency was ill-equipped to promote Ambisonics and negotiate licensing agreements with hardware manufacturers.

Still, Ambisonic titles are readily available today, and Meridian Audio incorporates Ambisonic decoders in all its digital controllers. It is possible to encode Ambisonics hierarchical data in multichannel datastreams, which may breathe new life into the technology. Just as it took 28 years for Alan Blumlein's invention of stereo to become a widespread commercial reality, Ambisonics may one day become the dominant multichannel music format. More information on Ambisonics, including a comprehensive list of titles and decoding hardware, can be found at ambisonics.net.

14 System Set-Up Secrets Part One: Loudspeaker Placement and Room Acoustics

Introduction

The room in which music is reproduced, and the positions of the loudspeakers within that room, have a profound effect on sound quality. In fact, the loudspeaker/listening room interface should be considered another component in the playback chain. Because every listening room imposes its own sonic signature on the reproduced sound, your system can sound its best only when given a good acoustical environment. An excellent room can help get the most out of a modest system, but a poor room can make even a great system sound mediocre.

Fortunately, you can greatly improve a listening room with a few simple tricks and devices. The possibilities range from simply moving your loudspeakers—or even just your listening seat—a few inches, to building a dedicated listening room from scratch. Between these two extremes are many options, including adding inexpensive and attractive acoustical treatment products.

This chapter covers optimizing your system acoustically, beginning with the most effective and least expensive acoustical technique of all—loudspeaker placement. Don't be intimidated by the technical information toward the end of the chapter—it's included for those wanting to know some of the theory behind the practical information presented in the beginning sections. (If you find this next section difficult, I've summarized it in "Acoustical Do's and Don'ts" later in the chapter.)

Loudspeaker Placement

The most basic problem in many listening rooms is poor loudspeaker placement. Finding the right spot for your loudspeakers is the single most important factor in getting good sound in your room. Loudspeaker placement affects tonal balance, the quantity and quality of bass, soundstage width and depth, midrange clarity, articulation, and imaging. As you make large changes in loudspeaker placement, then fine-tune placement with smaller and smaller adjustments, you'll hear a newfound musical rightness and seamless harmonic integration to the sound. When you get it right, your system will come alive. Best of all, it costs no more than a few hours of your time.

Before getting to specific recommendations, let's cover the six fundamental factors that affect how a loudspeaker's sound will change with placement. (Later we'll look at each of these factors in detail.)

1) The relationship between the loudspeakers and the listener is of paramount importance. The listener and loudspeakers should form a triangle; without this basic setup, you'll never hear good soundstaging and imaging.

2) Proximity of loudspeakers to walls affects the amount of bass. The nearer the loudspeakers are to walls and corners, the louder the bass.

3) The loudspeaker and listener positions in the room affect the audibility of room resonant modes. Room resonant modes are reinforcements at certain frequencies that create peaks and dips in the frequency response, which can add an unnatural "boominess" to the sound. When room resonant modes are less audible, the bass is smoother, better defined, and midrange clarity increases. (Room resonant modes are described in detail later in this chapter.)

4) The farther out into the room the loudspeakers are, the better the soundstaging—particularly depth. Positioning loudspeakers close to the wall behind the speakers destroys the impression of a deep soundstage.

5) Listening height affects tonal balance. With some loudspeakers, how high your ears are in the listening seat relative to the loudspeakers can affect the amount of treble you hear.

6) Toe-in (angling the loudspeakers toward the listener) affects tonal balance (particularly the amount of treble), soundstage width, and image focus. Toe-in is a powerful tool for dialing in the soundstage and treble balance.

Let's look at each of these factors in detail.

1) Relationship between the loudspeakers and the listener

The most important factor in getting good sound is the geometric relationship between the two loudspeakers and the listener (we aren't concerned about the room yet). The listener should sit exactly between the two loudspeakers, at a distance away from each loudspeaker slightly greater than the distance between the loudspeakers themselves. Though this last point is not a hard-and-fast rule, you should certainly sit exactly between the loudspeakers; that is, the same distance from each one. If you don't have this fundamental relationship, you'll never hear good soundstaging from your system.

Fig.14-1 shows how your loudspeaker and listening positions should be arranged. The listening position—equidistant from the speakers, and slightly farther from each speaker than the speakers are from each other—is called the "sweet spot." This is roughly the listening position where the music will snap into focus and sound the best. If you sit to the side of the sweet spot, the soundstage will tend to bunch up around one speaker. This bunching-up effect will vary with the loudspeaker; some loudspeakers produce a wider sweet spot than others.

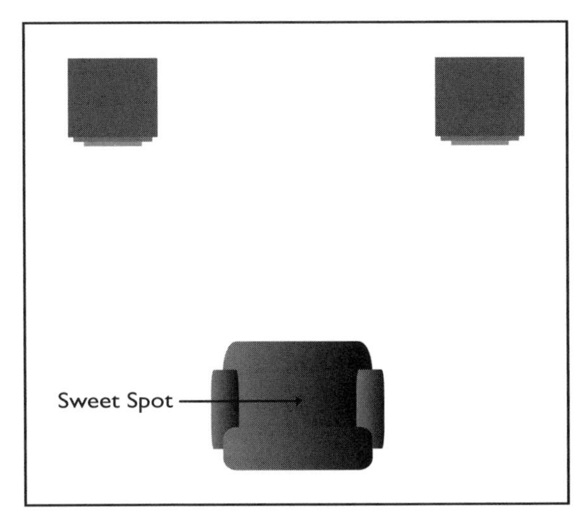

Fig. 14-1 The listener should sit between the loudspeakers and at the same distance from each loudspeaker.

Setting the distance between the loudspeakers is a trade-off between a wide soundstage and a strong center image. The farther apart the loudspeakers are (assuming the same listening position), the wider the soundstage will be. As the loudspeakers are moved farther apart, however, the center image weakens, and can even disappear. If the loudspeakers are too close together, soundstage width is constricted.

The optimum speaker width will produce a strong center image and a wide soundstage. You can experiment with angle simply by moving your listening chair forward and backward. There will likely be a position where the center image snaps into focus, appearing as a stable, pinpoint spot exactly between the loudspeakers. A musical selection with a singer and sparse accompaniment is ideal for setting loudspeaker spacing and ensuring a strong center image. With the loudspeakers fairly close together, listen for a tightly focused image exactly between the two loudspeakers. Move the loudspeakers a little farther apart and listen again. Repeat this move/listen procedure until you start to hear the central image become larger, more diffuse, and less focussed, indicating that you've gone slightly beyond the maximum distance your loudspeakers should be from each other for a given listening position. Note that you should wait until after you've completed the entire loudspeaker placement procedure to install the loudspeaker's spikes.

A factor to consider in setting this angle is the relationship to the room. You can have the same geometric relationship between loudspeakers and listener with the loudspeakers close together and a close listening position, or with the loudspeakers far apart and a distant listening position. At the distant listening position, the listening room's acoustic character will affect the sound more than at the close listening position. That's because you hear more direct sound from the loudspeaker and less reflected sound from the room's walls.. Consequently, the farther away you sit, the more spacious the sound. The closer you sit, the more direct and immediate the presentation. A technique that's helpful in small rooms, or those with fundamentally poor acoustics, is to sit very close to the loudspeakers, a condition called *near-field* listening. Some loudspeakers need a significant distance between the loudspeaker and the listener to allow the loudspeakers' individual drive units to integrate. If you hear a large tonal difference just by sitting closer, you should listen from a point farther away from the speakers.

2) Proximity to walls affects the amount of bass

The room boundaries have a great effect on a loudspeaker's overall tonal balance. Loudspeakers placed close to walls will exhibit a reinforcement in the bass (called "room gain"), making the musical presentation weightier. Some loudspeakers are designed to be near a rear wall (the wall behind the speakers); they need this reinforcement for a natural tonal balance. These loudspeakers sound thin if placed out into the room. Others sound thick and heavy if not at least 3′ from the rear and side walls. Be sure which type you're buying if your placement options are limited.

When a loudspeaker is placed near a wall, its bass energy is reflected back into the room essentially in phase with the loudspeaker's output. This means the direct and reflected waves reinforce each other at low frequencies, producing louder bass. Fig.14-2 shows the difference in a loudspeaker's frequency response when measured in an anechoic chamber (a reflection-free room) and in a normal room. A frequency-response graph plots amplitude (loudness) vs. frequency. As you can see in the graph, not only is the bass boosted, but the loudspeaker's low-frequency extension is also increased. Each surface near the loudspeaker (floor, rear wall, and side walls) will add to the loudspeaker's bass output. The closer to the corners the loudspeakers are placed, the more bass you'll hear.

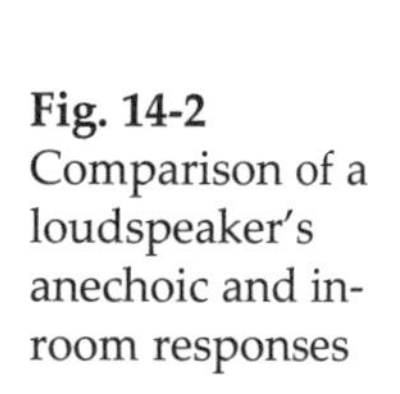
Fig. 14-2
Comparison of a loudspeaker's anechoic and in-room responses

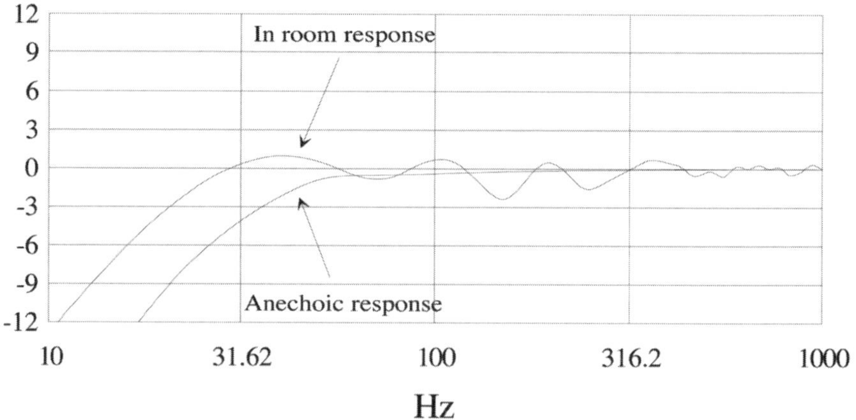

The loudspeaker's position in relation to the rear and side walls will also affect which frequencies are boosted. Correct placement can not only extend a loudspeaker's bass response by complementing its natural rolloff, but also avoid peaks and dips in the response. Improper placement can cause frequency-response irregularities that color the bass. That is, some frequencies are boosted relative to others, making the bass reproduction

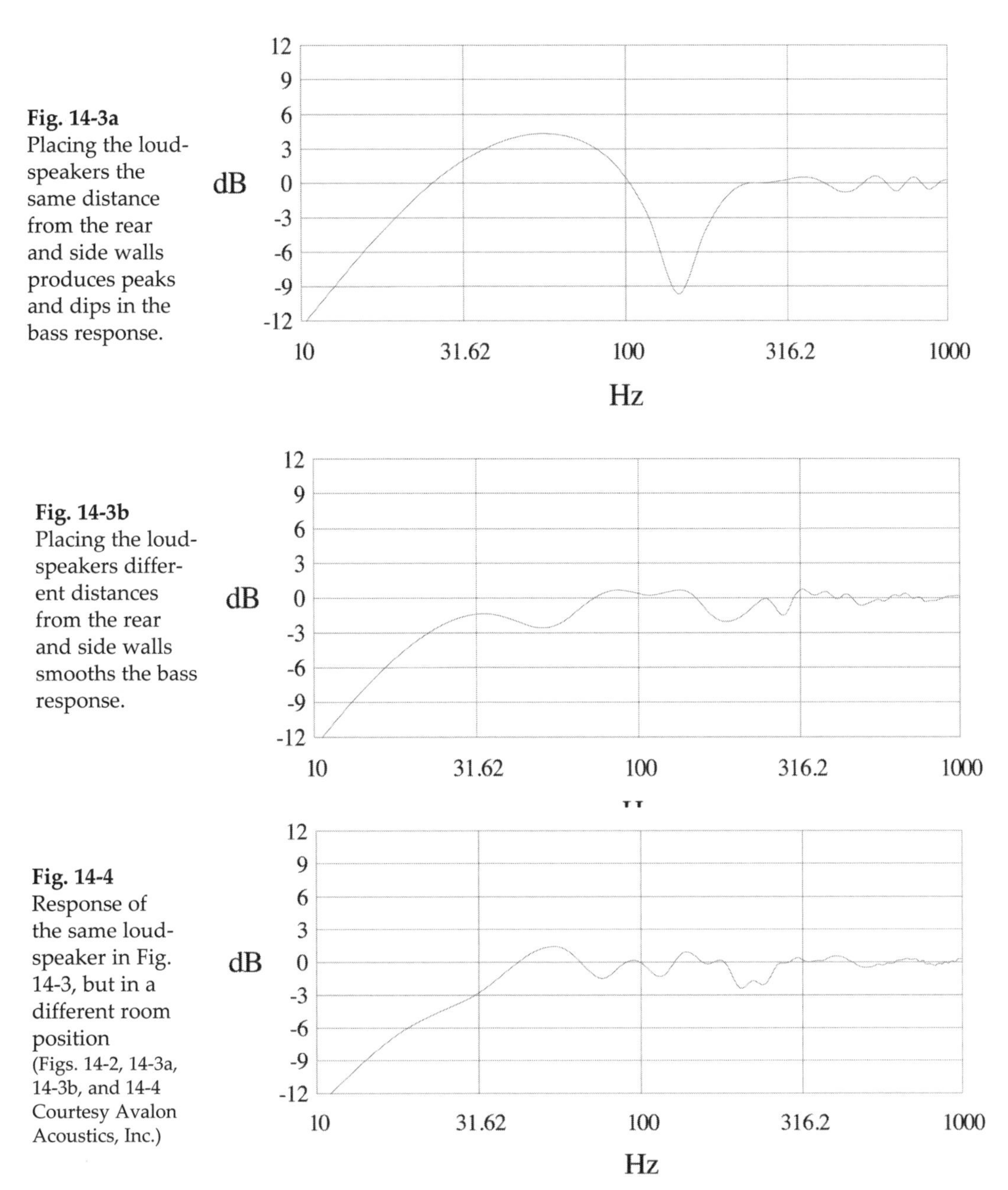

Fig. 14-3a Placing the loudspeakers the same distance from the rear and side walls produces peaks and dips in the bass response.

Fig. 14-3b Placing the loudspeakers different distances from the rear and side walls smooths the bass response.

Fig. 14-4 Response of the same loudspeaker in Fig. 14-3, but in a different room position (Figs. 14-2, 14-3a, 14-3b, and 14-4 Courtesy Avalon Acoustics, Inc.)

less accurate. The graph of Fig.14-3a is a loudspeaker's in-room response (the speaker's response as modified by the listening room) when placed equidistant from the rear and side walls. Note the 10dB notch (reduction in energy) at about 200Hz and the peak centered at 60Hz. The result will be a boominess in the bass and a leanness in the midbass. Moving the loudspeaker different distances from the rear and side walls can make the response much smoother (Fig.14-3b). The same loudspeaker's response at a different room position is shown in Fig.14-4.

These graphs illustrate that a loudspeaker's room position affects the bass response, and that the loudspeakers should be positioned at different distances from the rear and side walls. A rule of thumb: the two distances shouldn't be within 33% of each other. For example, if the loudspeaker is 3′ from the side wall, it should also be at least 4′ from the rear wall.

Many loudspeaker manufacturers will specify a distance from the rear and side walls. When a measurement is specified, the distance is between the woofer cone and the wall. Start with the loudspeakers in the manufacturer-recommended locations, then begin experimenting.

How close the loudspeakers are to the side walls affects the amplitude of the side-wall reflection. The closer the loudspeakers are to the side walls, the higher the level of the side-wall reflections reaching the listener—not a good thing. If you've treated the side walls as described later in this chapter, putting the loudspeaker closer to the side wall won't have as great an effect than if the side walls were left untreated.

A highly effective technique for finding the best distances from the rear and side walls was developed by David Wilson of Wilson Audio Specialties. I've watched him use this technique a number of times. Stand at the rear wall about the same distance from the sidewall as you expect the loudspeakers to end up. Begin speaking, moving in a line perpendicular to the rear wall. Listen to the timbre of your voice as you move. You'll hear different colorations as you move, and then suddenly, at one particular location, your voice will sound open, clear, and uncolored. There will be a small range (several inches) over which your voice sounds the most natural. Mark this area on the floor with masking tape. This is the best distance from the rear wall for your loudspeakers. Now repeat this exercise, but starting on the sidewall at the same plane at which you have placed the masking tape. The point at which your voice sounds the clearest is the best distance from the sidewall to locate your loudspeaker. The two points will intersect, indicating where you should position the loudspeaker. The intersection should be just behind the loudspeaker's front baffle and centered on the woofer. This technique is highly effective, and correlates very well with the placement suggested by computer modeling. The accuracy can be improved by having someone sit in the listening position as you move and speak, and confirm the point at which your voice sounds the least colored.

3) Loudspeaker and listener positions affect room-mode audibility.

In addition to deepening bass extension and smoothing bass response, correct loudspeaker placement in relation to the room's walls can also reduce the audible effects of your room's *resonant modes*. Room resonant modes, described later in this chapter, are reinforcements at certain frequencies that create peaks in the frequency response. Room modes also create *standing waves*, which are stationary patterns of high and low sound pressure in the room that color the sound. The standing-wave patterns in a room are determined both by the room's dimensions and by the position of the sound source in the room. By putting the loudspeakers and listener in the best locations, we can achieve smoother bass response.

A well-known rule of thumb states that, for the best bass response, the distance between the loudspeakers and the rear wall should be one-third of the length of the room (Fig.14-5). If this is impractical, try one-fifth of the room length. Both of these positions reduce the excitation of standing waves and help the loudspeaker integrate with the room. Ideally, the listening position should be two-thirds of the way into the room.

Starting with these basic configurations, move the loudspeakers and the listening chair in small increments while playing music rich in low frequencies. Listen for smooth-

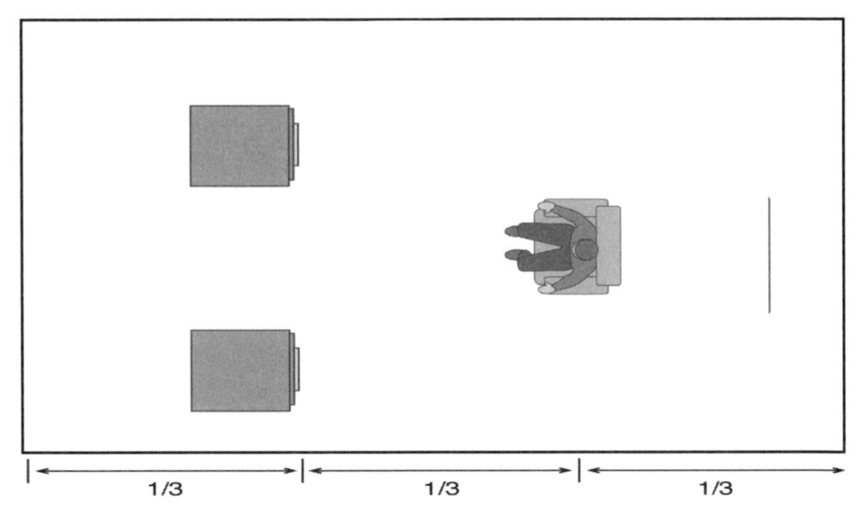

Fig. 14-5 The "rule of thirds" says to place the loudspeakers and listening position a third of the way into the room. If this isn't practical, try one-fifth of the way into the room.

ness, extension, and how well the bass integrates with the rest of the spectrum. When you find the loudspeaker placement where the bass is the smoothest, you should also hear an increase in midrange clarity and definition.

An excellent test signal for evaluating bass and midrange sound in a room is the Music Articulation Test Tone (MATT) developed by Acoustic Sciences Corporation (ASC). This special test signal is a series of tone bursts that rise in pitch, with silences between the bursts. Ideally, you should hear the bursts and silences as separate events. When heard through headphones or with your ear near the loudspeaker, each burst is clearly articulated. But when the sound is modified by the listening room, certain frequency bands of the ascending tone bursts become smeared or garbled, indicating that the listening room is storing, and then releasing, energy at those frequencies. By moving the loudspeakers and listening to the MATT, you can easily discover where your loudspeakers work best in the room. (The MATT is available on *Stereophile*'s *Test CD 2*, which also includes more detailed information about how to use this unique test signal. This disc is available at tubetrap.com.)

Another way of finding the right spots for your loudspeakers and listening chair with regard to reducing the influence of standing waves is a computer program called "Room Optimizer," created by RPG Diffusor Systems. The program asks for your room dimensions, loudspeaker type and woofer location within the cabinet, and other information. The software performs multiple calculations to determine the optimum loudspeaker and listener location."Room Optimizer" is highly recommended for this aspect of loudspeaker placement.

4) Distance from rear wall affects soundstaging

Generally, the farther away from the rear wall the loudspeakers are, the deeper the soundstage. A deep, expansive soundstage is rarely developed with the loudspeakers near the rear wall. Pulling the loudspeakers out a few feet can make the difference between poor and spectacular soundstaging. Unfortunately, many living rooms don't accommodate loudspeakers far out into the room. If the loudspeakers must be close to the rear wall, make the rear wall acoustically absorbent.

5) Listening height and tonal balance

Most loudspeakers exhibit changes in frequency response with changes in listening height. These changes affect the midrange and treble, not the bass balance. Typically, the loudspeaker will be brightest (i.e., have the most treble) when your ears are at the same height as the tweeters, or on the tweeter axis. Most tweeters are positioned between 32" and 40" from the floor to coincide with typical listening heights. If you've got an adjustable office chair, you can easily hear the effects of listening axis on tonal balance.

The degree to which the sound changes with height varies greatly with the loudspeaker. Some models have a broad range over which little change is audible; others can exhibit large tonal changes when you merely straighten your back while listening. Choosing a listening chair that sets your ears at the optimum axis will help achieve a good treble balance.

This difference in response is easily measurable. A typical set of loudspeaker measurements will include a family of response curves measured on various axes. The on-axis response (usually on the tweeter axis) is normalized to produce a straight line; the other curves show the *difference* between the on-axis response and at various heights. Chapter 16 ("Specifications and Measurements") includes a full description of how to interpret these measurements.

6) Toe-in

Toe-in is pointing the loudspeakers inward toward the listener rather than facing them straight ahead (see Fig.14-6). There are no rules for toe-in; the optimum amount varies greatly with the loudspeaker and the listening room. Some loudspeakers need toe-in; others work best firing straight ahead. Toe-in affects many aspects of the musical presentation, including mid- and high-frequency balance, soundstage focus, sense of spaciousness, and immediacy.

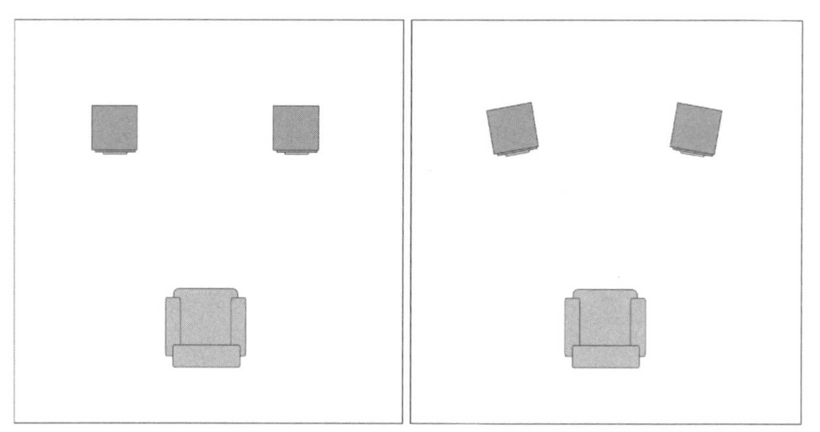

Fig. 14-6 Loudspeakers positioned with no toe-in (left) and with toe-in (right)

Most loudspeakers are brightest directly on-axis (directly in front of the loudspeaker). Toe-in thus increases the amount of treble heard at the listening seat. An overly bright loudspeaker can often be tamed by pointing the loudspeaker straight ahead. Some models, designed for listening without toe-in, are far too bright on-axis.

The ratio of direct to reflected sound increases with toe-in. That's because a toed-in loudspeaker will present more direct energy to the listener and project less energy into the room, where it might reach the listener only after reflecting from room surfaces. In a listening room with reflective side walls, toeing-in the loudspeakers can be a decided advantage. Moreover, the amplitude of side-wall reflections is greatly decreased with toe-in. Conversely, less toe-in increases the amount of reflected energy heard by the listener, adding to a sense of spaciousness and air. Reducing toe-in can open up the soundstage and create a feeling of envelopment.

Similarly, toe-in often increases soundstage focus and image specificity. When toed-in, many loudspeakers provide a more focused and sharply delineated soundstage. Images are more clearly defined, compact, and tight, rather than diffuse and lacking a specific spatial position. The optimum toe-in is often a trade-off between too much treble and a strong central image. With lots of toe-in, the soundstage snaps into focus, but the presentation is often too bright. With no toe-in, the treble balance is smoother, but the imaging is more vague.

Toe-in also affects the presentation's overall spaciousness. No toe-in produces a larger, more billowy, less precise soundstage. Instruments are less clearly delineated, but the presentation is bigger and more spacious. Toeing-in the loudspeakers shrinks the apparent size of the soundstage, but allows more precise image localization. Again, the proper amount of toe-in depends on the loudspeaker, room, and personal preference. There's no substitute for listening, adjusting toe-in, and listening again.

Identical toe-in for each loudspeaker is vital. Identical toe-in is essential to soundstaging because the speaker's frequency response changes with toe-in, and identical frequency response from both speakers is an important contributor to precise image placement within the soundstage. Achieving identical toe-in can be accomplished by measuring the distances from the rear wall to each of the loudspeaker's rear edges; these distances will differ according to the degree of toe-in. Repeat this procedure on the other loudspeaker, adjusting its toe-in so that the distances match those of the first loudspeaker. Another way to ensure identical toe-in is to sit in the listening seat and look at the loudspeakers' inside edges. You should see the same amount of each loudspeaker cabinet's inner side panel. You can also use a laser-alignment tool to assure identical toe-in. Substituting a piece of cardboard or similar material for the listener in the listening set, mount the laser on the top of one of the loudspeakers flush with one cabinet edge. Mark on the cardboard where the beam hits. Repeat the process with the other loudspeaker and adjust the toe-in so that the beam strikes the same spot as the first beam. The laser alignment tool comes in handy later to verify that each speaker has the same degree of backward tilt (or no tilt at all). The degree of tilt, called the rake angle, can vary if the loudspeaker's spikes are not screwed into the speaker at uniform depths.

Keep in mind that all loudspeaker placement variations are interactive with one another, particularly toe-in and the distance between loudspeakers. For example, a wide soundstage can be achieved with narrow placement but no toe-in, or wide placement with extreme toe-in.

Loudspeaker Placement in Asymmetrical Rooms

So far we've covered loudspeaker placement in an ideal room that is symmetrical and bounded on four sides. But what about those listening rooms in which one side is open to the rest of the house, or rooms in which proper loudspeaker placement is difficult? Fortunately, you can still get good sound in these odd-shaped rooms.

Let's first consider the room in which one wall is essentially missing, where the listening room opens to the rest of the house (Fig.14-7). The left loudspeaker will introduce more room gain than the right speaker because of its proximity to the corner. Remember that the closer the speaker is to walls, the greater the amount of bass. The right speaker has much less bass reinforcement, creating a bass imbalance.

The solution is two-fold. First, keep the speakers fairly far from the rear wall. This minimizes the room gain for the left loudspeaker, making its response more closely match that of the right speaker. Second, install a bass absorber (such as an ASC Tube Trap) in the corner behind the left loudspeaker. This trap will absorb bass energy rather than reflect it back into the room, and thus mimic the lack of room reinforcement from the right loudspeaker. You'll never be able to get a precise match, but these two techniques should get the two speakers close in bass output. I had a listening room for about two years that looked like Fig.14-7, and got good results with a 16" ASC Full Round Tube Trap in the left corner behind the loudspeaker.

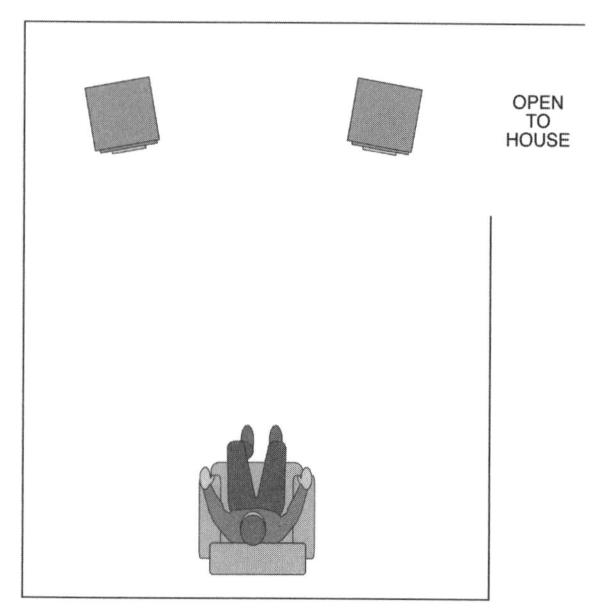

Fig. 14-7 A room in which one side is open can cause a bass imbalance.

If a room has non-parallel walls, try to position the loudspeakers at the room's narrow end. Figs.14-8a and b show incorrect and correct loudspeaker placement in such a room. The problem with the placement shown in Fig.14-8a (left) is that detrimental sidewall reflections are directed at the listening position. In Fig.14-8b (right) sidewall reflections are naturally directed away from the listener. Many recording-studio control rooms are narrower at the speaker location for just this reason.

Short-Wall vs. Long-Wall Placement

The examples shown have all assumed that the loudspeakers will be placed on the room's short wall. This traditional method provides the maximum space for bringing the loudspeakers out into the room and away from the rear wall. But putting the speakers on the long wall (Fig.14-9) also has some advantages.

Long-wall placement results in more direct sound at the listening position and less sidewall-reflected energy. This confers a large advantage in tonal purity and soundstage

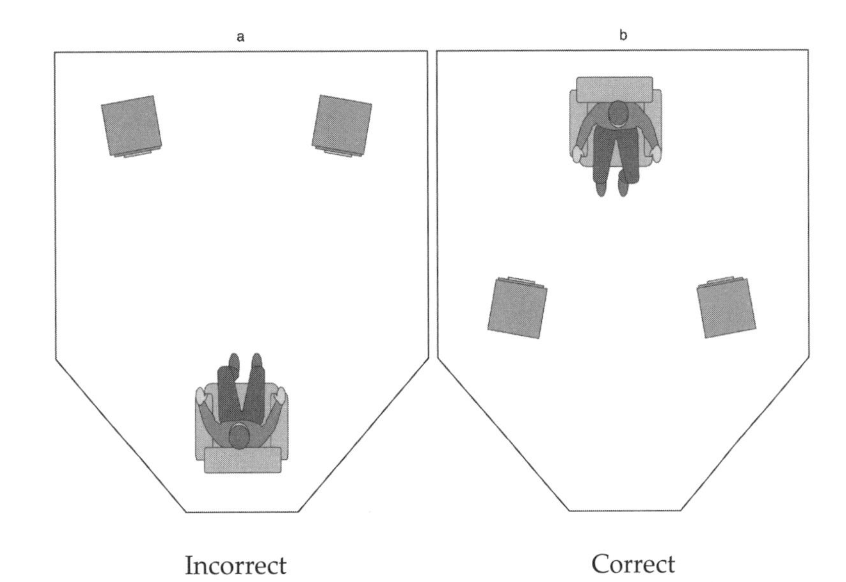

Fig. 14-8 In angled rooms, the loudspeakers should be placed so that sidewall reflections are directed away from the listener.

accuracy, for reasons described later in this chapter. You hear more of the speaker and less of the room. The sound tends to be more immediate, detailed, and present.

Long-wall placement works only when the room is wide enough to accommodate some space between the speakers and the wall behind them, space between the speakers and listening position, and some space between the listener and the wall behind the listening position. If you sit with your head very near the wall behind you, the sound will likely be boomy. Consequently, you end up sitting fairly close to the loudspeakers with long-wall placement in all but very large rooms. As described in Chapter 6, speakers with first-order crossovers require significant space between speakers and listener for the sound of the individual drivers to properly integrate. If you sit too close to a speaker with first-order crossovers, the tonal balance will be wrong. Consider the room's width and the type of crossovers in your loudspeakers when deciding if long-wall placement is right for you. The reduction in reflected sound at the listening position with long-wall placement is a considerable advantage.

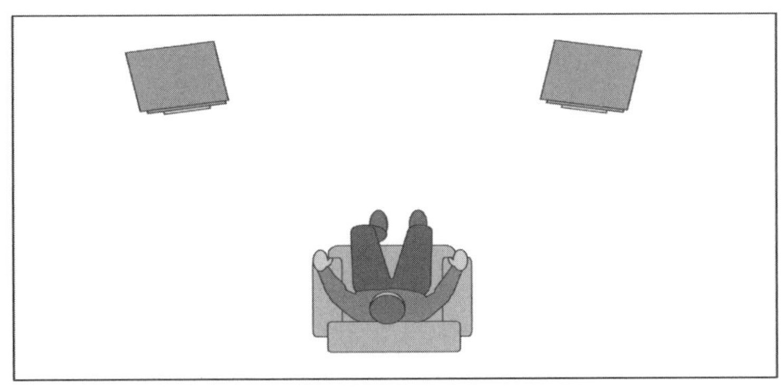

Fig. 14-9 Positioning the loudspeakers on the room's long wall reduces sidewall reflections at the listening position.

Finally, you may consider placing your loudspeakers at an odd angle in relation to the room, as shown in Fig.14-10. I've seen loudspeaker designers use this technique when trying to get good sound in a small hotel room at hi-fi shows. This placement, which works best for small speakers, has the advantage of reducing sidewall reflections at the listening position. The disadvantages are that it consumes much of the room's floor space and positions the speakers closer to walls, where the bass will be reinforced. That's why this technique is more successful with small speakers that have limited bass extension.

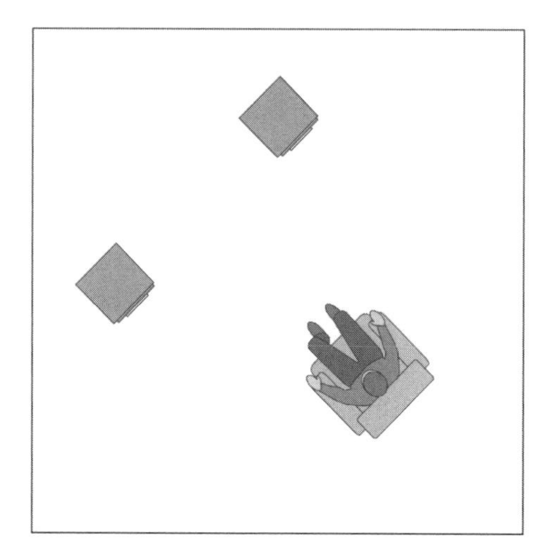

Fig. 14-10
Angled loudspeaker placement works best with small speakers.

Dipolar and Bipolar Loudspeaker Placement

Dipolar loudspeakers produce sound to the rear as well as to the front. An electrostatic speaker is a dipole because the vibrating diaphragm sits in open space rather than in a cabinet, launching sound equally to the front and rear. This rear wave from dipolar speakers is *out of phase* with the front wave; that is, when the diaphragm moves forward to create positive pressure in front of the diaphragm, it creates negative pressure behind the diaphragm.

A *bipolar* speaker typically uses arrays of conventional dynamic drivers on the front and rear of the loudspeaker enclosure. The front and rear waves from a bipolar speaker are *in phase* with each other. That's the difference between dipolar and bipolar: a dipole's rear wave is *out of phase* with the front wave, and a bipole's rear wave is *in phase* with the front wave.

Both of these speaker types are covered in detail in Chapter 6. What concerns us here, however, are the special placement requirements of bipolar and dipolar loudspeakers, along with the different ways in which they interact with the listening room.

The most important consideration when positioning dipoles is that the wall behind the speakers (the rear wall) has a much greater influence on the sound than it does with conventional point-source speakers (those that direct energy in only one direction). Conversely, how the sidewalls are treated is less important with dipoles because they radiate very little energy to the sides. (See Fig.6-4 in Chapter 6, which shows the dispersion patterns of point-source and dipolar speakers.)

Generally, dipoles like a reflective rear wall, but with some diffusing objects behind the speaker to break up the reflected energy. A highly absorbent rear wall defeats the purpose of a dipole; that reflected energy is beneficial, and you want to hear it. But if the wall is flat and lacks surfaces that scatter sound, the reflected energy combines with the direct sound in a way that reduces soundstage depth. Bookcases directly behind dipolar speakers help diffuse (scatter) the rear wave, as do rock fireplaces, furniture, and other objects of irregular shape. ASC Tube Traps can be positioned directly behind dipoles with their reflective sides out, or you can go all the way with RPG Diffusors. RPG's Skyline diffuser (Fig.14-11) is ideal for diffusing the rear wave from dipoles. These 2' by 2' squares are highly effective in scattering sound, and can be attached to a wall or mounted on a stand. A tall dipolar speaker benefits from two or three Skylines arrayed vertically directly behind each speaker.

Fig. 4-11 RPG's Skyline diffuser effectively scatters sound. (Courtesy RPG Diffusor Systems, Inc.)

You still want to treat the sidewalls to absorb reflections with dipoles, but their narrow radiation pattern makes treating the sidewalls less important than with point-source speakers. This is particularly true of dipoles that sound their best toed-in toward the listening position. Similarly, dipoles have very narrow vertical dispersion, meaning they direct very little sound toward the ceiling and floor.

Dipolar loudspeakers also need to be placed farther out into the room than conventional point-source speakers. You can't put dipoles near the rear wall and expect a big, deep soundstage. Be prepared to give up a significant area of your listening room to dipolar speakers.

Subwoofer Setup and Placement

It's relatively easy to put a subwoofer into your system and hear more bass. What's difficult is making the subwoofer's bass *integrate* with the sound of your main speakers. Low bass as reproduced by a subwoofer's big cone can sound different from the bass reproduced by the smaller cones in the left and right speakers. A well-integrated subwoofer produces a seamless sound, no boomy thump, and natural reproduction of music. A poorly integrated subwoofer will sound thick, heavy, boomy, and unnatural, calling attention to the fact that you have smaller speakers reproducing the frequency spectrum from the midrange up, and a big subwoofer putting out low bass.

Integrating a subwoofer into your system is challenging because the main speakers may have small cones, and the subwoofer has a large and heavy cone. Moreover, the subwoofer is optimized for putting out lots of low bass, not for reproducing detail. The main speakers' upper bass is quick, clean, and articulate. The subwoofer's bass is often slow and heavy. The term "slow" is not technically correct, but vividly describes how such a subwoofer sounds. More precisely, a "slow" subwoofer is one that has excessive *overhang*; the cone keeps moving long after the drive signal has stopped, which conveys the impression that the bass is "slow."

Achieving good integration between small speakers and a subwoofer is easier if you buy a complete system made by one manufacturer. Such systems are engineered to work together to provide a smooth transition between the subwoofer and the main speakers. Specifically, the crossover network removes bass from the left and right speakers, and removes midrange and treble frequencies from the signal driving the subwoofer. If all these details are handled by the same designer, you're much more likely to get a smooth transition than if the subwoofer is an add-on component from a different manufacturer.

If you do choose a subwoofer made by a different manufacturer, several controls found on most subwoofers help you integrate the sub into your system. One control lets you adjust the *crossover frequency*. This sets the frequency at which the transition between the subwoofer and the main speakers takes place. Frequencies below the crossover frequency are reproduced by the subwoofer; frequencies above the crossover frequency are reproduced by the main speakers. If you have small speakers that don't go very low in the bass and you set the crossover frequency too low, you'll get a "hole" in the frequency response. That is, there will be a narrow band of frequencies that aren't reproduced by the woofer *or* the main speakers.

Setting the subwoofer's crossover frequency too high also results in poor integration, but for a different reason. The big cone of a subwoofer is specially designed to reproduce low bass. When it is asked to also reproduce upper-bass frequencies, those upper-bass frequencies are less clear and distinct than if they were reproduced by the smaller main speakers. Finding just the right crossover frequency is the first step in achieving good integration. Most subwoofer owner's manuals include instructions for setting the crossover frequency. As a rule of thumb, the lower the subwoofer's crossover is set, the better.

Some subwoofers also provide a knob or switch marked *Phase*. To understand a subwoofer's phase control, visualize a sound wave being launched from your subwoofer and from your main speakers at the same time. Unless the main speakers and subwoofer are identical distances from your ears, those two sound waves will reach your ears at different times, or have a phase difference between them. In addition, the electronics inside a subwoofer can create a phase difference in the signal. The phase control lets you delay the wave generated by the subwoofer so that it lines up in time with the wave from the main speaker. When the sound waves are in-phase, you hear a more coherent and better-integrated sound. One way of setting the phase control is to sit in the listening position with music playing through the system. Have a friend rotate the phase control (or flip the phase switch) until the bass sounds the smoothest.

But there's a much more precise way of setting the phase control that guarantees perfect phase alignment between the subwoofer and main speakers. First, reverse the connections on your main loudspeakers so that the black speaker wire goes to the speaker's red terminal, and the red speaker wire goes to the speaker's black terminal. Do this with both speakers. Now, from a test CD that includes pure test tones, select the track whose frequency is the same as the subwoofer's crossover frequency. Sit in the listening position and have a friend rotate the subwoofer's phase control until you hear the *least* amount of bass. The subwoofer's phase control is now set perfectly. Return your speaker connections to their previous (correct) positions: red to red, black to black.

Here's what's happening when you follow this procedure. By reversing the polarity of the main speakers, you're putting them out of phase with the subwoofer. When you play a test signal whose frequency is the same as the subwoofer's crossover point, both the sub and the main speakers will be reproducing that frequency. You'll hear minimum bass when the waves from the main speakers and subwoofers are maximally out of phase. That is, when the main speaker's cone is moving in, the subwoofer's cone is moving out. The two out-of-phase waves cancel each other, producing very little bass. Now, when you return your loudspeakers to their proper connection (putting them back in-phase with the subwoofer), they will be maximally *in-phase* with the subwoofer. This is the most accurate method of setting a subwoofer's phase control becasuse it's easier to hear and identify the point of maximum cancellation than the point of maximum reinforcement.. Unless you move the subwoofer or main speakers, you need to perform this exercise only once.

The best integration comes from adding two (or more) subwoofers to your system. Two subwoofers drive the air in the room more uniformly, with fewer extremes of standing waves. The result is smoother bass throughout the room and better integration with the main speakers.

You can also get more dynamic impact and clarity from your subwoofer by placing it close to the listening position. Sitting near the subwoofer causes you to hear more of the sub's direct sound and less of the sound after it has been reflected around the room. You hear—and feel—more of the low-frequency wave launch, which adds to the visceral impact of owning a subwoofer. Bass impact is more startling, powerful, and dynamic when the subwoofer is placed near the listening position.

Subwoofer placement also has a large effect on how much bass you hear and how well the sub integrates with your main speakers. When a subwoofer is correctly positioned, the bass will be clean, tight, quick, and punchy. A well-located subwoofer will also produce a seamless sound between the sub and the front speakers; you won't hear the subwoofer as a separate speaker. A poorly positioned subwoofer will sound boomy, excessively heavy, thick, lacking detail, slow, and have little dynamic impact. In addition, you'll hear exactly where the front speakers leave off and the subwoofer takes over.

The simplest, most effective way of positioning a subwoofer is to put it near the listening position. Raise the subwoofer off the floor, if possible, so that it's close to where the listeners' ears would normally be. Play a piece of music with an ascending and descending bass line, such as a "walking" bass in straight-ahead jazz. Crawl around the floor on your hands and knees (make sure the neighbors aren't watching) until you find the spot where the bass sounds the smoothest, and where each bass note has about the same volume and clarity. Avoid positions where some notes "hang" longer, and/or sound slower or thicker, than others. When you've determined where the bass sounds best, put the subwoofer there. Now, when you're back in the listening seat, the bass should sound smooth and natural.

Some general guidelines for subwoofer placement: As with full-range speakers, avoid putting the subwoofer the same distance from two walls. For example, if you have a 20'-wide room, don't put the subwoofer 10' from each wall. Similarly, don't put the subwoofer near a corner and equidistant from the side and rear walls. Instead, stagger the distances to each wall. Staggering the subwoofer's distance from each wall smooths the bass because the frequencies being reinforced by the wall are randomized rather than coincident.

Multichannel Loudspeaker Placement

So far, we've discussed the placement of two loudspeakers for stereo music reproduction. With multichannel music and home theater becoming increasingly common, let's expand on these loudspeaker-placement principles to include positioning more than two loudspeakers.

As described in Chapter 13 ("Multichannel Audio"), the optimum loudspeaker type (point source, dipolar) and configuration differ for multichannel music reproduction and home theater. For multichannel music, the ideal loudspeaker array is five identical full-range loudspeakers placed equidistant from the listener. For film-soundtrack reproduction, the center loudspeaker is typically smaller and lacks bass extension, and the surround speakers are dipolar types mounted on the sidewalls. Loudspeaker arrays optimized for home theater also include a subwoofer.

Let's first look at the ideal multichannel music loudspeaker array. The center loudspeaker should be positioned on the room's center-line directly in front of the listening position, and slightly behind the plane of the left and right loudspeakers. This placement creates a gentle arc, and puts the center loudspeaker at the same distance from the listener as the left and right loudspeakers. If the three front loudspeakers were lined up, the sound from the center speaker would reach the listener before the sound from the left and right speakers.

The rear-channel speakers should be located at 135–150° as shown in Fig.14-12 and at the same distance from the listener as the front three loudspeakers. This placement isn't always practical, however, so many multichannel products provide a rear-channel delay for those situations in which the listener must sit closer to the rear loudspeakers. Delaying the signals to the rear channels causes the sound from the rear speakers to reach the listener at the same time as sound from the front speakers.

With this array, the front left and rear left loudspeakers can produce phantom images between them along the left sidewall, and the front right and rear right loudspeakers create phantom images along the right sidewall. Correct loudspeaker placement helps to achieve a soundfield that appears to be continuous from front to back, rather than as two separate soundfields at the front and rear of the room.

A multichannel music system can also employ seven speakers, with the two additional speakers located at the side positions. This array produces the greatest spatial realism because the side images are now "hard" rather than phantom, and the feeling of being immersed in a continuous soundfield is enhanced. Multichannel recordings made in a large acoustic and played back through a properly set up 7-channel system can be staggering in their ability to create a feeling of actually being inside the original acoustic space. It is even possible to hear discrete sidewall reflections localized along the sidewalls, just as one would hear at the original acoustic event.

Loudspeaker arrays optimized for film-soundtrack reproduction usually employ dipolar surround loudspeakers rather than point-source loudspeakers. Dipolar speakers produce sound equally to the front and rear; when positioned on the sidewalls, the listener hears sound from the surround speakers only after it has been reflected from the room's boundaries. This simulates the array of multiple surround speakers in a movie theater from just two surround speakers, and creates a greater feeling of envelopment. (Chapter 13 includes more detailed information on multichannel loudspeaker placement.)

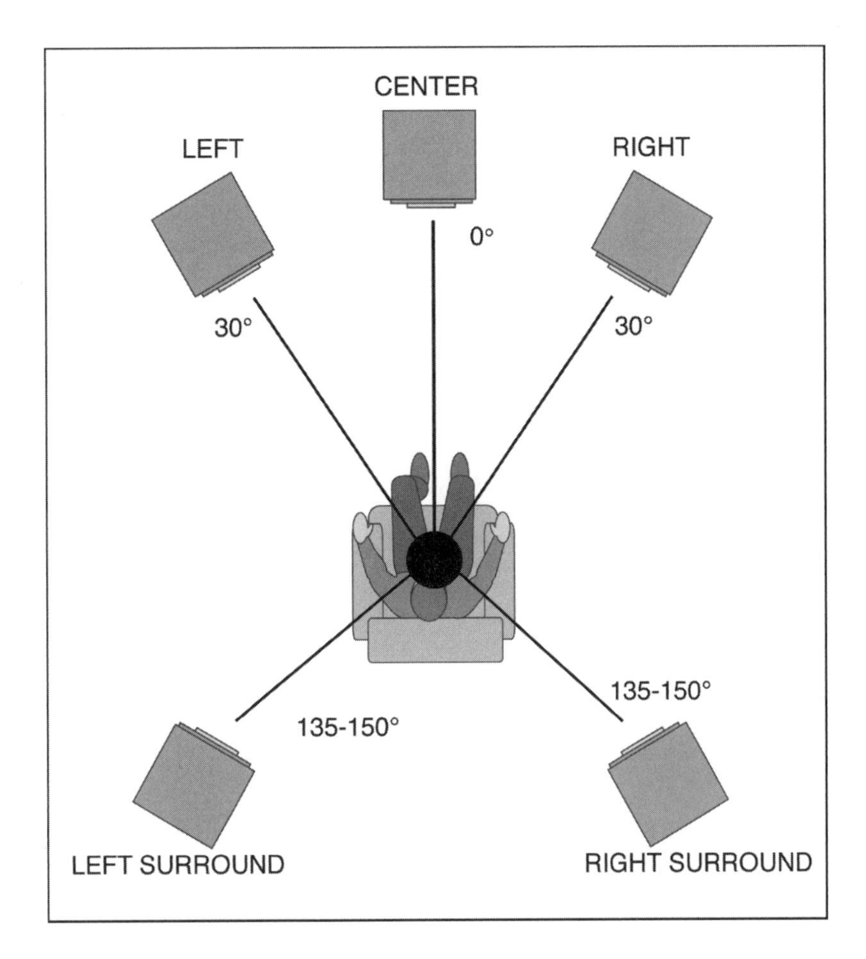

Fig. 14-12 A typical multi-channel music loudspeaker array employs five identical full-range loudspeakers positioned equidistant from the listener at the locations shown.

Loudspeaker Placement Summary

Loudspeaker placement is the single most important thing you can do to improve your system's sound. It's free, helps develop listening skills, and can make the difference between mediocre and spectacular sound with the same electronics and loudspeakers. Before spending money on upgrading components or acoustic treatments, be sure you've realized your system's potential with correct loudspeaker placement.

After you've found the best loudspeaker placement, install the carpet-piercing spikes (if any) supplied by the manufacturer. Level the spikes so that the loudspeaker doesn't rock: the loudspeaker's weight should be carried by all four (or three) spikes. If you have wood floors that you don't want to mar with spikes, place the round metal discs that are often supplied with the loudspeakers beneath the spikes. To be sure that the spikes are installed at the same depth into the loudspeaker, use a bubble level or laser

You've seen how loudspeaker placement gives you precise and independent control over different aspects of the music presentation. You can control both the quantity and the quality of the bass by changing the loudspeakers' distances from the rear and side walls. The audibility of room resonance modes can be reduced by finding the best spots for the loudspeakers and listening chair. Treble balance can be adjusted by listening height and toe-in. The balance between soundstage focus and spaciousness is easily changed just by toeing-in the speakers. Soundstage depth can be increased by moving the speakers farther out into the room.

I've had the privilege of watching some of the world's greatest loudspeaker designers set up loudspeakers in my listening room for review. At the very highest levels of the art, tiny movements—half an inch, for example—can make the difference between very good and superlative sound. The process can take as little as two hours, or as long as three days. I've often had the experience of thinking the sound was excellent after half a day of moving the loudspeakers, only to discover that the loudspeaker was capable of much greater performance when perfectly dialed-in. Loudspeaker positioning is a powerful tool for achieving the best sound in your listening room, and it doesn't cost a cent. Take advantage of it.

Common Room Problems and How to Treat Them

Treating your listening room can range from simply hanging a rug on a wall to installing specially designed acoustic devices. Large gains in sound quality can be realized just by adding—or moving—common domestic materials such as carpets, area rugs, and drapes. This approach is inexpensive, simple, and often more aesthetically pleasing than installing less familiar acoustic products. You can improve your room with existing household materials, or take the next step of installing dedicated acoustic-control devices. Here are some of the most common room problems, and how to correct them.

1) Untreated parallel surfaces

Perhaps the most common and pernicious of room problems is that of untreated parallel surfaces. If two reflective surfaces face each other, *flutter echo* will occur. Flutter echo is a "pinging" sound that remains after the direct sound has stopped. If you've ever been in an empty, uncarpeted house and clapped your hands, you've heard flutter echo. It sounds like a ringing that hangs in the air long after the clap has decayed. Flutter echo is a periodic repetition caused by the uncontrolled reflection of a sound back and forth between two surfaces. Imagine two mirrors facing each other, the reflections bouncing back and forth between the reflective surfaces to create the illusion of an infinitely receding distance. Flutter echo can blur transient attacks and decays and add a hard, metallic character to the upper midband and treble.

Try clapping your hands in various rooms of the house—particularly the bathroom or hallway. If your listening room has a pinging overhang similar to what you hear in the bathroom, you need to correct this problem.

Flutter echo is easy to prevent. Simply identify the reflective parallel surfaces and put an absorbing or diffusing material on one of them. This will break up the repeated reflections between the surfaces. The material could be a rug hung on a wall, a carpet on the floor (if the flutter echo is between a hard floor and ceiling), drapes over a window, or an acoustically absorbent material applied to a wall. Keep in mind that you need to treat just one of the parallel surfaces to prevent flutter echo. Whatever solution works best for you, killing flutter echo is of utmost importance.

2) Uncontrolled floor and side-wall reflections

It is inevitable that loudspeakers will be placed next to the room's side walls and near the floor. Sound from the loudspeakers reaches the listener directly, in addition to being reflected from the side walls, floor, and ceiling. Side-wall reflections are the music signal delayed in time, colored in timbre, and spatially positioned at different locations from the direct sound. All these factors can degrade sound quality. Moreover, floor and side-wall reflections interact with the direct sound to further color the music's tonal character. Fig.14-13 shows how the sound at the listening seat is a combination of direct and reflected sound.

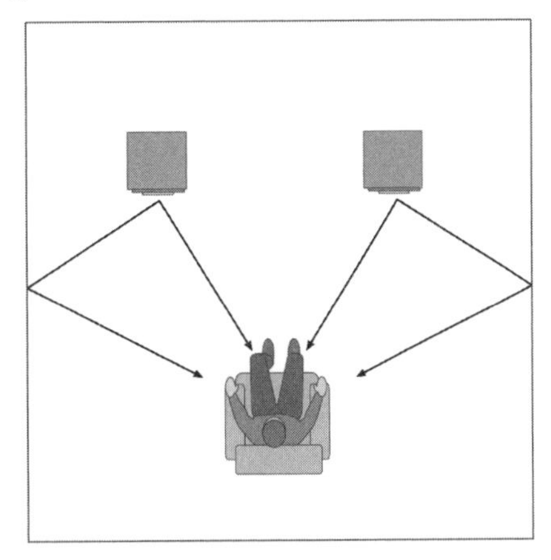

Fig. 4-13 The listener receives a combination of direct and reflected sound.

Side-wall reflections color the music's tonal balance in three ways. First, virtually all loudspeakers' *off-axis responses* (frequency response measured at the side of the loudspeaker) are much less flat (accurate) than their on-axis responses. The sound emanating from the loudspeaker sides (the signal that reflects off the side wall) may have large peaks and dips in its frequency response. When this colored signal is reflected from the side wall to the listener, we hear this tonal coloration imposed on the music.

Second, the side wall's acoustic characteristics will further color the reflection. If the wall absorbs high frequencies but not midband energy, the reflection will have even less treble.

Finally, when the direct and reflected sounds combine, the listener hears a combination of the direct sound from the loudspeaker and a slightly delayed version of the sound reflected from the side wall. The delay is caused by the additional path-length difference between the sound source (the loudspeaker) and the listener. Because sound travels at 1130 feet per second, we can easily calculate the delay time. If the additional path-length difference in Fig.14-13 was 4', the side-wall reflection will be delayed by 3.5ms (3.5 milliseconds, or 3.5 thousandths of a second) in relation to the direct sound.

The result is a phenomenon called *comb filtering* (see Fig.14-14), a sequence of peaks and notches (hence its similarity to a comb) in the frequency response caused by constructive and destructive interference between the direct and reflected sounds. The phase difference between the two signals causes cancellation at certain frequencies and reinforcement at others, determined by the path-length distance. It all adds up to coloration of the signal at the listening position.

Comb filtering is used as a special effect in the recording studio. Adding a slight delay to a signal and mixing the direct and delayed signals is called *phasing*, characterized

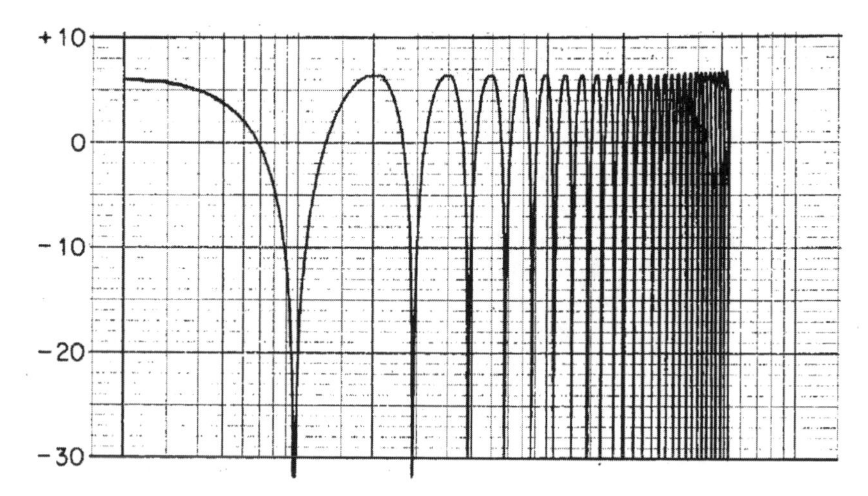

Fig. 14-14 Comb filtering is a sequence of peaks and notches in the frequency response caused by combining direct and reflected sound. (Courtesy *Audio Engineering Handbook*, K. Benson, McGraw-Hill, Inc., 1988)

by a hollow sound. Darth Vader's voice in the *Star Wars* series undoubtedly includes some phasing. Continuously varying the delay causes the frequencies at which the reinforcements and cancellations occur to sweep the audio band for an effect called *flanging*. It's called flanging because in the days before digital-delay lines, the engineer used a tape machine to create the delay, and would put his thumb on the tape reel's flange to vary the delay time and create the characteristic sweeping sound. The most widely heard example of flanging is perhaps the Doobie Brothers' song "Listen to the Music."

The result of these mechanisms—the loudspeaker's colored off-axis response, the sidewall's acoustic properties, and comb filtering—is a sound with a very different tonal balance from that of the direct signal from the loudspeakers. Side-wall reflections are one reason the same loudspeakers sound different in different rooms.

Side-wall reflections not only affect the perceived spectral balance, they also destroy precise image placement within the soundstage. The reflections present "virtual" images of the loudspeakers' signal that appear on the side walls. Although some degree of side-wall reflection adds spaciousness and size, strong reflections increase the *apparent* distance between the loudspeakers. This blurs the spatial distinction between individual images and makes the soundstage less focused and precise. When we hear a center-placed image as partially emanating from positions beyond the left and right loudspeaker boundaries, the tight image focus we seek is destroyed.

Sound also reflects from the floor and ceiling. Floor reflections tend to cause a reduction in mid bass energy, making the presentation a little leaner. The ceiling reflection affects the sound less than the side-wall reflections because of its greater path-length difference. Note that the sound of dipolar loudspeakers, which direct very little energy toward the ceiling, is less affected by ceiling reflections than is the sound of conventional loudspeakers. Finally, a sloped ceiling is advantageous with conventional loudspeakers when those speakers are placed at the short end of the room. The ceiling slope will tend to direct ceiling reflections away from the listening position.

Reflections reaching the listener from just one room boundary are called *first-order reflections*. These reflections have the greatest amplitude, and have the most effect on the reproduced sound. Fortunately, treating side-wall reflections is simple: just put an absorbing or diffusing material on the side walls between the loudspeakers and the listening position. Drapes are highly effective, particularly those with heavy materials and deep folds. The floor reflection is even easier to deal with: carpet or a heavy area rug on the floor will absorb most of the reflection and reduce its detrimental effects. Low frequencies, however, won't be absorbed by a carpet or rug, leading to a cancellation in the midbass caused by interference between the direct and reflected waves. This is the so-called "Allison Effect," named after loudspeaker designer Roy Allison, who first publicized the phenomenon.

Interestingly, the type of carpet between you and the loudspeakers can affect sound quality. Specifically, wool carpet produces a more natural tonal balance than does synthetic carpet. That's because the fibers in wool carpet all have slightly different lengths and thicknesses, which makes them absorb different frequencies. Synthetic carpet, by contrast, is composed of identically sized and shaped fibers that absorb only a very narrow band of frequencies. You can demonstrate this to yourself at a carpet store by speaking into sample pieces of wool and synthetic carpet and listening to the sound of your voice. The wool carpet will produce a more natural timbre.

Side-wall reflections should be diffused (scattered) or absorbed. Diffusion turns the single discrete reflection into many lower-amplitude reflections spread out over time and reflected in different directions (see Fig.14-15).

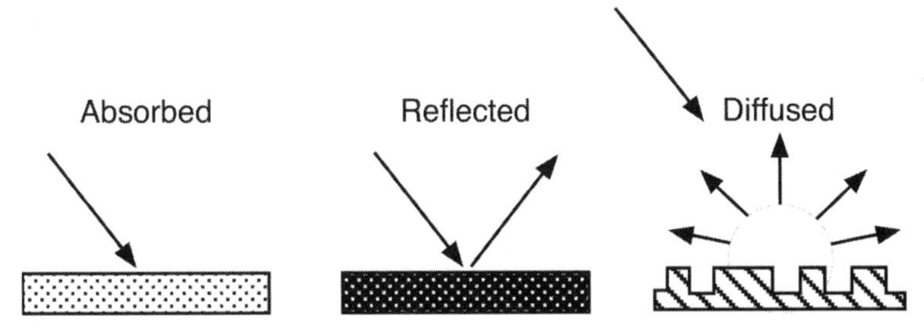

Fig. 14-15 Sound striking a surface is absorbed, reflected, or diffused (or a combination of the three).

Diffusion can be achieved with specialized acoustic diffusers such as those made by RPG Diffusors (shown in Fig.14-16, and described later in this chapter), or an irregular surface. A open-backed bookcase full of books makes an excellent diffuser, particularly if the books are of different depths, or are arranged with their spines sticking out at different distances..

The second option is to absorb the side-wall reflection with an acoustically absorbent material. The acoustical foams described later in this chapter will work, but completely absorbing the side-wall reflection with an aggressive foam can make the room sound lifeless and constrict the presentation's sense of size and space.

There is some debate in the high-end community as to whether side-wall reflections should be absorbed or diffused. Diffusion proponents argue that the reflected energy is beneficial if converted to many lower-amplitude reflections spread out over time and space, the diffused reflections increasing the presentation's spaciousness and air. Absorption proponents suggest that any reflections within the first 20ms of the direct sound degrade the signal from the loudspeakers. Most recording-studio control rooms are designed to provide a "Reflection Free Zone" (RFZ) where the engineer sits, so that he or she hears only the direct sound from the studio monitors. My experience suggests that absorbing side-wall reflections is better than diffusing them. Diffusing sidewall reflections tends to increase the spaciousness of the sound; absorbing sidewall reflections reduces soundstage width, but results in tighter image focus and a more spatially coherent presentation. There is *no* debate, however, about whether or not uncontrolled side-wall reflections degrade a room's sonic performance. They do.

Fig. 14-16 An RPG Diffusor. (Courtesy RPG Diffusor Systems, Inc.)

An excellent product for controlling side-wall reflections is the Tower Trap, also made by the Acoustic Sciences Corporation. This is a tall, cylindrical device with absorptive and reflective (diffusive) sides. Absorption or diffusion can be selected simply by turning the device. When placed near the side wall with the reflective side to the room rear, the absorptive side prevents the first reflection from reaching the listener directly. Some of the energy striking the side wall is reflected into the Tower Trap's rear (diffusive) side. Most of the side-wall reflection is absorbed, while some is delayed in time, attenuated, and diffused—exactly what we want.

Note that it isn't necessary to treat a listening room's entire side-wall area; the reflections come only from small points along the wall. At mid- and high frequencies, sound waves behave more like rays of light. We can thus trace side-wall reflections to the listening seat and put the treatment in exactly the right location. As with light rays, a sound wave's angle of incidence equals its angle of reflection. That is, the angle at which a sound wave strikes a reflective surface is the same as the angle at which it bounces off that surface.

The technique for tracing side-wall reflections is shown in the series of photographs and illustrations in Figs.14-17, 14-18, and 14-19. First, mount a reflective Mylar strip on the side wall between the listener and the loudspeaker. The strip's center should be at the height of your ears when you're sitting in the listening chair. Next, put light sources (two lamps with their shades removed is ideal) where the loudspeakers are normally placed, as shown in Fig.14-17. When sitting in the listening chair, you'll see the two lamps reflected in the Mylar strip (Fig.14-18). The points along the Mylar strip where you see the lamps' bulbs are exactly the points where sound is reflected from the side walls to the listening seat. This is where to put the acoustic treatment. The photograph in Fig.14-19 shows how strategically placed acoustical materials (in this case, ASC Tower Traps) can kill side-wall reflections from both loudspeakers. Compare Fig.14-18 to Fig.14-19.

Repeat the process for the left side wall. If your listening room is symmetrical and the listening position is in the middle of the room, you need use this technique on only one side wall, then duplicate the acoustic treatment on the other. To maintain acoustical symmetry in the room, both side-wall treatments should be the same.

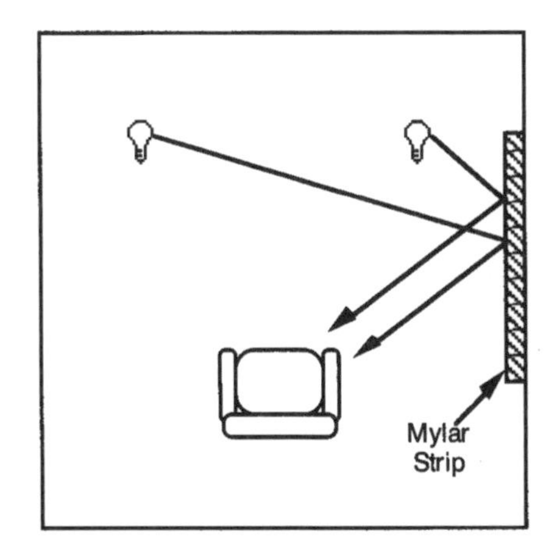

Fig. 4-17 Replace the loudspeakers with light sources and install a Mylar strip.

Fig. 14-18 The points of visual reflection are also the points of acoustic reflection.

Fig. 14-19 Acoustic treatments placed exactly at the reflection points kill side-wall reflections.

Treating the side-wall reflections from *both* loudspeakers on *each* side wall improves imaging. The right side wall will reflect sound from both right *and* left loudspeakers. The reflection of the left loudspeaker's signal from the right side wall confuses image placement and constricts soundstage width. This reflection can be thought of as a kind of "acoustic crosstalk"; we don't want left-channel information reflecting from the side wall into the right ear.

Note that a treatment placed away from the wall creates a larger apparent surface area than a treatment attached to the wall. The distance between the treatment and the wall causes the treatment to cast an acoustic shadow on the wall, widening the effective absorption area.

This technique can be applied to all reflections in the listening room. If you put a Mylar strip around the entire listening room, any point where you see a reflection of the light bulb itself is also a sound-reflection point. Additional absorptive or diffusive surfaces can then be placed and oriented exactly where they do the most good. You can also simply sit in the listening position and have a friend move a mirror along the sidewall. The point where you see the speaker's drivers in the mirror is the point where you should position absorbing or diffusing materials. This technique works for floor reflections; the point on the floor at which the mirror reflects the tweeter to the listening position is the place to add absorbing material.

Even if you don't go to the trouble of putting up a Mylar strip and lamps, you should do something to treat side-wall reflections. Bookcases, rugs, and drapes are all better than bare walls. If you really want to get the best from your system, however, there's no substitute for professionally designed acoustic treatments.

3) Thick, boomy bass

Thick, boomy bass is a common affliction that can be difficult to control. It often results from room resonance modes, poor loudspeaker placement, poor loudspeakers, or not enough low-frequency absorption in the listening room. As we will see in the later section on standing waves, listening-seat position can also exacerbate bass bloat.

If thick and boomy bass persists even after minimizing it with careful loudspeaker placement (the most effective method of alleviating the problem), you may want to consider different loudspeakers. If, however, the boominess is minor and you want to keep your loudspeakers, you can make the presentation leaner and tighter by adding low-frequency absorbers. These devices soak up low frequencies rather than reflecting them back into the room.

Passive low-frequency absorbers simply convert acoustic energy into another form—usually heat within a fibrous material. Low-frequency absorbers can be bought ready-made (such as ASC's Tube Traps and Tower Traps), built from common materials, or incorporated into an existing room structure. The easiest and most effective low-frequency absorber for the audiophile is ASC's 16" Full Round Tube Trap. These devices are extremely powerful, soaking up bass below 125Hz—the region below which most domestic materials stop absorbing. A pair of 16" Full Round Tube Traps in the corners behind the loudspeakers significantly tightens up the bass and removes boom and bloat.

Tube Traps employ a plastic sheet beneath the fabric exterior that covers about half the Trap. By rotating the trap, you can adjust the amount of high- and mid-frequency absorption; with the reflective side facing the listener, the Trap acts as a mid- and high-frequency diffusor. With the reflective side facing the wall, the Trap absorbs a broader range of frequencies. This variability is useful when tuning a room; rotating the Traps so that the reflective side faces the listener keeps the sound from becoming overdamped in the midrange and treble. (Tube Trap construction is shown in Fig.14-20.)

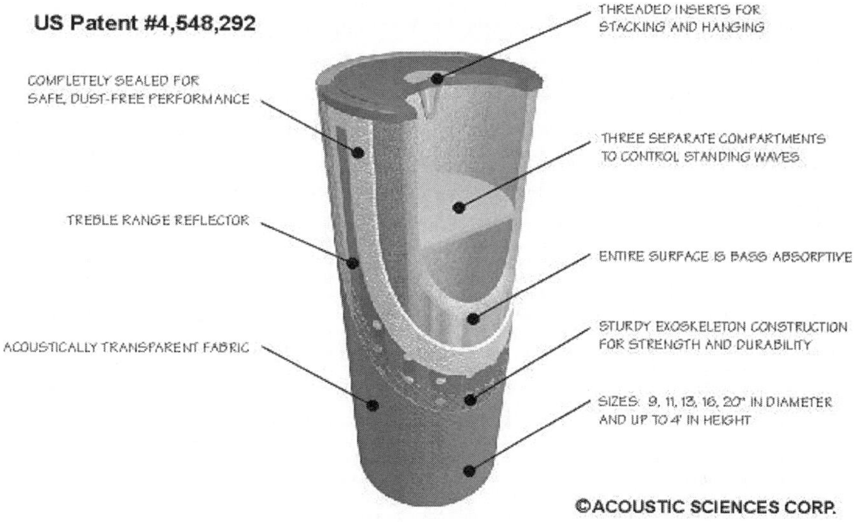

Fig. 14-20 Tube Trap construction. (Courtesy Acoustic Sciences Corporation)

A very inexpensive and effective low-frequency absorber can be made in a few hours for less than $20. These devices, called *panel absorbers,* have very high absorption at low frequencies, and can be tuned to the exact frequency and bandwidth required.

Panel absorbers can stand free of or be built right into an existing wall. In a typical built-in absorber, a 4′ by 8′ frame of 2x4s is nailed on its edge to the existing wall. The inside is sealed with caulking where the wood meets the wall to make it airtight, then filled with Fiberglas insulation. Next, a sheet of Masonite or plywood is nailed over the frame. Many tiny holes are then drilled in the panel. *Voilà*—a low-frequency panel absorber. (This type of perforated panel absorber is also called a Helmholtz resonator.)

Some panel absorbers have no holes, instead using a very thin sheet of material that flexes when struck by sound. The structure's absorption frequency is a function of the airspace depth (2x4s, 2x8s, 2x10s, or 2x12s may be used) and panel thickness. The Fiberglas inside the structure broadens the absorption peak. By changing the airspace depth and (in perforated absorbers) hole size, panel absorbers can be tuned to any low to mid frequency. Most rooms need broadband absorption in the bass, but panel absorbers can be narrowly tuned to absorb room resonance problems. A few panel absorbers, tuned to major resonance modes in the low-frequency spectrum, can greatly reduce bass problems in small rooms. To make these rather ugly contraptions palatable in a home, cover them with fabric. (Note that Helmholtz-type resonators shouldn't be covered directly with fabric: they need at least 1/4″ of airspace to allow the holes to "breathe" freely.)

Free-standing panel absorbers are built the same way, but with a sturdy backing (3/4″ particleboard, for example) instead of being attached to a wall. Specific details on building panel absorbers—material thickness, perforation size, spacing, etc.—can be found in F. Alton Everest's *The Master Handbook of Acoustics.*

Yet another way to get much-needed low-frequency absorption is by building a bass trap into an existing structure—a closet, say. Simply hanging an absorptive material such as acoustical foam or ordinary Fiberglas over the closet opening provides low-frequency absorption. This type of structure is called a *quarter-wavelength trap*. The trap will have peak absorption at the frequency whose 1/4 wavelength equals the distance between the closet back wall and the absorptive material, and also at odd multiples of the 1/4 wavelength. In fact, any absorptive material hung in front of a reflective surface will form a quarter-wavelength trap. Drapes in front of a window will work, but the distance between the window and drapes is so short that the lowest absorption frequency will be in the midband.

Let's calculate the absorption frequency for a closet 2' deep with absorptive material hung at the front. The formula is F = 1130/4D, where F is the lowest absorption peak, 1130 is the speed of sound in feet per second, and 4D is 4 times the distance between the reflective closet wall and the absorptive material. The lowest peak absorption frequency for a 2'-deep closet is 141Hz. This structure will also have peak absorption at odd multiples of a quarter wavelength: for instance, at three (423Hz), five (706Hz), and seven (989Hz) times 141Hz, and so on. Drapes hung 6" from a window will provide peak absorption at 565Hz, 1695Hz, 2825Hz, 3955Hz, and so on. Slanting the absorptive material will skew these values along a continuum, making the absorption more even with frequency.

The quarter-wavelength trap is typically limited to providing midbass absorption because of the large dimensions needed to achieve deep-bass absorption. I must stress, however, that proper loudspeaker placement is the easiest and most effective technique for reducing bass thickness. These acoustic treatments should be tried only *after* exhausting every possible option in loudspeaker placement.

4) Reflective objects near the loudspeaker

Reflective objects near the loudspeakers—equipment racks, windows behind the loudspeakers, subwoofers or furniture between the loudspeakers, even power amplifiers on the floor—can cause poor image focus and lack of depth. The best solution is to remove the offending object. I gained a huge increase in image focus and soundstage depth in my system just by moving my equipment racks from between the loudspeakers to the side of the room. If this is impractical, move the reflective objects as far behind the loudspeakers as possible. Power amplifiers should not extend in front of the loudspeakers' front panels, for example. And for the best musical performance, don't put a big television monitor between the loudspeakers. This is one reason why a music system should ideally be separate from a video system: imaging is degraded by a large piece of reflective glass near the loudspeakers. Having said that, however, there are a few tricks you can use to minimize the deleterious effect a video monitor has on music reproduction. (These techniques are described in Chapter 12.) Systems that employ a front-projector system and separate retractable screen don't suffer this problem.

If you can't move the offending object, cover it with an absorbing material such as Sonex. Windows behind the loudspeakers should be draped, with the drapes closed when listening. There's a large window directly behind my loudspeakers; I hear a significant increase in soundstage depth with the drapes pulled closed.

A good test of whether imaging is degraded by reflective objects near the loudspeakers is the Listening Environment Diagnostic Recording (LEDR) found on the *Chesky Records Jazz Sampler & Audiophile Test Compact Disc, Vol.1* (Chesky JD37). The test consists of a synthetic percussion instrument processed to sound as if it is moving in an arc between and above the loudspeakers. A system and room with good imaging will make the sound appear to move smoothly and consistently in the intended direction. Any holes in the image will be characterized by the sound jumping from one spot to the next rather than a smooth, unbroken motion. Such holes could be due to poor loudspeaker quality or placement, but can often be corrected just by moving reflective objects away from the loudspeakers. Best of all, a potentially large improvement in imaging and soundstaging can be had for a few minutes of your time.

Acoustical Do's and Don'ts

I've summarized this chapter into a few simple guidelines for improving your listening room. This next section is a short course for those readers who don't want to read about absorption coefficients and off-axis response. If you just want some practical tips for getting the most out of your system, this is the section to read. More detail on each of these points can be found throughout the chapter.

1) Loudspeaker placement

Just as real estate agents chant "location, location, location" as the three most important things about the desirability of a house, the three most important ways to improve the sound in your room are loudspeaker placement, loudspeaker placement, and loudspeaker placement. Follow the suggestions in this chapter and spend a few hours moving your loudspeakers around and listening. You'll not only end up with better sound, but become more attuned to sonic differences. All acoustic treatments should be built on a foundation of good loudspeaker placement.

2) Start with good ratios between the room's length, width, and height.

This isn't always possible, but when you're out house-hunting, take your tape measure and the table on good room ratios later in this chapter. If you can build a room from scratch, or convert a garage or basement into a listening room, choosing optimum dimensional ratios gives you a significant head start in getting great sound from your system.

3) Avoid untreated parallel surfaces.

If you've got bare walls facing each other, you'll have flutter echo. Kill the flutter echo by facing one wall with an absorbent material. Use one of the acoustical foams described, RPG Flutterfree, a rug, or the carpet-like material described later in this chapter. Note that a product such as ASC Tube Traps can provide low-frequency absorption, sidewall diffusion, and reduce flutter echo all at the same time.

4) Absorb or diffuse side-wall and floor reflections.

Bare floors should be covered with carpet between the listening seat and the loudspeakers. Treat the side walls between the loudspeakers and the listening position with an absorbing or diffusing material. Avoid having reflective surfaces, such as bare walls and windows, next to the loudspeakers. ASC Tower Traps, the various acoustical foams, and RPG Diffusors are all effective in treating side-wall reflections. Bookcases on the side walls make good diffusers.

5) Keep reflective objects away from loudspeakers.

Equipment racks, power amplifiers, furniture, and other acoustically reflective objects near the loudspeakers will degrade imaging and soundstaging. Move them behind the loudspeakers, or cover them with an absorbing material.

6) Choose a room with a high, sloped ceiling.

This isn't always possible, but a high, sloped ceiling skews the floor-to-ceiling resonance modes and reduces the early reflections at the listening position, resulting in smoother bass and a more spacious and open presentation. The loudspeakers should be positioned in the low-ceilinged portion of the room

7) Balance high-frequency absorbing materials with low-frequency absorbing materials.

Most rooms have lots of high-frequency absorption but little low-frequency absorption. Carpet, drapes, and soft furniture should be complemented with low-frequency absorbers (ASC Tube Traps or home-made panel absorbers) to keep the reverberation time flat across the band. Thick, slow bass results from such imbalances in reverberation.

8) Move the listening seat for best low-frequency balance.

Standing waves create stationary areas of high and low pressure in the room. Move the listening chair for best balance. Avoid sitting against the rear wall; the sound will be bass-heavy.

9) Break up standing-wave patterns with irregular surfaces or objects.

Strategically placed furniture or structures help break up standing waves. Large pieces of furniture behind the listening position diffuse waves reflected from the wall behind the listener.

A Short Course in Acoustical Theory

You don't need to know any acoustical theory to get good sound from your system. However, I've included the following survey of basic acoustical principles for those who want to take the next step in understanding how sound behaves in a room.

Listening-Room Resonance Modes

Resonance is the vibration of an object at its natural frequency, determined by the object's material and dimensions. Resonances surround us all the time—from a bell ringing at a certain pitch, to the sound made by blowing across the opening of a soda bottle. A singer breaking a glass with her voice is another example of resonance. It's not that the singer is that loud; when she hits the resonant frequency of the air in the glass, the reinforcement caused by the resonance adds energy to her voice.

As music lovers, we're interested in acoustic resonance: the reinforcement of certain frequencies within an enclosed volume of air, such as a listening room. When excited by the sound from the loudspeakers, the air in the listening room will resonate at particular frequencies, determined by the distances between the room's walls. These resonant points, called *room resonance modes,* can severely color the bass by creating large peaks and dips in the frequency response. Room resonance modes impose a sonic signature on the reproduced sound.

The room acts as an equalizer between your loudspeakers and ears, boosting certain frequencies and attenuating others. The result is degraded sound quality. In a poor room, the bass might lack specific pitch definition or sound slow and tubby, certain notes might be more prominent than others, dynamic impact could be reduced—or all of these at once. Resonance modes also cause the room to selectively store energy at the resonant frequencies, and then release that energy more slowly than other frequencies. The result is a blurring and muddying of the sound tonally, along with a distortion of music's dynamic structure and reduction in transient fidelity.

Let's consider what happens along the length of a listening room when the air inside it is excited by sound from the loudspeakers. For now, we'll ignore the fact that the room has height and width, and consider only the resonance between the two walls that define the room's length.

A listening room's resonant frequencies are determined by the distance between the room's walls. The farther apart the walls are, the lower the resonant frequency. Specifically, the lowest resonant frequency, called the *fundamental* resonance, occurs when the room's length equals half the wavelength of the sound. Put another way, a resonant mode will occur when the sound's wavelength is twice the length of the room.
Other resonant modes occur at twice this frequency, three times this frequency, and so on. Whenever the length of the room is a multiple of half of the sound's wavelength, a resonant mode will occur. This phenomenon, along with the formula for calculating resonant mode frequencies, is shown in Fig.14-21.

Calculating your listening room's resonant modes is easy. (But remember: we're still considering only the room's length.) The formula is F1 = 1130/2L. F1 is the first resonant mode, 1130 is the speed of sound in air (in feet per second), and 2L is two times the room's length (in feet). If the room is 21′ long, its first resonant mode will be 27Hz (1130/2 x 21).

We know that the next mode will occur when the wavelength equals the room's length, at 54Hz (2 x F1), then again at the next multiple of half a wavelength (1-1/2 wavelengths) at 81Hz (3 x F1), again at 108Hz (4 x F1), and so forth. It is only necessary to consider room modes up to about 300Hz, after which they tend to become so dense that they don't cause problems.

The room's height and width will also create their own resonant modes. If we have an 8′ ceiling, the resonant modes will occur at 71Hz (1130/2 x 8), 141Hz, 212Hz, and so on. If the width is 13′, the resonant modes will be at 43Hz (1130/2 x 13), 87Hz, 130Hz, 174Hz, 217Hz, 261Hz, etc.

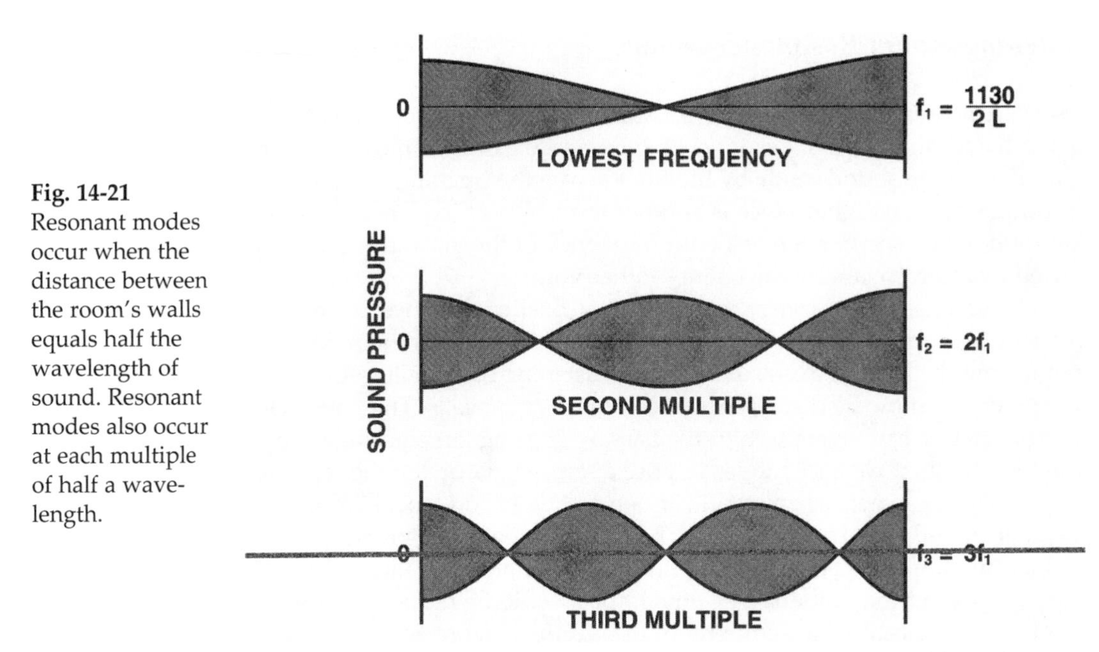

Fig. 14-21 Resonant modes occur when the distance between the room's walls equals half the wavelength of sound. Resonant modes also occur at each multiple of half a wavelength.

What this means to audiophiles wanting the most accurate reproduction possible is that those frequencies will be emphasized in a 21′ by 8′ by 13′ room because of the room's resonance modes. These emphases, or peaks, in the frequency response color the sound. In fact, room resonance modes can cause bass peaks and dips as large as 30dB. Resonance modes are a significant source of coloration in listening rooms, causing poor bass articulation, boominess, and bass thickness.

You can hear (and identify) room resonance modes most easily by playing the swept sine wave test from the *AVIA Guide to Home Theater* test DVD. This test signal sweeps from a high frequency to a low frequency, accompanied by a readout from the video portion of the disc (when the DVD player is connected to a video monitor) showing the descending frequency of the test signal at any given moment. You can simply listen and hear how some frequencies get much louder than others, or use a sound-pressure level (SPL) meter to measure the effect.

We can see how resonances store energy at certain frequencies, and then release that energy over time, in Fig.14-22. This is an actual measurement of my previous listening room before and after acoustic treatment. The MATT test signal is played through the loudspeakers, a measurement microphone picks up the signal, and the result is plotted graphically. The MATT signal is a swept sine wave, spanning the range from 20Hz–1950Hz, and interrupted with silence every sixteenth of a second. The signal sounds like a series of blips of ascending frequency, as the signal is on, and then off, for that duration. The measurement result shows how the room modifies the signal, specifically the frequencies at which the room stores energy, and how severely the transient nature of the test signal is distorted. Acoustic energy "hangs" in the room during the silences between blips, filling in the nulls in the chart. The deeper the nulls in the trace, and the flatter the peaks, and the more regular the trace's shape, the better.

The upper trace looks almost like a square wave, with deep silences between blips, indicating that the room isn't storing and releasing much energy. The lower trace, the room's measured performance before treatment, shows that the nulls are much less deep.

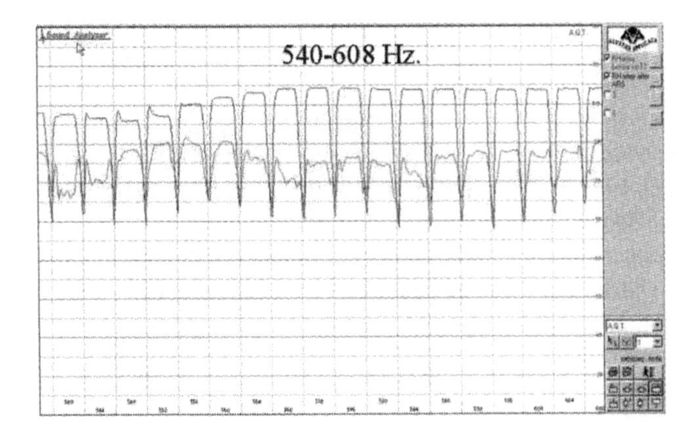

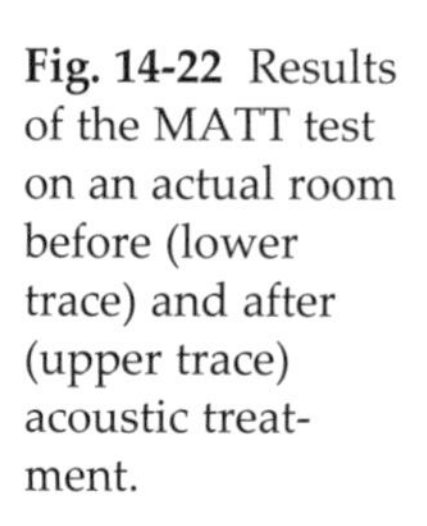

Fig. 14-22 Results of the MATT test on an actual room before (lower trace) and after (upper trace) acoustic treatment.

This indicates that the room is smearing the energy from the test signal over time, reducing articulation and degrading transient fidelity.

Optimizing Dimensional Ratios

We can minimize the frequency-response peaks and dips caused by room resonance modes by choosing a listening room with dimensional ratios that more evenly spread the resonant modes over the low-frequency band. Let's examine the resonance modes of two rooms of about equal volume. The first room has poor dimensional ratios, while the second room's dimensional ratios are excellent.

The first room is 24′ long and 16′ wide, with an 8′ ceiling. If we calculate and plot the resonant modes for each distance (length, width, height), we end up with a chart that looks like this:

Mode	Length	Width	Height
	24′	16′	8′
F1	24Hz	35Hz	71Hz
F2	48Hz	70Hz	142Hz
F3	72Hz	105Hz	213Hz
F4	96Hz	140Hz	284Hz
F5	120Hz	175Hz	355Hz
F6	144Hz	210Hz	
F7	168Hz	245Hz	
F8	192Hz	280Hz	
F9	216Hz	310Hz	
F10	240Hz		
F11	264Hz		
F12	288Hz		
F13	312Hz		

Notice that the third length mode (F3) at 72Hz coincides with the second width mode (F2) and the first height mode (F1). These three modes combine, piling up at 70Hz to create a huge peak in the response at this frequency. This undesirable situation occurs again at 140Hz, 213Hz, and 284Hz. The resonant modes coincide because the room's length, width, and height are multiples of each other. Actually, the three modes are perfectly coincident at 70.6Hz, 141.2Hz, etc. I've rounded them so they are slightly skewed, making the pileup more visible in Fig.14-23 (below). Moreover, there are no modes between 105Hz and 140Hz, which makes the pileup at 140Hz much more audible. The result will be a thick, peaky, and very colored bass reproduction.

We can also plot these modes graphically to better visualize room-mode distribution. This graph of mode vs. frequency is shown in Fig.14-23.

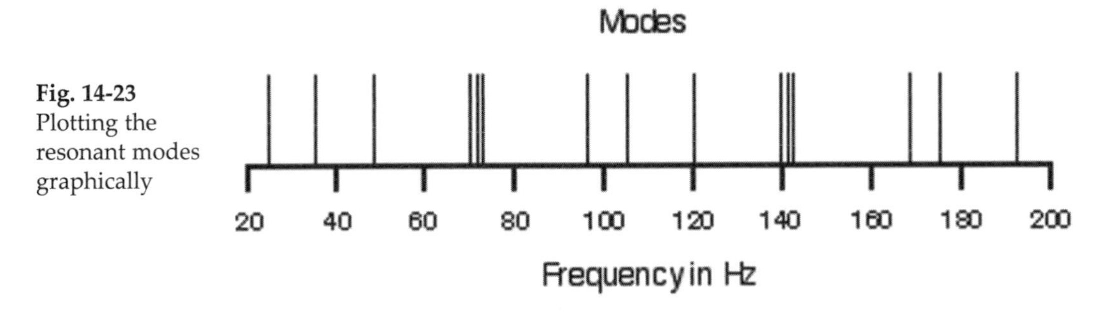

Fig. 14-23 Plotting the resonant modes graphically

As we can see, the modes are bunched together, with large spaces between them. The gaps between modes are just as detrimental as the coincident modes; they make the adjacent modes much more audible. Moreover, notes falling between modes are reduced in amplitude, making them sound weak and less prominent.

Let's see how this situation can be avoided by a room with better dimensional ratios. Our second room has approximately the same volume, but different distances between its walls. This room is 21′ long and 14′ wide, with a 10′ ceiling. Calculating the resonant modes for each dimension, we see that they are much better distributed:

Mode	Length	Width	Height
	21′	14′	10′
F1	27Hz	40Hz	56Hz
F2	54Hz	81Hz	113Hz
F3	81Hz	121Hz	169Hz
F4	108Hz	161Hz	226Hz
F5	135Hz	201Hz	282Hz
F6	161Hz	242Hz	339Hz
F7	188Hz	282Hz	
F8	215Hz	322Hz	

Fig.14-24 shows these data in graphical form.

We can see by comparing Fig.14-23 with Fig.14-24 that the resonant modes are more evenly distributed in the second room. Although there is some coincidence of modes, they are much less severe than in the first room. This second room will have smoother bass response, less coloration, and a more taut, punchy low end than the first room. Note that larger rooms have inherently more modes, and that the modes will therefore be more smoothly distributed—assuming good dimensional ratios. The small-

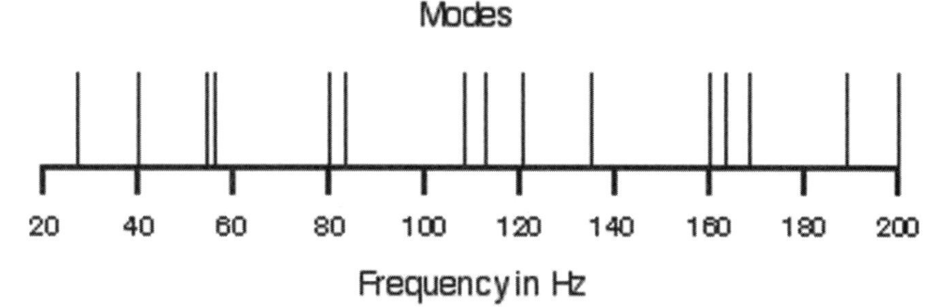

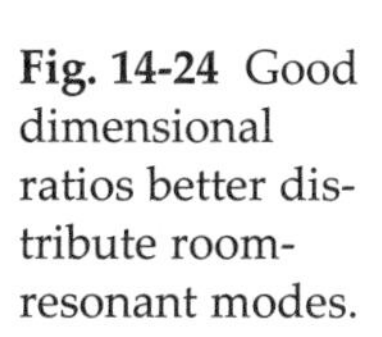
Fig. 14-24 Good dimensional ratios better distribute room-resonant modes.

er the room, the greater the challenge in distributing room resonance modes. Ideally, the room will have 20 or more axial modes below 325Hz, and the average spacing between those modes will be less than 15Hz. Distribution of axial modes along the width, length, and height can also be plotted as in Fig.14-25.

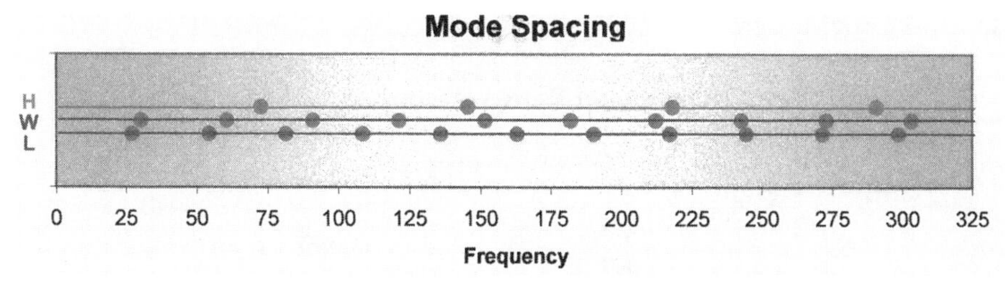

Fig. 14-25 Another way of graphically representing modal distribution. (Courtesy A/V RoomService, Ltd.)

A sloped ceiling helps greatly in skewing the height modes along a continuum. The resonant modes are naturally spread out rather than occurring at the same frequency as with a flat ceiling. If a ceiling starts at 8′ and slopes up to 12′, calculate its modes assuming a height of 10′ (the average of the two heights). The sloped ceiling doesn't eliminate resonance modes, but does help to distribute them. Similarly, splayed walls tend to skew the resonance modes, but don't eliminate them. Splayed walls make calculating room modes more difficult, and don't help much in reducing the audibility of resonance modes.

Room resonant modes can be made less audible by avoiding square or cubical rooms—i.e., rooms of equal width and height, or width and length, or length and height, or whose measurements are equal in all three dimensions—or rooms whose long dimensions are multiples of its short dimensions. If the ceiling is 8′, we want to avoid another dimension at 8′, 16′, 24′, etc. Note that resonance modes will always occur; good dimensional ratios merely distribute them evenly rather than allowing them to coincide. The chart on the next page shows, in descending order of quality, the best dimensional ratios (according to acoustician M.M. Louden):

According to this chart, the ideal length and width of a room with an 8′ ceiling is 15.2′ by 11.2′. These dimensions are calculated by multiplying each dimension by 0.8. For a 12′ ceiling, the optimum length and width are 22.8′ by 16.8′ (multiplying 1.2 by 19′ and by 14′). Note that, although these dimensions are different for the three ceiling heights, the *ratio* between the length, width, and height are identical. It's the ratio that determines how smoothly the resonance modes are distributed.

Note also that a larger room with good dimensional ratios will sound better than a smaller room employing those same ratios. The larger room will produce more resonance modes, which can be more closely spaced in frequency and thus achieve a smoother distribution. Remember that good dimensional ratios don't prevent resonances; they instead distribute those resonances more smoothly over the audio band. Good ratios prevent the coincidence of resonance modes, and the attendant gaps in resonance modes that make the modes more audible.

If you don't have the luxury of choosing the dimensions of your listening room, you can use the techniques described earlier in this chapter to minimize the audibility of room resonant modes.

Computer software for calculating room resonance modes is available at a moderate price. For example: Given your room's dimensions, RPG's Room Optimizer software will run a full modal analysis and recommend loudspeaker and listener placement.

This explanation of room resonance modes and the examples given are greatly simplified. I've just covered what are called *axial modes*—those that exist between one pair of surfaces. Other modes resulting from two surface pairs (*tangential modes*) and three surface pairs (*oblique modes,* from all six walls of a room) aren't discussed. *The Master Handbook of Acoustics* provides greater depth on this subject.

The 20 Best Dimensional Ratios (according to M.M. Louden)

Quality	Height	X	Y
1	1	1.9	1.4
2	1	1.9	1.3
3	1	1.5	2.1
4	1	1.5	2.2
5	1	1.2	1.5
6	1	1.4	2.1
7	1	1.1	1.4
8	1	1.8	1.4
9	1	1.6	2.1
10	1	1.2	1.4
11	1	1.6	1.2
12	1	1.6	2.3
13	1	1.6	2.2
14	1	1.8	1.3
15	1	1.1	1.5
16	1	1.6	2.4
17	1	1.6	1.3
18	1	1.9	1.5
19	1	1.1	1.6
20	1	1.3	1.7

(X can be length and Y width, or vice versa)

Standing Waves

If you've seen a cup of coffee sitting on a vibrating surface, you've seen standing waves. The vibration excites the liquid, causing waves to spread out from the center and reflect or bounce off of the cup's perimeter. At some points, the waves reinforce each other; at other points, the waves cancel. The result is a stationary pattern of peaks and dips in the coffee. Although the two waves are in motion, they create a stationary wave that is formed by the interaction between the two moving waves. This is exactly what happens to sound in a listening room.

Standing waves are stationary areas of high and low sound pressure in a room. They are caused by constructive and destructive interferences between the incident (direct) sound and the reflected sound from a room boundary. For example, when the reflected waveform's positive pressure phase is superimposed on the incident waveform's positive pressure phase, the two waves will reinforce each other (constructive interference), producing a *peak*. If the reflected wave's negative pressure phase meets the incident wave's positive pressure phase, the two waves will cancel (destructive interference), producing a dip or *null*. These interactions produce a particular pattern of stationary areas of more and less bass for each resonant frequency in the room.

Let's look at how a standing wave is set up along the length of a 20′-long listening room. If we excite the air in this room at its lowest resonant frequency, the pressure maxima will occur at the room walls, the pressure minima in the center. (See Fig.14-26. This drawing is analogous to Fig.14-21, which showed how a resonant mode occurs when half a wavelength equals the room's length.)

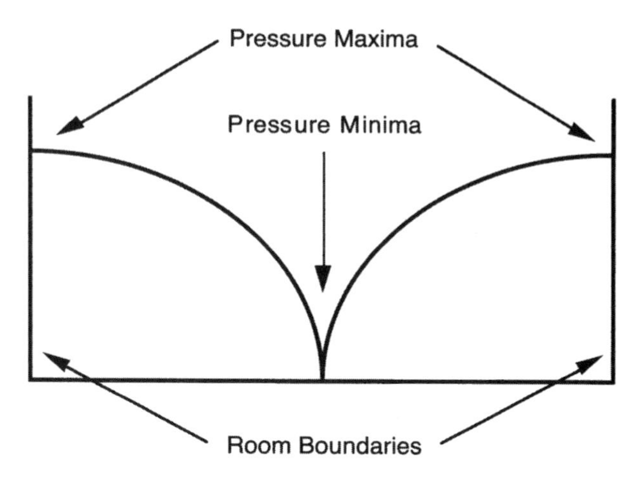

Fig. 14-26 Maximum pressure occurs at room boundaries.

By playing a tone at the room's resonant frequency and walking around in the room, we can hear the sound get louder and softer. In practice, the model in Fig.4-26 isn't so neat; the effects of other room boundaries and objects (furniture) in the room skew this theoretical representation. In addition, the associated resonant modes at multiples of the modal frequency tend to make the pattern of standing waves more complex.

Nevertheless, standing waves create areas in the room in which the bass sounds heavier, and some in which it sounds thinner. Play music with continuous bass energy (organ music works well) and move your listening chair forward and backward; the presentation will be heavier in some areas, leaner in others.

Although this unequal pressure distribution is undesirable, we can use standing waves to our advantage. If your loudspeakers and room tend to be thick and heavy in the bass, try moving your listening chair forward or backward until the bass is a little smoother and leaner. Or use the same technique to add fullness and weight to a thin-sounding loudspeaker.

Reverberation

The sound in a room is a combination of three components: 1) direct sound from the loudspeakers, 2) discrete early reflections from the floor, ceiling, and side walls, and 3) later, more diffuse reflections, called reverberation. This concept of three components to sound is illustrated in Fig.14-27

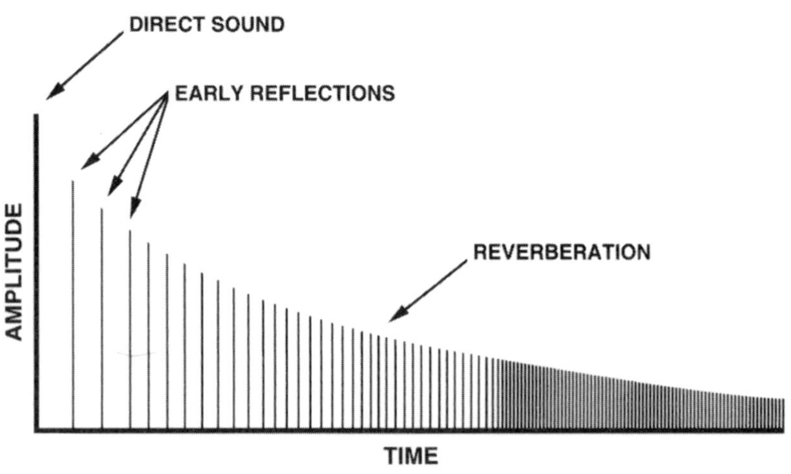

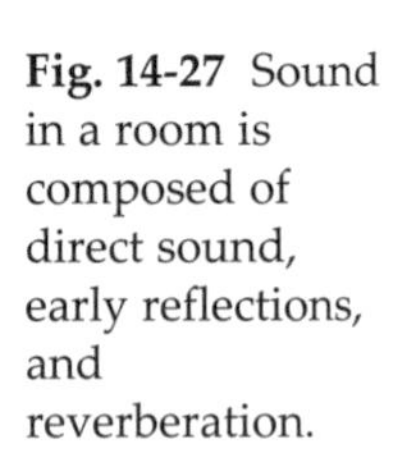

Fig. 14-27 Sound in a room is composed of direct sound, early reflections, and reverberation.

Reverberation is such a fundamental part of sound that we take it for granted. Although we don't hear it as a separate component of sound, reverberation adds warmth and space to music. If reverberation were completely missing from a sound, we would immediately identify the sound as being unnatural. You can hear how strange the absence of reverberation is on the *Denon Anechoic Orchestral Recording* (Denon PG 6006). The orchestra, recorded in an anechoic chamber (a room in which all sound reflections are completely absorbed), sounds completely different from even a recording made in a small, "dry" concert hall.

Reverberation in a room takes a finite time to decay into inaudibility.. The time it takes for sound to decay by 60dB is called the room's RT_{60} (the RT stands for "reverberation time"). A small, acoustically dead rock recording studio may have an RT_{60} of 0.1 second; a cathedral may have an RT_{60} of six or seven seconds. A room's RT_{60} quantifies how acoustically "live" or "dead" the room is. In practice, sound in a room doesn't decay the full 60dB because the decaying sound becomes swamped by the room's ambient noise.

Although the concept of reverberation is less applicable to small spaces such as hi-fi listening rooms than to concert halls, it is nonetheless helpful in balancing sound-absorbing materials in the room. A room with reflective surfaces will have a longer reverberation time than a room with absorptive surfaces such as carpets, soft furniture, and drapes. These materials absorb sound rather than reflect it back into the room. You want to choose materials in the listening room that contribute to an optimum RT_{60}, and balance those materials so that the reverberation time is constant over the audio band.

But what is the ideal reverberation time in a music listening room? And what combination of reflective and absorptive materials will produce the optimum reverberation time?

The quick answer is that optimum reverberation time in a listening room varies with the room's volume. For a 3000-cubic-foot room, a good reverberation time is about 0.4 second. For a 20,000-cubic-foot room, about 1 second is ideal. A surprisingly accurate measure of a room's reverberation time can be made with a handclap or balloon pop and a stopwatch. Have someone clap his hands loudly, or pop a balloon. Start the stopwatch when you hear the sound, and stop it when the sound has completely died away. Repeat the measurement several times and average the results. The measured result in seconds is a fairly reliable indicator of a room's RT_{60}.

Unfortunately, an optimum overall reverberation time isn't enough: we must ensure that the room's reverberation time is roughly the same at all audio frequencies. A room with thick carpet and heavy draperies—materials that absorb high frequencies, yet do nothing to low frequencies—will have a longer reverberation time at low than at high frequencies. This condition, shown in Fig.4-29, can make a hi-fi system sound slow and thick in the bass, or dead in the treble. Reverberation time is thus specified at six frequencies: 125Hz, 250Hz, 500Hz, 1kHz, 2kHz, and 4kHz. Our goal is to make the reverberation time roughly equal at these frequencies by choosing the appropriate mix of acoustical treatments. We do this by mixing different materials, each having absorption characteristics that vary with frequency. Let's look at how these surfaces affect the sound impinging on them.

Every surface in a listening room will either absorb, reflect, or diffuse (scatter) sound. We'll deal first with absorption and reflection. The degree to which a surface absorbs sound is called its *absorption coefficient*. This is a measure of the percentage of sound it absorbs, specified at the six frequencies listed in the previous paragraph. An absorption coefficient of 1.0 means that 100% of the sound is absorbed. This condition is likened to an open window—no energy is reflected back into the room. An absorption coefficient of 0.1 means that 10% of the sound is absorbed and 90% is reflected back into the room. Absorption coefficients of more than 1.0 are possible with a material that presents a surface area to the sound field that is greater than its area of wall contact, such as some wedge-shaped foams.

The absorption coefficients for two common materials—heavy carpet over felt, and drywall on studs (1/2" thick on 2x4s, spaced 16" on center)—are as follows:

	125Hz	**250Hz**	**500Hz**	**1kHz**	**2kHz**	**4kHz**
Carpet	0.08	0.24	0.57	0.69	0.71	0.73
Drywall	0.29	0.10	0.05	0.04	0.07	0.09

We can quickly see that carpet absorbs virtually no energy in the bass, but a significant percentage of the energy in the treble—nearly 75% at 4kHz. Conversely, drywall absorbs (and leaks out of the room) a moderate amount of energy in the bass (29% at 125Hz), and reflects virtually all midband and treble energy. Drywall on studs absorbs low frequencies by diaphragmatic action: sound striking the drywall causes it to flex. This flexing converts mechanical motion into a minute amount of heat.

Now, suppose we built a room completely covered in thick carpeting—like a rock band's garage studio. The carpet would absorb nearly all the treble and reflect nearly all the bass. The result would be a very short reverberation time at high frequencies and a very long reverberation time at low frequencies. The room would sound thick, heavy, and congested in the bass, and be generally unpleasant. Moreover, the room would store energy in the bass through resonance and release that energy over time, smearing transient signals. A short reverberation time in the midband and treble, accompanied by boomy bass, causes a loss of intelligibility. In such a room, each band member keeps turning up his volume so that he can hear himself over the other band members. Turning up the volume only makes the problem worse.

At the other extreme, a room of bare drywall, a tile floor, and no high-frequency absorbing materials would sound bright, hard, and thin. The trick is to balance absorptive materials so that the amount of absorption is roughly the same at all six frequencies. In practice, it's difficult to get enough low-frequency absorption; most rooms have longer reverberation times in the bass. Fortunately, a slightly longer reverberation time at low frequencies is desirable to give the room warmth.

Reducing a room's reverberation time by adding absorbing materials also has the added benefit of broadening the resonances, making them less audible.

Fig.14-28 shows a computer model of my former listening room's reverberation time. The upper plot is the reverberation time before installation of a complete room-acoustics package. The two lower plots are the ideal reverberation time and the actual reverberation time after room treatment.

Room Length =	21 ft.	0.5 in.	Room Volume =	2746	cf.
Room Width =	14 ft.	6.0 in.	Wall Area =	640	sf.
Ceiling Height =	9 ft.	in.	Ceiling/Floor Area =	305	sf.

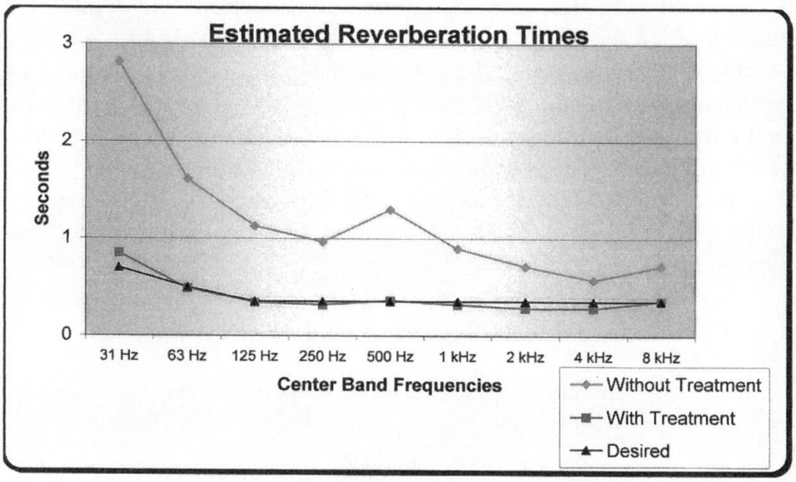

Estimated Reverberation Times

Seconds: 0, 1, 2, 3

Center Band Frequencies: 31 Hz, 63 Hz, 125 Hz, 250 Hz, 500 Hz, 1 kHz, 2 kHz, 4 kHz, 8 kHz

Without Treatment — With Treatment — Desired

	31	63	125	250	500	1000	2000	4000	8000	Average	
W/O	2.81	1.61	1.13	0.97	1.30	0.90	0.72	0.58	0.72	0.93	sec.
With	**0.85**	**0.49**	**0.34**	**0.32**	**0.36**	**0.32**	**0.29**	**0.28**	**0.36**	**0.32**	**sec.**
Ideal	0.7	0.5	0.35	0.35	0.35	0.35	0.35	0.35	0.35	0.35	sec.

Room Mode Bandwidths W/O ARS Treatment . 2.4 Hz.
Room Mode Bandwidths with ARS Treatment **6.9 Hz.**

Fig. 14-28 Computer model of my former listening room showing reverberation time without treatment, with treatment, and the ideal. (Courtesy A/V RoomService, Ltd.)

You can calculate your room's reverberation time. First, multiply the absorption coefficient of every surface in the room by the square footage of that surface at the six frequencies. For example, if you have 294 square feet of carpet, multiply 294 x 0.73 (absorption coefficient of carpet at 4kHz) to get 215 absorption units. (These absorption units are called "sabines," after Wallace Clement Sabine, the American physicist who founded the science of architectural acoustics. Boston Symphony Hall, designed in 1900 by Sabine, was the first concert hall to be designed using the new field of architectural acoustics.)

Multiplying each surface area's square footage by the absorption coefficients at the six frequencies gives us a chart of every material in the room and how many sabines of absorption that material provides at the six frequencies: 125Hz, 250Hz, 500Hz, 1kHz, 2kHz, and 4kHz.

To convert the total sabines at each frequency into a reverberation time at that frequency, we use the following formula:

$$RT_{60} = \frac{(0.049)(V)}{(S)\ (a)}$$

where

RT_{60} = reverberation time, seconds
V = volume of room, cu ft
S = total surface area of room, sq ft
a = average sabine absorption coefficient

This method of calculating reverberation time works best with large rooms (and assumes that the materials are uniformly distributed throughout the room), but nevertheless will give you an idea of the balance between low- and high-frequency absorption. If you perform these calculations on your own room and find a large difference in the total sabines between 125Hz and 4kHz, you can add low-frequency absorbers to even out the reverberation time. The acoustical materials described earlier in this chapter can be used to achieve a smooth reverberation time with frequency. A full listing of materials and their absorption coefficients is included in *The Master Handbook of Acoustics*, mentioned earlier in this chapter.

Building a Listening Room from Scratch

This chapter cannot guide someone through the long and complicated process of building a dedicated listening room from the ground up. Instead, I recommend two approaches: The first is to study the subject and design the room yourself, bringing in an acoustician for consultation two or three times during the design process. You can end up with a good room without spending a fortune. The acoustician can be a freelance contractor, or be associated with a company that sells acoustic treatments.

The second approach is to commission a full-fledged design from an architectural acoustics design firm. This way is ideal for those with large budgets or no time to learn acoustic design themselves. The design firm will often oversee construction, providing you with a completely finished room. This can be a *very* expensive proposition. A custom room built with this approach is shown in Fig.14-29 (photo courtesy of Pure Audio, Scotts Valley, California; www.balabo.com).

Isolating the Listening Room

Many music listeners want to play music at high levels without disturbing neighbors or other family members, or to listen to music in a quiet environment without outside sounds intruding on the experience. This brings us to a specialized area of acoustics called isolation. Acoustic isolation prevents sound in one area from escaping into another area. A good example of successful isolation is the midtown Manhattan recording studio that records during business hours.

Keeping sound in the listening room from getting into the rest of the house is a daunting challenge. It requires good design, precision construction, and lots of money. There are, however, a few principles and techniques that can help minimize sound transmission.

Very briefly, some of the sound escapes from the listening room into the rest of the house through doors and other openings. Even small cracks, such as the space between the floor and door, allow large amounts of acoustic energy through. This phenomenon is called "flanking." To reduce flanking, use a solid-core door and seal its perimeter with weather-stripping or refrigerator seals. You can also buy acoustic doors, complete with jambs and seals, from acoustic supply houses. Look for the ads in the back of *Mix* magazine or other professional audio publications.

Second, isolating a room requires lots of mass between the sound source and the area to be protected from sound. Thick walls, double walls with unbridged air cavities, lead sheeting, and double solid-core doors are all fundamental prerequisites to achieving high isolation. Most audiophiles don't need such isolation, but if you do, it's very expensive. Half-hearted attempts at isolation just don't work; it takes a serious effort to achieve adequate isolation.

The effectiveness of a wall's ability to reduce the amount of sound it passes is expressed in its Sound Transmission Classification (STC). For example, a standard wall built with 2x4 studs and Sheetrock has an STC of 34dB without insulation (as you would typically find in a home's interior walls) and 36dB with 3.5" of Fiberglas insulation. Using a 2x6 plate with 2x4 studs staggered (studs alternating on each side of the wall) increases the STC to as high as 52dB (with insulation). A full double 2x4 wall with an unbridged air cavity has an STC of 58dB. As you can see, the more elaborate (and expensive) the construction method, the greater the wall's ability to attenuate sound passing through it.

A related subject is the room's *noise criterion*, a numerical rating of the background noise level present in the listening room. The lower the NC rating (NC20 is excellent, for example), the lower the room's ambient noise. A quiet room makes low-level detail more audible, contributes to the sense of space (reverberation decay is better resolved), and fosters the impression of a black background from which the music emerges.

The Walldamp Technique

The modern Western construction technique of drywall over a wooden stud frame introduces audible problems not found in other parts of the world that use thicker and denser building materials. Specifically, sound striking a drywall-over-stud wall causes the wall to resonate at about 70Hz. Sound is converted to mechanical energy (the wall moving), then back to acoustic energy (the vibrating wall acts just like a loudspeaker's diaphragm). The wall hums away along with the music, introducing a drone-like sound of a specific pitch. What's worse, the acoustic energy produced by the wall is released relatively slowly over time, changing the music's dynamic characteristics.

This phenomenon, called *wall-stud resonance*, introduces a specific coloration to the reproduced sound. You can demonstrate this by hitting a drywall-over-stud wall with the ball of your fist and listening to the resulting thud. The sound produced by the motion of the wall has virtually the same character, whether the initial sound was a double bass or an explosion in a film soundtrack. Consequently, the music is overlaid with a common signature that degrades fidelity. Homes built with thick plaster walls or concrete block don't suffer from this problem.

While acoustic absorbers can stabilize the air in a listening room, they do nothing to address wall-stud resonance. Keeping the room's walls stabilized under dynamic conditions (loud transient signals) requires careful attention to wall construction. If you're building or remodeling a home, you can greatly reduce wall-stud resonance and isolate the listening room at the same time. One method is to buy dual-layer drywall that has damping material between the layers. Unfortunately, this is a specialized material that must be purchased by the truckload.

The second option is ASC's Walldamp material and construction technique, shown in Fig.14-30. The first step is to install a resilient channel between the studs and first layer of drywall. The resilient channel acts as a metal spring, isolating the wall from the studs and preventing the 70Hz wall-stud resonance from occurring. If left like this, however, the room would sound horrible; the entire wall would act as a drumhead, resonating whenever acoustic energy struck it.

This is where Walldamp and the second layer of drywall come in. The Walldamp in Fig.4-30 consists of squares of tarpaper coated on both sides with a special glue. When installed between two layers of drywall, the glue absorbs energy by a mechanism called *sheer stress damping*. That is, when the two pieces of drywall try to move relative to each other, the glue absorbs the energy on a molecular level. Specifically, the glue converts the mechanical energy of the wall's motion into a tiny amount of heat. Consequently, the acoustic energy striking the wall is absorbed by the damping action of the Walldamp instead of being reflected back into the room. Rather than shuddering when struck by sound, the wall behaves more like a thick plaster wall.

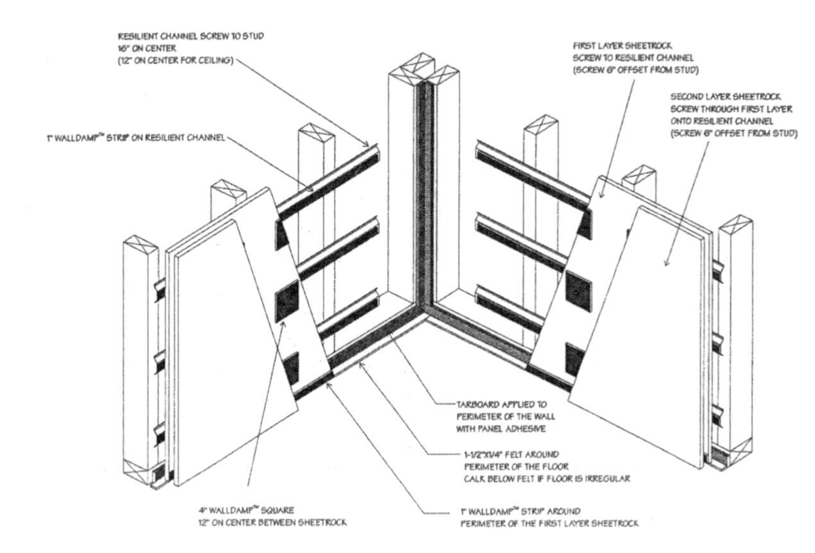

Fig. 14-30 ASC's Walldamp technique damps wall vibration and increases isolation. (Courtesy Acoustic Sciences Corporation)

In addition, the wall now acts as a large bass absorber at very low frequencies—just what most listening rooms need. Fewer freestanding bass absorbers are needed because now the entire *room* is acting as a bass absorber. Another benefit of this technique is that less of the sound inside a listening room escapes into the rest of the house: You can play music louder and later, without disturbing family members. Best of all, this construction technique adds very little to the cost of building or remodeling.

DSP Room Correction

We've seen in this chapter that the listening room is a crucial link in the music playback chain. Every listening room, no matter how good, radically alters the sound produced by a pair of speakers. The room acts as an equalizer, boosting some frequencies and attenuating others. This change in frequency response can be as much as 30dB in the bass, adding large amounts of coloration to the music. Moreover, the room briefly stores and releases energy at certain frequencies, causing some bass notes to "hang" in the room longer. The result is a thickness and lack of clarity in the bass, along with a reduced sense of dynamic drive.

It's been said that a loudspeaker designer has 100% control over his product's sound above 700Hz, 50% control from 300Hz to 700Hz, and only 20% control below 300Hz.

This aphorism reflects the increasing influence of the listening room on sound quality at lower and lower frequencies.

Although the listening room imposes far more severe colorations than any electronic component, we've grown to accept this musical degradation as an inevitable part of music listening. Even custom-built, acoustically designed rooms housing perfectly positioned loudspeakers will introduce significant colorations. All rooms introduce colorations: the better ones simply do less damage. Attempts to reduce room-induced colorations with analog equalizers have met with limited success. Analog equalizers add their own audible problems, and are limited in resolution. Digital equalizers can have greater resolution than analog ones, but are still a crude tool for dealing with room problems.

The advent of Digital Signal Processing (DSP) provides a powerful new opportunity to remove the effects of the listening room on reproduced music. DSP is the manipulation of audio signals by performing mathematical calculations on the numbers that represent the music. DSP chips can be programmed to perform as filters or equalizers with many hundreds of frequency bands, and with characteristics impossible to realize with analog equalizers. The trick is putting this advanced technology to work to surgically correct room-induced colorations.

Far more sophisticated than a digital equalizer, a DSP room-correction system analyzes the response of your loudspeaker/room combination with high resolution, then creates digital filters to remove the room-induced problems. The room-correction system not only removes the room's effects, but also corrects for speaker colorations. If your speakers have a peak of energy at 2kHz, for example, the room-correction system can remove this coloration. In addition, the room-correction system will make the speaker/room response identical for each speaker. This technique removes the room's effects on reproduced sound *electronically*, before it is reproduced by the loudspeakers. Put another way, the room-correction system distorts the signal in a way that counteracts the distortion imposed by the listening room. The result is flat response at the listening position.

Here's how room correction works: You (or your dealer) position the supplied measurement microphone at the listening seat, connect the room-correction system hardware to a personal computer, and run the supplied software on your PC. The room-correction system hardware puts out a series of pulses that are reproduced by your loudspeaker, then analyzed by the software running on your PC. The software then configures filters in the DSP chips that form an inverse of your room response. Peaks in the measured response become dips in the filter, and vice versa. The filter response is a mirror image of your room response. Newer, more sophisticated room-correction systems eliminate the need for the PC, and have greatly simplified the user interface. DSP room correction has become a mainstream technology through its inclusion in home-theater receivers.

The filter parameters calculated by the software are downloaded into the room-correction system's memory during the setup. The room-correction system hardware stays in your rack and performs the filtering on all signals running through it. If you change speakers, speaker location, or listening spot, you'll need to re-measure the room to create new filter coefficients (about a 15-minute job, once you've been through it a few times). The filter parameters are unique to your speakers, room, speaker positions, and listening position.

This measurment process is repeated at several locations around the primary listening seat, and the results averaged to provide flat response over a larger area.

At first glance, you might think that the room-correction system should produce perfectly flat response at the listening position. In practice, a flat response sounds thin, bright, and bass-shy. The ideal response has a rising response in the bass and a falling response in the treble. Some room correction systems don't give you the option of choosig the curve; you are stuck with whatever the designer thought was ideal. Such systems often improve the bass performance, but add excessive brightness to the overall sound. Other DSP room-correction systems allow you to choose one of several "target curves," or even to create your own curve. This target curve will be the response at the listening position.

An important technical factor in a room-correction system is its *frequency resolution*. This specification indicates the correction system's ability to correct very narrow peaks and dips in the frequency response. For example, a room-correction system with a frequency resolution of 10Hz wouldn't be able to correct a peak or dip in your room that was 8Hz wide. Some room-correction systems have a frequency resolution of less than 1Hz, which is fine enough to follow any peaks and dips.

Room-correction systems can also reduce the detrimental effects of sidewall reflections described earlier in this chapter. The music signal driving the loudspeakers is pre-conditioned to contain signals that have the same delay as the wall reflections at the listening position, but of opposite polarity. The wall's reflections are thus canceled by a series of delayed waves from the loudspeakers that have opposite polarity from the wall reflections. The result is cancellation of reflections and removal of room resonance modes.

Note that because room-correction systems operate in the digital domain, analog signals from an LP front end must be digitized, then converted back to analog. That's why most room-correction systems include an integral analog-to-digital converter. Digital signals, such as from a music server, are input directly to the room-correction system, which either converts that digital signal to analog for driving a power amplifier, or outputs a digital signal for conversion to analog with an outboard digital processor. Although room correction can produce profound improvements in bass quality and image focus, some audiophiles will be reluctant to digitize LPs. The purity of an analog source is inevitably degraded by the digital conversion process. You must consider, however, whether the loss of quality from digitizing an analog signal is a greater or lesser evil than the degradation imposed by the listening room.

My experience with room correction suggests that the technology is extremely effective in removing bass colorations and restoring clarity to reproduced music. With room correction, the sound has a "lightness," agility, and clarity that are difficult to achieve in any other way. The overall presentation seems to have less midbass boxiness, and extreme low frequencies have greater dynamic impact. In addition, soundstaging improves because identical response from both loudspeakers is essential to precise image focus. (That's why each speaker must be toed-in by exactly the same amount.) Without room correction, image size and placement can change as an instrument moves between musical registers because the response of each speaker/room is different. With identical output from each speaker, image focus becomes tighter and the soundstage is better delineated.

On the downside, DSP room correction can radically alter your system's sound. If you like your system's overall tonal balance, adding a room-correction system can be disappointing. Although room correction nearly always results in better bass, it often comes at the price of an unnatural midrange and treble balance.

If your loudspeakers are of the subwoofer/satellite variety, you can use room correction to improve the bass performance without affecting the midrange and treble. Room-correction devices are available that operate only on the signal driving the subwofoer, leaving the signal driving the satellite speakers unaltered. This approach gives you the best of both worlds.

15 System Set-Up Secrets Part Two: Accessories, Racks, AC Power Conditioners, and Expert Tuning Techniques

Introduction

I've been privileged to be a student in the art of audio system set up in the best classroom imaginable. As a full-time reviewer since 1989, I've watched the world's top high-end equipment designers install their products in my home for review. As you can imagine, these designers are highly motivated to get the best possible sound from their products. Consequently, they have pulled out of their bag of tricks all the techniques and insights gained over a lifetime in the field for getting better sound.

I realized early on the unique position I was in, and dedicated myself to learning these set-up techniques. Although there are many well-known standard set-up procedures, many of these elite designers had developed their own techniques for getting maximum performance out of an audio system. It has been fascinating to watch these masters in action and to listen to the results of every change. Frequently, the system sounds quite good after an hour of tweaking, but becomes spectacular after another couple of hours of experimentation.

In this chapter I'll share with you the culmination of what I've learned in this "classroom." You can use a few of the techniques to improve your own system, or study this section and practice the craft to become a master in the art of system setup.

I've divided this chapter into three sections. The first describes the wide range of accessories available to the audiophile. The second details specific procedures for improving a system's performance. Some of these procedures require buying an accessory (usually inexpensive), and some are purely techniques that require nothing more than your time, patience, and listening skill.

The third section goes into some depth on AC power conditioners (and AC cords) and equipment racks (and vibration isolation). Although these products are traditionally categorized as accessories, AC conditioners and equipment racks are fully equal in importance to any other component. In fact, good AC power and a vibration-resistant rack should be considered the foundation on which the rest of your system is built.

I can't overstate the need to base the tweaks in this chapter on a solid platform of correct loudspeaker placement. Correctly positioning your loudspeakers is the single most important thing you can do to get the best sound from your system. Only after you have correctly positioned your speakers and avoided major acoustical problems (described in the previous chapter) should you begin the fine-tuning process described here. Without proper loudspeaker and listener placement, none of the other set-up techniques will matter. Loudspeaker placement and acoustics are so important that I've devoted an entire chapter to them.

Accessories

The vast array and wide availability of audio accessories bears witness to audiophiles' need to squeeze out that last little bit of musical performance from their systems. Accessories can not only make your system sound better, they're also fun to try. It's quite rewarding to discover some simple trick or product that elevates your system's sound at little cost. This section surveys available accessories, describes what they do, and explains how they work.

How to Choose Accessories

Some of the accessories described in this section can make your system sound better than you thought it could—and at a modest cost. Other accessories can not only fail to improve the sound, but can actually degrade your system's musical performance. To make matters even more complicated, many accessories are completely worthless and are nothing more than "snake oil" sold by less-than-honest promoters. Finally, the effectiveness of an accessory can vary from system to system.

Fortunately, there's a simple and effective way of deciding which accessories are worth the money and which aren't: *Listen to the accessory in your system before you buy*. Most dealers will let you either borrow the accessory overnight, or return it for a refund if you don't hear an improvement. If you hear an improvement that seems commensurate with the product's asking price, buy it. If you don't, return the product. It's that simple. And if the accessory is inexpensive but not returnable, you can try it at minimal cost. Some CD treatments cost only a few dollars; if they don't work, it hasn't cost you much. But by all means try some of the devices described in this section. They can often make the difference between merely good and superlative sonic performance.

In any pursuit there are those who abandon reason, and high-end audio has its share of nonsensical, even bizarre accessories. Most of these products are completely worthless. Examples include an alligator clip attached to the loudspeaker cable that prevents the loudspeaker's "gravitational influence" from affecting the audio signal. Another claims to "energize" the electrons in an audio system with a combination of lithium salts and cobra venom.

We must be careful, however, in branding certain products as fraudulent. Many accessory products whose value is now without question—AC power cords and Tiptoes, for examples—were once dismissed as worthless by those who had never listened to their effects. In the early days of CD, it was assumed that all CD transports and digital interconnects sounded the same because they all produced the same stream of binary ones and zeros. It was only after much critical listening and closer scientific scrutiny that other mechanisms for creating differences in the sounds of transports and digital interconnects (primarily timing variations called "jitter") were discovered. It is undisputed today that transport and digital-interconnect jitter affects sound quality. We must listen and trust our ears, and not rely solely on pure theories, no matter how elegant or well-argued.

The bottom line with any accessory, as with any audio component, should sound familiar by now: listen before you buy. If the device makes an audible improvement, buy it—regardless of whether or not there's a scientific explanation.

Cable Enhancers

A cable enhancer is a device that breaks-in line-level interconnects and loudspeaker cables faster and more completely than running music through the cables. Both ends of the interconnect or cable are connected to the enhancer, which puts out a high-level signal to produce current flow through the cable. Cable enhancers cost about $200. Many dealers, however, will break in your new cables on an enhancer at no charge. Note that the improved sound of a cable after break-in is largely due to charging the dielectric, which dissipates over time. It may thus be necessary to periodically put your cables on a cable enhancer.

RF Filters

RF filters fit around cables and AC cords to prevent high-frequency noise from passing through the cable to the audio component connected by the cable. The filter is, in essence, a "choke" that turns the cable or power cord into a low-pass filter. RF filters can be put around interconnects at the preamp side to prevent RF from getting into the preamp, around interconnects at the power amplifier, or around AC cords. They cost about $25 per pair, are easy to install and remove, and can be an inexpensive improvement to your system. I've had mixed results with RF filters; well-designed AC power cords and interconnects are a better solution in my view.

Burn-in CDs

These discs contain special signals that reportedly improve your system's sound by demagnetizing the circuits inside your components. There's no scientific basis for the efficacy of these discs, but they do seem to work.

Contact Cleaners

Contact cleaners remove the residue and oxidation that builds up on RCA plugs and jacks, interconnect terminations, and loudspeaker and power-amplifier terminals. The fluid comes in a small bottle and is usually supplied with pipe cleaners or a brush for applying the liquid. Some cleaners come in spray form, but the contacts should still be rubbed to remove oxidation. Caig Laboratories makes a product called DeoxIT that is inexpensive and effective. The procedure for cleaning contacts and vacuum-tube pins and sockets is described later in this chapter.

Contact Enhancers

This is a fluid applied to electrical contacts in your system. Contact enhancers promote better conductivity between a plug and jack, improving the sound. Contact enhancers work especially well on phono cartridge pins and vacuum-tube pins. They are also effective on the blades of AC power cords. (Fig.15-1)

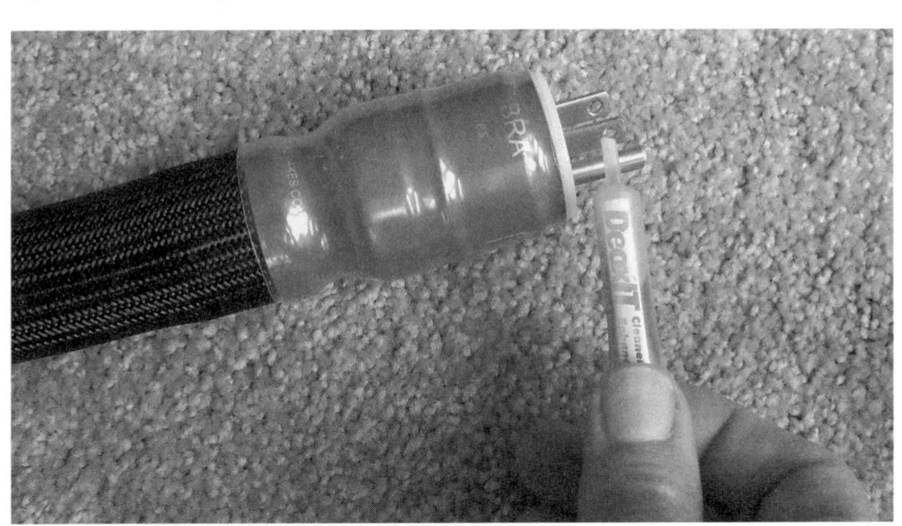

Fig. 15-1 Applying contact enhancer to the blades of AC cords is a simple and effective tweak.

Tube Dampers

Tube dampers are devices that fit over vacuum tubes to make them less microphonic. The damper is made from a rubber-like material—often Sorbothane—that damps a tube's vibration and improves sound quality. Some dampers are also claimed to extend tube life by acting as heatsinks, radiating heat away from the tube. Tube dampers cost between $4 and $25 each. You can see tube dampers on a tube in Fig.15-10 later in this chapter.

AC Polarity Testers and AC "Cheater" Plugs

An AC plug's orientation in a wall socket can affect a component's sound quality. Most audio components have three-pronged plugs, with the rounded lower pin connected to the component's chassis. This ground lead ultimately ends up at a copper rod buried in the ground outside the electrical service panel. If, for some reason, the AC voltage touches a component's chassis, the chassis will short that voltage to ground, blowing the breaker fuse and protecting anyone who may touch the chassis. The two other conductors are connected to the component's power transformer, which tends to leak some current to ground. The amount of current leaked depends on the orientation of the AC plug. One direction may produce a lower chassis voltage than the other. The system will sound best with the plug oriented in the direction producing the lower chassis voltage.

Determining which way the AC plug should go can be determined with an AC polarity tester. Just put the polarity tester near the chassis of the component and read the tester's meter. Reverse the AC plug and see if the voltage is lower. Repeat this process on each component until the entire system has been connected for best AC polarity. You can do the same thing with a voltmeter, but the polarity meter avoids direct contact with the AC line. A polarity meter costs about $50.

Reversing a three-pronged AC plug can be accomplished by adding a "cheater" plug (available at any hardware store) that allows the plug to be inserted upside down. The cheater plug's ground wire (which replaces the rounded pin) must be connected to the screw holding the AC wall cover in place to maintain the safety aspect of a grounded plug and receptacle. If the ground wire isn't attached, the cheater plug "lifts" the ground connection. When this happens, the component is grounded only through the interconnects connecting the ungrounded component to another component that is grounded at the wall. For example, the ground can be lifted on a preamplifier provided the power amplifier is grounded to the wall and the two are linked by interconnects. Ground lifting is sometimes required to eliminate hum and noise in the system created by ground loops. A ground loop occurs when there are two paths to ground of unequal resistance. A slight voltage difference is thus created that causes a small current to flow in the ground conductors, producing a hum. Lifting the ground at one component breaks the ground loop and removes the hum.

Note that the safety feature of a grounded system is defeated if one component is plugged into a cheater plug and not connected to a grounded component through the interconnects. Further, lifting the ground on all the components is extremely dangerous and could be fatal; if for some reason a high voltage appears on the chassis, you'll have no way of knowing it and could receive a lethal shock. But if the component is properly grounded, the chassis will cause a breaker fuse in your AC wall panel to blow, preventing electric shock.

Ground Isolators

A ground isolator may be needed when a cable TV connection is made to a component audio system, or in a system that includes a tuner for cable-delivered digital radio. The cable TV ground is likely to be different than your audio system's ground, producing hum and buzz when the signal from the digital radio tuner is connected to your preamp. A ground isolator is a transformer that physically separates the grounds yet passes the signal. If you have a multi-purpose audio/video system, slowly moving horizontal bars on your video display (called "hum bars") are a sure sign that you need a ground isolator.

AC Noise Filters

These small devices plug into unused AC outlets in your home and reportedly remove noise from the AC line (Fig.15-2). The price is generally about $150 for four devices.

Fig. 15-2 Noise filters plug into unused AC outlets and reduce noise on the AC line. (Courtesy PS Audio and Music Direct)

Replacement AC Wall Outlets

These outlets, some of them designed specifically for audio, replace the ones in your listening room to provide a better connection to your AC cables than is possible with the cheap outlets builders use. The quality outlets have more copper in them, and the receptacle provides a greater contact area between the receptacle and the plug's blades.

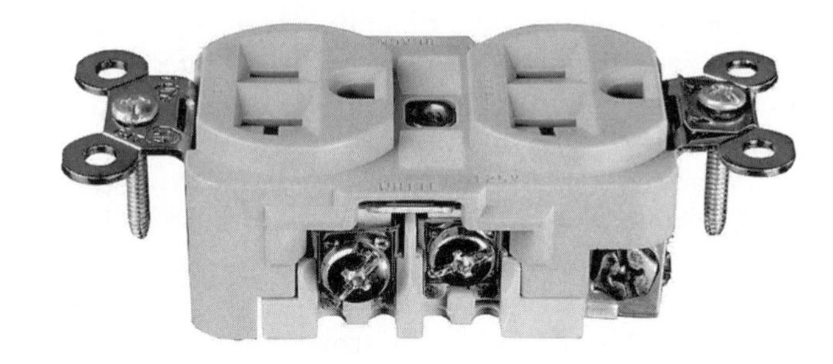

Fig. 15-3 AC outlets designed for high-performance audio offer an audible improvement over stock AC outlets. (Courtesy Shunyata Research)

Hi-Fi Tuning Fuses

These are the same size and current rating as stock fuses, but reportedly deliver better sound than conventional designs. They are made from ceramic bodies for low resonance along with silver internal wire. The end caps are made from pure silver with a layer of 24k gold for better signal transfer and protection from corrosion. Many power amplifiers have fuses in the DC supply to the output transistors, which would conceptually benefit the most from replacement with fuses of higher quality.

Shorting Plugs

A shorting plug is an RCA plug or XLR connector with no cable attached. When inserted into a preamplifier's unused inputs it shorts the signal conductors to ground. The shorting plug reduces noise inside the preamplifier.

Bi-Wire Jumpers

Loudspeakers that can be bi-wired are supplied with small plates or wires ("jumpers") that connect the two positive terminals together, and the two negative terminals together, when the speaker is connected by a single run of cable. (The jumpers are removed for bi-wiring.) Replacing these stock jumpers, which are often made from gold-plated brass, with high-quality jumpers can improve performance. Custom jumpers are often made from solid high-purity copper plated with rhodium (Fig.15-4).

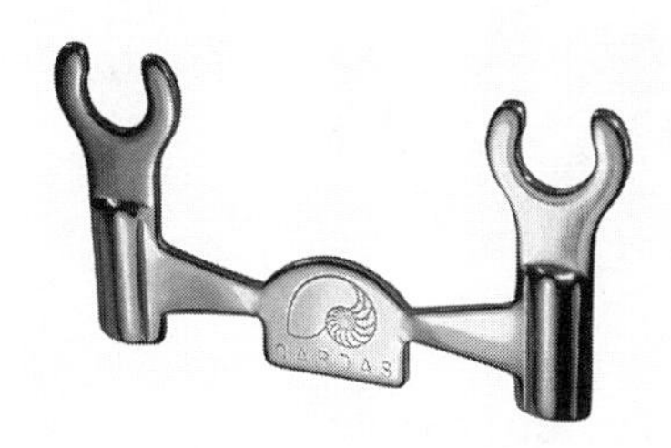

Fig. 15-4
Replacement bi-wire jumpers often sound better than stock brass jumpers.
(Courtesy Cardas and Music Direct)

Anti-Static Sprays

This spray reduces static on your equipment and cables, reportedly improving the sound.

Amplifier Stands

These are platforms just bigger than an amplifier that lift an amplifier off the floor a few inches, providing a more stable support as well as better ventilation.

Cable Elevators

These devices lift interconnects, loudspeaker cables, and AC cords off the floor. For some reason, getting cables off the floor improves the sound. One theory is that floors, particularly carpeted floors, store an electrostatic charge that adversely affects audio signals flowing through cables. Whatever the reason, cable lifters do work. Cable elevators can also be used to separate cables behind an equipment rack or at the power amplifier where an AC cord, interconnects, and loudspeaker cables meet.

Binding Post Wrenches

This $10 accessory should be in the toolkit of every audiophile. It is essentially a double-ended nut driver, with one side for 7/16" posts and one for ½" posts. It makes connecting loudspeaker cables easier, and also tends to ensure just the right amount of torque on the binding posts.

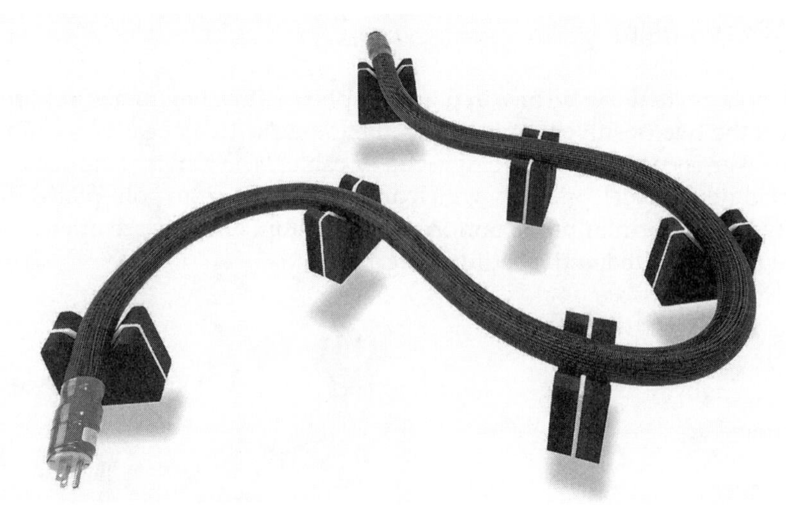

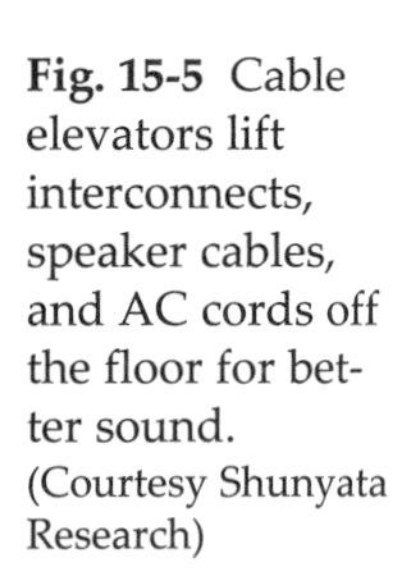

Fig. 15-5 Cable elevators lift interconnects, speaker cables, and AC cords off the floor for better sound. (Courtesy Shunyata Research)

Vibration Absorbers for Loudspeakers

Several products are said to reduce a loudspeaker's cabinet vibrations and improve sound quality. One device that I've found particularly effective is a cloth bag of small but very heavy damping beads—the finished product looks like a bean bag. When put on top of the loudspeaker, the bags absorb vibration, and are especially effective on small, lightweight, stand-mounted loudspeakers. The weight also helps couple the loudspeaker to the stand, and the stand to the floor.

Other vibration-control devices address the problem with small discs that reportedly absorb vibration through some internal mechanism, or are made of solid wood. The discs are placed on components, loudspeakers, listening-room walls, or equipment racks. These products tend to be quite expensive, however, and reports of their effectiveness are mixed.

CD Treatments

A wide variety of devices are available that claim to improve CD sound quality. These run the gamut from rubber rings that fit around a CD to liquids applied to a CD's surface. Some CD treatments make a worthwhile improvement to the sound; others are of dubious value.

The main category of CD tweaks is the damping device. This can take the form of a rubber ring stretched around the CD's edge, a flat adhesive ring stuck to the outer perimeter of the CD, or a mat that covers the entire label side of the CD. The theory is, the less disc vibration, the better the sound. One idea holds that a more stable CD results in less servo activity on the part of the transport's electromechanical systems, which maintains focus, tracking, and rotational speed. This means less current demand on the power supply.

Be careful about applying rings that cannot be removed. If they degrade the sound—as some may—you're stuck. In addition, some transports simply won't play discs that are a different size or shape. If you want to experiment with CD rings, use the removable kind. Rings generally cost between $1 and $3 each. Try a few: If they don't make an audible improvement, you're out only a few dollars.

CD mats can improve sound quality, but also stress the CD player's or transport's motor system; the rotational motor was designed to spin the mass of a CD, not a CD and a heavy mat. Some transport manufacturers have devised lightweight magnetic clamps that hold the CD firmly in place, yet add very little mass.

The second main category of CD treatments comprises those products that are intended to improve the CD system's optical performance. These include liquids applied to the CD surface, and colored paint around the disc edge. How these potions work is a mystery, but some of them do improve the sound. Again, you must be careful about what you apply to a CD surface: some chemicals can attack the CD's polycarbonate substrate and ruin the disc. Further, wiping the liquid off the disc can create scratches that may cause mistracking on playback.

Finally, CD lens cleaners are available to remove dust and dirt from a player's lens. They are the size of a CD, with tiny brushes that wipe the lens as the disc spins. Lens cleaners cost about $20. I've never found the need to clean a CD player's lens.

Analog Accessories

Vertical Tracking Force Gauges

An essential accessory for turntable setup, the vertical tracking force gauge (also called a "stylus force gauge") indicates the amount of tracking force at the stylus. A tracking force gauge is shown on page 259 in Chapter 9.

Azimuth Meters

Azimuth in LP playback is the perpendicularity of the stylus to the groove. If the azimuth is incorrect, one channel will be louder than the other, degrading imaging. You can't accurately judge azimuth by looking at the cantilever in the cartridge body; they are often misaligned at the factory. An azimuth meter is the easiest way to accurately set azimuth.

Replacement Tonearm Cables

Some tonearms allow replacement of the stock cable with an aftermarket model. Older tonearms, or lower-cost models, often benefit from a high-quality tonearm cable.

Record-Cleaning Machines

The best way to keep your valuable LPs sounding good and lasting longer is to clean them with a record-cleaning machine. Unlike a brush that removes only some of the dirt, or redistributes it to other parts of the record, a record-cleaning machine uses a motor-driven vacuum to extract dirt and grime from the record grooves. Although some record brushes can help remove dirt, none approaches the effectiveness of a cleaning machine. A record cleaned with a vacuum cleaner will have far fewer ticks and pops, and also have a cleaner, more open midrange. Removing the junk in the grooves allows the cartridge to extract more musical information from the record. A clean record will also last much longer than a dirty one; the stylus won't constantly be grinding dirt into the groove walls.

Record-cleaning machines vary in price according to their level of automation. Some machines, like the Record Doctor II by Nitty Gritty, require you to apply the cleaning fluid to the record surface, turn the record over (so the fluid side faces the vacuum slot), and spin the record yourself. Semi-automatic machines spin the record with a motor, but still require manual application of the cleaning fluid and turning over the record. Fully automatic machines apply the fluid, spin the record, and even clean both sides of the record at the same time.

Expect to pay about $80 for a bare-bones manually operated cleaner with no vacuum action, and up to $5000 for a state-of-the-art automatic system. A mid-priced vacuum-based record-cleaning machine is shown in Chapter 9.

Cartridge Demagnetizers

Cartridge demagnetizers are often referred to generically as "fluxbusters," after the Sumiko Fluxbuster, the first commercially available demagnetizer. Demagnetizers are about the size of a cassette box and cost less than $100. They generally have two RCA jacks that you connect to your tonearm cable's RCA plugs. When the demagnetizer's power switch is engaged, the demagnetizer puts out a high-frequency signal that slowly diminishes in intensity, removing unwanted residual magnetism in the phono cartridge.

Moving-coil phono cartridges need demagnetizing because the core around which the coils are wound gradually becomes magnetized by the powerful permanent magnets surrounding the coils. Air-cored moving-coil cartridges are not susceptible to accumulated residual magnetism and don't need demagnetizing.

Moving-magnet cartridges build up magnetism on the pole pieces, degrading sonic performance. Note that you must remove the stylus from a moving-magnet cartridge before demagnetizing. Otherwise, the demagnetizer will demagnetize the magnets, reducing the cartridge's output level. If you accidentally demagnetize the magnets, replace the stylus: the stylus assembly contains new magnets.

Cartridges should be demagnetized after each 50 hours of playing time. Note that demagnetizers designed to bulk-erase tapes or degauss televisions should not be used to demagnetize cartridges.

Record Brushes

Often made from carbon-fiber bristles, the record brush removes surface dust from LPs and should be used before playing every LP side. The record brush is not a substitute for periodic record cleaning with a liquid.

Replacement Record Sleeves

Paper record sleeves leave small fibers on your records, and plastic sleeves can leech chemicals onto the record surface. Special rice-paper sleeves suffer from neither of these problems and keep your records sounding good for years.

Platter Mats

A platter mat rests on top of your turntable platter and is designed to reduce record vibration. Platter mats might be better than no mats on some turntables, but higher-quality turntables are designed so that the least vibration is realized with no mat.

Cartridge Alignment Protractors

This device allows you to correctly set the offset and overhang of a cartridge (these terms are explained in Chapter 9). A protractor is shown on page 260 in Chapter 9.

Record Flatteners

This device, which looks like a large waffle iron but with flat inside surfaces, gently heats your LP while applying even pressure over the entire LP surface to remove record warp.

Stylus Cleaners

Cleaning your stylus before every record is essential not only to good sound, but to preserving record life. A dirty stylus will grind junk into the groove walls where it cannot be removed, even by a record-cleaning machine. Stylus cleaners resemble nail polish, with a brush in a bottle of cleaning fluid. The Discwasher brush, by contrast, has short, stiff bristles on which a fluid is applied. Whichever brush you use, the stylus should be cleaned with a gentle stroke from the back of the cartridge to the front. A stylus-cleaning kit sells for about $20.

Anti-Static Guns

This device removes the static charge from LPs by creating an opposite charge when the piezoelectric crystal in the "gun" is squeezed by the trigger.

LP Lubricants

This spray-on treatment reduces friction between stylus and groove, reportedly resulting in both improved sound and less record wear. One application lasts for the life of the record.

Headphones and Headphone Amplifiers

Headphones are essential for portable listening, but can also be used at home when you don't want to disturb family members. Headphones vary not only in sound quality, but in comfort; it's a mistake to buy headphones that you want to take off after just a short time. A significant factor in long-term comfort is the headphones' weight; the lighter models, of course, enjoy an advantage.

Some very-lightweight headphones sound excellent and are comfortable, but don't provide much isolation from outside sounds. If you plan on listening in a noisy environment, choose closed-ear headphones over the open-ear variety. In addition to keeping outside sounds out, closed-ear 'phones keep your music in, so it is less likely to disturb others.

High-quality headphones range in price from about $59 to as much as $2000 for electrostatic "earspeakers." The sound quality of the best under-$100 units is surprisingly good. Superlative models are available for about $500. The better-quality headphones provide a smoother tonal balance, greater resolution of detail, and a more open and transparent sound. It's worth shopping for headphones; just as with loudspeakers, less expensive models often outperform more costly units.

A recent trend has been toward the "ear bud" style, in which two small transducers fit inside your ears. Ear buds are much less bulky and obtrusive than traditional headphones, and are more portable. Although ear buds are commonly supplied with low-quality MP3 players, high-quality units are available that deliver superb sound.

For the ultimate performance, you can have in-ear buds created specifically for your ear canal. This ensures a tight fit and much better performance than is possible from "one-size-fits-all" in-ear 'phones.

Most listeners have never heard headphones at their full potential. That's because the headphone jack on the front of some components outputs a severely compromised signal. The headphone amplifier is usually an afterthought considered unworthy of serious design effort or expense. After all, how many consumers make purchasing decisions based on the quality of the headphone amplifier?

Audiophiles wanting a better headphone-listening experience can choose an aftermarket headphone amplifier. This small device, which works for portable or home use, takes in a line-level signal from a portable music player and amplifies it enough to drive headphones. Most headphone amplifiers operate on both AC and battery power.

The headphone amplifier is a miniature power amplifier that provides perhaps half a watt of output. The amplifier should be stable into reactive loads, and be built with audiophile-quality components. High-quality headphone amplifiers have a discrete output stage (made from separate transistors) rather than an operational amplifier chip, and are fed from power supplies designed with high-end techniques and circuits. They also incorporate audiophile-quality volume controls (Fig.15-6).

Another way to get great headphone sound at home is to drive the 'phones with a single-ended tubed amplifier. Some SET amplifiers drive the headphone output not with a cheap op-amp, but with the amplifier's tubed circuitry. These low-powered amplifiers make ideal headphone drivers, and can produce superlative sound. The high impedances of many headphones, along with their low current draw, make them the perfect load for SET amplifiers. Check with the manufacturer to find out if the headphone jack is driven by the same tubed circuitry that powers the loudspeakers.

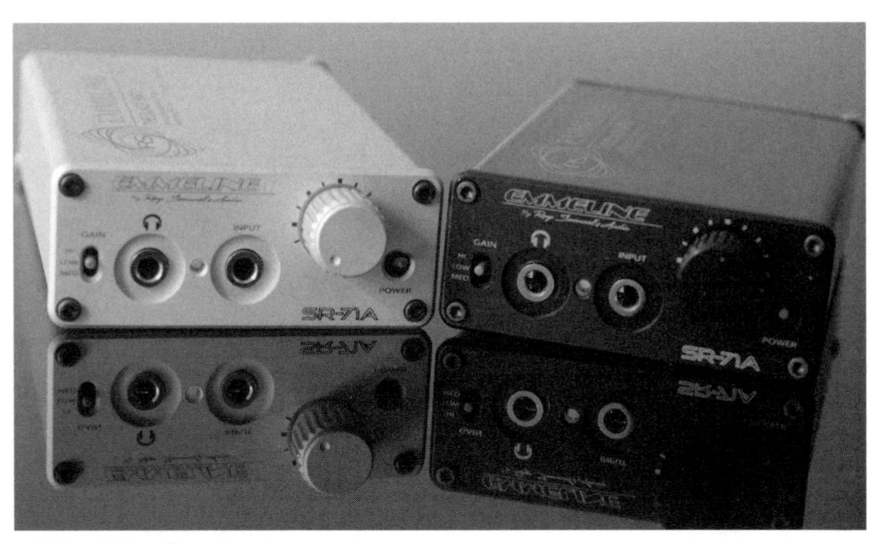

Fig. 15-6 A high-quality headphone amplifier can greatly improve the sound of portable music players as well as that of most home-audio components. (Courtesy Ray Samules Audio)

System Set-Up Techniques

I've organized set-up techniques into specific categories, such as tips for equipment placement, cables, tubed amplifiers, loudspeakers, etc.

Equipment Placement

• Position components for adequate ventilation. Overheating will shorten product life. In addition, some digital products are fine-tuned to operate at a specific temperature. If an outside heat source such as a tubed preamplifier is placed beneath it, the DAC's sound will be degraded.

• Maintain adequate distance between the preamplifier and power amplifier. The power amp's large transformer can radiate 60Hz hum. This is more crucial in preamplifiers with phono stages or with separate phono preamps.

• Don't position your power amplifier (or amplifiers) on the floor in front of the loudspeaker plane. That is, the amplifier's front panel should be behind the loudspeakers' baffles. Acoustic reflections from the amplifier can degrade imaging.

• Install the equipment on a sturdy rack. Vibration can affect system performance, particularly that of turntables. (Equipment racks are covered in detail later in this chapter.)

• Avoid placing the equipment rack in a corner where acoustic pressure builds up and could cause components in the rack to vibrate.

• If you place amplifiers on a carpeted floor, use spikes, cones, or an amplifier stand to raise the amplifier above the carpet to improve ventilation. If the amplifier's bottom plate sinks into thick carpet, the convection cooling essential to keeping the output stage from overheating is compromised.

Fig. 15-7
Spikes or cones beneath an amplifier or integrated amplifier on a carpeted floor helps with vibration as well as providing much-needed ventilation.

Cables and Interconnects

• Because all wire degrades the signal passing through it, the less wire you have in your system, the better. Keep analog interconnects and loudspeaker cables as short as possible, but always the same length between left and right. Note that this doesn't apply to digital interconnects, which sound best when they're between 1 meter and 1.5 meters in length.

• Keep interconnects away from AC power cords. If they must meet, position them at right angles to each other instead of parallel.

• If you must choose between long interconnects and short loudspeaker cables (such as with a preamplifier in an equipment rack, and power amplifiers on the floor next to the loudspeakers) or short interconnects and long loudspeaker cables (such as with the power amplifier in the equipment rack), go with the shorter loudspeaker cables. There's some controversy on this point, but my experience suggests that loudspeaker cables are more detrimental to sound quality than interconnects.

• At the back of an amplifier where an AC power cord, interconnects, and loudspeaker cables converge, separate the cables as much as possible. If they do meet, cross the cables at right angles rather than allowing them to run in parallel. Cable elevators designed to lift cables and interconnects off the floor are ideal for separating cables behind an amplifier (Fig.15-8).

• Lift cables and interconnects off the floor with cable elevators.

• If you have excess cable or interconnect, don't wind it into a neat loop behind the loudspeaker or equipment rack. This will make the cable more inductive and change its characteristics. Instead, drape the cable so it crosses other cables at right angles.

• Periodically disconnect all interconnects and loudspeaker cables for cleaning (Fig.15-9). Oxidation builds up on jacks and plugs, degrading the audio signal passing through the connections. Use a contact cleaner (available at most high-end stores). It works. In fact, switching interconnects sometimes cleans the jacks, making the system sound better even

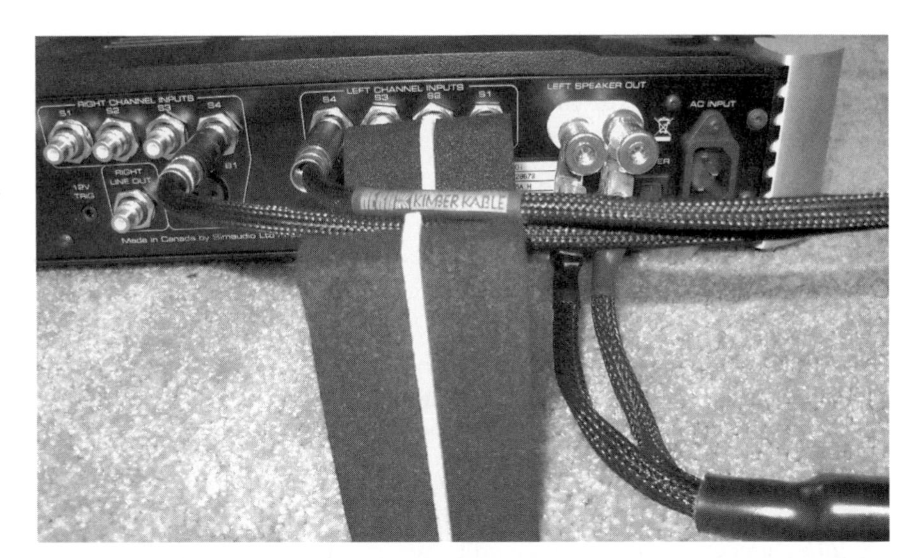

Fig. 15-8 Separate AC cords, interconnects, and loudspeaker cables where they meet behind an amplifier. Use the same technique for interconnects and power cords behind an equipment rack.

Fig. 15-9 Periodically clean spade lugs, binding posts, RCA plugs, RCA jacks, XLR jacks, XLR plugs, tube pins, tube sockets, and other connections with contact cleaner.

though the interconnect may not be intrinsically better. Start by cleaning all the RCA and XLR jacks on your equipment and cables. Treat the outer ring as well as the unseen junction inside the jack that contacts the RCA plug's center pin. Pay special attention to phono jacks and plugs; connectors that carry the tiny phono signals to your preamplifier are the most sensitive to degradation from dirt and oxidation. Next, clean the binding posts on loudspeakers and power amplifiers, as well as the spade lugs on speaker cables. Be sure not to touch any cleaned surface with your fingers—body oils will trap contaminants. It's a good idea to wear gloves when cleaning contacts. You may also consider cleaning the pins on phono cartridges and the tonearm's cartridge sleeve connectors. Periodic cleaning of all electrical contacts in a system can keep your system sounding its best. Once every two months is a good interval. Make sure, however, that the contact cleaner doesn't leave a residue that could attract dust and end up degrading the connection.

• Follow up a contact cleaning by using a contact enhancer on all jacks and plugs.

• When connecting and disconnecting RCA plugs, always grip the plug, never the cable. Remember to push the tab when disconnecting XLR plugs.

• Ensure tight connection of all RCA plugs, and particularly spade lugs on power amplifiers and loudspeakers. Make sure you get lots of contact surface area between the spade lug and post, then tighten down the binding posts.

• Avoid sharp bends in cables and interconnects.

• High contact pressure is more important than a large contact area when connecting loudspeaker cables. A tight connection—enough to slightly distort a thin spade lug—is ideal. Be careful not to overtighten the binding post, which could cause the entire post to turn inside the loudspeaker.

• Once cables and interconnects are in place, don't disturb them. It seems as though cables and interconnects "settle" and sound better over time, a process that starts anew if the cable is moved. Cables and interconnects should be moved only for cleaning.

• If you have the choice between using spade lugs and banana plugs, opt for spade lugs which provide better contact with the binding post.

Tubed Equipment

• Tubes are microphonic, meaning that they act as microphones, creating a signal when vibrated. Combat tube microphonics by applying specially made damper rings around each tube.

• Keep the tubes biased according to the manufacturer specifications.

• When installing or changing vacuum tubes, wear gloves to avoid depositing body oils on the tube's glass envelope (Fig.15-10).

• Clean the contact points between the tubes and the socket. Disconnect the power to the component, remove the tubes, and then clean the pins on the tube sockets using an audio contact cleaner and an electrically isolated cleaning tool. Then clean the tube's pins. Treat the pins with contact enhancer and reinsert the tubes.

• Tubed amplifiers often offer different transformer taps that allow you to select the taps that best match your loudspeaker's impedance. Maximum power transfer occurs when the correct tap for a given loudspeaker's impedance is selected. However, the tap that provides the greatest power transfer might not produce the best sound. The surest way to decide which tap to use is to ignore the loudspeaker's nominal-impedance specification and just listen to all the taps.

• Tubes' performance degrades over time, but so slowly that you don't notice it from one week to the next. Consequently, your tubes might be degrading the sound of your system. The solution is to replace the tubes.

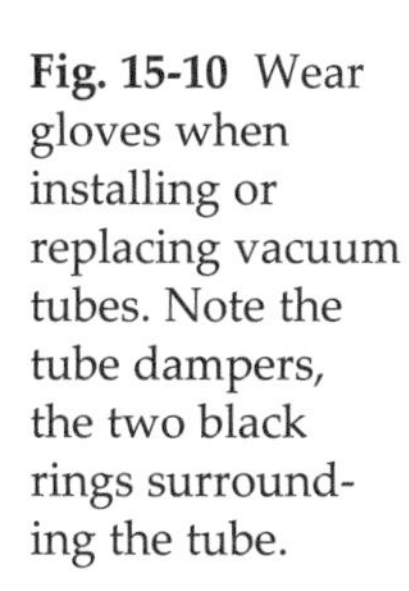

Fig. 15-10 Wear gloves when installing or replacing vacuum tubes. Note the tube dampers, the two black rings surrounding the tube.

• Tubed components are particularly susceptible to sonic degradation when vibrated. Follow the vibration-isolation guidelines later in this chapter.

LP Playback

• Turn off your digital components (CD players, DACs, music servers)) when playing LPs. This applies to any component in your system that has a microprocessor in it. Digital circuits radiate noise that can contaminate the audio signal.

• Demagnetize your phono cartridge after every 50 hours of playing time. Be sure to remove the stylus from a moving-magnet cartridge before demagnetizing.

• Clean the stylus before every record side with a commercially available brush.

• Periodically give your stylus a deep cleaning; Hold a matchbook from which all the matches have been removed underneath your stylus and very gently scrape the striking area over the stylus in a back-to-front motion (the same direction as the stylus moves through the record). This abrasive action removes the accumulated coating of vinyl on the stylus.

• Clean LPs with a liquid cleaning solution, even new ones. LPs have a mold-release compound that should be removed before playing. Cleaning machines with a vacuum system are optimum, but you can still clean records with a brush and liquid solution.

Digital Playback

• Keep digital data interconnects (between a separate CD transport and digital processor) and analog interconnects between source components and your preamplifier separate from each other. The very high frequencies carried by digital cables can radiate noise and pollute analog signals.

• Digital interconnects seem to deliver the best sound when they are between 1 meter and 1.5 meters in length.

• If your CD player or DAC has a digital-domain volume control, run it with 3dB of attenuation rather than at maximum level. Although attenuating the volume with a digital-domain volume control theoretically reduces resolution, most digital filters sound better when they are not processing full-scale signals. The very slight reduction in resolution is more than made up for by the greater sense of ease when a few dB of attenuation is used.

• Warm up your digital components for a minimum of two hours before listening. Longer is better; I leave my DAC powered up continuously. It can take a full day for a DAC to sound its best.

Loudspeakers

• Periodically check the tightness of the driver mounting bolts (Fig.15-11). These bolts can come loose over time, degrading performance. If the driver's frame isn't tightly coupled to the baffle, the frame will vibrate and radiate a sound of its own. Don't overtighten the bolts. Because these bolts can loosen during shipping, check their tightness when you first unbox new loudspeakers.

Fig. 15-11 Periodically tighten the bolts holding the loudspeaker drivers to the baffle.

• Ensure good contact between loudspeaker cables and binding posts. Get the spade lug fully on the binding post and tighten it with a binding-post wrench.

• Be sure that the loudspeaker doesn't rock on its spikes. All four (or three) spikes should be fully coupled to the floor.

• If you don't want to damage hardwood or tile floors with spikes, put coins between the spikes and the floor.

• When placing small speakers on stands, couple the speaker to the stand with four small balls of Bostik Blue-Tak.

• Using a bubble level, check that each speaker has the same degree of tilt-back (also called "rake angle"). This is particularly important with loudspeakers using first-order crossovers.

• When bi-wiring loudspeakers, be very careful if using two different types of cable. Every cable has capacitance and inductance, which interacts with the capacitance and inductance of the speaker's crossover network. Using cables with different characteristics can change the crossover's performance.

• To break in loudspeakers more quickly, play them at a somewhat loud level and note the volume setting on the preamplifier. Face the speakers directly toward each other about an inch apart. Reverse the red and black leads on one speaker. Throw a blanket over the speakers. Play a test CD that has wideband pink noise at a level of –10dB and put that track on repeat. This technique exercises the loudspeakers with minimal noise. By facing the speakers toward each other and putting them out of phase, their outputs largely cancel, reducing the sound level. The blanket damps any remaining sound. Notice that we set the volume level before wiring the speakers out of phase; when out of phase, it's impossible to judge how loudly they are playing and it's easy to overdrive and damage the speakers.
Remove speaker grilles. Although some manufacturers claim that their grilles don't degrade fidelity, virtually all of them do by creating diffraction.

AC Power

• Don't use light dimmers or fluorescent lighting in the listening room. Dimmers put lots of noise in the AC line, which gets into your components through their AC cords. If the listening room has dimmers, turn the dimmer all the way off. Dimmers work by chopping up the 60Hz AC sinewave, which puts noise on the AC line.

• When evaluating AC power cords, replace every cord in your system with the cord under evaluation, not just on one component. The degradation imposed by all the other cords can easily swamp the benefits of a single superior cord.

• Replace the stock AC wall sockets with audio-grade models. Use contact enhancer on the blades of AC plugs.

• Have an electrician perform a tune-up of your home's AC power. 1) Tighten the screws holding the breakers to the panel and to the buss bars. 2) Replace breakers that supply your audio system; they have carbon deposits that can add noise to the line. While you're at it, have the electrician replace 15-Amp breakers with 20-Amp units. 3) Clean and reseat the home's main ground wire to the grounding rod.

General

• With the system turned off, periodically turn your preamplifier's volume control up and down several times. This action cleans the volume control's wiper.

• Warm up your system before a listening session. An hour is a good rule of thumb, although different components have different warm-up times. Large solid-state amplifiers seem to take as much as two hours to sound their best, and digital gear needs at least 12 hours (preferably 24) to reach optimum performance.

• Turn off all fluorescent lights when listening. This includes the newer compact fluorescent types, all of which employ a small switching circuit that radiates noise.

Equipment Racks and Isolation Devices

Equipment Racks

The most important accessory for any audio system is the equipment rack. A good rack presents your components in an attractive way, makes your system functional and easy to use, and, most importantly, helps you get the best sound possible from your system. If your system includes a turntable, a high-quality rack is absolutely essential to realizing the LP front end's musical potential.

Apart from providing convenient housing for your equipment, the rack's main job is isolating the equipment from vibration. There is no question that vibrations degrade the sonic performances of preamplifiers, DACs, CD transports, and particularly turntables. This vibration is generated by transformers in power supplies, motors in turntables and CD transports, and from acoustic energy impinging on the electronics.

Vibration affects electronic products through a phenomenon called microphony. A microphonic component emits a small electrical signal when vibrated; in other words, it acts as a microphone, converting mechanical energy into electrical energy. This electrical energy pollutes the audio signal, resulting in degraded musical performance.
Tubed products are particularly susceptible to vibration. In fact, you can shout into some tubed preamplifiers and hear your voice come out of the loudspeakers. The preamplifier converts the acoustic energy of your voice into an electrical signal. When the tubed component is vibrated by the sound from the loudspeakers, or by transformer or motor vibration, noise is added to the music.

The sound qualities of digital products are also degraded by vibration, but through a different mechanism. The crystal oscillators in many transports and DACs change their frequencies slightly when vibrated. Although this frequency shift is slight, even small timing variations can wreak havoc in digital audio systems. Some manufacturers of digital products go to extreme lengths to ensure accurate and stable crystal frequencies, and build some form of vibration isolation around the oscillators.

The LP playback system is the most sensitive to vibration; the vibrational energy is transmitted to the tonearm and cartridge, adding sonic garbage to the tiny signal recovered from the record groove. This vibration comes from the turntable motor, from resonances in the tonearm, and from acoustic energy impinging on the turntable, arm, and cartridge. Clearly, isolating your system from vibration is essential to realizing the best musical performance.

Fig. 15-12 A high-quality equipment rack reduces vibration in your components. (Courtesy Harmonic Resolution Systems)

A good equipment rack fights vibration with rigidity, mass, and careful design. The massive, inert structure of a high-quality rack is much less likely to vibrate when in the presence of sound pressure generated by the loudspeakers. Moreover, the equipment rack can absorb, or damp, the vibration created by power transformers and motors. Many equipment racks have built-in vibration-damping mechanisms in their shelves.

One technique for vibration control is the pneumatic rack that floats the rack shelves on a cushion of air. An air bladder in the rack is filled from a bicycle-tire pump, which isolates the components in the rack from vibration. Pneumatic platforms are devices that hold a single component; the component sits on the platform, and the platform rests on a shelf in the equipment rack. One pneumatic platform that works particularly well for audio was actually designed to isolate a scanning electron microscope from vibration. If you want to experiment with pneumatic isolation without spending much money, try putting a bicycle inner tube underneath a component and pumping it half full of air. It's surprising how much of an improvement pneumatic isolation can render in a high-quality music system.

When choosing an equipment rack, look for a thick steel frame (14-gauge steel is excellent), welded construction, the ability to be filled with sand, vibration-damping shelves (Medium Density Fiberboard, or MDF, for example), bracing posts around the shelves, and floor spikes. Spikes between the rack and a floor act as a mechanical diode, draining vibrational energy from the rack and providing the best coupling between the rack and floor. For a spiked rack to be effective, the floor must be sturdy and flat so that the rack doesn't rock. The heights of most spikes are adjustable; you can level the rack and get good contact between the floor and all four (or three) spikes. Rack spikes are usually much heavier duty, with more rounded points than loudspeaker spikes. Note, however, that both types of spike will damage wood floors. Some racks are supplied with small plates to hold the spikes, preventing floor damage.

Not all good racks will have each of these features, but most will have at least some of them. The ability to fill the rack's posts with sand is especially important. Sand is an excellent vibration-absorbing material, and it adds mass to the rack. It's also cheap.

Since the previous edition of this book a number of manufacturers have entered the market with extremely high-tech racks designed using aerospace engineering. These racks are elaborately designed using the same computer modeling programs used in vibrational analysis in other fields. These products, while expensive, have taken vibration isolation to an unprecedented performance level (Fig.15-12).

The preceding discussion of high-mass racks, floor spikes, and intimate coupling between components, the rack, and the floor is only one school of thought. The second theory holds that audio components should be *decoupled* from the support surfaces and the floor with lightweight racks and compliant feet. Vibration in the rack and components is thus damped rather than transmitted to the floor. Generally, mass coupling works better for heavy components, and decoupling for lightweight components. The best way to determine which method is best for your system is to try both before you buy, or ask your dealer's advice.

Before installing a heavy rack—particularly one to be filled with sand—make sure the floor structure can support the weight. Suspended wood floors may need additional bracing. Bracing also makes the floor more solid and less prone to vibrating. A concrete slab is the ideal foundation for an equipment rack; you don't have to worry about the weight, and it provides a solid surface for the rack. You should set up the rack exactly where you want it before filling it with sand.

These features—sand-fill, spikes, vibration-absorbing shelves, thick steel construction, and bracing—are found together in the most elaborate and expensive racks. Many racks without all of these features still provide good performance. Avoid racks with large, unsupported shelves, flimsy construction, low mass, and poor vibrational damping. And don't even consider the generic "stereo stands" sold in furniture and department stores. Get a rack specially designed for high-performance audio systems. Consider a good equipment rack an essential part of your hi-fi system; it will help your system achieve its full musical potential.

Vibration Control Accessories: Spikes, Feet, and Cones

Vibration-isolation accessories have the same goals as the specialized audio equipment rack: reduced vibration in the audio component. Isolation devices can be used on their own or in conjunction with an equipment rack.

The field of vibration-control accessories for audio systems was established with a product called Tiptoes. Invented by Steve McCormack of The Mod Squad, Tiptoes are machined aluminum cones with sharp points. When used point-side down to support pieces of audio equipment, they act as mechanical diodes to remove mechanical energy from the components. When first introduced, Tiptoes were greeted with scorn—until their effects were heard. Tiptoes spawned an entire industry of cones, spikes, and similar devices of all manner of construction, made from a wide range of sometimes exotic materials (Fig15-13).

A different approach to vibration control is supporting the component with compliant feet, or mounting the component on a compliant platform. These devices are made of a rubber-like material (usually Sorbothane) chosen for its excellent damping characteristics. Compliant feet isolate the audio product from external vibration by absorbing the energy before it can get into the product. These devices are also excellent for stacking one component atop another. The feet provide airspace for ventilation, and the additional space reduces potential magnetic interaction between components.

Fig. 15-13 Spikes, cones, and compliant feet. (Courtesy AudioPoints, Stillpoints, Vibrapod, and Music Direct)

The most effective place for Tiptoes or other vibration-control devices is underneath tubed power amplifiers, particularly if the amplifiers are placed near loudspeakers. The acoustic energy from the loudspeakers impinges on the tubes, which convert that energy into an electrical signal that degrades sound quality. These devices also provide a small airspace between the floor and amplifier, increasing ventilation around the amplifier.

One designer of extremely high-quality turntables—who is also a physicist and mechanical engineer—considers vibration control so important that he has had a local tombstone-maker cut huge slabs of granite on which to support his equipment. (He also told me that if we could hear our audio systems with zero vibration, the sound quality improvement would be shockingly large.) You probably won't go to that extreme, but you should certainly try some of the available vibration-control devices.

AC Power Conditioners and AC Cords

Over the past ten years, the audio world has learned much about the role AC power plays in an audio system's sound quality. We had AC power conditioners before that time, of course, but many of those early AC conditioners (and indeed, many of today's models) take a "hit and miss" approach to AC conditioner design. The designers of these products attempted to address the problems of AC power without a full understanding of the phenomena involved, and consequently created products that often did as much harm as good (particularly by compressing music's dynamics). But some of today's AC conditioners and AC power cords are much more sophisticated and effective than those of a decade ago, and are based on a firmer understanding of how AC power interacts with the audio signal and the resulting effect on the system's sound. AC power conditioner design has, in some cases, been transformed from a "black art" to a serious engineering exercise based on real scientific research. These conditioners are vastly improved over those of the previous generation, and are capable of rendering significant sonic improvements.

An AC power conditioner plugs into the wall outlet and provides multiple AC outlets for plugging in your audio equipment. But before we talk about what a power conditioner does, let's look at the AC power line and its relationship to an audio system.

The AC power from the wall outlet is a 60Hz, 120V sinewave that powers the audio system (see Appendix B for more detail on the power line). All your components are connected via the power line. In fact, your audio components are connected to every other electrical device in your house, and to every home and factory also using the power grid.

Power equipment on the AC line generates noise that travels back into the line where it enters your audio components. This noise is called electromagnetic interference, or EMI. Light dimmers, refrigerators, and other household appliances put high-frequency junk in the AC line. Vacuum cleaners and electrical power tools are a major source of power-line noise because the fibers in the motors' brushes are continually making and breaking contact. The line is also polluted by AM radio stations; power lines act as antennae, superimposing the AM signal over the 60Hz line.

The designers of the first-generation of power conditioners assumed that this noise on the AC line was the most significant problem in AC power supplying audio systems. Consequently, those conditioners attempted to filter incoming noise from the line, and did so effectively. But the more significant source of noise on the AC line comes not from the outside world but from your audio system itself.

How do your audio components create noise on the AC line? It's natural to assume that AC is continuously pulled from your wall when your system is running. In this view, AC power is a low-frequency event (50Hz or 60Hz). Indeed, this is the case with fans, motors, and other constant-draw devices. But an audio system makes very different demands on the AC power system. The capacitors in an audio component's power supply only charge (drawing current from the wall) in very short bursts at the tops and bottoms of the 60Hz AC sinewave. This means that 120 times per second, the AC power system delivers a current transient that refills the audio component's capacitors. Specifically, the diodes in the bridge rectifiers (rectifiers turn AC into DC) are tuning on and off, radiating noise every time they do so. These short bursts of current flow put a large amount of noise back onto the AC line—noise that is coupled to all the other audio components in your system.

Today's best power conditioners reduce this noise-coupling between components, while simultaneously providing the least possible resistance to the flow of current from the AC wall outlet to the component's AC input jack. Power conditioners are often designed in conjunction with AC cords as a complete engineered system to minimize noise and maximize power transfer. The design of the conditioner and cord greatly affect the amount of radiated noise, the nature of that noise, and the amount of instantaneous current delivery possible. Fig.15-14 is a pictorial diagram showing how an AC line conditioner prevents noise in one component (in this case, the CDP or compact disc player) from getting into other components. Fig.15-15 is the back panel of a high-quality AC conditioner.

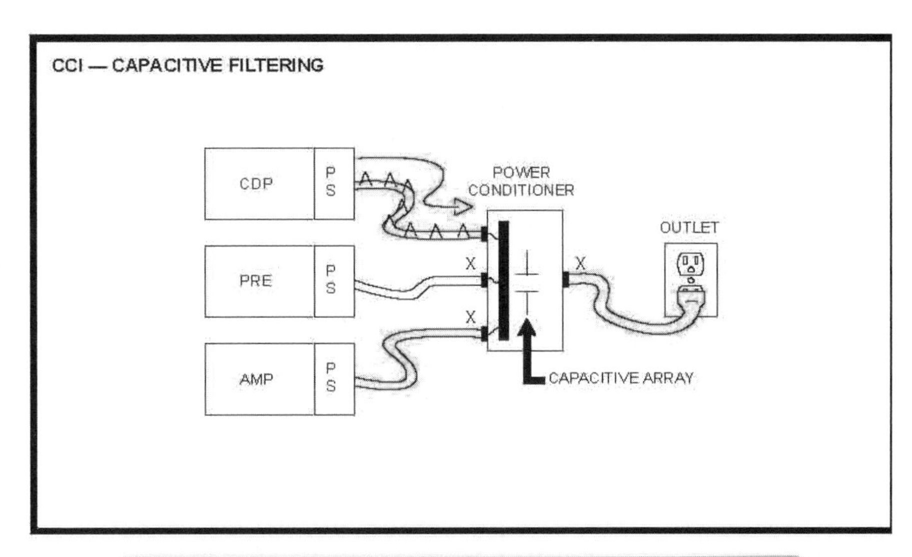

Fig. 15-14 Pictorial diagram showing how an AC conditioner isolates each component from noise. (Courtesy Shunyata Research)

Fig. 15-15 A high-quality AC power conditioner. (Courtesy Shunyata Research)

Noise isolation is only half of the AC conditioner's job. The AC conditioner and the AC power cord should also provide a low-impedance current path. Consequently, some AC conditioners are built with large solid-copper buss bars, custom AC sockets that provide more contact area and greater contact pressure with the plug, and other techniques to provide maximum current delivery. These parts are often cryogenically treated by subjecting them to very low temperatures and then slowly bringing them up to room temperature. Cryogenic treatment alters the metal's molecular structure and, for some reason, results in better sound quality. (I've heard comparisons of cryogenically treated and non-treated interconnects, and have been surprised by the difference.) Earlier-generation power conditioners, and many of today's models, often compressed musical dynamics, probably because they restricted current flow. High-quality AC sockets designed for audio can also be installed in your wall, replacing stock sockets. This is a crucial interface that benefits from the audio-grade sockets' greater ability to transfer current from the wall wiring to the AC cord. The AC sockets in your home, and the generic black power cords that are supplied with audio components, have become commoditized; there's no incentive for their manufacturers to improve quality. That's why you should use AC sockets, AC conditioners, and AC power cords designed specifically for high-performance audio.

It's a common mistake to evaluate the quality of an AC cord by putting it into your system on one component. The new AC cord could be spectacularly great, but you might not know it if the rest of the AC cords in your system are compromised. That's why it's essential to replace all the AC cords at the same time before reaching any judgments about AC cord quality.

If you can afford it, there's no substitute for having separate, dedicated AC lines installed to supply your audio system. Ideally, your front-end components would be run off one line, and each monoblock power amplifier would be supplied by its own dedicated line. Running dedicated lines to your listening room can have a dramatic sonic effect.

Because the AC line voltage varies according to the time of day and the load on the line, one may expect a line conditioner to regulate the voltage and provide a constant 120VAC to your system. Regulation, however, doesn't improve the sound of an audio system, and can actually degrade it if the input voltage moves around the threshold at which a separate transformer tap kicks in. Moreover, most high-end audio equipment is designed to work within the tolerances of the AC line supplied by the electric company. This is why power conditioners for computers that incorporate line regulation shouldn't be used for audio systems.

When choosing a line conditioner, make sure its power capability exceeds the power consumption of the components you'll be plugging into it. Choose a conditioner with a sufficient number of outlets for your present and anticipated needs. As with all accessories, try the power conditioner in your system before you buy. Expect to pay a minimum of $250 for a conditioner with just a few outlets, and several thousand dollars for a state-of-the-art system.

A radical approach to AC line conditioning is to create an entirely new 60Hz sinewave by amplifying the output from a small 60Hz oscillator. The power conditioner is essentially a power amplifier. The device converts the incoming 60Hz, 120V to DC; a low-level oscillator generates a clean 60Hz sinewave, which is amplified to 120V. As you can imagine, the output power is limited, making such a conditioner best suited to front-end components rather than power amplifiers. These products are large, heavy, run hot, and—unlike a power amplifier whose output depends on the audio signal's amplitude—consume just as much power when your audio system is running as when it is idle. Although this technique creates perfectly clean power, it nonetheless does not address noise put onto the local (within your audio system) AC line by your components themselves.

A power-line conditioner can't make poor audio components sound good; instead, it merely provides the optimum AC environment for those components so that they may realize their full potentials. The sonic benefits of a good line conditioner include a "blacker" background, with less low-level grunge and noise. The music seems to emerge from a perfectly quiet and black space, rather than a grayish background. The treble often becomes sweeter, less grainy, and more extended. Soundstaging often improves, with greater transparency, tighter image focus, and newfound soundstage depth. Midrange textures become more liquid, and the presentation has an ease and musicality not heard without the conditioner.

AC Power Cords

At first glance, it may seem that the several short pieces of wire carrying AC to the components in your audio system—the AC power cords—couldn't affect the sound of the system. After all, the AC cord is just another piece of wire at the end of a miles-long transmission system; how could it make any sonic difference?

From the perspective of your audio components, the AC cord isn't the last few feet of a transmission system, but the first few feet. The AC cord is directly connected to the primary windings in the transformer of your audio component and thus is an integral part of the AC delivery system. The noise problems described earlier can be reduced with careful AC cord design. Magnetic interaction and coupling between the conductors is reduced with dedicated audio power cords, resulting in better transfer of AC to the components. This magnetic interaction, addressed in interconnects and loudspeaker cables, affects AC transmission to a greater degree because of the high current flow in an AC cord.

Even an entry-level audio-grade power cord ($95) can make a significant difference in the sound of a system. Replacing all the stock black AC cords with entry-level cords takes you much of the way toward the state of the art in AC power. More expensive cords are generally better-sounding in some manufacturers' lines, but even the basic cords can render a significant improvement over stock cords.

A Final Note

The accessories and techniques described in this chapter are designed to improve the sound of your system so that you can more easily forget about the equipment and enjoy the music. When listening for pleasure (as opposed to critical listening to evaluate an accessory or technique), forget everything you've read in this chapter. The last thing that should be on your mind when enjoying music is whether the accessory you just installed is an improvement, what accessory you are going to try next, and whether the system would sound better with more tweaking. It would be a supreme irony if the very products and techniques created to take you a step closer to the music ended up distracting you from the real reason we own high-performance audio systems.

16 Specifications and Measurements

An audio component's technical specifications and measured performance can tell us much about the component. Understanding what the specifications mean, and interpreting laboratory-test results is useful—though not essential—to the audiophile. Specifications and measurements have their place in audio, but they are not the beginning and ending of the story. Rather, they are just one facet of a multidimensional puzzle.

What specifications and measurements cannot tell us is how the product communicates musical expression. To discover that, we must listen. Although understanding this chapter is not essential to enjoying high-quality music reproduction in your home, I've included an explanation of specifications and measurements for each component category for those interested readers, and to make this book complete.

Preamplifier Specifications and Measurements

An important preamplifier specification is a line-stage's *gain,* or the amount of amplification the preamp provides. Gain is expressed either in decibels (dB), or as a number representing the ratio between the input and output voltages.

The preamp's gain is important to consider in relation to the output voltage of source components. If you have a very-high-output CD player or DAC feeding a high-gain preamplifier, you'll barely crack open the volume control before you get loud listening levels. This has several disadvantages. First, it makes setting a "just right" volume difficult: The volume control is rendered so insensitive by such high outputs that even small adjustments make large differences in listening level. Second, volume controls are often less well matched between channels at the low end of their range. This means that one channel becomes louder than the other when the volume control is turned down. The effect is the same as if you'd turned a balance control off-center. This difference can be as much as 3dB—enough to shift the image to one side. A DAC putting out, say, 5V RMS at full scale and driving a line-stage preamp with 25dB of gain would produce this problem.

High gain can be an advantage when using a line-stage preamplifier with an external phono stage. Many phono stages have just enough gain of their own to feed a line-stage, but no more. The line-stage therefore requires a fair amount of gain.

Line-stage gain varies between 5dB and 25dB. The 5dB figure is very low—perhaps too low for some outboard phono stages, and even for tuners and tape decks. This will be especially true if the power amplifier has low gain or the system has insensitive loudspeakers, and could require using the volume control near the maximum setting. For most CD sources, about 10dB of gain is ideal.

As described in Chapter 4, phono-stage gain should be matched to your phono cartridge. Moving-magnet cartridges need about 35dB of gain, while moving-coil types require up to 65dB. If the phono preamp has less than adequate gain, a line-stage preamp with high gain can further amplify the phono signal for adequate listening levels.

A preamplifier's gain can be expressed indirectly as *input sensitivity*. This is the voltage required to produce some reference level (usually 0.5V) at the preamplifier output with the volume turned all the way up. A preamplifier with an input sensitivity of 0.05V has a gain of 10 (which can also be expressed as 20dB, because 20dB represents a 10-to-1 voltage ratio).

A preamplifier's *input impedance* specification describes the electrical resistance to current flow "seen" by the source component (such as a CD player) driving the preamp. Most preamps have input impedances between 10,000 and 50,000 ohms (10–50k ohms). The preamp's input impedance is the *load* the source component drives. As described earlier, the higher the input impedance, the easier it is for the source to drive. As the input impedance increases, however, current flow though the interconnects and preamplifier input stage decreases. Some designers believe moderate rather than very high input impedance results in the best sonic performance. If the preamplifier's input impedance is too low, it can *load down* the source component, resulting in poor technical and sonic performance.

A preamplifier is also specified according to its *output impedance*. Output impedance is best thought of as a resistor inside the preamplifier that couples its output stage to the rear-panel jacks. Output impedance describes the preamplifier's ability to deliver current into a load (the power amplifier). A preamplifier's output impedance is much lower than its input impedance; where input impedance may be 50k ohms or more, output impedance may be less than a thousandth of that value.

By having a high input impedance and a low output impedance, the preamplifier acts as a buffer between the source components and the power amplifier. The sources drive a high input impedance—which is very easy for them—and the preamp takes on the burden of driving the interconnects and power amplifier. The preamplifier's low output impedance makes it ideal for this job. (The advantages of low output impedance are discussed in more detail in the section on passive level controls earlier in this chapter.)

A preamplifier's *input overload* specification refers to the maximum input voltage the preamp can accept without distorting (with distortion defined as 1% Total Harmonic Distortion, or THD). The higher the input overload figure, the better. A full-function preamplifier will have two input overload specifications: one for the line-level inputs (typically more than 10V RMS), and one for the phono inputs (typically several hundred millivolts). *Maximum output level* is the maximum output voltage the preamplifier can swing from its output stage. Maximum output level is often as much as 50V—far higher than any audio signal. Because most power amplifiers have an *input sensitivity* (the input voltage required to produce maximum power output) of under 1.5V, virtually any preamp has enough output voltage to drive any power amplifier.

Signal-to-noise ratio, or *S/N ratio*, is a measurement of how quiet the preamplifier is. It expresses in decibels the ratio between a signal level of 0.5V output and the residual noise floor. The higher the signal-to-noise ratio, the quieter the preamp. Because the ear isn't equally sensitive to noise at all frequencies, a *weighting curve* is sometimes applied to the noise spectrum to more closely approximate the noise's audibility. (Weighting curves are described in more detail in Appendix A.) A signal-to-noise ratio of 90dB unweighted (no weighting curve applied) is good; higher signal-to-noise ratios are common when "A"-weighted (some of the noise is removed from the measurement to approximate the audibility of the noise). A preamp with lots of 60Hz hum in the audio signal (caused by AC from the wall getting into the audio signal path) will exhibit a very large increase in S/N ratio (as much as 25dB) when "A"-weighted. If the preamplifier's S/N ratio is referenced to 1V output instead of 0.5V, subtract 6dB from the S/N to find the equivalent 0.5V figure.

Preamps rarely have frequency-response errors; virtually all modern preamps have flat response from below 20Hz to above 50kHz. Phono stages, however, often have errors in their RIAA response, producing the equivalents of frequency-response errors. These errors are caused primarily by capacitor tolerances in the RIAA circuit. An example of a preamp with RIAA errors is shown in Fig.16-1a. The solid line is the left channel, the dotted line is the right channel. A phono preamplifier with no RIAA errors would produce a perfectly flat line. The rolloff in the upper midrange and treble seen in Fig.16-1a will make this phono stage sound dark, closed-in, and lacking immediacy. A phono preamp with more accurate RIAA equalization is shown in Fig.16-1b.

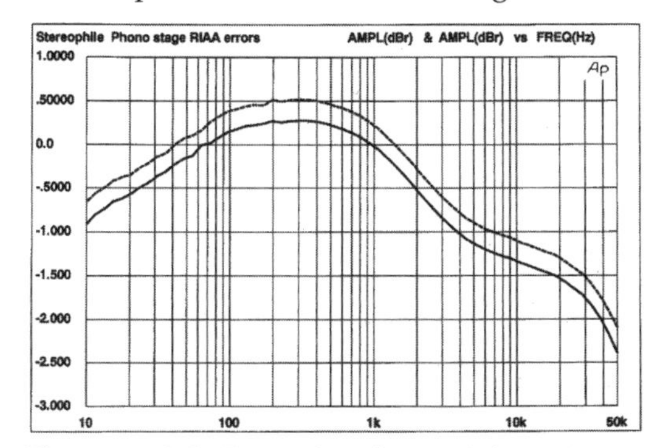

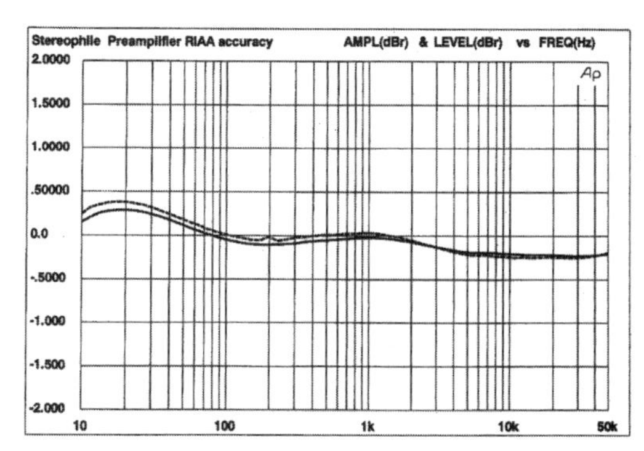

Fig. 16-1a & b Examples of poor (left) and typical (right) RIAA accuracy (Courtesy *Stereophile*)

A preamplifier's *Total Harmonic Distortion* (THD) is often measured and expressed as THD+N (Total Harmonic Distortion Plus Noise). This name reflects the technique that measures the sum of harmonic distortion and noise. Here's how THD+N is measured: The preamp under test is driven by a swept sinewave signal, a band-reject filter removes the test signal, and whatever is left (distortion and noise) is plotted as a function of frequency. Note that very-low-distortion preamps can have a higher THD+N figure than their intrinsic harmonic distortion; the noise, rather than distortion, dominates the measurement.

The THD+N vs. frequency plot of a typical preamplifier is shown in Fig.16–2. The vertical scale is THD+N in percent. The upper pair of traces is measured through the phono inputs, the lower pair through the line inputs. The apparently higher distortion through the phono inputs is undoubtedly caused by its higher noise level, not increased distortion.

Interchannel crosstalk, also called *channel separation,* is a measure of how much the signal in one channel leaks into the other channel. The higher the channel separation number, the better, with 100dB being excellent and 50dB being on the poor side. A crosstalk specification should include the measurement frequency; crosstalk usually increases as frequency rises. Crosstalk will generally increase at the rate of 6dB per octave, due to capacitive coupling between channels.

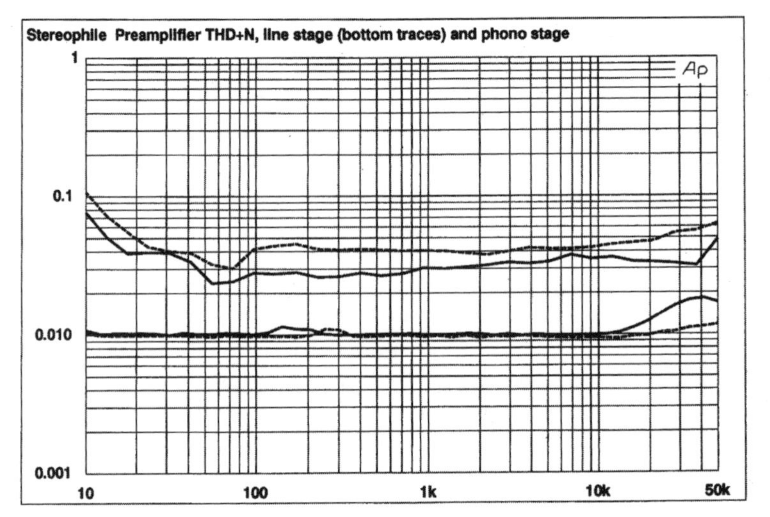

Fig. 16-2 A typical preamplifier THD+N measurement (Courtesy *Stereophile)*

When expressing this phenomenon as *crosstalk*, the *lower* the specified crosstalk, the better; this figure indicates how much signal has leaked into the other channel. Conversely, when expressing *channel separation*, the *higher* the figure, the better; the number indicates the degree of isolation between channels. Both terms, however, describe the same phenomenon. Fig.16-3a shows mediocre crosstalk performance; Fig.16-3b shows excellent crosstalk performance. The measurement is made by driving one channel of the preamplifier with a swept sinewave at a reference output level (the 0dB reference on the graph's vertical scale), then plotting the undriven channel's output signal as a function of frequency. The measurement is repeated with the channels switched, and both traces are combined on the same graph. In the example of poor crosstalk performance, the crosstalk in the undriven left channel was 60dB below that of the driven right channel at 1kHz. In Fig.16-3b, the signal in the undriven channel was down by 100dB at 1kHz.

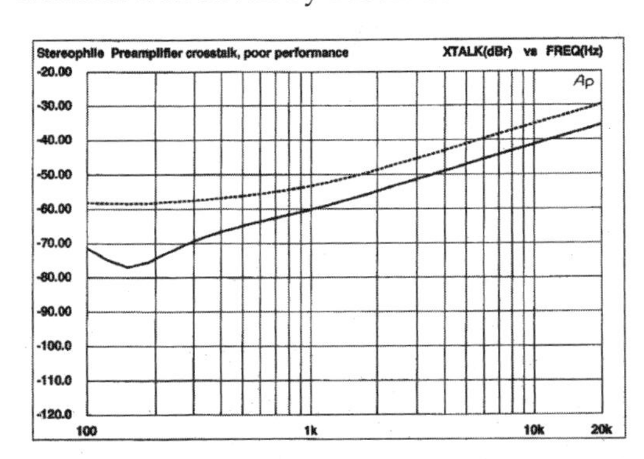

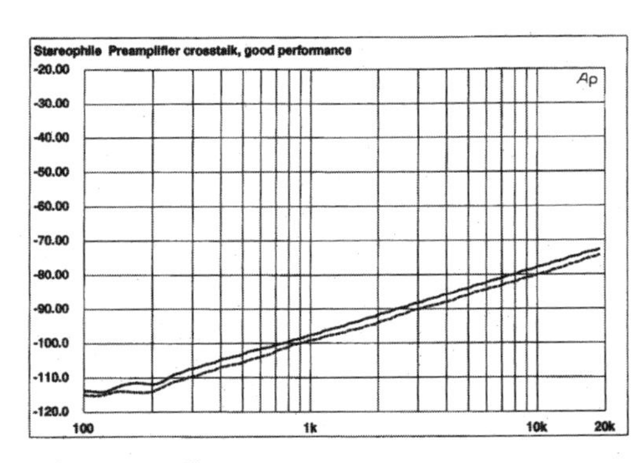

Fig. 16-3a & b Examples of poor (left) and excellent (right) crosstalk (Courtesy *Stereophile*)

Channel balance is a measure of how closely in level the two channels are matched. Some preamplifiers have slightly more gain in one channel than the other, creating a channel imbalance. Volume controls can also introduce channel imbalance, particularly at the low end (counterclockwise) of their range. Channel balance is often measured at three volume-control positions, and should be within 0.2dB at all volume settings. Large channel imbalances can cause the center image to shift to one side. Tubed preamps sharing a single dual tube for left and right channels are particularly susceptible to channel imbalances; the two elements within the tube may have slightly different gains. Fortunately, the problem can be corrected simply by changing the tube to one with matched gain.

Preamplifier specifications should be used to ensure electrical compatibility with other components, not as indicators of sound quality. Many superb-sounding preamps measure poorly, and some preamps that look great on paper disappoint in the listening room. Let your ears—not the specification sheet—be your guide in choosing a preamplifier.

Power Amplifier Specifications and Measurements

Power amplifiers share many specifications (signal-to-noise ratio, crosstalk, etc.) with other components, yet have some unique specifications and measurements.
Power amplifiers typically have input impedances of between 20 and 150k ohms, with 47k ohms being a standard for unbalanced inputs. This is a high enough value to ensure that the power amplifier's input stage won't load down the preamplifier's output and put a strain on the preamplifier's output stage.

A power amplifier's output impedance is usually a very low value—between 0.05 and 0.5 ohms for solid-state amplifiers. Tubed power amplifiers have a higher output impedance—about 0.5 ohms to as high as 2 ohms. Single-ended tubed amplifiers can have output impedances as high as 3 ohms. Think of output impedance as a resistor in series with the output transistors (or output transformer in a tubed amplifier) connected to the load. Low output impedance is a desirable design goal; when the power amplifier's output impedance becomes a significant fraction of the loudspeaker's input impedance, the loudspeaker and amplifier interact. The result can be peaks and dips in the amplifier's frequency response caused by the loudspeaker's impedance change with frequency. This is a serious consideration when contemplating a single-ended tubed amplifier; the amplifier will have a different tonal balance when driving different loudspeakers. A high output impedance is also characterized by soft bass, lack of bass extension and dynamics, and less bass control. A power amplifier with a high output impedance is less able to control woofer motion. Output impedance is related to an amplifier's damping factor—its ability to control cone motion in the loudspeaker. The higher the damping factor, the better. Note that damping factor is specified at the amplifier's output terminals; resistance in loudspeaker cables can dramatically decrease the effective damping factor, particularly in long cable runs.

Gain refers to the ratio between the input and output voltages. Gain can be specified as the ratio between the voltages, or as a figure in decibels (dB). Typical gain is about 26dB, or a ratio of 20:1 (i.e., an input voltage of 1V will produce 20V at the output). Input sensitivity is the input voltage required to drive the amplifier to clipping. A typical value is between 0.775V and 2V, with about 1.2V being standard. If the power amplifier has a very high input sensitivity (less than 0.5V), the preamplifier's volume control will be extremely insensitive—you'll get plenty of volume with the knob barely turned up. Conversely, a low-output preamp driving a low-sensitivity power amplifier (2V, say) will require the preamp's volume control to be turned up very high to get adequate volume.

We've talked a lot about maximum output power; now let's see how it's measured. With a dummy load (a huge power resistor) connected to the amplifier's output, the power amplifier is driven with a gradually increasing voltage. Distortion in the output is monitored and plotted as a function of output power. When the amplifier clips, the distortion shoots up instantaneously. The clipping point, defined as 1% THD, is the amplifier's maximum output power. This test is performed with the amplifier driving loads of 8, 4, and 2 ohms. The results are combined and plotted on a single graph (Fig.16-4a). The farther to the right the "knee" (bend) in the distortion trace, the higher the output power at clipping. Note that some plots of THD vs. output power have only two traces, at 8 and 4 ohms. These amplifiers cannot be tested into 2 ohms at full output power: the power-supply fuses blow, or the amplifier simply shuts down.

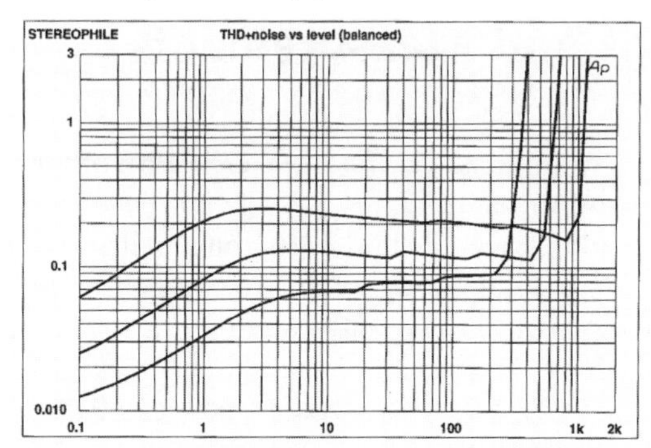

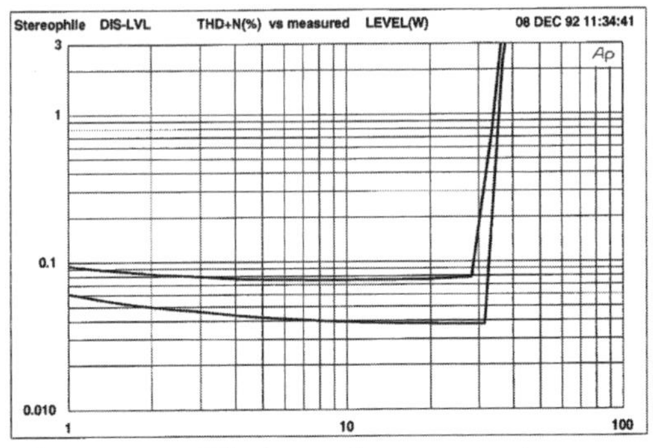

Figs. 16-4 a and b The power output vs. THD curves show an amplifier's maximum output power, and how well it can increase its power into low-impedance loads. (Courtesy *Stereophile*)

By combining the three traces on one graph, we can quickly assess the amplifier's ability to increase its output power into low impedances. The greater the distance between the curves' knees, the greater the amplifier's ability to drive current into low-impedance loads. Fig.16-4b shows the measurement of an amplifier with very little ability to deliver current into a 4 ohm load: It produced 15.6dBW at 8 ohms, but only 12.5dBW at 4 ohms.

By contrast, Fig.16-4a shows an amplifier behaving as a virtually perfect voltage source, maintaining its voltage and doubling its current output each time the load impedance is halved. This graph, the result of an actual measurement, reveals exemplary—and extraordinary—performance. The performance would have been even better had the AC line voltage not drooped under the heavy load, decreasing the raw power available to the amplifier—the amplifier pulled so much current out of the wall that the AC line voltage dropped from 118V to 106V. This amplifier put out 325W into 8 ohms (25.1dBW), 635W into 4 ohms (25dBW), 1066W into 2 ohms (24.3dBW), and 1548W into 1 ohm (22.9dBW). Moreover, it was perfectly stable driving this tremendous amount of current into such a low impedance.

These methods are used for most tubed and solid-state push-pull amplifiers. Single-ended tubed amplifiers produce so much distortion that a different criterion must be used to determine the clipping point. Rather than using 1% THD as the point at which we say the amplifier has run out of power, 10% THD is often called the clipping point of a single-ended amplifier. That's because a single-ended tubed amplifier may produce 2 or 3% THD when operated at just a few watts of output power.

Another way to look at a power amplifier's output capability is the power-bandwidth test, which determines the widest bandwidth over which the amplifier can maintain full output power. The amplifier is driven to just below clipping, and the output power is plotted against frequency.

How much peak power an amplifier can deliver on transient signals (musical peaks) above its continuous power rating is called its dynamic headroom, expressed in dB. A power amplifier rated at 100Wpc (RMS) that can deliver 200Wpc for brief periods (measured in milliseconds) without severe distortion is said to have 3dB of dynamic headroom. You may remember from earlier in the chapter that doubling the power represents an increase in volume of 3dB. An amplifier's dynamic-headroom specification says a lot about its subjective ability to play loudly.

In addition to plotting THD against output level to find the clipping point, a power amplifier's distortion can be measured at low signal levels and plotted as a function of frequency. Fig.16-5 is such a graph. The lower trace is THD+N (THD plus Noise) measured at 1W into 8 ohms. The next higher trace is at 2W into 4 ohms, and the upper trace (highest distortion) is with 4W driving a 2 ohm load. The lower the distortion and noise, the better, with less than 0.2% constituting good performance.

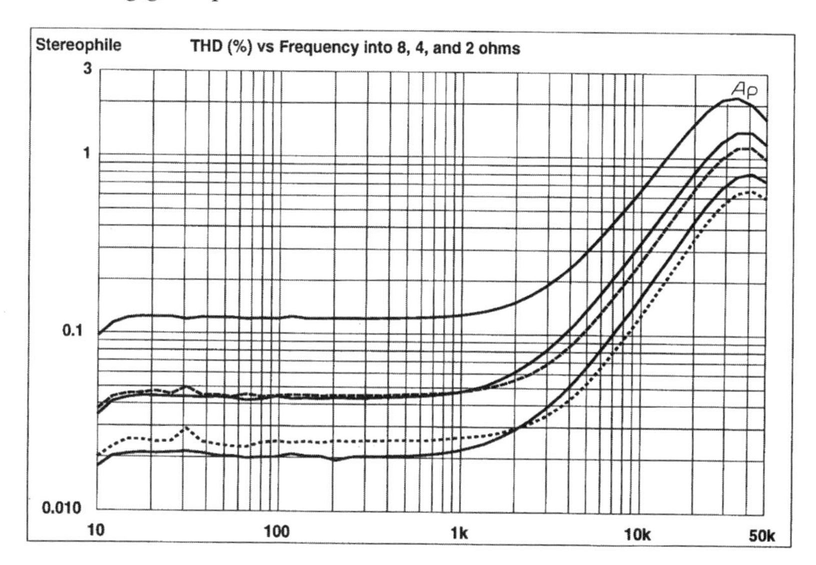

Fig. 16-5 THD vs. frequency measurements show how the amplifier's distortion changes into low-impedance loads (Courtesy *Stereophile)*

Although knowing how much total distortion a power amplifier generates is important, an examination of the harmonic-distortion spectrum is more instructive. We want to know which harmonics are present; some harmonic products are much more audible and annoying than others. As stated earlier in the discussion of tubed power amplifiers, 10% second-harmonic distortion is less musically objectionable than 0.5% seventh-harmonic distortion. We thus need to characterize the distortion spectrum rather than use a single THD figure.

To do this, the power amplifier is driven at two-thirds rated power with a 50Hz sinewave, and a Fast Fourier Transform (FFT) is performed on the output signal. The FFT captures the amplifier's output signal and transforms it into its constituent parts. (Specifically, the FFT can convert a time-domain signal into the frequency domain.) An FFT produces a plot showing energy vs. frequency, shown in Fig.16-6a. We can see that this particular amplifier generates an entire series of harmonic-distortion products. The distortion spectra of Fig.16-6b, however, show a much lower level of distortion products.

Although harmonic distortion has been touted as the definitive performance parameter of power amplifiers and receivers, this specification in itself isn't that significant a

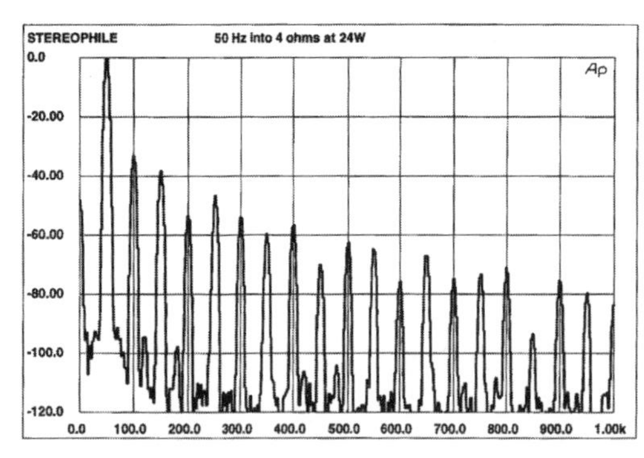

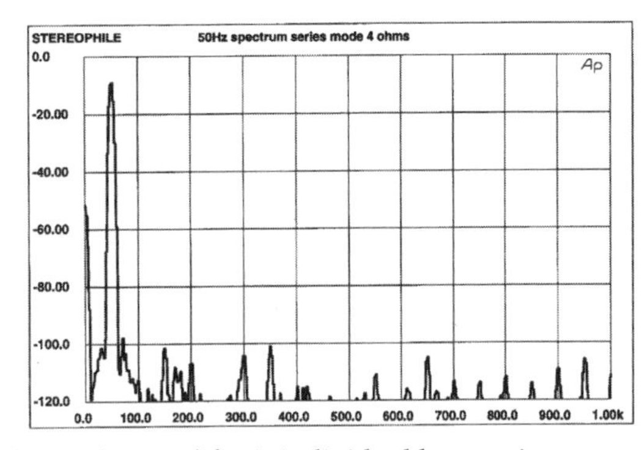

Figs. 16-6a and b An FFT of an amplifier's output shows the amplifier's individual harmonic-distortion components. (Courtesy *Stereophile*)

criterion in the evaluation of power amplifiers. Many amplifiers with 3% distortion sound just fine, while others with 0.01% sound awful. As described in Chapter 1, THD can be reduced by increasing the amount of negative feedback in the amplifier. Unfortunately, negative feedback introduces other problems that make the amplifier sound worse. Although it's interesting to know how much harmonic distortion an amplifier produces and the spectral distribution of those distortion products, I wouldn't let THD figures influence my purchasing decision.

Intermodulation Distortion (IMD) is distortion that occurs when two signals of different frequencies are amplified simultaneously—something that happens all the time in music. IMD creates sum-and-difference signals related to the frequencies being amplified. For example, if we drive a power amplifier with 10kHz and 1kHz sinewaves, the amplifier produces an output signal containing the original test tones at 10kHz and 1kHz, but also intermodulation products at 11kHz (the sum of the two signals) and at 9kHz (the difference between the two signals). Though the 9kHz and 11kHz sum-and-difference IMD products are usually the highest in amplitude, IMD creates an infinite series of distortion products. Intermodulation distortion is expressed as a percentage of the IMD products in relation to the test-signal amplitude, with typical values falling between 0.01% and 0.2%. IMD test signals are often 60Hz and 7kHz combined at equal amplitude, or with the higher-frequency tone at one-quarter the amplitude (–12.04dB), or 19kHz and 20kHz at equal amplitude. To test with the latter signals, the amplifier is driven at two-thirds rated power and an FFT is performed on the amplifier's output. We look for the 1kHz difference signal (20kHz minus 19kHz); the lower its amplitude, the better. This 1kHz intermodulation product is usually between 55dB down (55dB below the test signal amplitude) and 75dB down. We also sometimes see a family of sideband signals around the test tones.

Some of these measurements are unique to power amplifiers. Other, more universal measurements include those of frequency response, crosstalk, and signal-to-noise ratio. Except for single-ended amplifiers, however, it's rare for a modern amplifier to depart from flat frequency response. Note that the sonic characteristics of "bright," "extended," "forward," etc. are not functions of the amplifier's frequency response.

Finally, a power amplifier is said to be inverting if it reverses absolute polarity between its input and output. A non-inverting amplifier maintains the same polarity between input and output. In an inverting amplifier, the positive- and negative-going parts of the wave are flipped over: positive at the input becomes negative at the output. This can slightly improve some recordings and degrade others. Because most gain stages invert absolute polarity, an inverting amplifier usually has an odd number of gain stages, while a non-inverting amplifier has an even number; one isn't necessarily better than the other. (See "Absolute Polarity" in Appendix A.)

Loudspeaker Specifications and Measurements

There is more correlation between a loudspeaker's measured performance and its sound quality than there is with any other component in an audio system. A loudspeaker that measures as having a treble rolloff will likely sound dull; a loudspeaker with a measured rising top end will likely sound bright and forward. Having said that, however, I must stress that you can't judge a loudspeaker's performance by reading its specifications or measurements. Although good measured performance is valued, the loudspeaker's musical abilities in the listening room are what count. To find that out, we must listen.

Following is an explanation of each of the measurements found in some loudspeaker reviews or manufacturers' specification sheets.

Impedance is the electrical resistance the loudspeaker presents to the power amp. As described in detail in Appendix B, a loudspeaker's impedance is a combination of its simple DC resistance, inductive reactance, and capacitive reactance. The higher the impedance, the easier it is for the amplifier to drive. A low-impedance loudspeaker requires the amplifier to drive more current through the loudspeaker, which puts more stress on the amplifier.

Even though a loudspeaker's impedance may be specified as "6 ohms," the actual impedance almost always varies with frequency. Fig.16-7a shows an impedance curve (solid line) for a loudspeaker of rather low impedance. The impedance dips to less than 4 ohms (the "40.0m" horizontal division) through the bass and most of the midrange. The impedance minima is just over 1 ohm at 30Hz—an extremely low value.

The dotted trace in the impedance plot shows the loudspeaker's phase angle, which indicates how inductive or capacitive the loudspeaker looks to the amplifier. If the loudspeaker presented a pure resistance (no capacitance or inductance), the phase-angle plot would be a straight line. The farther away from the center line, the more reactive the loudspeaker is at that frequency. (Phase angle is discussed in detail in Appendix B.)

A combination of high reactance and low impedance is especially difficult for a power amp to drive. The severe phase-angle swings seen in Fig.16-7a, coupled with its low impedance, will be more demanding of the power amplifier. This doesn't mean that the loudspeaker won't work, only that it will work better with a power amp that can deliver a substantial amount of current into low-impedance loads. For example, a power amp that nearly doubles its power into a 4 ohm load compared to an 8 ohm load would be a better choice than one that barely increases its power into 4 ohms.

A much easier load to drive is illustrated in Fig.16-7b. The impedance is much higher over a wider band, and the phase-angle swings are less severe. The impedance never dips below 6 ohms, and is well above that figure over wide bands. This loudspeaker will demand much less current from the power amplifier.

A loudspeaker's impedance magnitude and phase angle must be considered in relation to its sensitivity. A loudspeaker with a high sensitivity is much less demanding of a

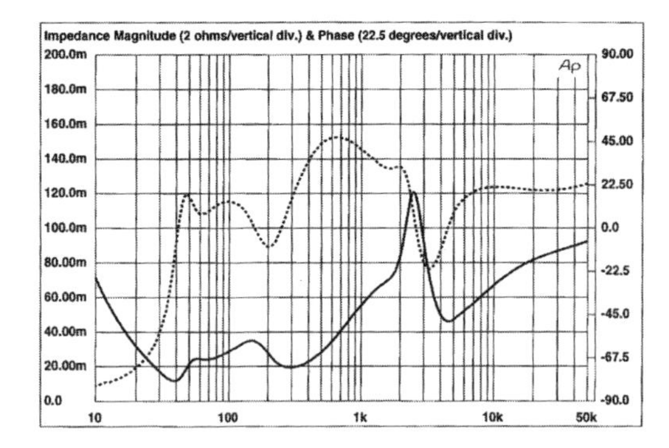

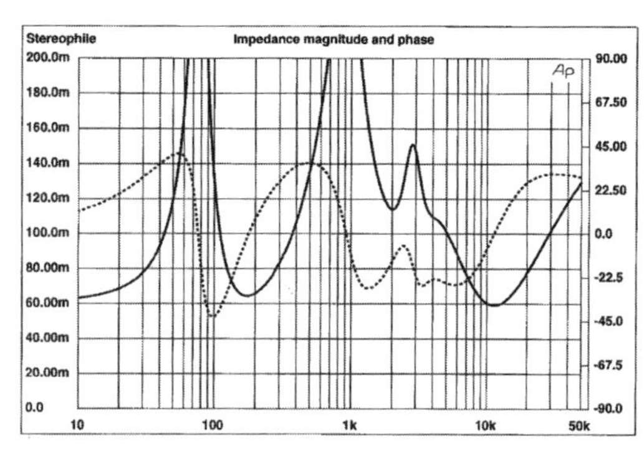

Figs. 16-7a and b A loudspeaker's impedance magnitude (solid line) and phase angle (dotted line) indicate how easy or difficult it is for a power amplifier to drive. (Courtesy *Stereophile)*

power amp than one with a low sensitivity, regardless of the loudspeaker's impedance. The combination of low sensitivity, low impedance, and severe phase angle requires an amplifier of the highest quality to achieve musically acceptable results.

Sensitivity can be expressed as a sound-pressure level with 1 watt input measured 1 meter away. A loudspeaker's sensitivity may be expressed as 88dB/1W/1m. High sensitivity is generally considered any value above 90dB/1W/1m, while low sensitivity is a value below 85dB/1W/1m.

Note that sensitivity can also be expressed with a drive signal of 2.83V, which corresponds to 1W into an 8 ohm load. Loudspeaker manufacturers can cheat on sensitivity figures by using the 2.83V figure into a 4 ohm loudspeaker, which increases the loudspeaker's sensitivity rating by 3dB. The loudspeaker isn't 3dB more sensitive, it's just drawing twice as much current from the power amp at the same 2.83V. If you see the 2.83V figure, make sure the impedance is 8 ohms. Note that the term efficiency is often used incorrectly in place of sensitivity. Technically, efficiency is the percentage of electrical power converted by the loudspeaker into acoustical power.

Measuring a loudspeaker's impedance and phase angle can be done electrically on a test bench. Measuring a loudspeaker's acoustical characteristics, however, requires putting a microphone in front of the loudspeaker and driving the loudspeaker with test signals. Consequently, the measured response will be not the loudspeaker's intrinsic characteristics, but the loudspeaker's response as modified by the room in which the measurements are taken. As we know, a loudspeaker's performance is greatly dependent on the room in which it is used. We could end up measuring more of the room's characteristics than the loudspeaker's.

One way of solving this problem is to measure the loudspeaker in an anechoic chamber—a room whose surfaces are covered with highly absorbent material. This material absorbs virtually all the sound energy striking it, reflecting no energy back into the room. In fact, the term anechoic literally means "no echo." Consequently, the measurement microphone "sees" only the loudspeaker's response.

Anechoic chambers are expensive to build and consume lots of real estate. Engineers have therefore developed methods of measuring loudspeakers in a normal room to exclude the room's influence from the measured results. One of these techniques, invented by Richard Heyser, is called time-delay spectrometry (TDS). In TDS measurements, the loudspeaker under test is driven by a swept sinewave. The signal picked up by the measurement microphone is filtered by a narrow filter that allows only a narrow range of frequencies to pass. The frequency the filter allows to pass tracks the swept sinewave driving the loudspeaker, but is delayed in time by a few milliseconds (ms). The filter's delay is exactly the amount of time it takes for sound from the loudspeaker to reach the measurement microphone. The sound reaching the measurement microphone is the loudspeaker's direct acoustic output, followed a few milliseconds later by room reflections; or, the same sound after it has been reflected/bounced off of the room's walls, floor, and ceiling.

This is where the swept bandpass filter comes in. The filter tracks the sinewave stimulus so that when a reflection reaches the microphone, the filter has already moved past the reflection's frequency and rejects it. The result is the loudspeaker's intrinsic response, unmodified by the room.

A second, more popular technique is implemented in a personal-computer–based hardware and software package called the Maximum Length Sequence System Analyzer (MLSSA). The loudspeaker under test is driven by a stimulus containing a specific pattern of noise sequences. The measurement microphone a few feet in front of the loudspeaker picks up this signal, which is compared to the noise-sequence pattern stored in memory. From the way the loudspeaker alters this sequence, the system computes the loudspeaker's impulse response. The period between the impulse and the first reflection is free from reflections and is indistinguishable from an anechoic measurement. The impulse response is then "windowed" to remove the information from the first reflection onward, leaving only the anechoic window.

The impulse response, which is a time-domain signal, is transformed into the frequency domain using Fourier analysis. Fourier analysis is a mathematical technique of converting an impulse response (a time-domain signal) into a frequency response (a frequency-domain signal), or vice versa. The term Fast Fourier Transform (FFT) describes this conversion from one domain to another.

From the loudspeaker's impulse response we can derive the system's frequency response. The result is the loudspeaker's response minus the room's contribution. Note, however, that the anechoic window's length determines the low-frequency resolution of the measurement. The longer the anechoic portion of the impulse response that is transformed into the frequency domain, the lower the frequency that can be accurately measured. When we measure loudspeakers with this technique, the loudspeaker is placed on a tall stand so that it is halfway between the floor and ceiling, maximizing the anechoic window. The room should therefore also be as large as possible.

Despite these techniques, the low-frequency resolution of MLSSA measurements isn't satisfactory unless you have a huge room (a long anechoic window). To better measure a loudspeaker's low-frequency performance, a measurement microphone is placed so that it nearly touches the woofer cone. At this position, the room's contribution is negligible. This separately measured nearfield woofer response is appended to the MLSSA-derived response to provide an accurate wideband measurement.

With that background, let's go through a set of measurements that you might find in a typical loudspeaker review and examine what they mean.

A loudspeaker's impulse response reveals its time behavior, including how time-coherent the loudspeaker system is, how much the drivers exhibit ringing, and whether the drivers are connected in the same polarity. I described time coherence in the section on crossovers; a time-coherent loudspeaker has no phase lag between drivers. That is, they all move in step in response to the input signal. In a loudspeaker that isn't time-coherent, the output from the individual drivers is slightly staggered in time. Ringing is driver motion after the input signal has stopped.

Fig.16-8a is the impulse response of a loudspeaker that has poor time coherence. Notice the negative-going spike (the tweeter's output), which indicates that the tweeter's output occurs before the other drivers' outputs, and that it is connected in the opposite polarity. The impulse response in Fig.16-8b is much more coherent; all the drivers respond to the impulse in an identical manner. Ideally, the impulse response should have a fast rise-time (indicated by the steepness of the impulse) and minimal driver ringing after the impulse has stopped. (Ringing can be seen as squiggles after the impulse.)

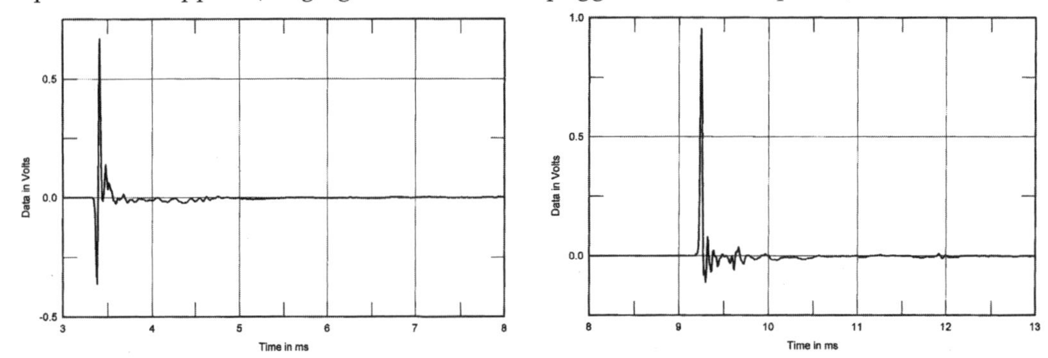

Figs. 16-8a and b A poor impulse response (left) and an excellent impulse response (right). (Courtesy *Stereophile)*

Another way of looking at a loudspeaker's time behavior is the step response, or how the loudspeaker behaves when presented with direct current (DC). Derived from the impulse response, the step response plots the loudspeaker's output vs. time. Fig.16-9a is the step response of the same loudspeaker whose impulse response we saw in Fig.16-8a. The lag between the tweeter (the negative-going spike) and the other drivers is apparent, as well as a slow risetime (the time it takes for the loudspeaker to reach full amplitude in response to the input signal). For contrast, Fig.16-9b is a virtually perfect step response. The drivers are time-coherent, and the risetime is extremely fast (the step is a straight vertical line). This step response was derived from the impulse response in Fig.16-8b.

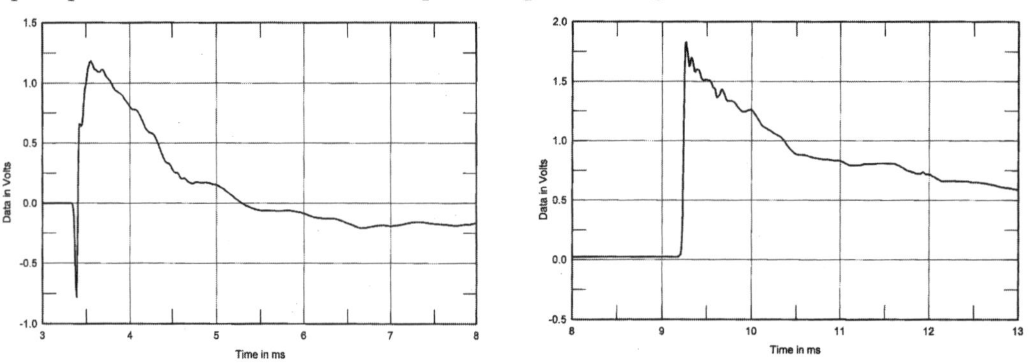

Figs. 16-9a and b A poor step response (left) and an excellent step response (right). (Courtesy *Stereophile)*

Fig.16-10 is a loudspeaker's overall frequency response as measured over a 30° horizontal window. This simply means that the frequency response was measured at various angles from the loudspeaker's front baffle and averaged to produce the curve. Averaging the response gives a better representation of the listener's subjective response to the loudspeaker; some of that side energy will reach the listener after being reflected from the listening room's sidewalls. This particular loudspeaker has some excess treble energy between 5kHz and 15kHz, which may make it sound tizzy in the treble. Notice how the response drops rapidly above 15kHz. The broad dip between 500Hz and 5kHz may give the loudspeaker a distant, rather than present, perspective. The rising bass indicates that the low-frequency alignment may be underdamped, giving the bass a generous but fat quality. In short, these deviations from a flat line indicate colorations in the loudspeaker.

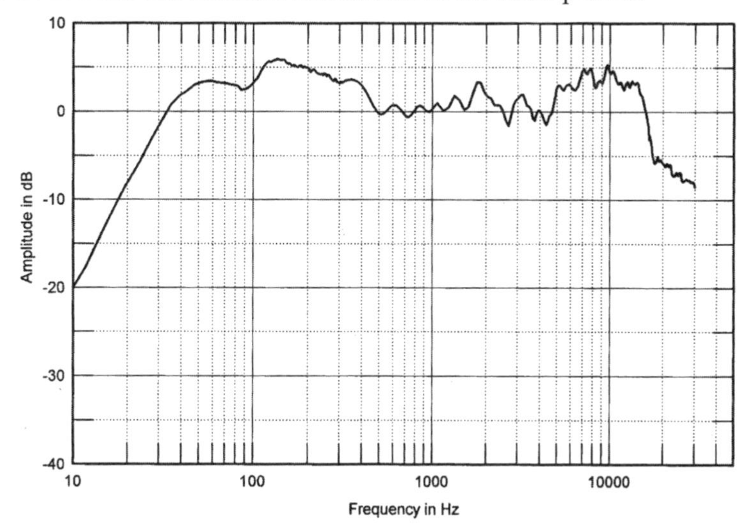

Fig. 16-10 Frequency response averaged over a 30° horizontal window. (Courtesy *Stereophile)*

Fig.16-11 is a loudspeaker's "horizontal family" of response curves. These show how a loudspeaker's response changes as the measurement microphone is moved around the loudspeaker. The frequency-response measurement just described is an average of these individual curves.

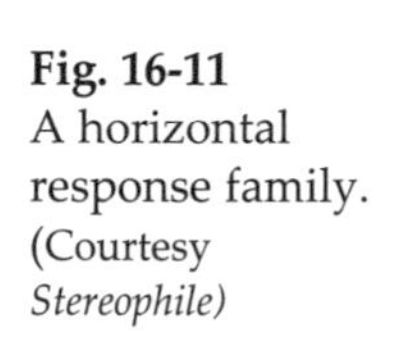

Fig. 16-11 A horizontal response family. (Courtesy *Stereophile)*

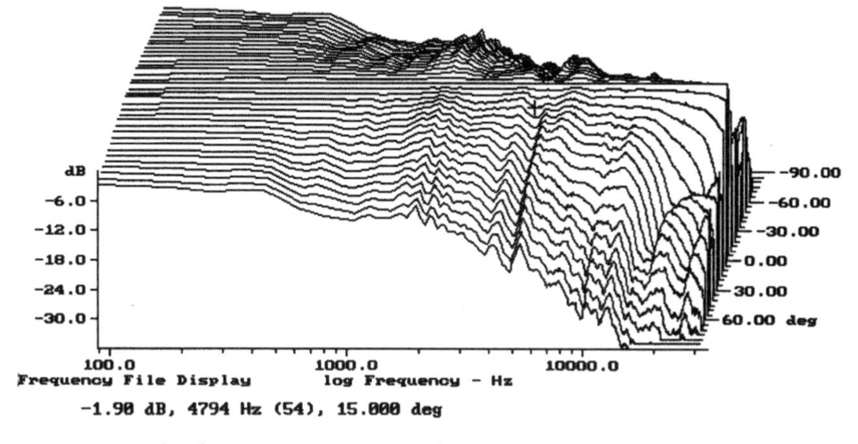

The trace representing the frequency response directly in front of the loudspeaker (on-axis) has been normalized to be a straight line so that the other curves show only the difference from the on-axis response. The horizontal family shows the loudspeaker's dispersion, or how it radiates acoustical energy into the room as a function of frequency. This particular loudspeaker shows a typical treble rolloff as the microphone is moved to the sides of the cabinet, but also excess treble energy on-axis. These curves suggest that this loudspeaker should be listened to without toe-in; the excess treble on-axis will be reduced if the listener sits slightly off-axis (no toe-in). A loudspeaker that exhibits radical changes in off-axis response can sound colored, even though the on-axis response may be fairly flat.

The next measurement (Fig.16-12) shows how the loudspeaker's response changes with listening height. Again, the flat line has been normalized so that only the differences from vertical on-axis response are shown. The on-axis point is usually the tweeter axis, or a typical listening axis (the text accompanying the measurement will usually indicate which). As you can see, the frequency response changes radically with listening height. The loudspeaker has severe peaks and dips in the response (caused by interference effects) above and below the optimum listening axis. This means the loudspeaker will sound very different depending on listening height. This example is extreme; most loudspeakers don't exhibit such radical response changes along the vertical axis. All loudspeakers, however, evince some change in tonal balance at different listening axes. This measured response underscores the importance of listening height when auditioning a loudspeaker.

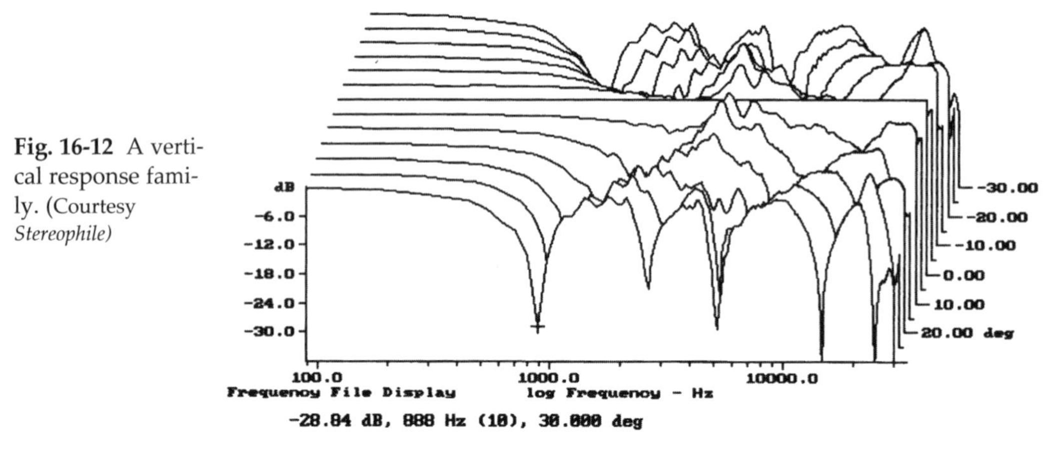

Fig. 16-12 A vertical response family. (Courtesy *Stereophile*)

The ideal loudspeaker would stop generating acoustical energy as soon as the input signal stopped. In practice, some acoustical energy, however small the amount, is always stored in the drivers and enclosure. This energy is radiated over time, like a bell ringing. To look at a loudspeaker's response over time, its frequency response is plotted in discrete slices of time. As described previously, the frequency response is derived from the impulse response using the Fast Fourier Transform. By moving the window in the impulse response on which the FFT is performed, the loudspeaker's frequency response can be derived at different points in time. Specifically, the FFT-derived frequency response is calculated perhaps 100 times over 4 milliseconds (ms) to create the plots of Figs.16-13a and b. This graph is properly called a cumulative spectral-decay plot, but is more commonly called a waterfall plot because of its similarity to cascades of falling water. It shows how the loudspeaker's output changes over time and with frequency. The topmost curve is the loudspeaker's frequency response at time zero. The next lower line is the frequency response 40 microseconds (40µs) later. The next lower line is 40µs after that, and so forth.

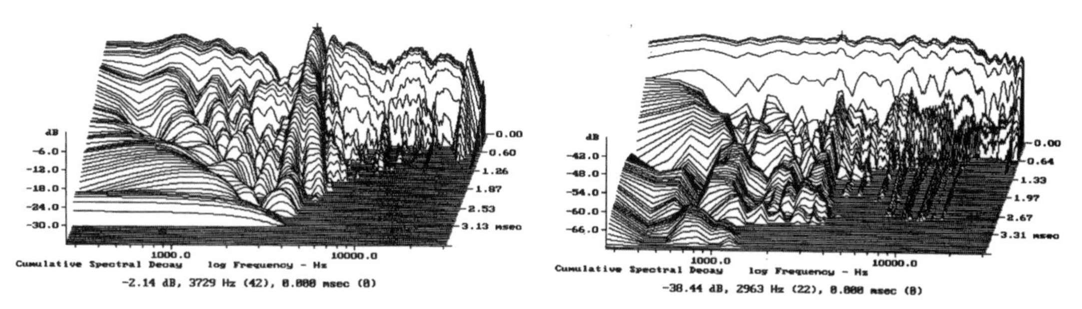

Figs. 16-13a and b Poor (left) and excellent (right) cumulative spectral-decays. (Courtesy *Stereophile)*

By plotting the response over time, we can easily see cabinet resonances, driver ringing, driver breakup, and other problems. A driver resonance is seen as a vertical ridge in the plot; the loudspeaker continues to radiate energy at the ridge's frequency long after the impulse has stopped. The perfect loudspeaker would produce a flat line at the top, and instantly decay to no energy output.

Examples of poor and excellent performance on the cumulative spectral-decay test are illustrated in Figs.16-13a and b. In Fig.16-13a we can see a ridge at 3729Hz, indicating that the loudspeaker continues to produce acoustic energy at this frequency for 2ms after the drive signal has ceased—a long time in loudspeaker terms. By contrast, Fig.16-13b shows exemplary behavior: the acoustic output drops very quickly (the white space between the top few traces and the lower traces), and almost no ridges are apparent. The one minor resonance at 2963Hz is minuscule compared to the resonances seen in Fig.16-13a, which will be audible as a change in timbre when this frequency is excited by the music.

The same technique can be used to examine the behavior of loudspeaker cabinets. An accelerometer is attached to the loudspeaker, which is driven by the MLS signal. The accelerometer output is drawn in a waterfall plot, which shows the frequency and amplitude of cabinet vibrations. For example, Fig.16-14 shows an enclosure with a severe resonance at 187Hz, seen as the narrow ridge at that frequency. This loudspeaker stores, then slowly releases, energy at 187Hz. This will undoubtedly color the bass when the loudspeaker is excited by music with energy at this frequency.

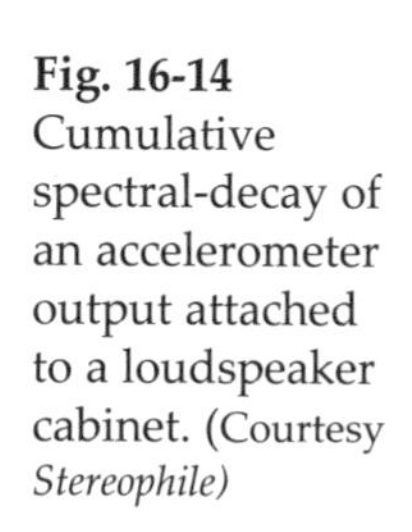

Fig. 16-14 Cumulative spectral-decay of an accelerometer output attached to a loudspeaker cabinet. (Courtesy *Stereophile)*

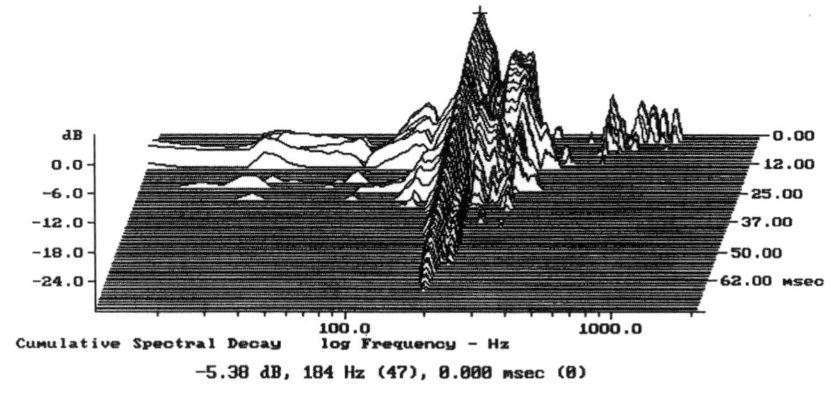

Those readers wanting to delve deeper into loudspeaker design are directed to Martin Colloms's definitive reference book on the subject, *High Performance Loudspeakers* (Halsted Press, ISBN 0-470-21721-9).

Digital Specifications and Measurements

Many tests and measurements exist to quantify a DAC's technical performance, but few—if any—measurements tell us how the DAC sounds. Nonetheless, good technical performance indicates that the DAC is well-engineered, and is more likely to have good sonic performance as well.

Let's start with a DAC's maximum output level. (I'll use the term "DAC," although I'm also referring to CD players—the measurements are the same.) The standard analog output voltage is 2VRMS, with units varying between 1.75V and a whopping 7.2V. Most CD players and DACs put out between 2.2V and 3.5V. Note that this value is the highest RMS output voltage possible from the player, measured by playing a full-scale, 1kHz sinewave.

Knowledge of a DAC's output level is useful in several ways. First, a DAC of very high output wouldn't be ideal for driving a preamplifier that had lots of gain. Although it is unlikely that the DAC would overload the preamplifier's input, you would end up using the preamp's volume control at the very low end of its range. Most volume controls exhibit their greatest channel imbalances (i.e., one channel is louder than the other) when turned down. In addition, setting a "just right" volume is more difficult with high-output DACs and high-gain line-stage preamplifiers; the volume control becomes overly touchy to small movements. This is particularly true of preamps with detent level steps instead of a continuously variable volume control. A high output level is an asset when driving a passive level control, but a drawback with a high-gain line-stage preamplifier.

Another important measurement to consider when determining if a particular DAC or CD player is a good candidate for a passive level control is output impedance. If a passive level control is used between a DAC and power amplifier, the burden of driving the cables and power amplifier falls on the DAC's output stage. If the DAC has a high output impedance, dynamics can become compressed and the bass may get mushy. In extreme cases, a DAC with a very high output impedance can even become current-limited and flatten the waveform peaks if it is asked to drive a low-input-impedance preamplifier. DACs with the highest output impedances have an impedance of nearly 2k ohms at 20kHz and 1.5k ohms at 1kHz. This high output impedance causes the DAC's output stage to clip when asked to drive impedances of less than about 15k ohms.

Further, the output impedance of the passive level control itself must be added to the equation. The passive control's output impedance (which varies depending on which resistor is switched in, or the position of the potentiometer) must be added to the driving component's source impedance to find the total source impedance driving the power amplifier.

DAC output impedances vary from less than 1 ohm to 5.6k ohms (the highest value I've measured in a product). Generally, if the output impedance is less than a few hundred ohms, the processor should have no trouble driving a passive level control. Between 50 and 500 ohms is a typical value.

Bottom line: Don't drive a low-input-impedance power amplifier (less than about 20k ohms) with a passive level control, with a high-output-impedance (greater than about 800 ohms) DAC through long, high-capacitance interconnects. These aren't hard-and-fast numbers but general guidelines. Appendix B includes a full analysis of what can happen with a passive level control and high-output-impedance sources.

The next measurement, frequency response, doesn't reveal a processor's overall tonal balance as it often does with loudspeakers, but is included in specifications and measurements for the sake of completeness. Most processors are down a few tenths of a dB at 20kHz, the result of rolloff in the digital filter.

More revealing, however, is a CD player's or DAC's de-emphasis error. As previously described, some CDs have had their trebles boosted (emphasized) during recording, and carry a flag that tells the CD player to switch in its de-emphasis circuit to restore flat response. These de-emphasis circuits are often inaccurate (due to resistor and capacitor tolerances), producing frequency-response irregularities when the player is decoding emphasized discs. De-emphasis error should be less than ±0.1dB across the band. De-emphasis errors can cause tonal-balance irregularities (tizzy upper treble, for example) when playing pre-emphasized CDs. De-emphasis errors of 0.2dB over an octave of bandwidth are audible. Poor and good de-emphasis accuracy is shown in Figs.16-15a and b. Both of these measurements were made on digital processors using analog de-emphasis circuits; processors with digital-domain de-emphasis have no errors. The positive error in Fig.16-15a will make the processor sound bright when playing pre-emphasized CDs.

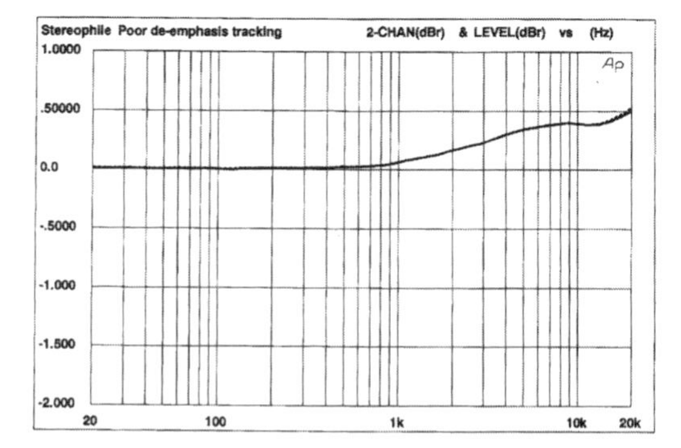

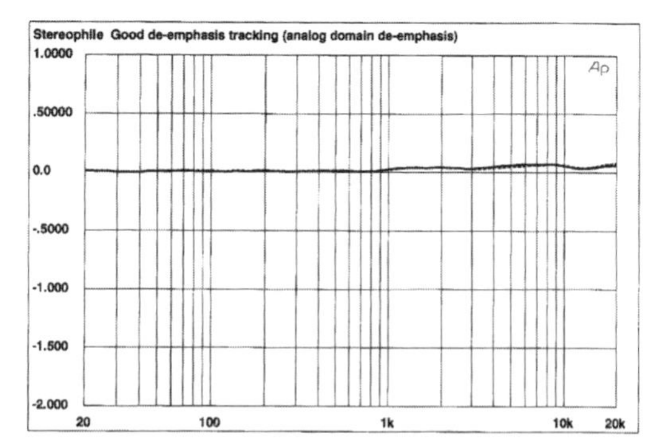

Figs. 16-15a and b De-emphasis errors (left) and accurate de-emphasis tracking (right) (Courtesy *Stereophile)*

Interchannel crosstalk, also called channel separation, is a measurement of how well the left and right channels are isolated from one another. Ideally, when a DAC is fed a right-channel signal, that signal should not appear in the left channel, and vice versa. The lower the crosstalk (or, put another way, the greater the channel separation), the better. Many DACs (and other products) have decreasing channel separation as frequency increases, a result of capacitive coupling between channels. Typical CD players and DACs have about 90dB of channel separation at 1kHz, this decreasing to about 70dB at 20kHz. Although there is no correlation between low crosstalk and soundstage width (an intuitive link), high channel separation indicates good overall engineering. Mediocre and good crosstalk performance are shown in Figs.16-16a and b. At 1kHz, the processor shown in Fig.16-16a has 67dB of channel separation, compared with the 112dB of channel separation at 1kHz measured in the processor shown in Fig.16-16b.

A very revealing test of a processor's ability to re-create low-level signals looks at the processor's output vs. frequency (spectral analysis) when the processor under test is driven with the digital code representing a –90dB, dithered 1kHz sinewave. A spectral analysis plots energy level vs. frequency. Peaks in the trace at 60Hz, 120Hz, 180Hz, and related frequencies indicate that power-supply noise is getting into the audio circuitry. A spectral analysis can also hint at how well the unit's DACs (digital-to-analog converters) are performing; the signal should peak at the –90dB horizontal division, and the left and right traces should overlap. This indicates that the DACs are linear and have similar performance

If the linearity trace dips below the center line, this indicates a negative linearity error. If it rises above the center line, the processor has a positive linearity error (the analog output is higher than it should be). The digital input level is read across the bottom, with the highest level at the graph's right-hand side. Some processors with excellent low-level linearity appear to have a positive error at very low signal levels (–115dB). This is actually noise swamping the DAC's output. The practical lower limit of measuring linearity is about –112dBFS (in quiet processors). Linearity is determined by the digital-to-analog converter chip inside the DAC.

Another test is the FFT-derived spectral analysis of the DAC's output when decoding a full-scale mix of 19kHz and 20kHz. This test reveals intermodulation components such as the 1kHz difference tone (20kHz minus 19kHz), and interactions between the sampling frequency (44.1kHz) and the test signals.

As with the spectral analysis we saw earlier of the –90dB, 1kHz sinewave, this spectral analysis also plots energy against frequency. But this technique does it a little differently. In an FFT-derived spectral analysis, the DAC's output is captured in the time domain, creating a waveform which is then transformed into the frequency domain using the Fast Fourier Transform (FFT).

The result of this technique is shown in Figs.16-20a and b, the FFT-derived spectral analysis of two processors' outputs when decoding this unusual test signal. Ideally, there should be no spikes in the noise floor; Fig.16-20b shows exemplary performance on this test.

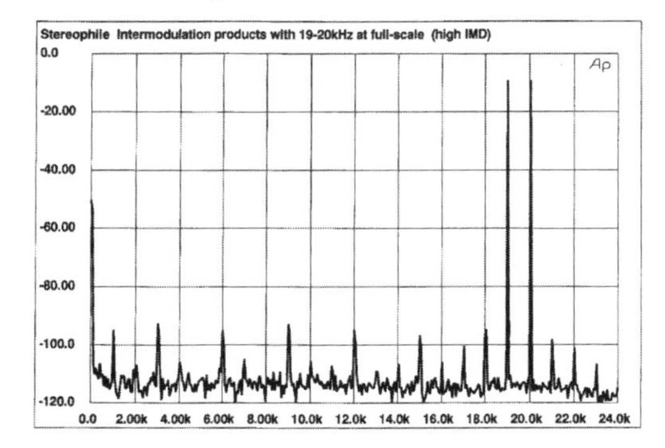

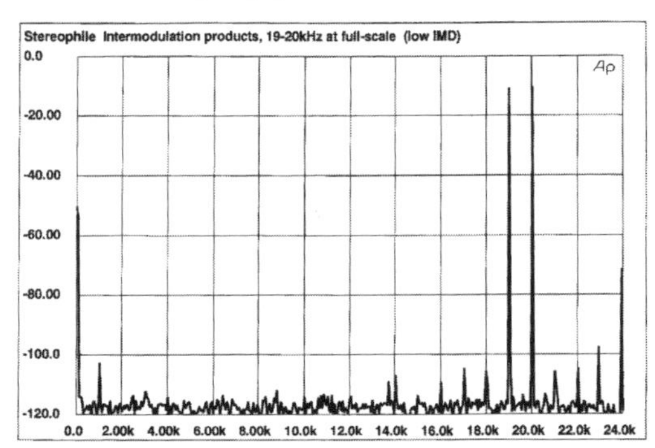

Figs. 16-20a and b Some processors produce high intermodulation distortion (left); others generate very little IMD (right). (Courtesy *Stereophile)*

Note the greater number of spikes in Fig.16-20a. The higher the spikes' amplitude and the greater their number, the worse the intermodulation distortion performance. The trace in Fig.16-20b reveals a 1kHz difference component, but very few other intermodulation products.

DACs run into problems when converting very-low-level signals to analog. One way to examine a DAC's behavior on small signals is to capture its analog output waveform when reproducing a –90dB, undithered 1kHz sinewave. This test signal produces three quantization steps: 0, +1, and –1. The three levels should be of equal amplitude, and the signal should be symmetrical around the center horizontal division. A poor-quality reproduction of this signal is shown in Fig.16-21a, a high-quality one in Fig.16-21b. Note the absence of noise in Fig.16-21b, and the three clearly defined quantization steps. By contrast, Fig.16-21a is so noisy it's difficult to make out the wave's shape.

A DAC's reproduction of a 1kHz, full-scale squarewave (not shown) reveals its time-domain behavior. Most measurements show the overshoot and ringing typical of the

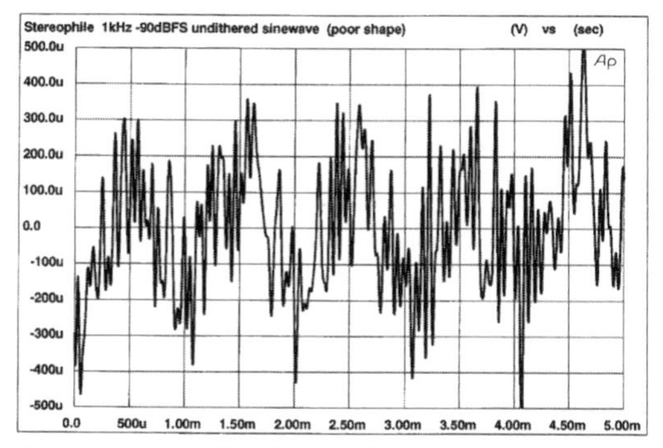

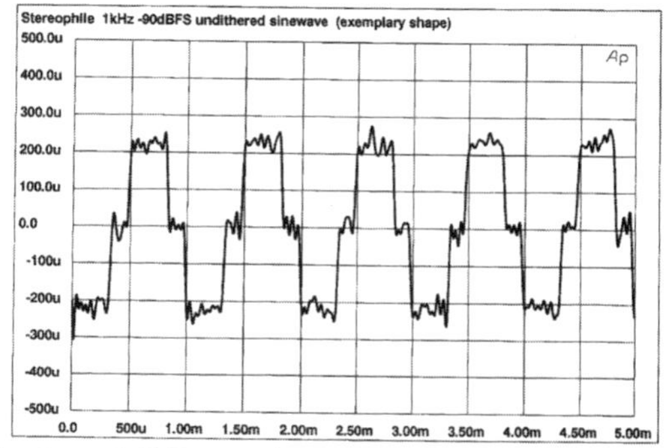

Figs. 16-21a and b Comparison of a noisy and poorly shaped 1kHz –90dB undithered sinewave (left), and nearly perfect reproduction (right). (Courtesy *Stereophile)*

digital filters used in most processors and CD players. The digital filter determines the waveform shape; the time distortion occurs in the digital filter. Some digital processors with custom DSP-based filters don't suffer from this type of distortion.

The amount of DC at the DAC's analog outputs is measured and noted. The lower the DC level, the better. DC may cause a thump through the system when the preamplifier's input selector is switched to select the DAC's output. If the preamplifier and power amplifier are both direct-coupled (no DC blocking capacitors in the signal path) and use no DC servo correction, the DC from the DC could be amplified and appear at the loudspeaker terminals—not a good condition. Low DC—less than, say, 20mV—is therefore desirable.

The measurements section of a digital product review will often mention that the DAC is polarity-inverting or non-inverting. A non-inverting DAC's analog output will be of the same absolute polarity as the digital data on the CD. Most processors don't invert polarity, but if one does, that's no reason for concern. Without knowing the absolute polarity of the recording and the rest of your system, it's a fifty-fifty chance that the absolute polarity will be wrong. Knowing that a DAC inverts absolute polarity is useful when comparing DACs;: all DACs under audition should maintain the same polarity for the listening comparisons to be valid.

Measuring a DAC's word-clock jitter (inaccuracy in the crucial timing signal) was first made possible in 1992 by an instrument called the LIM Detector, developed by Ed Meitner of Museatex. LIM stands for Logic Induced Modulation. Although other designers have looked at word-clock jitter, the LIM Detector was the first commercially available jitter-analysis tool. When used with an FFT machine and an RMS-reading voltmeter, the LIM detector provides the digital processor's overall RMS jitter level (over a specified bandwidth) and a spectral analysis of that jitter.

To use the LIM Detector, the word-clock pin on the DAC is found, and a probe is connected between the word clock and the LIM Detector. The LIM Detector output is only the jitter component of that word clock. A wide range of word-clock frequencies is accommodated. The LIM Detector output is fed to the input of an FFT analyzer, which displays the jitter's spectral content. The analyzer's RMS reading meter displays an overall RMS voltage, which is converted to jitter in picoseconds (ps). We thus end up with a number indicating the overall amount of jitter on the digital processor's word clock, and FFTs showing the jitter's spectral distribution.

The FFT of Fig.16-22a shows a jitter spectrum in which the jitter is periodic; that is, the jitter energy is concentrated at specific frequencies, seen as spikes in the plot. Where you see a spike, it means the digital processor's word clock is being jittered at that frequency. Note that the spikes occur at 1kHz and multiples of 1kHz; the test tone driving the digital processor is a 1kHz sinewave. As described earlier in this chapter, jitter is often correlated with the audio signal—something we can readily see in Fig.16-22a. The RMS jitter level, measured over a 400Hz–20kHz bandwidth, is 350ps.

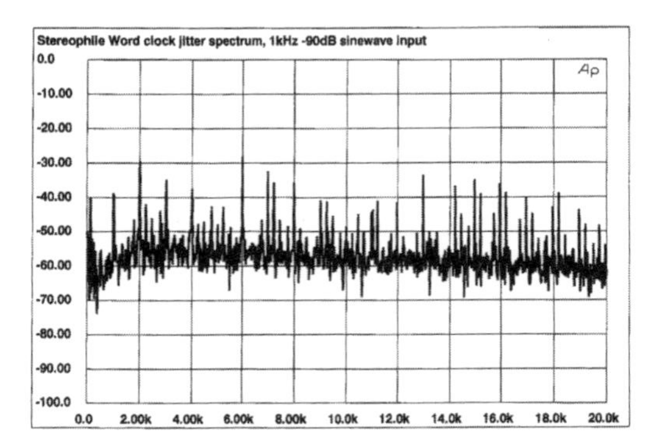

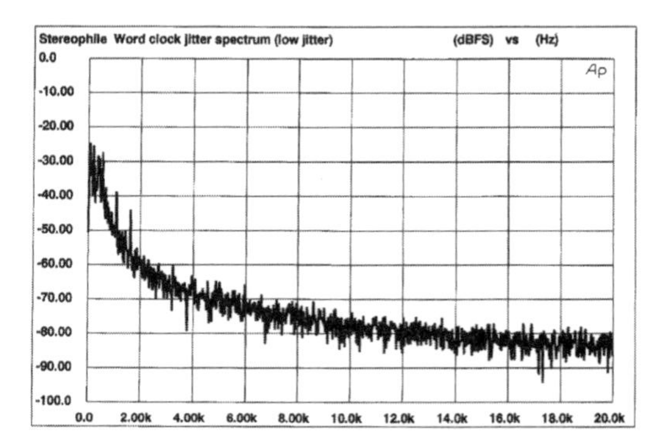

Figs. 16-22a and b High word-clock jitter in a digital processor (left), and low jitter with few periodic jitter components (right). (Courtesy *Stereophile)*

For contrast, Fig.16-22b is the jitter spectrum of a very-low-jitter digital DAC. Not only is the overall RMS level lower (less than 80ps), but the jitter spectrum is nearly free from signal-correlated jitter. In short, the lower the trace on the FFT, and the smoother the line (lack of spikes), the better the DAC's jitter performance.

This method of measuring jitter is direct in the sense that we're looking at clock jitter where it matters, but indirect because it doesn't tell us the effect of that jitter on the analog output (what really matters). More recently, measurement systems have been developed that measure the amount of clock jitter by analyzing the DAC's analog output. These analyzers better correlate with listening impressions. Moreover, other factors influence how much jitter is actually present at the conversion to analog; many DAC chips add jitter to the clock inside the chip where it is impossible to measure with the LIM technique.

There are few measurements available that even attempt to characterize a transport's performance. One of them, however, can be made by anyone, and gauges how well the transport's tracking and error-correction systems work. All you need is the Pierre Verany Digital Test CD (Pierre Verany PV.788031/788032). Disc 2 of this two-CD set contains a section in which the spiral track of pits is interrupted by blank areas (no pits). This simulates data dropouts caused by poor-quality CDs, dirt, scratches, or other media contamination. A continuous tone is recorded on the tracks. The blank areas gradually increase in length, and each increase is represented by a different track number. The higher the track number in this series, the longer the dropout. By listening for when the test tone is interrupted and noting the track number, the dropout length the player can track and correct can be determined. The higher the track number the transport will play flawlessly, the better. Most transports will play up to track 33. Skip-free reproduction of track 37 or higher is excellent.

I must reiterate that a DAC's bench performance doesn't predict its sound quality. There is no way to predict a component's musical character from looking at its test results.

Appendix A: Sound and Hearing

Introduction: What is Sound?

Sound is a series of physical disturbances in a medium such as air. When an object vibrates, it sends out a series of waves that propagate through the air. These waves are composed of fluctuations in air pressure above and below the normal atmospheric pressure of 14.7 pounds per square inch. When a loudspeaker cone moves forward, for example, it compresses the air in front of it, creating an area in which the air molecules are denser than normal atmospheric pressure. This portion of the sound wave is called a *compression*.

When the loudspeaker cone moves backward, it creates an area in front of it in which the air molecules are less dense than atmospheric pressure. This portion of the sound wave is called a *rarefaction*. Sound is made up of a series of alternating compressions and rarefactions moving through the air.

Sound waves are transmitted through the air by passing on the moving molecules' momentum to adjacent molecules. The original compressed molecules return to their original positions because of the elastic properties of air. The air molecules don't move very far; instead, their "bumping into each other" is what transmits sound energy.

When these compressions and rarefactions strike our eardrums, we perceive the phenomenon as sound. The greater the change in air pressure above and below normal atmospheric pressure, the greater the *amplitude* of the sound.

Because most objects vibrate with a periodic back-and-forth motion, or oscillation, most sound waves (and nearly all musical sounds) have a periodic repetition. The sound wave is a replica of the object's motion. Consequently, sound waves have a regular, periodic pattern of compressions and rarefactions.

This periodic pattern is illustrated in Fig.A-1. The loudspeaker at one end of the tube of air is driven by a periodic signal, such as that made by the cyclic back-and-forth motion of a tuning fork. Fig.A-1 (b) shows the movement of the loudspeaker diaphragm as a function of time. When the cone moves forward, a compression is created; when it moves backward, a rarefaction is formed. This pattern moves down the tube at 1130 feet per second, the speed of sound in air. The resulting pattern of pressure change, shown in Fig.A-1 (c), is called a *sinewave*, the most basic periodic repetition. Fig.A-1 (c) shows the instanta-

of greater amplitude. Specifically, two waves of equal amplitude, equal frequency, and the same phase will yield an increase of 6 decibels (dB) when combined. (Decibels are explained later in this appendix.) These waves are said to *reinforce* each other. If, however, one of these waves is phase-shifted by 180°, the two waves will *cancel* each other and produce no signal. When one wave is at peak positive, the wave shifted by 180° is at peak negative, producing the cancellation.

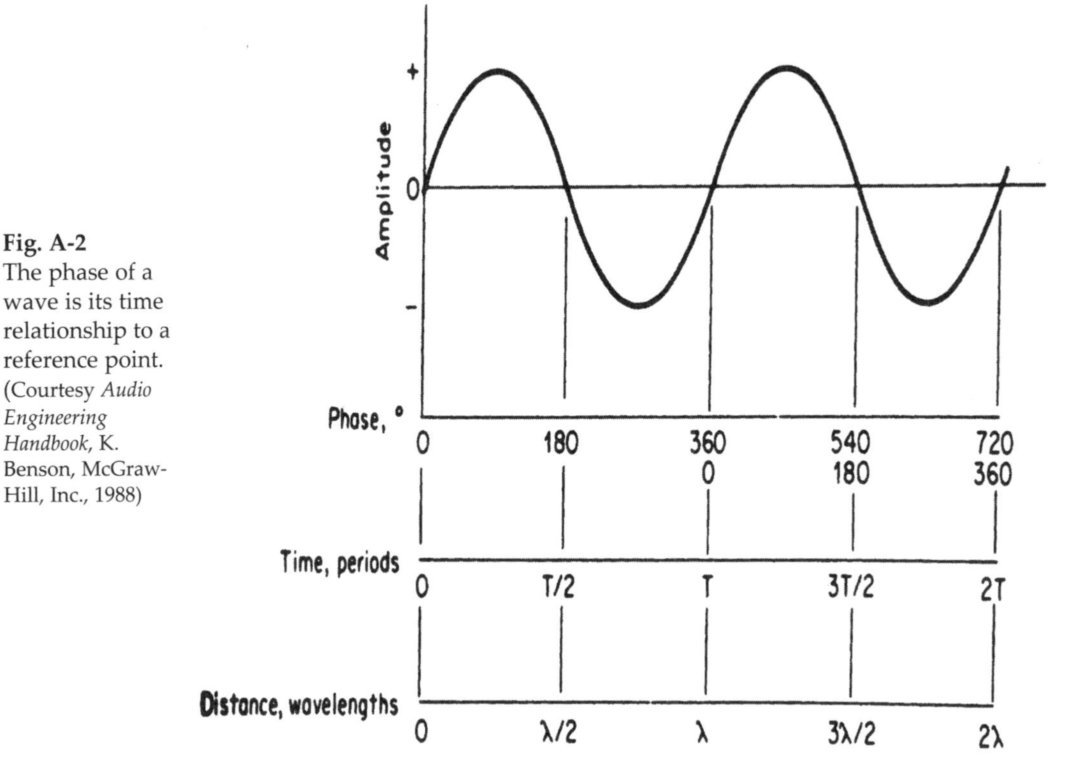

Fig. A-2
The phase of a wave is its time relationship to a reference point. (Courtesy *Audio Engineering Handbook*, K. Benson, McGraw-Hill, Inc., 1988)

This is exactly what happens in a hi-fi system if *one* loudspeaker is connected with the loudspeaker cables reversed (the red amplifier terminal is connected to the loudspeaker's black terminal, and the black amplifier terminal is connected to the loudspeaker's red terminal). Reversing the leads to one loudspeaker "flips" that signal over, making positive into negative and negative into positive. This is similar to a phase shift of 180°. Now, when one woofer cone pushes forward to create a compression, the second loudspeaker's woofer pulls back to make a rarefaction. When the compressions and rarefactions are combined, they cancel; the compression from one loudspeaker "fills in" the rarefaction from the second loudspeaker, and less sound is heard. Because the two waves don't perfectly coincide, they don't completely cancel, but they do greatly reduce the acoustic output in the bass. If these two signals were combined electrically, however, complete cancellation would occur. Note that if you reverse the red and black leads on *both* loudspeakers, there is no phase shift between the left and right signals, and thus no cancellation occurs.

Absolute Polarity

Some audio products (primarily preamplifiers and DACs) have switches marked "polarity," "phase," or "180°." These invert the polarities of *both* audio channels. The switch "flips" the signals over, making positive into negative and negative into positive. The switch inverts the stereo signal's *absolute polarity.*

Absolute polarity shouldn't be confused with *phase* reversal *between* channels (described earlier). In that case, there was a 180° phase shift in one channel with respect to the other—with severe audible consequences.

Absolute polarity, however, describes a polarity reversal of *both* channels. If musical waveforms were completely symmetrical—that is, identical in the positive- and negative-going halves—absolute polarity would make no difference. You could flip the signal upside-down and never be able to detect the difference, either by listening or by measuring.

But most musical signals aren't perfectly symmetrical, and are degraded if reproduced with inverted absolute polarity. When a recording is played back with correct absolute polarity, a compression from the musical instrument or voice at the original acoustic event causes a compression from the loudspeaker; a rarefaction from the instrument causes a rarefaction from the loudspeaker. Conversely, playing a recording with inverted polarity results in a compression from the instrument or voice to be reproduced as a rarefaction from the loudspeaker.

It may seem obvious and intuitive that all recordings and playback systems should maintain correct absolute polarity, but that isn't the case. An audio signal undergoes many reversals of absolute polarity during the recording and reproduction chain—reversals that no one keeps track of. Some recordings have correct absolute polarity and others don't—there's about a 50-50 chance that it's right. Similarly, some playback systems invert absolute polarity and others don't. The only thing that matters is that the sum total of all the polarity reversals results in correct polarity at the loudspeaker output. If a recording with inverted absolute polarity is played on a system that inverts absolute polarity, the result is correct polarity at the loudspeaker end of the playback chain.

The only way to ensure correct absolute polarity for each recording is by throwing the polarity switch and listening for which position sounds more natural. You can mark your records and CDs as to their polarity, and put the polarity switch in the appropriate position for each recording. Reversing the red and black leads on the cables going to *both* loudspeaker will also invert absolute polarity.

Note that it doesn't do any good to have a non-inverting playback system: roughly half your records and CDs will be reproduced with incorrect polarity whether your system is inverting or not. Moreover, many recordings have no single correct polarity; the disc often contains a mix of inverted and non-inverted tracks. Finally, music recorded with multi-track techniques often contains a mix of inverted and non-inverted signals in the same piece of music, making moot the entire issue of absolute polarity.

Some listeners report radical degradation of the sound when absolute polarity is reversed; others never notice the difference. The audibility of absolute polarity is highly variable, depending on the listener's polarity sensitivity, the instruments on the recording (some instruments produce a less symmetrical wave shape than others), whether or not the recording has a mix of inverted and non-inverted signals (from different microphones), and the phase coherence of the loudspeakers. Loudspeakers with poor time-domain performance can obscure the difference between correct and incorrect polarity.

saying "80dB SPL," we mean that the sound is 80dB greater in magnitude than the threshold of hearing. Note that 0dB SPL isn't absolute silence, but the softest sound an average person can hear in a very quiet environment.

The decibel scale is logarithmic, meaning that as the number of dB increases arithmetically, the value that number expresses increases exponentially. For example, each doubling of sound pressure level represents an increase of 6dB, as shown in Table 1.

Sound or electrical power ratio	Decibels	Sound pressure, voltage, or current ratio	Decibels
1	0	1	0
2	3.0	2	6.0
3	4.8	3	9.5
4	6.0	4	12.0
5	7.0	5	14.0
6	7.8	6	15.6
7	8.5	7	16.9
8	9.0	8	18.1
9	9.5	9	19.1
10	10.0	10	20.0
100	20.0	100	40.0
1,000	30.0	1,000	60.0
10,000	40.0	10,000	80.0
100,000	50.0	100,000	100.0
1,000,000	60.0	1,000,000	120.0

Let's say you were in a very quiet room and measured the sound of a fly buzzing at 40dB SPL. We can see in Table 1 that 40dB represents a pressure ratio of 100; the sound of the buzzing fly produces a pressure 100 times greater than a sound at 0dB SPL, the threshold of hearing. The threshold of hearing, represented by the reference level of 0dB SPL, has been established as a pressure of 0.0002 dynes per square centimeter (dynes/cm^2). A dyne is a unit of force. The letters "SPL" after the decibel notation tell us that 0.0002 dynes/cm^2 is the reference level. Knowing this, we can calculate that the sound a buzzing fly produces is a pressure 100 times that of the pressure at 0dB SPL, or 0.02 dynes/cm^2 (100 x 0.0002).

Two flies would produce a sound pressure level of 46dB—double the pressure, and an increase of 6dB over the 40dB one fly produces. Because 40dB represents a pressure of 0.02 dynes/cm^2, 46dB represents a pressure of 0.04 dynes/cm^2. (For the purposes of this illustration, we'll assume the flies produce the same sound pressure, and that their sounds are perfectly in-phase with each other. In reality, the phase relationship between the two sounds will be random, producing a sound *power* increase of 3dB.)

Now consider another example: Let's say that a jet taking off produces an SPL of 120dB. Looking at Table 1, we see that 120dB represents a ratio of 1,000,000—in other words, the jet creates a pressure on our eardrums a million times greater than the pressure at the threshold of hearing. Multiplying 1,000,000 by the reference pressure (0.0002 dynes/cm^2), we know that the jet creates a pressure of 200 dynes/cm^2 (0.0002 x 1,000,000).

If we add a second jet taking off, we know that the number of dB SPL will increase from 120dB to 126dB, and also that the pressure will double from 200 dynes/cm^2 to 400 dynes/cm^2. (Again, we'll assume for illustrative purposes that the jets produce equal loudness and are in-phase.)

Obviously, the sound of the second jet taking off is much greater than that of the second fly. But both increased the sound pressure level by the same *ratio*: 6dB. The same 6dB increase represented an increase of only 0.02 dynes/cm^2 in the case of the buzzing fly, but a whopping 200 dynes/cm^2 with the jet.

This logarithmic nature of the decibel makes handing very large numbers easier. These very large numbers are needed to express the vast range of sound pressure levels we can accommodate, from the threshold of hearing to the threshold of pain. The threshold of pain, 140dB SPL, is a pressure on your eardrum of 2000 dynes/cm^2, or ten million times the pressure at the threshold of hearing. Rather than saying that the threshold of pain is 10,000,000 times louder than the threshold of hearing, it's easier to represent this value as 140dB.

Moreover, this exponential increase parallels our subjective impression of loudness. If we asked a person listening to music at 60dB SPL to turn up the volume until it was "twice as loud," the person would increase the volume by about ten dB. If the person were listening at 100dB SPL and was asked to turn up the volume until it was twice as loud, he would also increase the volume by 10dB. As we can extrapolate from the fly and jet examples, the increase in pressure from 60dB to 70dB is far less than the pressure increase from 100dB to 110dB. Both increases, however, produce similar subjective increases in loudness.

This phenomenon is analogous to the way we perceive pitch: We hear each doubling of frequency as an octave, yet each ascending octave spans a frequency difference twice that of the next lower octave. The octave span between 20Hz and 40Hz is perceived as the same musical interval as the octave between 10kHz and 20kHz, although the latter is a much larger change in frequency.

We can calculate the number of dB between two values with the following formula: $NdB = 20 \log P1/Pr$ where NdB = the number of dB, $P1$ is the measured pressure, and Pr is the reference pressure. Expressing this formula verbally, we say, "The number of dB equals twenty times the logarithm of the ratio between the two pressures (or voltages, or currents)." This formula works for voltage, electrical current, or sound pressure level. Just as a doubling in SPL is represented by an increase of 6dB, a doubling of voltage is represented by an increase of 6dB.

Using this formula, we can calculate the difference in dB between two CD players having different output voltages. We'll say CD player A puts out 3.6V when playing a full-scale, 1kHz test tone, and CD player B puts out 2.8V with the same test signal. How much difference in dB is there between the two players?

We first find the ratio between the two voltages by dividing 3.6 by 2.8. The ratio is 1.2857. By pressing the "log" key on a calculator with 1.2857 in the display, we get the logarithm of the ratio (0.0109). Then we multiply the result by 20, giving us the answer in dB. CD player A's output voltage is 2.18dB higher than that of CD player B.

Calculating power ratios is slightly different. With acoustic or electrical power, an increase of 3dB represents a doubling of power. This is given by the following formula: $NdB = 10 \log P1/Pr$ where $P1$ is the measured power and Pr is the reference power. If we want to find the difference in power output in dB between two amplifiers, we multiply the logarithm of the ratio of the two values by 10, not by 20 as in voltage, current, and sound pressure.

Let's say one power amplifier has a maximum output power of 138W, and another has a maximum output power of 276W. What is the decibel difference between them? First find the ratio between the two powers by dividing 276 by 138. The answer is 2. Find the logarithm of 2 (0.301) and multiply by 10. The answer is that the second

The Dynamic Nature of Music

So far we've talked about sound only in the context of steady, continuous tones. But music is dynamic, constantly changing in level, and is composed of impulsive sounds as well as steady ones. A basic understanding of this dynamic quality is important to learning about music reproduction.

We can divide a dynamic sound into three components: *attack, internal dynamics,* and *decay* (see Fig.A-8). These three aspects of a sound are called the sound's *envelope.* Attack is the way in which a sound begins. The attack can be sudden (a snare drum, for example) or gradual (the build-up of an organ note in a large room). Decay is the way a sound ends. Internal dynamics are the main portion of the sound between the attack and decay. The envelopes shown in Fig.A-8 could be for a snare drum (left figure) or an organ (right). The snare drum has a very sudden attack; it takes very little time for the sound to reach its highest volume. It also has a very short duration and a rapid decay. By contrast, the organ has a very slow attack, taking a longer time to reach its full amplitude. The internal dynamics are very long, and define the note's sound. Similarly, the decay is slow and gradual.

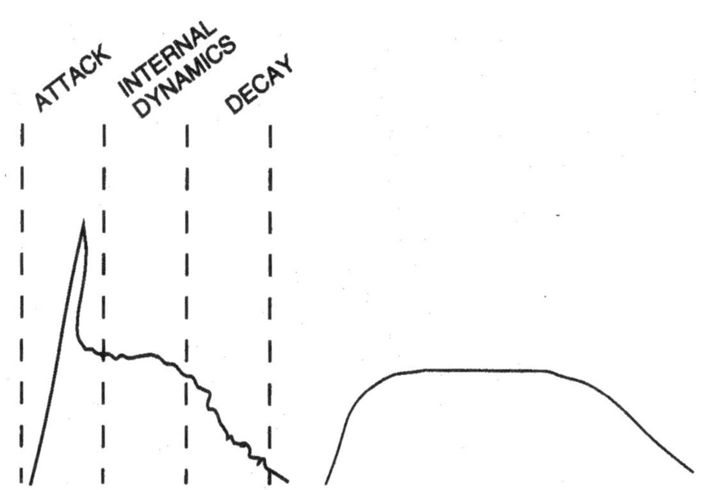

Fig. A-8 Musical sounds are composed of an attack, internal dynamics, and decay.

A sound characterized by sudden attack and rapid decay is a *transient* signal. *Transient response* refers to an audio component's or system's ability to reproduce the steep attack of transient signals. This type of signal has a high *peak-to-average ratio,* meaning that the peaks are much higher in level than the average level. Conversely, the organ has no peaks, so it has a very low peak-to-average ratio. A sound's peak-to-average ratio greatly influences the perceived volume; the ear tends to integrate the sound so that peaks add very little to the overall perceived volume.

Another way of expressing the dynamics of a musical signal is the term *dynamic range*—the difference between the loudest and softest sounds. Technically, a piece of audio equipment's dynamic range is the difference in dB between the device's noise floor and its maximum undistorted signal level. Subjectively, dynamic range is the contrast between a piece of music's quietest and loudest passages.

Localization

Localization is our ability to identify the spatial position of a sound. This ability to localize sounds derives in part from the fact that we have two ears, and that sounds arriving at each ear differ slightly from one another. The brain uses these differences (and other cues) to construct a three-dimensional representation of where a sound source is located.

The most basic mechanism for localizing a sound source is attenuation of high frequencies by the head. Low frequencies can diffract around the head more easily than high frequencies, causing one ear to receive a signal with a gradual attenuation of the midrange and treble. The brain interprets this high-frequency rolloff at one ear as a cue that the sound originates closer to the other ear.

At low frequencies (below about 700Hz), there is virtually no amplitude difference between the left and right ears for a sound emanating from one side. The sound diffracts around the head and the treble-rolloff localization cue doesn't exist. At these low frequencies, a second mechanism takes over in which the phase difference between the left and right ears is analyzed. Because low frequencies have long wavelengths compared to the size of the head, one ear receives a slightly different phase of the sound wave. This information is another cue as to the location of a sound source.

The head, upper body, and outer ear (called the *pinna*) also modify the acoustic signals before they reach the ear. The pinna's unusual shape reflects sounds from its structures into the ear. These reflections combine with sounds directly striking the eardrum, causing comb-filter effects. The brain processes the resultant cancellations and reinforcements to determine a sound source's direction. The modification of the sound by the human body is called the head-related transfer function. Recordings have been made by putting microphones inside a dummy head; the microphones "hear" the signal as modified by the dummy head. In one interesting experiment, recordings were made by placing microphones inside the ear canals of a variety of human subjects. These recordings were then played back by ear buds placed inside the ear canals of the listeners. This technique in essence replaced the head-related transfer function of one listener with that of another. The listening subjects reported wildly different tonal balances and spatial perceptions when listening to another person's head-related transfer function.

The Listening Environment Diagnostic Recording (LEDR) on the first Chesky Test CD (Chesky JD37) is a synthesized percussion sound that appears in one loudspeaker, then slowly moves across to the other loudspeaker. Another LEDR test moves the image from the loudspeaker up to a plane above the loudspeaker. The third LEDR test moves the image in an arc from one loudspeaker to the other. These special signals, used to evaluate a listening room, were created by simulating the frequency and amplitude cues imposed by the pinna. By creating notches in the signal's spectrum at precisely calculated frequencies, the signal simulates the effect of a sound source's movement.

You can conduct a simple experiment to demonstrate the pinna's effect on localization. Close your eyes and have a friend jingle keys about four feet away from you, either at head level, waist level, or knee level. Point to where the sound is coming from. Now repeat the experiment with your hands folding your outer ears forward. This eliminates the pinna cues that aid in localization, making it much more difficult to tell where the sound of the keys is originating.

Other Psychoacoustic Phenomena

An interesting phenomenon occurs when we hear a direct sound and a delayed version of that sound (as we would hear an echo). If the echo occurs within about 30ms (milliseconds) of the first sound, the echo isn't perceived as a discrete sound, but instead slightly modifies the sound's timbre. This first 30ms is called the *fusion zone,* and the phenomenon is called the *Haas Effect.*

Similarly, when exposed to a direct and a delayed sound that originate from different directions, we hear the sound as coming from the direction of the first sound. The ear ignores the second sound. Dolby Surround decoding exploits this phenomenon to increase the apparent channel separation between front and rear channels.

Masking is a psychoacoustic phenomenon that renders one sound inaudible when accompanied by a louder sound. Our inability to hear a conversation in the presence of loud amplified music is an example of masking; the music masks the conversation. Similarly, we hear tape hiss or record-surface noise only during quiet passages, not during most of the music. Masking is particularly effective when the masking sound has a similar frequency to the masked component. This kind of masking is called *simultaneous masking.*

Masking also occurs when two sounds are close together in time, an effect called *temporal masking.* The ear/brain tends to hear only the louder of the two sounds—oddly, within a limited time window, it doesn't matter if the louder sound occurred slightly earlier or slightly later than the softer sound. Although it seems logically impossible that a later sound could mask an earlier sound, this occurs because it takes time for the auditory system to process the signal and send it to the brain. When confronted by a louder sound slightly later in time, the louder sound takes precedence.

This appendix is meant only as a cursory survey of the basics of sound and hearing for the interested audiophile; the subject can be studied in great depth. For those interested in learning more about sound and hearing, I recommend the superb introduction to the subject by Dr. Floyd Toole in *Audio Engineering Handbook,* published by McGraw-Hill, ISBN 0-07-004777-4. Further references are included in Dr. Toole's chapters.

Appendix B: Audio and Electronics Basics

A knowledge of electronics and audio technology isn't necessary to choosing and enjoying a high-end music system. You can just select a system, set it up, and not worry about what's going on inside those mysterious black boxes.

Some audiophiles, however, want at least a basic understanding of how audio products work. They have an inherent fascination with audio equipment and want to broaden their knowledge of high-end audio to include a basic conception of electronics and audio circuits. This Appen-dix is by no means a comprehensive treatment of the subject—that could fill many books. Instead, it provides an overview of audio technology for the interested audiophile.

Moreover, it bridges the gap between the technical descriptions of products published in some high-end audio magazines and the reader who feels intimidated by all the technical jargon. So next time you read that a particular power amplifier has a transformer with dual secondary windings, three full-wave bridge rectifiers, a discrete, direct-coupled JFET Class-A input and driver stage, with a push-pull output stage that switches to Class B operation at 20W RMS, you'll know what it means.

But before we get to all that, let's take a look at the basic vocabulary of electronics and audio technology.

Voltage, Current, Resistance, and Power

The most basic elements of electronics are voltage, current, and resistance. Let's look first at voltage.

A *voltage* exists between two points when one point has an excess of electrons in relation to the other point. A battery is a good example: the negative terminal has an excess of electrons in relation to the positive terminal. When this happens, we say the battery is *charged*.

Voltage is also called *potential difference*, because it has the *potential* to do work and because voltage is always the *difference* in the electrical charges between two points. This is why we say a voltage exists *across* two points, such as a pair of loudspeaker terminals (when driven by an amplifier), or the positive and negative terminals of a battery. One point is more negative (an excess of electrons) than the other point. The *volt* is the unit of measurement that expresses just how great the difference in electrons between two points is.

voltage source. Most amplifiers, however, can't maintain the same voltage across a low-impedance load as they can across a higher impedance, thus the power available to drive a low-impedance loudspeaker is reduced. Amplifiers can also run out of current, again limiting the power delivered to the load.

Series and Parallel Circuits

So far we've discussed a type of circuit called a *series* circuit. In a series circuit, there is only one path through which current can flow. A *parallel* circuit provides multiple paths for current flow. If we add another resistor to our series circuit, we've turned it into a parallel circuit (Fig.B-3). Some of the current flows through the first resistor, some through the second resistor.

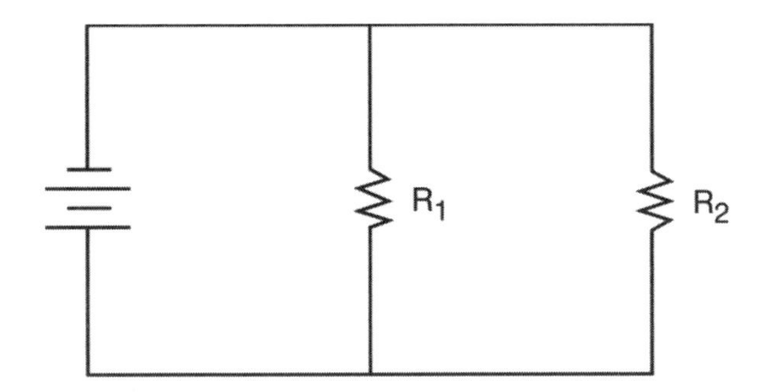

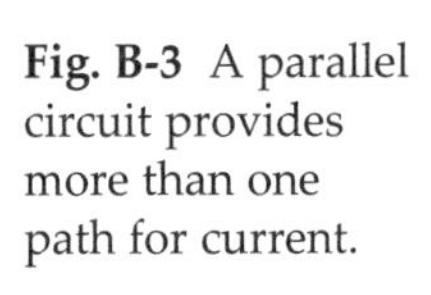

Fig. B-3 A parallel circuit provides more than one path for current.

Let's look at series and parallel circuits from the standpoint of a power amplifier driving loudspeakers. An amplifier driving a single loudspeaker is shown in Fig.B-4. The amplifier and loudspeaker form a series circuit; there is only one path for current. If we want to add another loudspeaker to this amplifier channel, we can do it two ways. The first is to put the second loudspeaker in series with the first one (Fig.B-5). The same current flows through both loudspeakers. In a series circuit, the individual resistances are added to find the total resistance. If each loudspeaker presents a resistance of 8 ohms, the total load the amplifier must drive is 16 ohms.

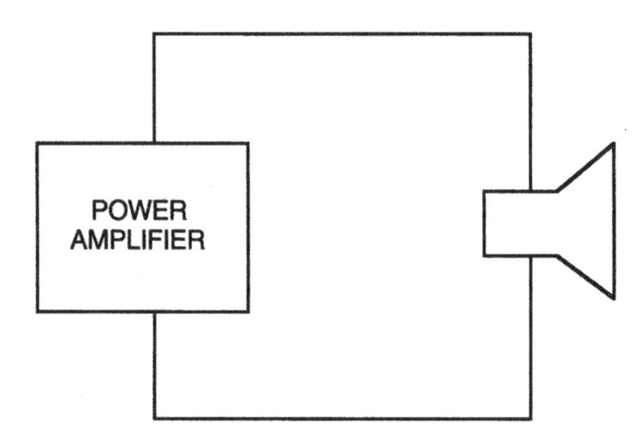

Fig. B-4 An amplifier driving a loudspeaker is a series circuit.

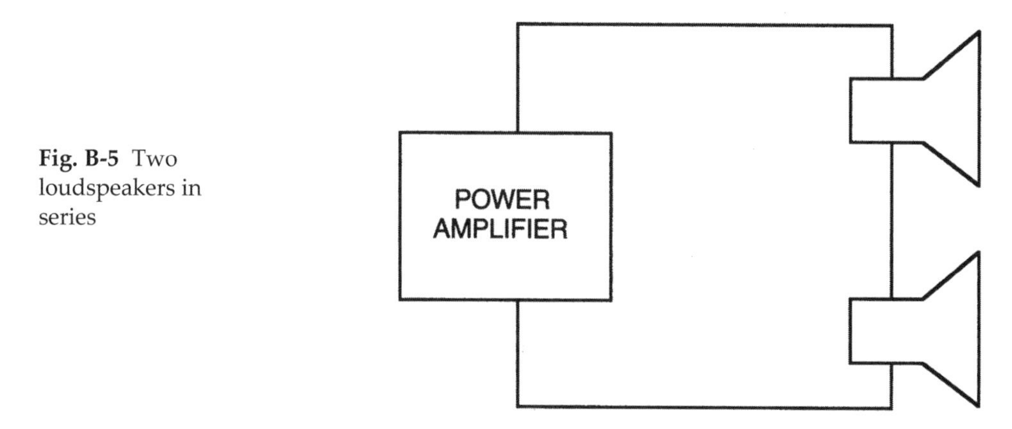

Fig. B-5 Two loudspeakers in series

Another way to connect the second loudspeaker is to wire it in parallel with the first one (Fig.B-6). We've now provided more than one path for current; some current flows through loudspeaker 1, some through loudspeaker 2. Unlike a series circuit, in which the resistances are added to find the total resistance, adding a parallel resistance *lowers* the total resistance (R_T). In the case of our two 8-ohm loudspeakers in parallel, the total resistance the amplifier must drive is 4 ohms. When two equal resistances are in parallel, the total resistance is half that of one resistance. The total resistance (R_T) in a parallel circuit can also be found with the formula $1/R_T = 1/R_1 + 1/R_2 + 1/R_3$, etc.

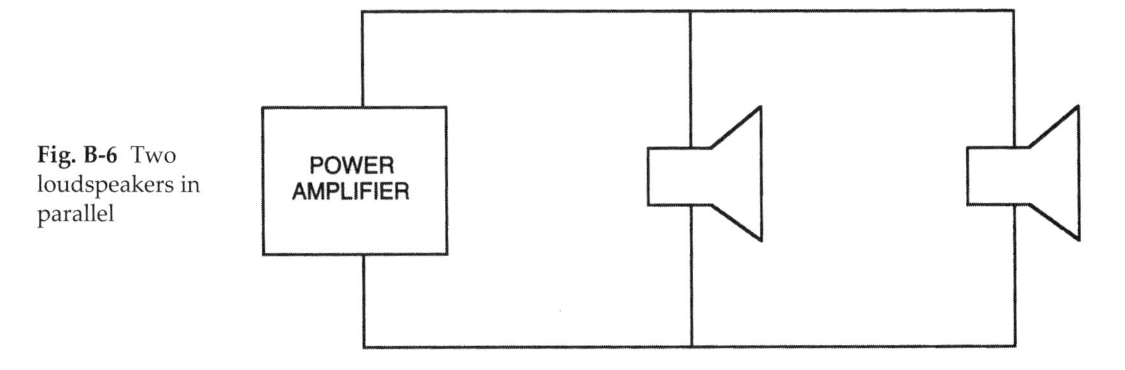

Fig. B-6 Two loudspeakers in parallel

Parallel circuits differ from series circuits in that the voltage across each branch of a parallel circuit is the same as the applied voltage. If we put ten resistors of different resistance values in parallel and apply a source voltage of 20V, that 20V will appear across all ten resistors. However, the current flow will be different through each resistor. It's the current that changes in parallel circuits, not the voltage.

Conversely, in a series circuit, the current remains the same while the voltage across each resistor is different. As current flows through each resistor in series, some of the applied voltage is dropped across each resistor. The amount of that voltage drop is determined by the resistor's resistance value. The sum of the voltage drops across each resistor equals the applied voltage.

Note that, at each resistor in the parallel circuit, the voltage stayed the same while the amount of current flow was different—the opposite of the series circuit.

Alternating Current (AC)

So far we've talked only about circuits using direct current (DC). Another type of current, called *alternating current,* or *AC,* is the basis for all audio circuits. An audio signal is an example of an AC voltage. An AC voltage reverses polarity, producing current that reverses direction. The principles of DC circuits and Ohm's law described earlier still apply, but AC circuits use additional elements we'll discuss later.

Let's use the AC power line from a wall outlet as an example of AC voltage and current. The 120 volts from the wall socket appear across the socket's two vertical slots. The voltage's polarity (which slot is positive, which negative) reverses itself at the rate of 60 times per second, or 60 Hertz (*Hz*). If we looked at this 120V, 60Hz voltage we'd see the waveform in Fig.B-7.

Starting at the left side of Fig.B-7, the voltage increases until it reaches its positive peak, then decreases to the zero crossing line (no voltage). The voltage then increases again, but in the opposite polarity. After reaching its negative peak, it decreases again. This represents one complete cycle of a sinewave, sixty of which occur every second. The reversing-polarity voltage causes current to reverse direction in whatever conductor is attached to it—usually an AC power cord and power transformer.

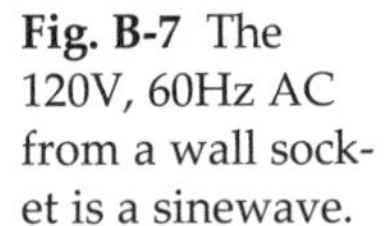

Fig. B-7 The 120V, 60Hz AC from a wall socket is a sinewave.

An audio signal is also an AC signal, but unlike our simple 60Hz sinewave, an audio signal is a *complex* AC voltage because it's composed of many frequencies. The AC audio signal is an electrical analog of the sound-pressure wave that created the original audio event. Just as a sound wave is composed of compressions and rarefactions, the audio signal is composed of positive and negative polarity phases.

AC voltage can be expressed in two ways: *peak-to-peak* (p-p) and *root mean square* (RMS). In a p-p expression, the voltage between the waveform peaks is measured. The more commonly used RMS value specifies the voltage at roughly 70% of the peak voltage. RMS better describes the "effective" voltage, or the ability of the voltage to do work. The 120V AC power line has a peak value of 170V, but an RMS value of 120V. When-ever you see an audio signal expressed as a voltage—a preamplifier output level, for example—it is virtually always an RMS value.

Manufacturers of power amplifiers sometimes exploit this difference between peak and RMS to overstate a product's power output. An amplifier can deliver more power on peaks than continuously. Chapter 5 ("Power Amplifiers") has a more complete discussion of power-amplifier output ratings.

Electromagnetic Induction, Inductance, and Capacitance

We've learned how voltage is equivalent to electrical pressure, current is the flow of electrons "pushed" by the voltage, and resistance is the opposition to current flow. Now let's look at three more electrical properties very important in audio: electromagnetic induction, inductance, and capacitance.

Electromagnetic induction is the relationship between moving electrons and magnetism; moving electrons produce a magnetic field, and a moving magnetic field causes electrons to flow through a conductor. Many audio devices, from phono cartridges to loudspeakers, rely on electromagnetic induction to work.

Let's first look at how a moving magnetic field causes electrons to flow. If you have a conductor, a magnetic field, and relative motion between them, a current (electron flow) will be *induced* in the conductor. For example, if we move a magnet upward through a coil of wire (the conductor), a voltage will be induced across that wire. A pointer on a voltmeter across that coil will swing in one direction when the magnet is moved. When the magnet stops moving, the pointer will return to its original position—no more voltage is being induced across the coil. Now, if we move the magnet in the opposite direction, a voltage will also be induced, but of opposite polarity to that of the voltage induced by moving the magnet up. The voltmeter pointer will now swing to the opposite side of the meter while the magnet is moving. When the magnet stops moving, the pointer will return to its center zero position.

This is exactly how a phono cartridge works. In a moving-coil cartridge, the coils (left and right audio channels) are attached to the cantilever; the cantilever is moved back and forth by the stylus as the stylus follows the modulations in the record groove. The coils sit in a fixed magnetic field. As the coil moves back and forth, a voltage is induced across the coil—the audio signal. The coil cuts through *magnetic lines of flux* surrounding the magnets, inducing a voltage across the coil. The faster the coil cuts these magnetic lines of flux, the higher the induced voltage. The more turns on the coil, the more lines of flux are cut, and the higher the induced voltage. High-output moving-coil cartridges have more turns on their windings than low-output moving-coils. A moving-magnet cartridge works on exactly the same principle, but with the roles reversed: the magnets move and the coils are fixed.

There's another interesting phenomenon of electromagnetic induction: electrons flowing through a conductor create a magnetic field around that conductor. A conductor connected across a battery's terminals will be surrounded by a magnetic field. If we make current flow in the opposite direction by reversing the battery terminals, the north-south magnetic field will also reverse polarity. The magnetic field's density can be increased by coiling the wire to concentrate the magnetic lines of flux, creating a device called an *inductor,* which creates a property called *inductance.* Inductance is measured in henrys (the unit of inductance) and is represented by the letter *L*. (The schematic symbol for an inductor is shown in Fig.B-8).

Fig. B-8 The schematic symbol for an inductor.

In another combination of a capacitor and an inductor, a *bandpass* filter is created. The bandpass filter will roll off frequencies on either side of the *passband*. A three-way loudspeaker crossover will have a high-pass filter on the tweeter, a low-pass filter on the woofer, and a bandpass filter on the midrange driver.

A filter's *cutoff frequency* is the frequency at which the amplitude is reduced by 3dB. Similarly, the *bandwidth* of an audio device or filter is the frequency span between the –3dB points. All the filters described here are called *passive filters* because they aren't built around an amplifier. Filters that use amplifying devices are called *active filters*.

Capacitors are also used to block direct current, yet pass alternating current. As we'll see later in this chapter, many audio circuits mix DC with the AC audio signal. Putting a capacitor in series with the signal blocks the unwanted DC component and allows the desired AC audio signal to pass. The capacitor looks like an infinitely high resistance at DC, but a low-value resistor to AC signals.

Capacitors can also store a voltage. In fact, the definition of capacitance is the ability to store a charge. If we put a battery across a capacitor, electrons flow from the battery's negative terminal to the capacitor's plate. The voltage source redistributes electrons from one plate of the capacitor to the other. The applied voltage stays across the capacitor even after the battery is removed.

This stored charge can be lethal. A power amplifier with large power-supply capacitors can kill you even if the amplifier hasn't been plugged in for years. Even small capacitors can give you a shock. The storage ability of capacitors can be demonstrated by turning off your power amplifier while it is reproducing music. It takes several seconds for the music to fade after the amplifier's power source has been removed. The amplifier is running off the charge stored in its capacitors.

Impedance

Impedance is the opposition to current flow in an AC circuit, specified in ohms and represented by the symbol Z. Impedance is to an AC signal what resistance is to DC. If we're dealing with DC, opposition to current flow is resistance. If the voltage is AC, opposition to current flow is impedance. Because loudspeakers are driven by AC signals, they have an impedance, not a resistance.

Impedance differs from resistance in that impedance implies that the load is not a simple resistance, but a combination of resistance, inductive reactance, and capacitive reactance.

An audio playback chain is a series of impedances. Every component with an input presents an impedance to the signal driving it. The impedance a component presents to the drive signal is called its *input impedance* (also called *load impedance*). A DAC's analog output drives the preamplifier's input impedance, the preamplifier drives the power amplifier's input impedance, and the power amplifier drives the loudspeaker's impedance. Impedance can be thought of as the load through which the component must force current.

Components in an audio chain also have an *output impedance*, or *source impedance*. If you think of a component—a CD player, for example—as an audio signal generator, its output impedance is like a resistance in series with that signal generator.

Generally, audio products have output impedances of between a few tens of ohms and a few hundred ohms. Input impedances, however, are very high, usually between 10,000 ohms (10k ohms) and 1,000,000 ohms (one Megohm, or one million ohms). A product with a 1M ohm impedance will draw less current through it than one with a 10k ohms input impedance. The higher the input impedance, the easier it is for the source component to drive; the input impedance is less of a load on the source component.

Let's combine our knowledge of Ohm's law, capacitance, and impedance to examine the interaction between a preamplifier and a power amplifier through a long run of interconnect. Fig.B-11 shows this circuit pictorially and schematically. Because all cables and interconnects have some inherent capacitance, we can represent the interconnect capacitance by the symbol for a capacitor between the signal conductor and ground. The preamp's output impedance is represented by the resistor in series with the signal generator. The power amplifier's input impedance is represented by the resistor inside the power amplifier outline.

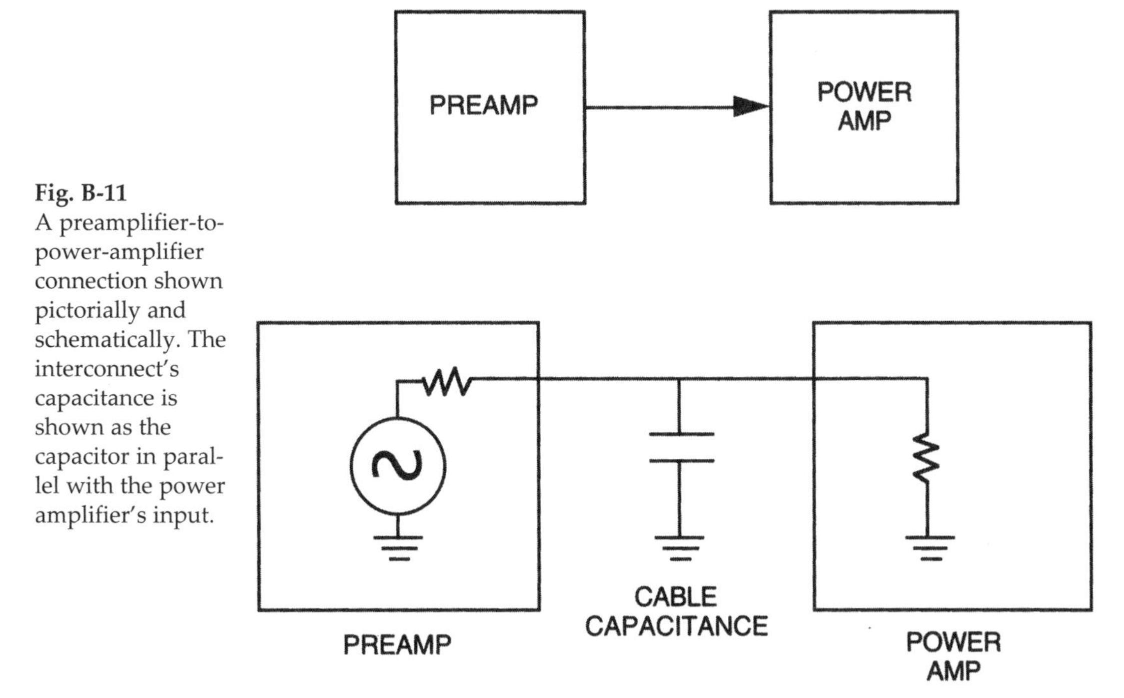

Fig. B-11
A preamplifier-to-power-amplifier connection shown pictorially and schematically. The interconnect's capacitance is shown as the capacitor in parallel with the power amplifier's input.

Let's say that the preamp has a very low output impedance, the power amplifier has a high input impedance, and the cable capacitance is low. All these are ideal conditions. The power amplifier's high input impedance means very little current will flow, causing very little voltage drop across the preamplifier's output impedance. Because the cable's capacitance is low, its reactance will remain high up to a very high frequency. Everything works just fine.

But if the preamplifier has a high output impedance and the cable is excessively capacitive, we run into problems. At low frequencies, the cable capacitance has no effect: its capacitive reactance is extremely high. In fact, it's almost as if the capacitor weren't there. But as frequency rises and the capacitive reactance drops, the capacitor looks like a resistor connected to ground. The higher the frequency, the lower the value of the "resistor" con-

The Power Supply

Every audio component that plugs into a wall outlet contains a power supply. The power supply converts 120V, 60Hz AC power from the wall outlet into the lower DC voltages needed by audio circuits. This is a critical job in high-end audio: the power supply has a large influence on the product's sound quality. Often, the power supply accounts for much of a component's size, weight, and cost. The general thinking in high-end product design is to have extremely simple audio circuits and extremely elaborate power supplies. We want a pure and simple path for the audio signal to follow, and a very clean and stable DC supply. The first component in the power supply is the *power transformer*. This device *steps down* the 120V AC powerline to a lower AC voltage, perhaps ±30V (we'll see how it does this in a minute). The transformer's schematic symbol is an accurate representation of its construction: it consists of two coils of wire placed next to each other (Fig.B-13).

Fig. B-13
The schematic symbol for a transformer represents its physical structure.

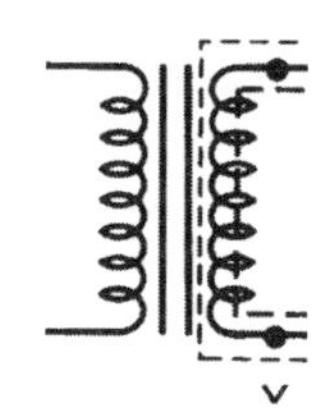

A little earlier we learned that whenever current flows through a conductor, a magnetic field is set up around that conductor. We also learned that whenever we have a conductor, a magnetic field, and relative motion between the two, a voltage is induced. Both principles are in operation in a transformer. When the 60Hz AC flows through the transformer's first coil, called the *primary winding*, an expanding and collapsing magnetic field is set up around this coil. This expanding and collapsing field created around the primary winding induces a voltage across the second coil, called the *secondary winding*. The 60Hz AC signal thus appears across the secondary winding, even though there is no physical connection between this winding and the AC wall outlet.

The voltage induced across the secondary is determined by the *turns ratio* between the primary and secondary. By changing the turns ratio, any value of output voltage can be obtained. Some transformers have multiple secondary windings to supply different circuits. The transformer thus converts 120V, 60Hz AC to a lower 60Hz AC voltage or voltages. The transformer also *isolates* the component from the power line.

Transformers are rated by power capacity, called a *VA* rating (Volt/Amperes). A transformer rated at 1kVA (1000VA), for example, can deliver 100V at 10A (assuming a 1:1 turns ratio). Transformers in low-signal components (preamps, CD players) are typically rated at a few tens of VA, while the largest power-amplifier transformers may reach a whopping 5kVA. Such a transformer may weigh more than 100 pounds.

Many audio products use *toroidal* transformers instead of the more conventional *laminated* type. A toroidal transformer is doughnut-shaped to concentrate its magnetic field in a tighter configuration. Toroidals are more expensive than laminated types, but are more efficient and radiate less noise into the surrounding circuitry. Just as the magnetic field surrounding the primary winding induces a voltage across the secondary winding, audio circuitry near the transformer can pick up 60Hz *hum*. Toroidal transformers reduce this risk. The stepped-down voltage from the transformer's secondary winding is still AC, but we need DC to supply the audio circuits. The next element in the power supply is the *rectifier*, a device that turns AC into DC. Rectifiers are made from diodes, the most basic *semiconductor* element. A diode allows current to flow in one direction but not the other. Fig.B-14 shows the effect of putting a single diode on a transformer's output. The diode allows one polarity to pass, but blocks the 60Hz sinewave's other half. (Which half it passes—positive or negative—is determined by the diode's direction.) In effect, the diode removes one polarity of the signal. Because the resultant voltage never changes polarity, it is, technically, direct current. This method, called *half-wave rectification*, isn't very efficient. But by putting four diodes together, we can make a *full-wave bridge rectifier* (Fig.B-15). In effect, the other sinewave polarity is "flipped over." The full-wave rectifier produces smoother DC and delivers more power to the supply.

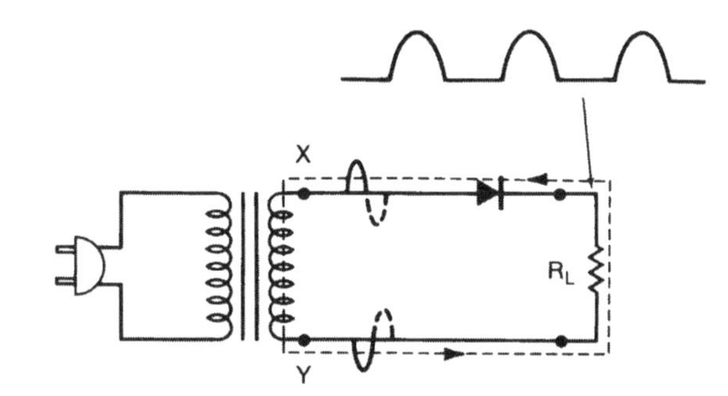

(A) Transformer feeding a half-wave rectifier

Fig. B-14 A half-wave rectifier passes only one phase of the AC powerline.

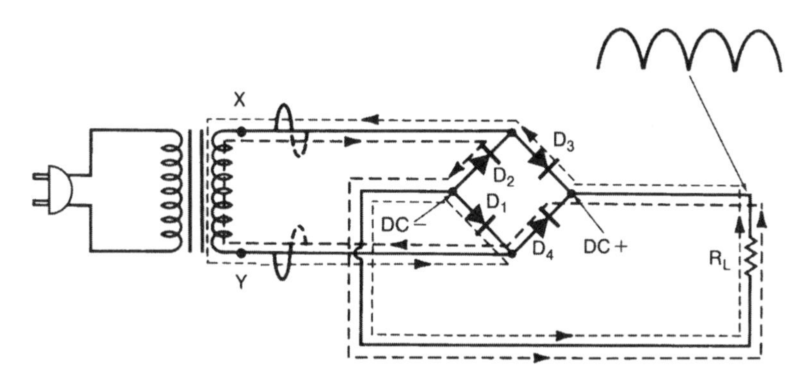

(C) Transformer feeding a bridge rectifier

Fig. B-15 A full-wave rectifier "flips over" one polarity of the AC powerline. (Figs. B-14 and B-15 copied with permission from *Audio Technology Fundamentals* by Alan A. Cohen)

The rectifier output is DC (it never changes polarity), but still has AC *ripple* riding on it. Ripple consists of vestiges of the AC signal superimposed on the DC; the DC shifts up and down at 120Hz. Note that a full-wave rectifier produces ripple at 120Hz, not 60Hz, because the negative phase of the sinewave is "flipped over" (shown earlier in Fig.B-15). Ripple is reduced by adding *filter capacitors*, whose storage function smooths the AC ripple riding on the DC voltage. In fact, filter capacitors are also called *smoothing* capacitors.

The filtered output is now DC with much less ripple, but it still isn't ready to supply our audio circuits. Any changes in the input line voltage will affect the DC voltages supplying the audio circuits—an unacceptable condition. The job of making this DC more stable and at the correct voltage falls on the *voltage regulator*. A voltage regulator maintains a constant output voltage regardless of changes to its input voltage. The voltage regulator's input

nation refers to the transistor's internal structure, which determines the voltage polarities needed to operate the transistor. In a complementary pair, an NPN transistor is connected to a PNP transistor (Fig.B-21). The "top" transistor will handle the positive half of the audio signal, and the bottom transistor will amplify the negative portion of the signal. When the top transistor is conducting, the bottom transistor is turned off, and vice versa. This technique keeps the transistors turned off half the time, producing less heat and reducing stress on the transistors.

The signal driving a complementary pair is processed by a *phase splitter* before the transistor bases. A phase splitter duplicates the audio signal so that the positive phase drives the NPN transistor of the complementary pair, and the negative phase drives the PNP half of the pair.

A Class-A/B amplifier is also called a "push-pull" output stage because one transistor "pulls" current while the other transistor "pushes." Current flow is indicated by the directions of the arrows in Fig.B-21.

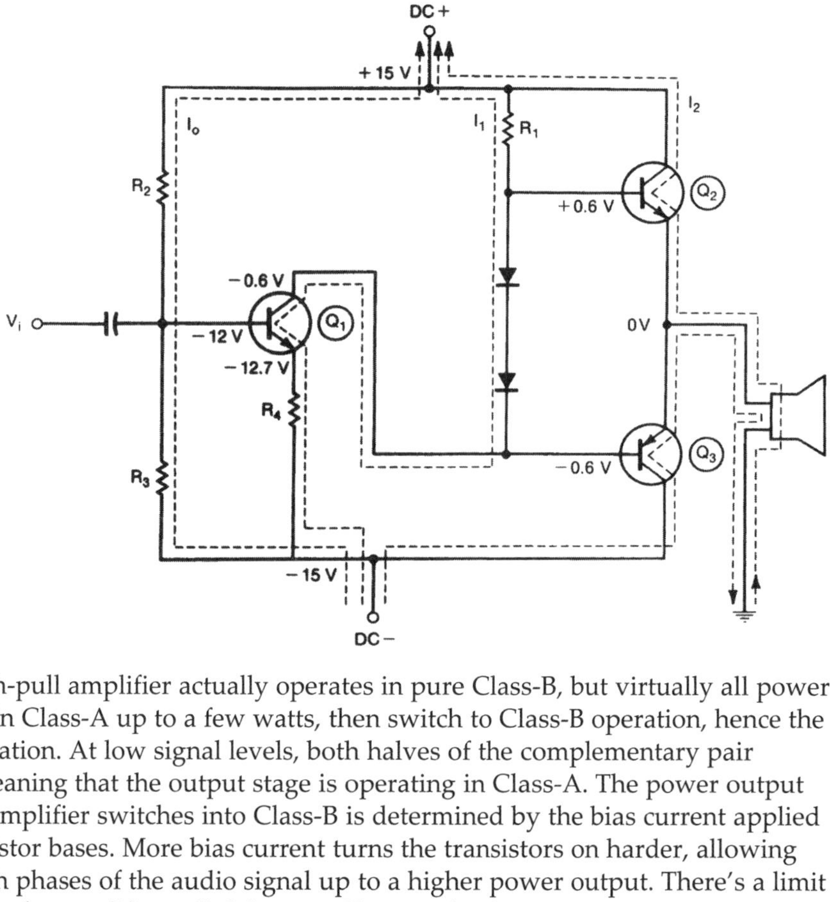

Fig. B-21 A complementary output pair of transistors driven by a Class-A driver stage. (Copied with permission from *Audio Technology Fundamentals* by Alan A. Cohen)

A true push-pull amplifier actually operates in pure Class-B, but virtually all power amplifiers operate in Class-A up to a few watts, then switch to Class-B operation, hence the "Class-A/B" designation. At low signal levels, both halves of the complementary pair always conduct, meaning that the output stage is operating in Class-A. The power output level at which the amplifier switches into Class-B is determined by the bias current applied to the output transistor bases. More bias current turns the transistors on harder, allowing them to handle both phases of the audio signal up to a higher power output. There's a limit to how much power the amplifier will deliver in Class-A; the designer just can't keep increasing the bias current. Maximum Class-A power output is determined by how much current the transistors can handle, how well heat is removed from the transistors, and how much current the power supply can deliver. Pure Class-A amplifiers are huge, very heavy (due to their massive power transformers), generate lots of heat, and are very expensive in terms of "dollars per watt."

Class-A operation produces less distortion than Class-A/B, and generally sounds better. A kind of distortion in Class-B amplifiers called *crossover distortion* can produce a waveform discontinuity at the zero crossing point where one transistor turns off and the other turns on. The two halves of the complementary pair aren't biased identically, producing crossover distortion. This condition can't occur in Class-A amplifiers. Class-A output stages have other another big advantage: the transistors are always at the same temperature regardless of the signal they are amplifying, making the output stage more stable and linear. (Chapter 6 contains a discussion of amplifier output stages and their musical qualities.)

The NPN and PNP transistors described previously are called *bipolar* devices because current flows through both polarities (P and N) of semiconductor material. Another type of transistor, called the *Junction Field Effect Transistor* (JFET), is often used in small-signal applications. The JFET is called a *unipolar* device because current can flow in only one direction. A JFET has a *gate*, a *source*, and a *drain*, which correspond to the base, emitter, and collector in our NPN and PNP transistors. JFETs are often used in amplifier input stages for their high input impedance. JFETs are also turned on (biased) by voltage rather than by current, as are bipolar transistors. JFETs are found only in small-signal circuits. In addition, JFETs are also used as electronic switches.

Another type of field-effect transistor is called the MOSFET (MOS stands for Metal Oxide Semiconductor). MOSFETs are sometimes used in power-amplifier output stages.

Amplifier Distortion

All the amplifier circuits we've talked about produce distortion—their output signals aren't exact replicas of their input signals. Instead, the signals are changed in certain ways. In *harmonic distortion,* harmonics of the input signal appear in the output. For example, if a 100Hz sine-wave is input to the amplifier, the output will be a 100Hz sinewave, but with the harmonics of 100Hz (the second harmonic at 200Hz, third harmonic at 300Hz, fourth at 400Hz, etc.) added to the signal.

Harmonic distortion is expressed as a percentage of the output signal. The term *Total Harmonic Distortion,* or *THD,* is an expression of the percentage of all the harmonic components combined. For example, if the amplifier puts out 10V and the harmonic distortion products are 10mV, the amplifier has 0.1% THD. Although THD is a common ampli-fier specification, a more instructive analysis of harmonic distortion involves examining *which* harmonics are produced by the amplifier. Second- and third-harmonic distortion components aren't nearly as objectionable as upper-order harmonic distortion components such as the fifth, seventh, and ninth. In fact, 10% second-harmonic distortion is less musically objectionable than 0.5% seventh-harmonic. We can quickly see that a single THD figure tells us very little about the audibility of an amplifier's distortion. All amplifiers produce different harmonic distortion spectra—one possible reason why they sound different.

Another distortion in amplifiers is called *intermodulation distortion,* or *IMD.* When a signal of two frequencies is input to an amplifier, the amplifier will output the two frequencies, plus the *sum* and *difference* of those frequencies. For example, if we drive an amplifier with a mix of 100Hz and 10kHz, the amplifier's output will be 100Hz and 10kHz, *plus* 9.9kHz (the *difference* between 100Hz and 10kHz) and 10.1kHz (the *sum* of 100Hz and

amplified and combined into one signal at the op-amp's output, but any noise, distortion, or spurious junk common to both polarities will be not be amplified by the op-amp. The phenomenon by which a differential amplifier rejects signals common to both channels is called *common-mode rejection*; the op-amp's ability to reject signals common to both channels is called its *common-mode rejection ratio*, or *CMRR*.

The op-amp driven with a differential signal naturally converts a balanced signal to an unbalanced signal. Many audio components with balanced inputs have a differential amplifier at the input, which converts the balanced signal to a single-ended (unbalanced) signal that the component then amplifies.

Digital Electronics

Digital electronics are becoming increasingly important in audio recording and reproduction. Not only CD players and digital processors use digital electronics: more and more control and switching functions in preamps and power amps are being handled by digital circuits.

Simply, digital circuits operate in one of two states: on or off. The "On" state is represented by a voltage (such as +5V), and "Off" is represented by no voltage (also called *ground*). The terms *high* and *low*, or *one* and *zero*, also describe the two conditions in digital circuits. Unlike an analog signal that has a continuously varying amplitude, digital signals are a train of high and low pulses. Digital circuits are also called *binary* circuits to describe the two-state nature of a digital signal.

In digital signals, information is represented as a series of high and low states. Each state carries one piece of information, or *bit*. (Appendix C explains how binary coding can represent numbers, which in turn can represent an audio signal. Binary coding can store virtually any sort of information; this book's text and graphics were stored as binary code.) The building blocks of digital electronics are called *logic gates*. A logic gate has more than one input, but only one output. Logic gates are made with tiny transistors that are either fully turned on or fully turned off. Simple logic gates can make basic decisions. For example, an *OR gate* will produce a high output if either input #1 *or* input #2 is high. A two-input *AND gate* will produce a high output if input #1 *and* input #2 are high. Digital electronics use many other types of logic gates.

Logic gates can be arrayed to perform nearly any mathematical task. In practice, many hundreds of thousands of gates are needed to make digital circuits. These gates are combined into *integrated circuits*, or ICs. Several hundred thousand gates can be put on a single IC.

Appendix C: Digital Audio Basics

Introduction

Throughout the history of recorded sound, the goal has been to capture an acoustic event by storing a mechanical or magnetic representation of the original acoustic waveform. The modulations in an LP groove, for example, are an analog of the acoustic wave that was originally heard as music. The more alike the groove modulations and the acoustic waveform, the higher the fidelity. Tiny changes in the shape of that squiggly line—inevitably introduced in record cutting, pressing, and playback—produce an almost infinite variability in sound quality.

The advent of digital audio has fundamentally changed the way audio is encoded, stored, and played back. Rather than attempting to preserve the acoustic waveform directly, digital audio converts that waveform into a numerical representation. The numbers representing the waveform are stored and later converted back into an analog signal. Because those discrete numbers can be stored and recovered more precisely than the infinitely variable analog signal, digital audio has the potential to exceed the musical performance of analog media.

I say "potential" because, in many respects, digital audio has yet to sound better than the best analog formats. Virtually anyone who's listened to a live microphone feed, then the same signal played back both from analog tape and after encoding into digital form, will attest that the analog tape produces a truer representation of the music. Similarly, a comparison of a properly played LP and the CD of the same music reveals musical virtues in the analog representation not heard from the digital medium.

This isn't to say that analog is without flaw; rather, it suggests that the distortions generated by digital audio are less sonically benign than those from analog media.

Fortunately, this state of affairs is only temporary; digital audio has taken a large step forward with the commercial availability of high-resolution formats such as SACD and high-res downloads. Only in the last decade did we discover that building digital audio products is a far more exacting process than theory suggests. Although some audiophiles shun digital, it's shortsighted to summarily reject all digital audio based on the crude sound provided by the technologies of digital's first decades. The future promises much more than that.

Binary Number System

All digital audio systems—and digital electronics in general—use the *binary* number system. "Binary" means "two," reflecting the two possible values in the binary, or base 2, number system: 0 and 1. A binary number of these two values strung together can represent any quantity we want, as can the ten values of the decimal system we're so familiar with.

When we count in decimal and reach the highest digit (nine), the ones place is reset to zero and the sum is carried to the next column, producing the number 10. This number is a code that represents a 1 in the tens column and a 0 in the 1s column—one ten, no ones. The position of a digit indicates the weight given that digit.

Just as there is no single digit greater than nine in decimal notation, there is no single digit greater than 1 in binary notation. Consequently, when we count in binary, we reset and carry with every count. For example, the number 01 in binary represents the decimal number 1. If we add the binary numbers 01 and 01, we reset the ones column and carry to the next column, producing the binary number 10. This represents a 1 in the twos column and a 0 in the 1s column, or the number 2 (expressed in decimal). Binary addition works like this:

```
  01      10      11
 +01     +01     +01
  10      11     100
```

Just as the position of a digit within a decimal number changes its value by a factor of ten (ones, tens, hundreds, thousands, etc.), the position of a binary digit changes its value by a factor of 2. For example, the binary number 0101 indicates a 1 in the 1s column, a 0 in the 2s column, a 1 in the 4s column, and a 0 in the 8s column. The binary number 0101 thus represents the decimal quantity 5. (Fig.C-1 shows binary and decimal equivalences.)

Fig. C-1 Binary and decimal equivalences

Binary	Decimal
0000	0
0001	1
0010	2
0011	3
0100	4
0101	5
0110	6
0111	7
1000	8
1001	9
1010	10
1011	11
1100	12
1101	13
1110	14
1111	15

Any decimal number can be represented by binary notation, provided the binary number is long enough. The highest number that can be represented in binary notation is 2^x, where x is the number of places. For example, a binary number with four places can represent any decimal number between 0 and 15 (2^4). The number 15 in binary is 1111: a one in the 8s column, a one in the 4s column, a one in the 2s column, and a one in the 1s column.

Each of these positions in a binary number is called a *bit*, short for *binary digit*. The greater the number of bits, the higher the number we can represent. The bit in the position of highest value in a binary number (usually at the far left) has the greatest effect on the number's value and is called the *Most Significant Bit*, or *MSB*. The bit in the position of lowest value in the binary number (usually the far right column) is called the *Least Significant Bit*, or *LSB*.

Binary notation makes computers and electronic circuits simpler and more practical. The electronic circuit or device need operate in only two states: one or zero, on or off, yes or no. If a voltage is present, we call that binary 1; if no voltage is present, that represents binary 0. These conditions relate to a transistor being fully biased on (binary 1) or fully biased off (binary 0). Similarly, a north-south polarization of a magnetic medium could represent binary 1, and a south-north polarization might indicate binary 0.

In digital audio, the analog signal is represented by a stream of these binary numbers. In the next sections, we'll see exactly how music can be represented by binary numbers.

Sampling and Quantization

Sampling and quantization are the cornerstones of digital audio. These two distinct but interrelated processes form the basis for all digital audio systems. To see how sampling and quantization can preserve an analog audio signal as a series of numbers, let's first look at an audio signal's characteristics.

An analog audio signal is a voltage that varies over time. The faster the variation over time, the higher the audio signal's frequency. The greater the amplitude swings, the louder the signal. The audio signal thus has two properties—time and amplitude—that must be encoded to correctly preserve that signal.

An LP record is a good example of how time and amplitude information are preserved. The side-to-side modulations in the record groove encode the amplitude information; the greater the modulation, the higher the signal amplitude. The time information is encoded by the LP's rotation, which must be the same on playback as when the disc was recorded. If we change the speed of the LP, we change the time relationship and thus the audio signal's frequency.

Digital audio must also preserve the time and amplitude information of an audio signal. But instead of encoding and storing those characteristics continuously as in an LP groove, digital audio preserves time and amplitude in discrete units.

A digital audio system encodes the signal's time information by *sampling* the audio signal at discrete time intervals. The amplitude information is encoded by generating a number at each sample point that represents the analog waveform's amplitude at sample time—a process called *quantization*. Sampling preserves the time information, quantization preserves the amplitude information.

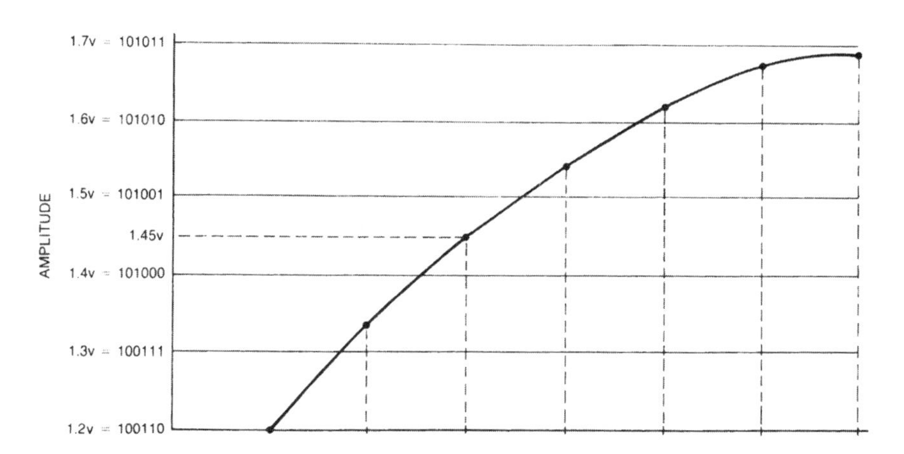

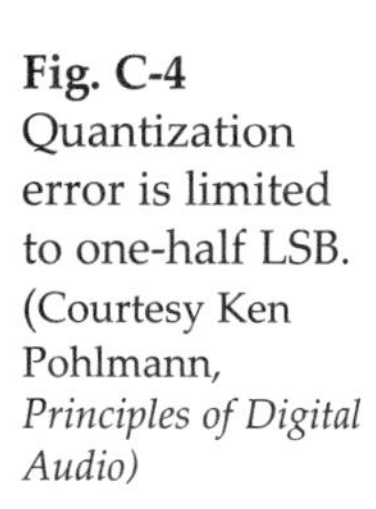
Fig. C-4 Quantization error is limited to one-half LSB. (Courtesy Ken Pohlmann, *Principles of Digital Audio)*

The worst case is when the analog amplitude falls exactly between two quantization levels, as occurs at the second sample in Fig.C-4; the difference between the analog amplitude and the quantization word representing that amplitude is the greatest.

Quantization error can be expressed as a percentage of the least significant bit (LSB). At the first sample, the quantization error was one-quarter of an LSB. At the second sample, the error was one-half of an LSB. Note that quantization error can never exceed half the amplitude value of an LSB. Consequently, the smaller the quantization step size, the less the quantization error. Adding one bit to the quantization word length doubles the number of quantization steps and halves the quantization error. Because such a halving represents a difference of 6dB, the signal-to-noise ratio of a digital system improves by 6dB for each additional bit in the quantization word. A digital system with 18-bit quantization will have 12dB less quantization noise than a system with 16-bit quantization.

We can also approximate the signal-to-noise ratio of a digital system by multiplying the number of bits in the quantization word by 6. Sixteen-bit quantization provides a theoretical dynamic range of about 96dB. A 20-bit digital audio system has a dynamic range of about 120dB, or 24dB higher than a 16-bit system.

Quantization error is audible as a rough, granular sound on low-level signals, particularly reverberation decay. Instead of hearing the sound decay into silence, the reverberation decay grows in coarseness and grain as the signal decays. This is because the quantization error becomes a greater percentage of the signal as the signal's amplitude decreases.

This increase in distortion as the signal level drops is unique to digital audio; all analog recording formats exhibit higher distortion at *high* signal levels, such as saturation in analog tape. This situation presents unusual challenges: an increase in distortion as signal level decreases makes that distortion much more audible. Increasing the word length to 20 from 16 bits significantly reduces this problem.

Note that signal-to-noise ratio and distortion figures specified for a digital audio system are with a full-scale signal. Most of the time, a music signal is well below full-scale and therefore closer to the noise floor. The distortion isn't a function of how many bits the system has available, but the number of bits used by the signal at any given moment. This is why distortion and noise are inversely proportional to signal amplitude, and why digital audio has problems with low-level signals.

These factors make setting recording levels in a digital system an art completely different from setting analog recording levels. Ideally, the highest peak over the entire audio program should just reach full-scale digital (using all the bits). If the recording level is set so that the highest peak reaches –6dB, this is equivalent to throwing away one bit of the quantization word, with the attendant 6dB reduction in signal-to-noise ratio. And if the analog signal amplitude is higher than the voltage represented by the highest number, the quantizer simply runs out of bits and repeats the highest number, making the waveform peaks flat. This highly distorted waveform produces an unacceptable "crunching" sound on peaks. If you have a digital-audio recording device with meters, you can look at the recording levels on compact discs by taking the digital output from a CD transport and plugging it into the digital recorder. The recorder's meters will show the exact levels recorded on the CD. If the highest peaks never reach full-scale, some resolution has been needlessly lost.

Note that an audio program with a very wide dynamic range will be closer to the quantization noise floor more of the time than a signal with limited dynamic range. The peaks on the wide-dynamic-range-signal will be set just below full-scale, and the much-lower-level signal will consequently be encoded with fewer bits. This is a particular problem in classical music, which has a very wide dynamic range. This is a strong incentive for recording or mastering engineers to compress the music's dynamic range. In addition, producers of pop music want their records to sound loud in relation to other songs over the radio. Severely limiting the music's dynamic range makes the music loud all the time, but at the expense of dynamics, life, and rhythmic power.

Digital signal levels are referenced to a full-scale signal, which occurs when all the bits are used and no higher signal can be coded. This reference level is called 0dBFS, with FS standing for "full-scale." We may refer to a signal as –20dBFS, meaning it is 20dB below full scale.

Dither

The most extreme case of quantization error occurs when the signal amplitude is less than one LSB (Fig.C-5). This low-level signal isn't encoded at all by the quantizer: the quantizer outputs the same code at each sample point and the information is completely lost (C and D). If the signal does traverse quantization steps, the encoded signal is a squarewave, which

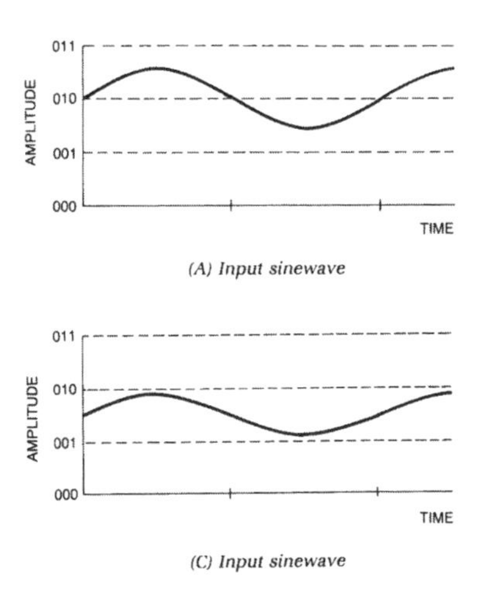

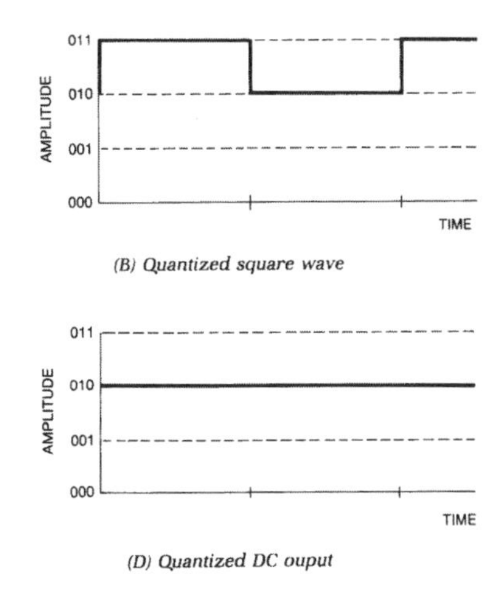

Fig. C-5 Quantization error is large relative to a signal traversing only a few quantization steps. (Courtesy Ken Pohlmann, *Principles of Digital Audio)*

represents a significant distortion of the original signal. This suggests that any information lower in amplitude than the LSB will be lost.

Fortunately, this limit can be overcome by adding a small amount of noise, called *dither*, to the audio signal. Dither allows a quantizer to resolve signals below the LSB, and greatly improves the sound of digital audio.

Fig.C-6 is identical to Fig.C-5, but this time we've added a small amount of white noise to the audio signal. This noise causes the audio signal to traverse quantization levels, allowing the signal to be encoded. The original sinewave is better preserved in the pulse-width modulation signal of Fig.C-6.

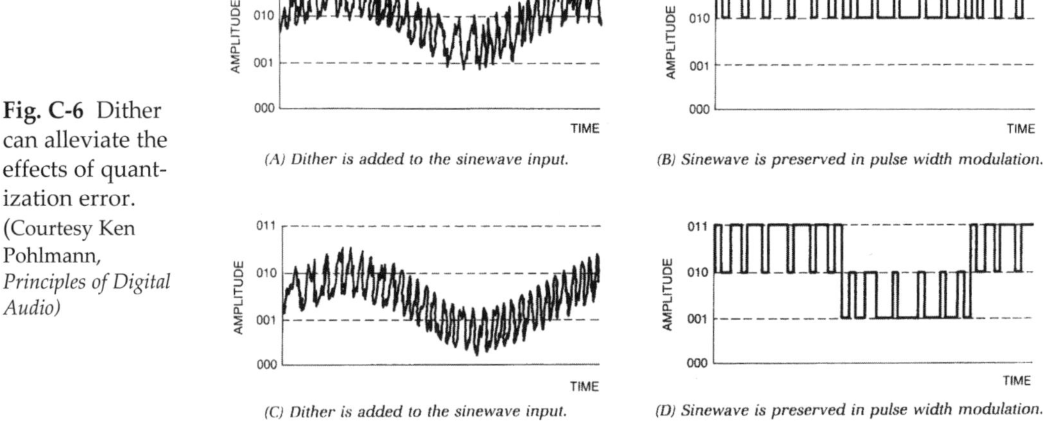

Fig. C-6 Dither can alleviate the effects of quantization error. (Courtesy Ken Pohlmann, *Principles of Digital Audio*)

Dither reduces quantization artifacts, allows the system to resolve information lower in amplitude than one-half the least significant bit, and makes a digital audio system sound more like analog. Among other benefits, dither improves low-level resolution and smooths reverberation decay. Without dither, reverberation decay becomes granular in texture, then seems to drop off into a black hole. It's ironic that a small amount of analog white noise can greatly improve the performance of digital audio. Still, despite the great advantages of dither, we do end up paying a small penalty in decreased signal-to-noise ratio. Dither is an odd concept to someone unfamiliar with digital audio; specifically, that adding noise to the audio signal can improve fidelity.

Finally, it should be pointed out that the preceding explanation of sampling and quantization describes a pulse-code modulation (PCM) digital-audio system, the standard encoding method in nearly all professional digital audio and in the compact disc and Blu-ray Disc formats. Other forms of encoding are possible such as Pulse-Width Modulation (PWM) used in the Super Audio CD format. This encoding method is explained in Chapter 7.

Digital Audio Storage

Once the audio signal is in the digital domain, we must store it on some medium. Common storage formats include the compact disc, solid-state memory, magneto-optical discs, and computer hard disks.

One might intuitively conceptualize a digital audio recording system as storing the binary information representing the audio signal directly on the medium. For example, a reflective spot on a CD could represent binary one, a non-reflective spot binary zero. In practice, however, digital audio is always encoded with a modulation scheme. Further, a significant amount of additional information is added to the raw audio data.

Every encoding scheme increases the storage capacity of the medium and facilitates data recovery. The Eight-to-Fourteen Modulation (EFM) encoding scheme used in the CD is a good example of encoding. The disc surface is a spiral track of small indentations called *pits* and flat areas called *land*. A transition from pit to land or land to pit represents binary one; all other surfaces (pit bottom or land) represent binary zero. The pattern of pit and land is created by the EFM encoding: 8 audio bits are converted into 14 bits for storage on the CD. These 14-bit words are linked by 3 "merging" bits to produce 17 bits for 8 audio bits. The result is a specific pattern of ones and zeros that follows certain rules. For example, each binary one is separate from other binary ones by a minimum of two zeros and a maximum of ten zeros. Although EFM encoding more than doubles the number of bits to be recorded, it actually increases the storage capacity of the CD compared to unmodulated coding. Note that both left and right audio channels as well as this additional information are encoded into a single serial datastream.

The compact disc is also a good example of how audio data are mixed with non-audio data and stored on the medium. The CD data-stream is formatted into structures called *frames*. A CD frame (Fig.C-7) contains a synchronization pattern to identify the beginning of a block, bits that contain information about the signal (pre-emphasis, elapsed time, data format, etc.), and error-correction information. Between EFM coding, error correction, subcode, and synchronization, only about a third of the data stored on a CD are actual audio data. All digital audio-storage formats add non-audio data and structure the data into clearly defined blocks.

For a full discussion of how audio data are stored and recovered from the compact disc, see the section in Chapter 7 on how CD transports work.

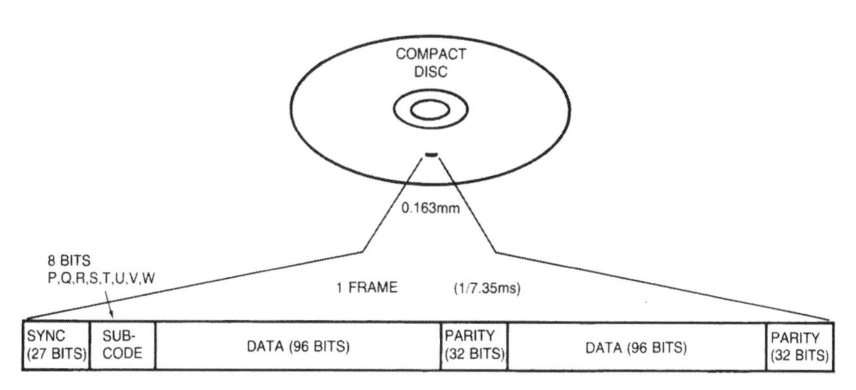

Fig. C-7 The CD frame format prior to EFM modulation. (Courtesy Ken Pohlmann, *Principles of Digital Audio*)

Error Correction

No storage or transmission medium is perfect. LPs have ticks and pops, and analog magnetic tape has imperfections in the magnetic coating that cause a momentary loss of high frequencies. In analog recording, nothing can be done about these errors; we must live with their sonic imperfections.

Digital audio, however, provides an opportunity to make the audio signal virtually impervious to media imperfections. Lost or damaged data can be reconstructed with error-correction techniques. Error correction doesn't merely approximate the missing data; in most cases it can precisely restore the lost information bit for bit. For example, the compact disc's error-correction system can completely correct up to nearly 4000 consecutive missing bits. These errors could result from poor-quality disc manufacturing, or scratches on the disc from mishandling. Without error correction, the CD simply wouldn't function. The official CD specification—called the "Red Book"—allows a Block Error Rate (BLER) of 220 errors per second.

The two principles on which error correction is based are *redundancy* and *interleaving*. Redundant data is information added to the signal that isn't necessary to decode the signal—unless that signal is corrupted. The crudest form of redundancy is simply to store the data twice. Data corrupted in one spot are unlikely to be corrupted at the second storage location. A more efficient technique transmits a *checksum* with the data that indicates if the data have been reliably retrieved. In practice, error-correction schemes are enormously complex.

Digital audio media are subject to two kinds of errors: *bit errors* and *burst errors*. Bit errors are short in duration and easy to correct. Burst errors, however, affect long, consecutive streams of data. Not only is a large chunk of data missing, but the redundant information accompanying that data is also lost and so the chance of correction is reduced.

A technique called *interleaving* increases the possibility that a burst error can be corrected. Interleaving scatters data to different areas of the media. When those data are deinterleaved (put back in the original order), the long burst error is broken up into many shorter errors that are more easily corrected.

If the error is so large that it cannot be completely corrected, the error is *concealed* by approximating the missing data. Some concealed errors may be audible as a *click* or *pop* sound. This is a rare occurrence; the CD's error-correction system is so good that very few discs—even those abused and scratched—have uncorrectable errors.

Digital-to-Analog Conversion

The digital-to-analog conversion process requires recovering the digital audio data from the medium and converting it to analog form.

The recovered datastream must first be demodulated. In the CD format, this involves converting the 14-bit EFM-coded words back into their original 8-bit form. The data are deinterleaved and error correction performed. Synchronization information—essential to data recovery—is no longer needed and is discarded. The subcode information (track number and timing, along with other housekeeping data) is stripped out of the datastream and handled separately from the audio data. All of these functions are performed by the CD transport, which outputs a formatted datastream to the outboard DAC (or internal digital-to-analog converters in a CD player), which actually converts the signal to analog. (A full description of how a DAC converts digital data to an analog signal is included in Chapter 7.)

Jitter Explained

In analog recording, speed variations in the recording media cause audible distortion. The classic case is a tape machine's "wow and flutter," speed fluctuations that cause the music's pitch to shift up and down slowly ("wow") or quickly ("flutter"). Digital audio doesn't suffer from wow and flutter; digital audio data can be read from the medium at a varying rate, yet be clocked out at a constant speed. But digital audio does suffer from a distortion-producing mechanism conceptually similar to wow and flutter, called *clock jitter*.

Jitter—timing variations in the clocks that serve as the time-base reference in digital audio systems—is a serious and underestimated source of sonic degradation in digital audio. Only recently has jitter begun to get the attention it deserves, both by high-end designers and audio academics. One reason jitter has been overlooked is the exceedingly difficult task of measuring such tiny time variations—on the order of tens of trillionths of a second. Another reason jitter has been ignored is the mistaken belief by some that if the ones and zeros that represent the music are correct, then digital audio must work perfectly. Unfortunately, getting the ones and zeros correct is only part of the equation. Those ones and zeros must be converted back to analog with extraordinarily precise timing to avoid degrading the signal.

As described in Chapter 7, the series of discrete audio samples are converted back into a continuously varying signal with a digital-to-analog converter (DAC) chip. A DAC takes a quantization word and converts it to a voltage—exactly opposite the function of the A/D converter. All that is required for perfect conversion (in the time domain) is that the samples be input to the DAC in the same order they were taken *and with the same timing reference* as when the samples were created. In theory, this sounds easy—just provide a stable clock to the A/D converter and a stable clock of the same frequency to the D/A converter.

It isn't that easy in practice. If the samples aren't converted back to an analog waveform with the identical timing with which they were taken, distortion of the analog waveform will result. These timing errors between samples are caused by variations in the clock signal that controls *when* the DAC converts the digital words to an analog voltage.

Let's take a little closer look at how the DAC decides *when* to convert the digital samples to analog. In Fig.C-8, the binary number at the left is the quantization word that represents the analog waveform's amplitude when it was first sampled. The bigger the number, the higher the amplitude. (This is only conceptually true—in practice, the data are in 2's complement form, a code that is easier to handle by circuits.) The squarewave at the top is the "word clock," the timing signal that tells the DAC *when* to convert the quantization word to an analog voltage. Assuming the original sampling frequency was 44.1kHz, the word clock's frequency will also be 44.1kHz (or, if the processor uses an oversampling digital filter, some multiple of 44.1kHz). On the word clock's *leading* edge, the next sample (quantization word) is loaded into the DAC. On the word clock's *falling* edge, the DAC converts that quantization word to an analog voltage. This process happens 44,100 times per second (without oversampling). If the digital processor has an 8x-oversampling digital filter, the word-clock frequency will be eight times 44,100, or 352.8kHz.

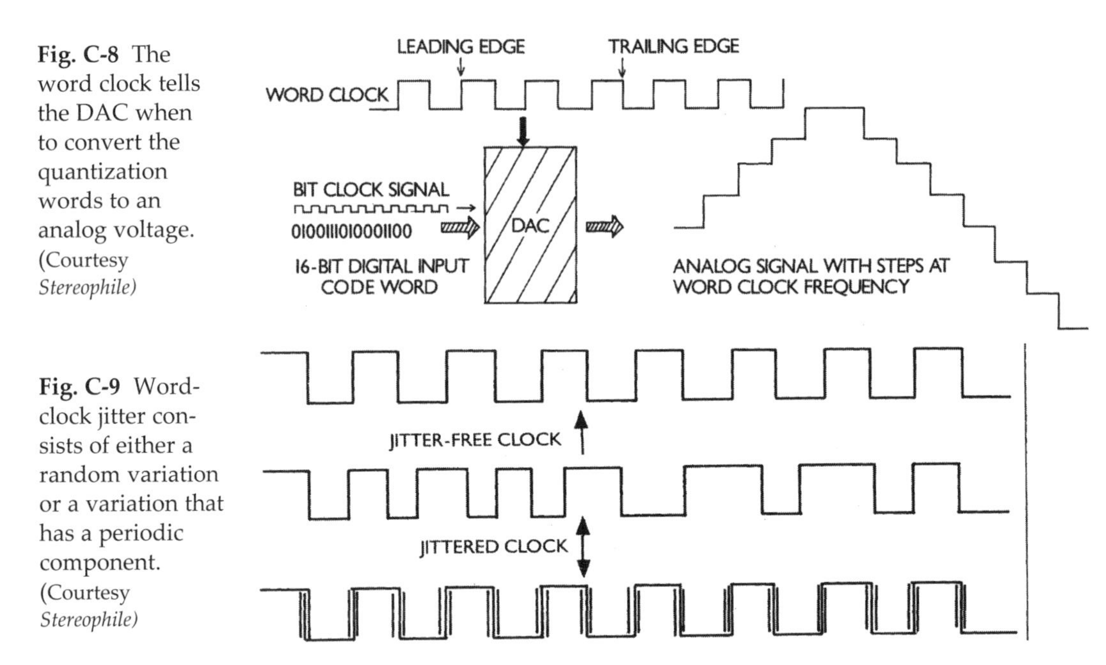

Fig. C-8 The word clock tells the DAC when to convert the quantization words to an analog voltage. (Courtesy *Stereophile)*

Fig. C-9 Word-clock jitter consists of either a random variation or a variation that has a periodic component. (Courtesy *Stereophile)*

It is here at the word clock that timing variations affect the analog output signal. Specifically, clock jitter is any time variation between the clock's trailing edges. Fig.C-9 shows perfect and jittered clocks (exaggerated for clarity).

Now, look what happens if the samples are reconstructed by a DAC whose word clock is jittered (Fig.C-10). The sample amplitudes—the ones and zeros—are correct, *but they're in the wrong places*. The right amplitude at the wrong time is the wrong amplitude. A *time* variation in the word clock produces an *amplitude* variation in the output, causing the waveform to change shape. A change in a waveform's shape is the very definition of distortion. Remember, the word clock tells the DAC *when* to convert the audio sample to an analog voltage; any variations in its accuracy will produce an analog-like variability in the final output signal—the music.

Clock jitter can also raise the noise floor of a digital converter and introduce spurious artifacts. If the jitter has a random distribution (called "white jitter" because of its similarity to white noise), the noise floor will rise. If, however, the word clock is jittered at a specific frequency, artifacts will appear in the analog output as sidebands on either side of the audio signal frequency being converted to analog. It is these specific-frequency artifacts that are the most sonically detrimental; they bear no harmonic relationship to the music, and may be responsible for the hardness and glare often heard from digital audio.

Fig.C-11a is a spectral analysis of a computer-simulated digital processor's output when reproducing a full-scale, 10kHz sinewave with a jitter-free clock. Fig.C-11b is the same measurement, but with 2 nano-seconds (two billionths of a second) of jitter with a frequency of 1kHz on the clock. The plot in Fig.C-11b reveals the presence of discrete-frequency sidebands on either side of the test signal caused by jitter of a specific frequency. The amplitude

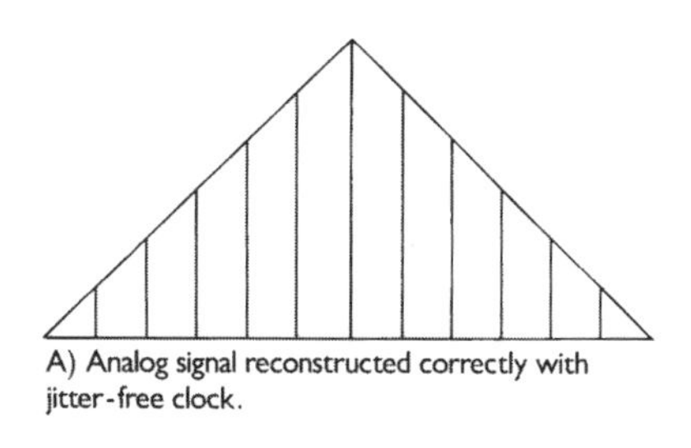

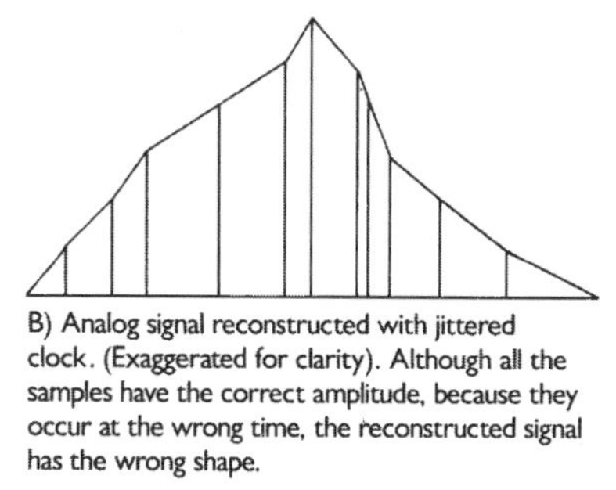

Fig. C-10 An analog waveform is reconstructed correctly with a jitter-free word clock (top); word-clock jitter results in a distortion of the analog waveform's shape. (Courtesy *Stereophile)*

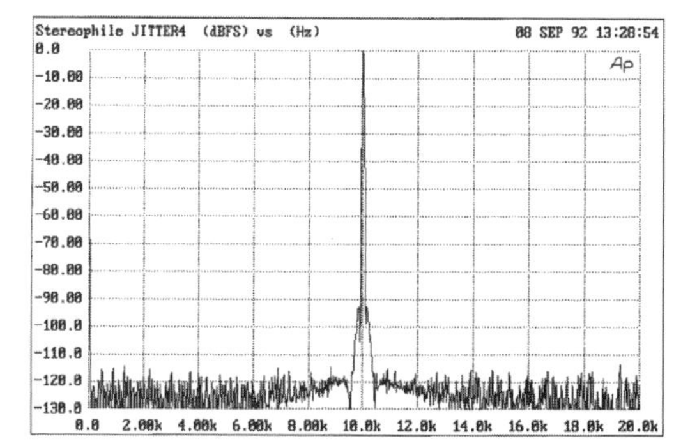

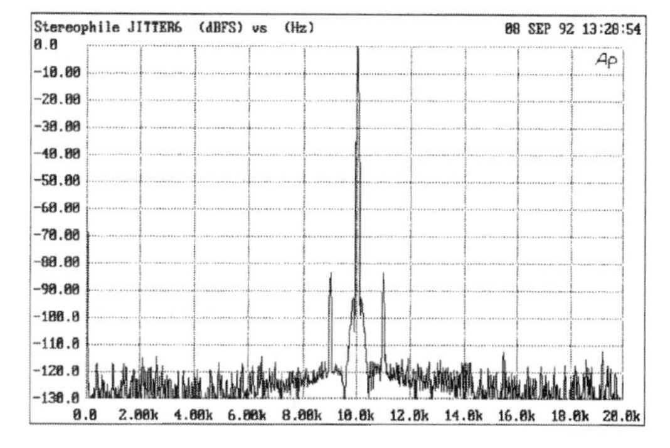

Figs. C-11a and b Computer-simulated spectral analysis of a digital processor's output with a jitter-free clock (left), and with 2ns of jitter (right). (Courtesy *Stereophile)*

of these artifacts is a function of the input signal level and frequency; the higher the signal level and frequency, the higher the sideband amplitude in the analog output signal.

How much jitter is audible? In theory, a 16-bit converter must have less than 100 picoseconds of clock jitter if the signal-to-noise ratio isn't to be compromised (there are 1000 picoseconds in a nanosecond). Twenty-bit conversion requires much greater accuracy, on the order of 8ps. One hundred picoseconds is one-tenth of a billionth of a second (10^{-10})—about the same amount of time it takes light to travel *an inch*. Moreover, this maximum allowable figure of 100ps assumes that the jitter is random (white), and does not have a specific frequency that would be less sonically benign. Clearly, extraordinary precision is required for accurate conversion.

Where does clock jitter originate? The primary source is the interface between a CD transport and a DAC. The master clock signal is embedded in the S/PDIF (Sony/Philips Digital Interface Format) signal that connects the two devices (though it's more accurate to say that the audio data are embedded in the clock). The DAC recovers this clock signal at the input receiver chip. The typical method of separating the clock from the data and creating a new clock with a *Phase Locked Loop* (PLL) produces *lots* of jitter. Even if the clock is recovered with low jitter, just about everything inside a digital processor introduces clock jitter: noise from digital circuitry, processing by integrated circuits—even a pcb trace.

It's important to note that the only point where jitter matters is at the DAC's word-clock input. A clock that's recovered perfectly but is degraded just before it gets to the DAC is no better than a high-jitter recovery circuit that's protected from additional jitter on its way to the DAC. Conversely, a highly jittered clock can be cleaned up just before the DAC with no penalty.

Digital Signal Processing (DSP)

DSP is the manipulation of an audio signal in the digital domain. DSP opens vast and exciting new possibilities for improving the quality of reproduced sound—possibilities unimaginable with analog techniques.

DSP is nothing more than the mathematical processing of the digital data representing an audio signal. This processing can take many forms, allowing DSP chips to be configured in a wide range of applications. For example, DSP chips can be programmed to be a high-pass and a low-pass filter, thus replacing the capacitors, resistors, and inductors in a loudspeaker crossover. Some loudspeakers accept a digital input, implement the crossovers with DSP, then convert the signal to analog inside the loudspeaker. The advantages of DSP crossovers are significant: the designer can make the crossover frequency and slope exactly as he or she wants it without regard for the limitations of analog crossover components—capacitors, resistors, and inductors. Moreover, the phase response can be precisely controlled, as can the time offset between drivers. In addition to such control, DSP crossovers get analog components out of the signal path for better sound. Such "digital" loudspeakers were pioneered by Meridian Audio. (DSP-based loudspeakers are covered in Chapter 6.)

Distortions in conventional loudspeakers can be corrected by DSP. If we know that a particular loudspeaker has amplitude errors and departures from linear phase response, those errors can be corrected in the digital domain by DSP.

Perceptual Coding

Storing or transmitting even standard-resolution (16-bit/44.1kHz) digital audio requires a large storage capacity of the recording medium, or fast Internet connection if downloading music. Large storage capacity and wide bandwidth translate to greater expense in delivering digital audio to consumers. This quandary is resolved by reducing the number of bits needed to encode an audio signal. This technique, called *perceptual coding,* attempts to radically reduce the data rate of digital audio signals with minimal effect on sound quality.

Dolby Digital is a good example of why perceptual coding is such a hot field. A 16-bit/44.1kHz digital audio signal (as is found on CD) consumes 705,600 bits per second per channel (16 times 44,100). With Dolby Digital's six discrete audio channels, a conventionally coded film soundtrack would consume a whopping 4,233,600 bits per second (4.23Mbs).

This is roughly half the maximum continuous bitrate possible from DVD, leaving very little data for digital video, which has an even more voracious appetite for data capacity. The Dolby Digital perceptual encoder produces a digital audio datastream with a bitrate of just 384,000 or 448,000 bits per second—for all six channels. That's less than one tenth the amount of data without perceptual coding.

Let's look at how this is possible. Perceptual coding exploits idiosyncrasies of human hearing to allocate the available bits to sounds we can hear, and to ignore sounds we're less likely to hear. For example, the psychoacoustic effect of *masking* (described in Appendix A) prevents us from hearing quiet sounds in the presence of louder sounds. Perceptual coders introduce severe noise and errors into the audio signal because the music is encoded with so few bits. But that noise and distortion are cleverly manipulated so that they are just under the *masking threshold*, or the point below which a sound is inaudible. For example, if most of the musical energy at a given moment is concentrated between 1kHz and 3kHz, the noise will also be concentrated in that frequency band. The noise spectrum is constantly shifting so that the encoding error (noise) "hides" beneath the correctly coded component of the audio signal. Frequency bands that contain more information at any given moment are allocated more bits; those bands with little information are allocated fewer bits. In practice, perceptual coding is vastly more complex.

I was once able to attend a remarkable demonstration of perceptual coding that highlighted how the technology works. I first heard a musical selection that had a signal-to-noise ratio of 13dB—a ridiculously low figure. For comparison, most electronics have signal-to-noise ratios of 80dB or greater. The sound was so noisy and distorted that the music was barely recognizable. I next heard the same music, again with a signal-to-noise ratio of 13dB, but this time it sounded fine. On an objective basis, the two presentations had exactly the same amount of noise. The difference was that in the second presentation, the noise was constantly shifting in spectral content so that it could "hide" beneath the correctly coded musical signal.

As powerful and useful as this technology can be, it has unfortunately been taken to extremes by some researchers in the field. They pronounced early perceptual coders as "transparent" (the perceptually coded signal was indistinguishable from the original), even though they sounded horrible. Moreover, in their zeal to further reduce the data rates, proponents of perceptual coding have sacrificed musical quality. The researchers speak of "psychoacoustic redundancy" and "informational irrelevance" with little regard for preserving the music with the highest fidelity. With this technological power comes hubris; perceptual coding has been promoted as suitable for master-tape archiving, a truly frightening thought. Our musical heritage should be preserved with the best technology, not the most economical.

Those readers interested in learning more about digital audio are directed to the excellent *Principles of Digital Audio* by Ken Pohlmann.

Glossary

A/B comparison A back-and-forth listening comparison between two musical presentations, A and B.

absolute polarity A recording with correct absolute polarity played back through a system with correct absolute polarity will produce a positive pressure wave from the loudspeakers in response to a positive pressure wave at the original acoustic event. Incorrect absolute polarity introduces a 180° phase reversal of both channels. Absolute polarity is audible with some instruments and to some listeners.

AC line-conditioner A device that filters noise from the AC powerline and isolates equipment from voltage spikes and surges. Prevents noise generated by one component in the audio system from getting into other components.

acoustic absorber Any material that absorbs sound, such as carpet, drapes, and thickly upholstered furniture.

acoustic diffuser Any material that scatters sound.

acoustic feedback Sound from a loudspeaker causes a turntable to vibrate; that vibration is converted by the phono cartridge into an electrical signal, which is reproduced by the loudspeaker, which causes the turntable to vibrate even more. This sets up a *feedback loop* in which the vibration feeds on itself, becoming louder and louder. You've probably heard acoustic feedback at an amplified concert as a howl or screech from the PA system.

acoustic panel absorber A device that absorbs low to mid frequencies by diaphragmatic action. That is, sound striking the panel absorber causes the panel to move, converting acoustic energy into a minute amount of heat in the panel.

acoustics The science of sound behavior. Also refers to a room; e.g., "This room has good acoustics."

AC synchronous A motor whose rotational speed is determined by the frequency of the AC voltage driving it. Used in most belt-drive turntables.

AC-3 Another name for Dolby Digital, the 5.1-channel discrete digital surround-sound format. (see "Dolby Digital")

active subwoofer A speaker designed to reproduce only low frequencies, and which includes an integral power amplifier to drive the speaker.

adaptive USB Type of USB connection in which the receiving device (usually a DAC) locks to the source device's clock. Contrast with "asynchronous USB" in which the receiving device's clock runs independently of the source device's clock, resulting in lower jitter.

ADC see "analog-to-digital converter"

adjacent-channel selectivity Tuner specification that describes a tuner's ability to reject radio stations adjacent to the desired station.

AES/EBU interface Professional system for transmitting digital audio. An AES/EBU cable is a balanced line and is terminated with XLR connectors. Also found on some consumer products. Named for the Audio Engineering Society and European Broadcasting Union.

agile Sonic description of bass that can follow quickly changing pitches and dynamics.

aggressive Sonic description of a forward presentation that seems to thrust the music on the listener.

AIFF "see Audio Interchange File Format"

air Sonic description of treble openness, or of space between instruments in the soundstage. Contrast with "dull," "thick."

air-bearing tonearm A tonearm in which the armtube floats on a cushion of air.

air-bearing turntable A turntable in which the platter floats on a cushion of air.

All Media Guide (AMG) Online service that downloads metadata to music servers.

alternate-channel selectivity Tuner specification describing a tuner's ability to reject stations two channels away from the desired station.

ambience Spatial aspects of a film soundtrack that create a sense of size and atmosphere, usually reproduced by the surround speakers.

Ambisonics A method of capturing a three-dimensional soundfield and reproducing that soundfield through loudspeakers. Although Ambisonics is compatible with monophonic and stereo reproduction, we generally think of Ambisonics as a multichannel music format.

AMG see "All Media Guide"

ampere Unit of electrical current, abbreviated A.

analog An analog signal is one in which the varying voltage is an analog of the acoustical waveform; i.e., the voltage varies continuously with the original acoustical waveform. Contrasted with a digital signal, in which binary ones and zeros represent audio or video information.

analog bypass Input on a digital controller that passes analog input signals to the output without analog-to-digital and digital-to-analog conversions.

analog-to-digital converter A circuit that converts an analog signal to a digital signal.

analytical Sonic description that describes a component that reveals every nuance in the recording, but in an unpleasant way. An analytical component is rarely musical.

anechoic Literally "without echo."

anechoic chamber An acoustically reflection-free room. An anechoic chamber's walls are covered in highly absorbent material so that no sound is reflected back into the room. Used in loudspeaker testing.

anti-skate adjustment Control on a tonearm that adjusts the amount of force applied to the tonearm to counteract the arm's natural tendency to skate (pull inward).

Apple Lossless Audio Codec Usually called just "Apple Lossless" or ALAC. File format that delivers perfect bit-for-bit accuracy to the source data with about a 40% reduction in file size.

articulate Sonic description of a component that clearly resolves pitches.

asynchronous sampling-rate converter Sampling rate converter in which the input and output sampling rates are free-running independently of each other.

asynchronous USB Type of USB connection in which the receiving device's clock runs independently of the source device's clock, resulting in lower jitter. Contrast with "adaptive USB" in which the receiving device locks to the source device's clock.

atmosphere see "ambience."

Audio Interchange File Format (AIFF) Digital audio file format in which the data are stored in raw PCM format.

audiophile A person who values sound quality in reproduced music.

Audiophilia nervosa Humorous name for a condition in which the listener is constantly fretting over the equipment at the expense of enjoying music.

Audio Return Channel Feature of HDMI 1.4a that sends the audio signal back from a television to a controller or AV receiver. Obviates the need for a separate audio cable when listening to audio through your theater system that was picked up by the television, such as when connected to an over-the-air antenna.

A/V Short for audio/video. Identifies a component or system as one that processes video as well as audio signals.

A/V input An input on an A/V receiver or A/V preamplifier that includes both audio and video jacks.

A/V loop An A/V input and A/V output pair found on all A/V receivers and A/V preamplifiers. Used to connect a component that records as well as plays back audio and video signals. A VCR is connected to a receiver's or preamplifier's A/V loop.

A/V preamplifier Also called by its more descriptive name of "A/V controller," the A/V preamplifier is a component that lets you control the playback volume and select which source you want to watch. A/V preamplifiers also perform surround decoding.

A/V preamplifier/tuner An A/V preamplifier that includes, in the same chassis, an AM or FM tuner for receiving radio broadcasts.

A/V receiver The central component of a home-theater system; receives signals from source components, selects which signal you watch and listen to, controls the playback volume, performs surround decoding, receives radio broadcasts, and amplifies signals to drive a home-theater loudspeaker system. Also called a "surround receiver."

azimuth In magnetic tape recording, the angle of the heads along the line of the tape, ideally 90°. In LP playback, the angle of the cantilever in relation to the record surface.

Baerwald alignment Cartridge alignment developed in 1941 by H.G. Baerwald. that specifies a certain cartridge offset and overhang for minimum tracking error One of two optimum alignments; the other is Löfgren.

baffle The front surface of a loudspeaker on which the drivers are mounted.

balanced connection A method of connecting audio components with three conductors in a single cable. One conductor carries the audio signal, a second conductor carries that audio signal with inverted polarity, and the third conductor is the ground.

banana jack A small tubular connector found on loudspeakers and power amplifiers for connecting speaker cables terminated with banana plugs.

banana plug A common speaker-cable termination that fits into a banana jack.

bandwidth The range of frequencies that a device can process or pass. In reference to an electrical or acoustic device, bandwidth is the range of frequencies between the –3dB points.

bass Sounds in the low audio range, generally frequencies below 500Hz.

bass extension A measure of how deeply an audio system or loudspeaker will reproduce bass. For example, a small subwoofer may have bass extension to 40Hz. A large subwoofer may have bass extension to 16Hz.

bass management A combination of controls and circuits in an A/V receiver, A/V preamplifier, or multichannel digital audio player (DVD-A and SACD) that determines which speakers receive bass signals.

bass reflex A speaker design with a hole or slot in the cabinet that allows sound inside the cabinet to emerge into the listening room. Bass reflex speakers have deeper bass extension than speakers with sealed cabinets, but that bass is generally less tightly controlled. Also called "vented" or "ported." Contrast with "infinite baffle."

B-Format In Ambisonics, the W (mono pressure signal), X (sides), and Y (front to back) signals. Also refers to these signals along with the Z (height) signal.

bi-amping Using two power amplifiers to drive one loudspeaker. One amplifier typically drives the woofer, the second drives the midrange and tweeter.

binaural recording Recording technique in which microphones are placed inside the "ear canals" of a dummy head. The recording contains spatial cues introduced by the dummy head's physical structure. When reproduced through headphones, a binaural recording re-creates the original three-dimensional acoustic space with uncanny realism.

binding post A connection on power amplifiers and loudspeakers for attaching loudspeaker cables.

bipolar speaker A speaker that produces sound equally from the front and the back. Unlike the dipolar speaker, the bipolar's front and rear soundwaves are in-phase with each other.

bipolar transistor Most common transistor type in audio electronics. Name "bipolar" (two poles) comes from fact that current flows through two types of semiconductor material, P and N. Bipolar transistors are either NPN or PNP types, which refers to the polarity of their operating voltages.

bit rate The number of bits per second stored or transmitted by a digital audio or digital video signal. For example, the bit rate of compact disc is 705,600 bits per second per channel. Higher bit rates generally translate to better audio quality. Also called "datarate."

bit transparency Ability of a music server to store and output a bitstream that is identical to the input data. A bit-transparent server does not corrupt the audio data.

bi-wiring Technique of running two cables to each loudspeaker. One cable is connected to the loudspeaker's woofer input terminals, one to the tweeter's input terminals. Bi-wiring is possible only with loudspeakers with two pairs of input terminals.

bleached Sonic description of a component that emphasizes the upper harmonics of an instrument's sound, while de-emphasizing the lower harmonics and fundamental. A bleached sound is thin, bright, and lacking warmth.

bloom Sonic description of a sense of air around instrumental images.

Blu-ray Disc 120mm optical disc format developed as the high-definition successor to DVD. Can store high-resolution multichannel audio in addition to HD video. An audio-only version can deliver up to eight channels of PCM audio at 192kHz/24 bit.

boomy Excessive bass, particularly over a wide band of frequencies.

boutique brand A product line that appears to be high-end, but actually puts inferior electronics in slick, eye-catching chassis.

break-in Initial period of use of a new audio component, during which time the component's sound improves.

bridging Amplifier-to-loudspeaker connection method that converts a stereo amplifier into a monoblock power amplifier. One amplifier channel amplifies the positive half of the waveform, the other channel amplifies the negative half. The loudspeaker is connected as the "bridge" between the two amplifier channels.

brightness In audio, an excessive amount of treble that adds shrillness to the sound.

brittle A midrange or treble character that makes instrumental timbres sound harsh. Contrast with "liquid."

buffer Electronic circuit that isolates one component or circuit stage from another. A preamplifier is a buffer because it acts as an intermediary between source components and a power amplifier. The preamplifier's buffering function relieves the source components of the burden of driving a power amplifier. A digital buffer stores data temporarily in solid-state memory. Digital buffers act as jitter-reduction devices because the output is controlled by a precision clock.

build quality The quality of electronic parts, chassis, and construction techniques of an audio or video component.

bypass test Listening test in which the component under evaluation is alternately removed from and inserted into the signal path in order to assess its sonic characteristics.

calibration The act of fine-tuning an audio or video component for correct performance. In an audio system, calibration includes setting the individual channel levels.

cantilever Thin tube protruding from a phono cartridge that holds the stylus.

capacitive reactance Property of capacitance that blocks low frequencies but passes high frequencies. Capacitive reactance makes a capacitor behave as a frequency-dependent resistor. Because of capacitive reactance, a capacitor connected to a tweeter allows treble to pass but blocks bass.

capacitor Electronic component that stores a charge of electrons. Reservoir capacitors are used for energy storage in power amplifiers; filter capacitors filter traces of AC from DC power supplies;

coupling capacitors connect one amplifier stage to another by blocking DC and allowing the AC audio signal to pass.

capture ratio Tuner specification: the difference in dB required between the strengths of two stations needed before a tuner can lock to the stronger station and reject the weaker one. The lower the capture ratio, the better the tuner.

cartridge demagnetizer Device that removes stray magnetic fields from metal parts inside a phono cartridge.

CD Compact Disc Plastic disc 120mm in diameter that stores up to 80 minutes of music.

CD-4 Compatible Discrete 4. A Quadraphonic format invented by JVC in which the two rear channels are encoded along with the front channels, and a high-frequency (30kHz–50kHz) difference signal allows the rear-channel signals to be demodulated.

CD Recordable Format for recording digital audio on a compact disc. CDR is a "write once" format, meaning that once recorded on, the disc cannot be erased.

CD-Rewritable Recordable CD format that can be rerecorded over multiple times.

center channel In a multichannel audio system, the audio channel that carries information that will be reproduced by a speaker placed in the center of the viewing room between the left and right speakers. The center channel carries nearly all of a film's dialogue.

center-channel mode A setting on A/V receivers and A/V preamplifiers that configures the receiver or preamplifier for the type of center-channel speaker in the system.

center-channel speaker The speaker in a home-theater system located on top of, beneath, or behind the visual image; reproduces center-channel information such as dialogue and other sounds associated with onscreen action.

channel balance The relative levels or volumes of the left and right channels in an audio system or individual component.

channel separation A measure of how well sounds in one channel are isolated from the other channels.

Chesky 6.0 An alternative multichannel audio technique proposed by Chesky Records using four front loudspeakers and two rear loudspeakers.

chesty A loudspeaker coloration that emphasizes the deep chest sounds in spoken or singing voice.

chuffing Sound created by the port of a bass reflex loudspeaker when reproducing low bass at high levels. Caused by large movement of air in the port.

Class-A Mode of amplifier operation in which a transistor or tube amplifies the entire audio signal.

Class-A/B Mode of amplifier operation in which the output stage operates in Class-A at low output power, then switches to Class-B at higher output power.

Class-B Mode of amplifier operation in which one tube or transistor amplifies the positive half of an audio signal, and a second tube or transistor amplifies the negative half.

clipping An amplifier that is asked to deliver more power than it is capable of will flatten the tops and bottoms of the audio waveform, making the peaks appear to be clipped off. Clipping introduces a large amount of distortion, audible as a crunching sound on musical peaks.

clock A signal present in all digital products that acts as a timekeeper. Can also refer to an outboard device that provides a precision clock to a DAC or disc transport.

coaxial cable A cable in which an inner conductor is surrounded by a braided conductor that acts as a shield.

coaxial digital output A jack found on CD transports and some CD players that provides a digital audio signal on an RCA jack for connection to another component through a coaxial digital interconnect.

coaxial driver A loudspeaker driver in which one drive unit (typically a tweeter) is mounted inside a second drive unit (typically a midrange).

coherence The impression that the music is an integrated whole, rather than made up of separate parts.

coincident microphones A stereo or multichannel recording technique in which microphones are placed very close to each other. Examples include Blumlein (also called crossed figure-of-eight), XY (crossed cardioids), and MS (middle-side; a forward-oriented omnidirectional or cardioid microphone and a figure-of-eight microphone).

coloration A change in sound introduced by a component in an audio system. A "colored" loudspeaker doesn't accurately reproduce the signal fed to it. A speaker with coloration may have too much bass and not enough treble, for example.

comb filtering A phenomenon that introduces a series of deep notches in the frequency response, usually caused by combining a direct signal with a slightly delayed version of that signal, such as the sidewall reflection from a loudspeaker in a listening room.

common-mode rejection When a balanced signal is input to a differential amplifier, only the difference between the two phases of the balanced signal is amplified. Any noise common to both phases (common-mode noise) is rejected by the differential amplifier.

compliance In phono cartridges, a number expressing the cantilever's stiffness. Specifically, compliance is the length of cantilever movement when a force of 10–6 dynes is applied, expressed in millionths of a centimeter. High compliance is any value above 20. Low compliance is any value below 10.

compression A component of a sound wave in which the air density is greater than normal atmospheric pressure. A woofer cone moving outward creates a compression. One compression and one rarefaction make one complete cycle of a sound wave.

cone The paper, plastic, metal, or exotic-material diaphragm of a loudspeaker that moves back and forth to create sound.

congested A thickening of the sound that makes instrumental images less separate and distinct.

constrained-layer damping Construction method in which a layer or layers of soft materials are placed between layers of harder materials to dissipate vibrational energy.

Consumer Electronics Show (CES) Annual trade show for the consumer electronics industry, held in Las Vegas in early January.

contact cleaner Fluid for removing oxide and dirt from audio jacks and plugs.

continuousness Quality of a musical presentation in which the sound seems a whole, rather than disjointed.

control components An audio system's preamplifier is a control component because it adjusts the volume and determines which source you listen to.

controller Another term for an A/V preamplifier.

critical listening The art of evaluating audio equipment quality by careful analytical listening for specific sonic flaws.

crossover A circuit that splits up the frequency spectrum into two or more parts. Crossovers are found in virtually all dynamic loudspeakers, and in A/V receivers and A/V preamplifiers.

crossover distortion see "zero-crossing distortion"

crossover frequency The frequency at which the audio spectrum is split. A subwoofer with a crossover frequency of 80Hz filters all information above 80Hz out of the signal driving the subwoofer, and all information below 80Hz out of the signal driving the main speakers.

crossover slope Describes the steepness of a crossover filter. Expressed as "*x*dB/octave." For example, a subwoofer with a crossover frequency of 80Hz and a slope of 6dB/octave would allow audio frequencies at 160Hz (an octave above 80Hz) into the subwoofer, but signals at 160Hz would be reduced in amplitude by 6dB. A slope of 12dB/octave would also allow 160Hz into the subwoofer, but the amplitude would be reduced by 12dB. The most common crossover slopes are 12dB/octave, 18dB/octave, and 24dB/octave. Crossover slopes are also referred to as "first-order" (6dB/octave), "second-order" (12dB/octave), "third-order" (18dB/octave), and "fourth-order" (24dB/octave). The "steeper" slopes (such as 24dB/octave) split the frequency spectrum more sharply and produce less overlap between the two frequency bands.

crosstalk see "channel separation"

current The flow of electrons in a conductor. For example, a power amplifier "pushes" electrical current through speaker cables and the voice coils in a loudspeaker to make them move back and forth.

current-to-voltage converter Circuit following the digital-to-analog converter chip that converts the DAC's current pulses to a voltage.

cutoff frequency see "crossover frequency"

DAC see "digital-to-analog converter"

damping factor A number that expresses a power amplifier's ability to control woofer motion. Related to the amplifier's output impedance.

DAS see "Direct Attached Storage"

data compression see "perceptual coding"

data rate see "bitrate"

dB see "decibel"

dBFS Decibels referenced to full scale in digital audio. Full scale is the maximum recorded level possible in digital. All digital levels are expressed in dBFS; a signal level 20dB below full scale is –20dBFS.

dBSPL A measurement of loudness (sound-pressure level) in decibels, referenced to the threshold of hearing (0dBSPL).

dBW A measurement of amplifier power in decibels referenced to 1W of output power.

DC (Direct Current) Flow of electrons that remains steady rather than fluctuating. Contrasted with alternating current (AC).

decibel The standard unit for expressing relative power or amplitude levels of voltage, electrical power, acoustical power, or sound-pressure level. Abbreviated dB.

de-emphasis In the compact disc format, a treble boost can be applied during recording (pre-emphasis), and a treble cut during playback (de-emphasis), increasing the signal-to-noise ratio.

depth The impression of instruments, voices, or sounds existing behind one another in three dimensions, as in "soundstage depth."

detail Low-level components of the musical presentation, such as the fine inner structure of an instrument's timbre.

dialogue intelligibility The ability to clearly hear and understand the dialogue in a movie without strain. Dialogue intelligibility is affected by the quality of components in a home-theater system, room acoustics, and how the system is set up.

diaphragm The surface of a loudspeaker driver that moves, creating sound.

dielectric Insulating material inside a capacitor or around a cable.

differential amplifier Electronic circuit that amplifies only the difference between the two phases of the balanced signal.

diffraction The bending of soundwaves as they pass around an object. Also: a re-radiation of sound caused by discontinuities in surfaces near the radiating device, such as the bolts that secure drivers to a speaker cabinet.

diffusion Scattering of sound. Diffusion reduces the sense of direction of sounds, which benefits sound produced by surround loudspeakers.

digital Calculation or representation by discrete units. For example, digital audio and digital video can be represented by a series of binary ones and zeros.

digital amplifier Amplifier that takes in PCM audio data and converts the PCM to a pulse-width modulated signal that turns the amplifier's output transistors on and off. Not to be confused with a Class-D amplifier that employs an analog signal path with a switching output stage.

digital loudspeaker Loudspeaker incorporating a digital crossover and power amplifiers. A digital loudspeaker takes in a digital bitstream, spits up the frequency spectrum with digital signal processing, converts each of those signals to analog, and amplifies them separately. The individual power amplifiers then power each of the loudspeaker's drive units.

digital preamplifier Audio component that performs source selection, volume control, and other functions in the digital domain.

Digital Satellite System (DSS) A method of delivering high-quality digital video into consumers' homes via an 18" roof-mounted dish.

Digital Signal Processing (DSP) Manipulation of audio or video signals by performing mathematical functions on the digitally encoded signal.

Digital Theater Systems (DTS) A discrete digital surround-sound format used in movie theaters and some home-theater systems. A better-sounding alternative to Dolby Digital. Also called DTS Digital Surround. The DTS format encompasses film soundtracks and music releases.

digital-to-analog converter A chip that converts digital audio signals to analog audio signals. Also: A stand-alone component in an audio system that converts digital audio signals to analog signals. Also called a digital processor or DAC.

Digital Transmission Content Protection (DTCP) Copy-protection system that allows content owners to control how the content may be copied. Also known as "5C" after the five companies that developed the standard.

digital volume control Digital circuit that adjusts the signal level by performing mathematical computation on the ones and zeros representing the audio signal.

diode Electronic component that allows current to flow in only one direction.

dip A reduction in energy over a band of frequencies. Contrast with peak.

dipolar speaker A loudspeaker that produces sound from the rear as well as from the front, with the front and rear sounds out-of-phase with each other.

Direct Attached Storage (DAS) Type of storage device (usually hard-disc drive) in which the drive is addressable only by the host computer. Contrast with Network Attached Storage (NAS) in which the storage device can be addressed by other devices on the network.

direct-coupled Electronic circuit in which no coupling capacitors are in the signal path.

Direct Stream Digital (DSD) Method of digitally encoding music with a very fast sampling rate, but with only 1-bit quantization. Developed by Sony and Philips for the Super Audio CD (SACD).

direct-to-disc Recording format in which the LP master lacquer is cut in real-time as the music is performed with no internediate storage medium.

discrete Separate. A circuit using separate transistors rather than an integrated circuit. A discrete digital surround-sound format contains 5.1 channels of audio information that are completely separate from each other; contrasted with a matrixed surround format such as Dolby Surround, which mixes the channels together for transmission or storage.

dispersion The directional pattern over which a loudspeaker distributes its sound.

dither A small amount of noise added to an audio signal that improves the low-level resolution of digital audio.

Dolby Digital A 5.1-channel discrete digital surround-sound format used in movie theaters and consumer formats. The surround format required on the DVD format and HDTV broadcasts.

Dolby Digital EX Surround format that matrix-encodes a third surround channel in the existing left and right surround channels of a Dolby Digital signal. This third surround channel (called "surround back") drives a loudspeaker or loudspeakers located directly behind the listening position for greater spatial realism. Called THX EX in THX-certified products.

Dolby Digital Plus Surround format that provides a higher bitrate than Dolby Digital along with more efficient coding, both of which contribute to higher sound quality. Dolby Digital Plus is an extension of Dolby Digital in that it appends additional data to a Dolby Digital signal. This feature makes Dolby Digital Plus backward compatible with Dolby Digital decoders. Dolby Digital Plus was developed for Blu-ray Disc

Dolby Pro Logic A type of Dolby Surround decoder with improved performance over standard Dolby Surround decoding. Specifically, Pro Logic decoding provides greater channel separation and a center-speaker output. A Dolby Pro Logic decoder takes in a 2-channel, Dolby Surround-encoded audio signal and splits that signal up into left, center, right, and surround channels.

Dolby Pro Logic II Pro Logic surround decoder with improved performance and added features compared with older Pro Logic circuits.

Dolby Pro Logic IIx Pro Logic surround decoder that generates the surround-back channel from 2-channel or 5.1-channel sources.

Dolby Pro Logic IIz Surround decoding format that provides outputs that drive additional "height" loudspeakers located above the left and right loudspeakers.

Dolby Surround An encoding format that combines four channels (left, center, right, surround) into two channels for transmission or storage. On playback, a Dolby Pro-Logic decoder separates the two channels back into four channels.

Dolby TrueHD Surround format developed for Blu-ray Disc. Unlike Dolby Digital which is a "lossy" compression format, TrueHD is "lossless" in that it delivers high-resolution multichannel audio with perfect bit-for-bit fidelity to the source data.

downconverted A digital audio signal that has been converted from a higher sampling frequency to a lower sampling frequency. For example, a digital master tape recorded at 96kHz is downconverted to 44.1kHz sampling for storage on a compact disc.

downmix converter A circuit found in DVD players that converts the 5.1-channel discrete Dolby Digital soundtrack into a 2-channel Dolby Surround-encoded signal. A DVD player's downmix converter lets you hear surround sound from DVD if you don't have a Dolby Digital decoder. Also refers to the circuit in all DVD-Audio players that creates, on the fly, a 2-channel mix from the disc's multichannel signal.

driver The actual speaker units inside a loudspeaker cabinet.

dry Sonic presentation in which the ambience or reverberation is reduced.

DSD see "Direct Stream Digital"

DSP see "Digital Signal Processing"

DSP room correction Technique of removing room-induced frequency-response peaks and dips with digital signal processing.

DTCP see "Digital Transmission Content Protection"

DTS see "Digital Theater Systems"

DTS ES Discrete Surround-sound format that delivers discrete 6.1-channel sound. The additional channel (compared with 5.1-channel sound) drives a loudspeaker or loudspeakers located directly behind the listening position.

DTS ES Matrix Surround-sound format that matrix-encodes a third surround channel into the existing left and right surround channels in a DTS signal.

DTS HD Surround format developed for Blu-ray Disc that provides a higher bitrate and more efficient coding than DTS, both of which improve sound quality.

DTS HD Master Audio Surround format developed for Blu-ray Disc that provides high-resolution multichannel audio with perfect bit-for-bit fidelity to the source data. The "Master Audio" appelation signifies "lossless" compression.

DTS Neo:6 Decoding circuit developed by Digital Theater Systems for creating 6-channel audio from 2-channel sources. Comes in two flavors: DTS Neo:6 Music and DTS Neo:6 Cinema.

dull Lacking treble energy.

DVD CD-sized disc that can contain digital video (DVD-Video), high-resolution digital audio (DVD-Audio), or computer data (DVD-ROM).

DVD-Audio Subset of the DVD standard that provides for high-resolution, multichannel digital audio on DVD.

DVD-Video Format created for storing movies on a CD-sized disc.

dynamic compression Phenomenon in loudspeaker drivers in which an increase in drive signal results in little increase in acoustic output. Occurs when the driver's voice coil heats.

dynamic range In audio, the difference in volume between loud and soft. In video, the difference in light level between black and white (also called contrast).

dynamic range compressor A circuit found in some Dolby Digital-equipped receivers and preamplifiers that reduces dynamic range. A dynamic range compressor can reduce the volume of peaks, increase the volume of low-level sounds, or both. Useful for late-night listening when you don't want explosions to disturb other family members, but still want to hear low-level sounds clearly.

dyne A unit of force. The threshold of hearing (0dBSPL) represents a pressure on your eardrum of 0.0002 dynes/cm2. A force of 10^{-6} dynes applied to a phono cartridge cantilever will push that cantilever approximately 15 millionths of a centimeter.

edge A harshness in transient musical events such as the initial attacks of percussion instruments.

effective tonearm mass The total mass of a tonearm's moving parts, and where along the tonearm that mass is distributed. Mass near the pivot point only slightly increases a tonearm's effective mass, but the same amount of mass near the tonearm's cartridge end greatly increases the effective mass. A tonearm's effective mass must be matched to the phono cartridge's compliance for optimum performance.

electromagnetic interference (EMI) Disturbance caused by radiation of electromagnetic waves.

electrostatic loudspeaker Loudspeaker in which a thin diaphragm is moved back and forth by electrostatic forces. Contrasted with a dynamic loudspeaker, in which electromagnetic forces move the diaphragm back and forth.

EMI see "electromagnetic interference"

envelopment The impression of being surrounded by sound, such as in a surround-sound system.

equalization In tape or LP record playback, a treble cut to counteract a treble boost applied during recording. Also describes modification of tonal balance by employing an equalizer.

equalizer A circuit that changes the tonal balance of an audio program. Bass and treble controls are a form of equalizer.

etched An unpleasant emphasis on transient musical information, accompanied by excessive brightness.

extension How high or low in frequency an audio component can reproduce sound.

Fast Fourier Transform (FFT) Mathematical technique that converts information from the time domain to the frequency domain.

feedback In amplifier circuits, taking part of the output signal and sending it back to the amplifier's input. Feedback reduces distortion and makes the circuit more stable. (see also "acoustic feedback")

FFT see "Fast Fourier Transform"

filter Electronic circuit that selectively removes or reduces the amplitude of certain frequencies.

filter capacitor Type of capacitor used in power supplies to filter traces of 60Hz AC from the DC supply.

FireWire A high-speed interface that can carry high-resolution digital audio. Officially called IEEE1394, but also known as i.LINK (Sony's trade name). An open standard developed by Apple Computer.

5C see "Digital Transmission Content Protection"

5.1 channels The standard number of channels for encoding film soundtracks. The five channels are left, center, right, surround left, and surround right. The ".1" channel is a 100Hz-bandwidth channel reserved for high-impact bass effects.

FLAC (Free Lossless Digital Audio Codec) Digital audio file format that provides perfect bit-for-bit accuracy to the source through lossless compression. A FLAC file is approximately 60% the size of a WAV file of the same data.

flat A speaker that accurately reproduces the signal fed to it is called "flat" because that is the shape of its frequency-response curve. Flat also describes a soundstage lacking in depth.

floorstanding speaker A speaker that sits on the floor rather than on a stand.

flutter echo Back-and-forth acoustic reflections in a room between pairs of surfaces. Think of a pair of facing mirrors, each reflecting light into the other. Flutter echo can be heard as a "pinging" sound after a handclap. Caused by untreated parallel surfaces.

forward A description of a sonic presentation in which sounds seem to be projected forward toward the listener.

frequency Number of repetitions of a cycle. Measured in Hertz (Hz), or cycles, per second. An audio signal with a frequency of 1000Hz (1kHz) undergoes 1000 cycles of a sinewave per second.

frequency response A graphical representation showing a device's relative amplitude as a function of frequency.

full-range speaker A speaker that reproduces bass as well as midrange and treble frequencies.

gain Number expressing the amount of amplification provided by an amplifier. An amplifier that converts a signal with an amplitude of 0.1V to 1V is said to have a gain of 10.

geometry In cables, the physical arrangement of the conductors and dielectric.

gimbal bearing Type of tonearm bearing in which the arm is held within a set of gimbals. Contrast with unipivot.

Gracenote CDDB Database identifying album, song title, and artist that can be embedded in digital audio servers, computer-based audio systems, and portable music players, or accessed via the Internet.

grainy Sonic description of a roughness to instrumental or vocal timbres.

half-speed mastering Technique of cutting an LP record in which the master tape and the lathe's turntable run at half the normal speed, resulting in correct pitch and tempo when played at normal speed. Half-speed mastering gives the cutting head more time to trace the signal, resulting in better sound.

harmonic distortion The production of spurious frequencies at multiples of the original frequency. A circuit amplifying a 1kHz sinewave will create frequencies at 2kHz (second harmonic), 3kHz (third harmonic), and so forth.

HDCD see "High Definition Compatible Digital"

heatsink Large metal device that draws heat away from the interior of an electronic device and dissipates that heat in air. The fins protruding from the sides of power amplifiers are heatsinks.

hierarchical encoding In Ambisonics, a multichannel signal that is scalable to any number of playback channels, including mono and stereo, ensuring compatibility with a wide range of playback systems.

Helmholtz resonator Acoustic device with a small opening that absorbs sound by causing the air in the device to resonate at a particular frequency.

Hertz (Hz) The unit of frequency; the number of cycles per second. Kilohertz (kHz) is thousands of cycles per second.

high-blend circuit FM tuner feature that automatically switches the signal to monaural when the broadcast signal strength falls below a certain level.

high-cut filter Filter that removes treble energy.

High Definition Compatible Digital (HDCD) Process for improving the sound quality of 16-bit/44.1kHz digital audio on compact disc. An HDCD-encoded disc will play on any CD player, but sounds best when played on a CD player or digital processor equipped with an HDCD decoder.

high-density layer The information layer in a hybrid Super Audio CD that contains high-resolution digital audio.

high-pass filter A circuit that allows high frequencies to pass but blocks low frequencies. Also called a "low-cut filter." High-pass filters are often found in A/V receivers and A/V preamplifiers to keep bass out of the front speakers when you're using a subwoofer.

high-resolution digital audio Generally regarded as digital audio with a sampling rate greater than 48kHz and a word length longer than 16 bits.

home theater The combination of high-quality sound and video in the home.

Home THX A set of patents, technologies, and playback standards for reproducing film soundtracks in the home. THX doesn't compete with surround formats such as Dolby Digital or DTS, but instead builds on them.

HRx Name for Reference Recordings' format in which high-resolution WAV files are stored on DVD. HRx files can be played directly by a few disc players, but more commonly the files are transferred from the DVD to a hard-disc drive for later playback.

hybrid An audio component that combines more than one technology, such as tubes and transistors in the same amplifier, or dynamic and ribbon drivers in the same loudspeaker.

hybrid disc SACD-based disc that is compatible with CD players as well as SACD players. Has two information layers; one layer contains CD-quality audio, the second contains high-resolution audio.

Hz see "Hertz"

IC Integrated circuit. Some products use ICs for processing and amplifying audio signals; higher-quality units use discrete transistors instead.

i.LINK Sony trade name for the IEEE1394 (FireWire) interface.

image specificity Soundstage attribute in which instrumental or vocal images are precisely positioned and clearly defined.

imaging The impression of hearing, in reproduced music, instruments and voices as objects in space.

IMD see "intermodulation distortion"

immediate An immediate musical presentation is somewhat vivid, lively, and forward. Contrast with "laid-back."

impedance Resistance to the flow of AC electrical current. An impedance is a combination of resistance, inductive reactance, and capacitive reactance.

inductive reactance The property of an inductor that increases its opposition to current flow as frequency increases.

inductor Electronic component that increases its opposition to current flow as frequency increases. Many loudspeaker crossovers use a large inductor on the woofer to filter out high frequencies.

infinite baffle A sealed loudspeaker cabinet. The cabinet wraps around the drive units, mimicking a baffle of infinite size. Also called air suspension or acoustic suspension. Contrast with a reflex, or ported, loudspeaker.

infrared (IR) The frequency of light emitted by wireless remote controls.

input impedance The resistance to current flow presented by a circuit or component to the circuit or component driving it. Impedance is a combination of resistance, capacitive reactance, and inductive reactance.

input overload A condition in which a component (typically a preamplifier) produces severe distortion because the signal driving the component is too high in level. For example, connecting a high-output moving-magnet cartridge to a preamplifier's moving-coil input could overload the preamplifier's input, introducing audible distortion.

input receiver Chip or circuit in a digital processor that takes in the S/PDIF datastream, generates a new clock based on that datastream, strips out the audio data, and sends the audio data to the digital filter. Converts S/PDIF to audio data, control data, and a clock.

integer upsampling Increasing the sampling frequency of a digital audio signal by an integer multiple; i.e. 44.1kHz to 88.2kHz (2x) or 176.4kHz (4x). Contrasted with non-integer upsampling; i.e. 44.1kHz to 96kHz.

integrated amplifier Audio product combining a preamplifier and power amplifier in one chassis.

interconnect A cable that carries line-level audio signals (audio interconnect), composite video signals (video interconnect), or S/PDIF-encoded digital audio (digital interconnect.).

intermodulation distortion (IMD) Distortion introduced by amplifier circuits that is the sum and difference of the input signal. For example, an amplifier driven with a mix of 1kHz and 5kHz generates intermodulation-distortion products at 6kHz (the sum of 1kHz and 5kHz) and at 4kHz (the difference between 1kHz and 5kHz). These IMD products then intermodulate with each other to create a nearly infinite series of distortion products.

interpolation Filling-in missing information with a "best-guess" estimate.

JFET (Junction Field Effect Transistor) A type of transistor often used in preamplifier input stages that differs in internal structure and operation with the more common bipolar transistor.

jitter Timing variations in the clock that synchronizes events in a digital audio system. The clock could be in an analog-to-digital converter that controls when each audio sample is taken. Of more interest to audiophiles is clock jitter in digital audio reproduction; the clock controls the timing of the reconstruction of digital audio samples into an analog signal. Jitter degrades musical fidelity.

jitter-reduction device A component inserted between a digital source such as a CD transport and a digital-to-analog converter. The jitter-reduction device takes in a jittered S/PDIF signal and outputs a low-jitter S/PDIF signal.

Kbs Kilo-bits per second. Thousands of bits per second; a measure of bit rate.

kHz see "kilohertz"

ladder DAC see "multibit DAC"

laid-back Sonic description in which the musical presentation has a sense of ease, with some distance between the listener and the soundstage. A laid-back presentation has been likened to sitting in a distant row in a concert hall. Contrast with "forward" and "immediate."

land The flat areas on a CD surface between the pits. Digital data are encoded in the land-to-pit and pit-to-land transitions.

lean Sonic description of a musical presentation lacking midbass. Synonyms are "thin," "lightweight," and "underdamped" (to describe loudspeakers). Contrast with "weighty," "full," and "heavy."

least significant bit (LSB) In binary notation, the bit that has the lowest value. In 16-bit digital audio, the 16th bit has the least effect on the numerical value of the 16-bit word. An example in decimal notation: in the decimal number 15,389, the digit in the ones column (9) has the least effect on the value of the number. Contrast with most significant bit (MSB).

LED (Light Emitting Diode) Solid-state device that emits light when current is passed through it. LEDs are used as indicators on the front panels of many audio components.

level matching Technique of ensuring that two musical presentations are reproduced at exactly the same volume so that more accurate judgments of the audio quality can be reached.

LFE see "Low Frequency Effects"

light-emitting diode see "LED"

lightweight Musical presentation lacking bass energy.

linear An output that varies in direct proportion to the input.

linearity Digital audio specification that describes the amplitude accuracy of a digital-to-analog converter. For example, a DAC driven with a digital signal at –90dBFS may reproduce that level as –88dB; the DAC has a 2dB positive linearity error. That is, the output level doesn't perfectly match the input level.

line level An audio signal with an amplitude of approximately 1V to 2V. Audio components interface at line level through interconnects. Contrasted with "speaker level," the much more powerful signal that drives speakers.

line-source loudspeaker Loudspeaker with a tall and narrow dispersion pattern. A tall ribbon driver is naturally a line source, as is a vertical array of point-source drivers. Contrasted with a point-source loudspeaker, which has a shorter, broader dispersion pattern.

liquid Sonic description of a musical presentation lacking shrillness. Usually applied to the midrange, liquidity implies correct reproduction of musical timbre.

litz cable Cable in which many small strands are grouped, with each strand insulated. This technique isolates each strand from the surrounding strands.

loading Method of enclosing a driver (usually the woofer) in a loudspeaker cabinet. Different loading methods include the infinite baffle, bass reflex, and transmission line.

localization The ability to detect the directionality of sounds.

Löfgren alignment Cartridge alignment developed in 1938 by E. Löfgren that specifies a certain cartridge offset and overhang for minimum tracking error One of two optimum alignments; the other is Baerwald.

loudspeaker A device that converts an electrical signal into sound. The loudspeaker is the last component of the playback chain.

low-cut filter A circuit that removes bass frequencies from an audio signal. Also called a "high-pass filter."

low-frequency cutoff The point at which a loudspeaker's output drops in the bass by 3dB.

Low Frequency Effects (LFE) A separate channel in the Dolby Digital and DTS formats reserved for low bass effects, such as explosions. The LFE channel is the ".1" channel in the 5.1-channel or 7.1-channel Dolby or DTS format.

low-pass filter A circuit that removes midrange and treble frequencies from an audio signal. Also called a "high-cut filter."

LP Long-Playing record. Double-sided 12" vinyl disc that stores approximately 25 minutes of audio per side.

LSB see "Least Significant Bit"

lush see "liquid"

luxury products Audio components built with lavish cosmetic design and expensive chassis. Contrasted with value products, which attempt to produce the best sound for the least money.

mass loading Technique of increasing the mass of loudspeaker stands or equipment racks by filling them with sand, lead shot, or other dense material.

matrix A method of encoding four audio channels into two channels for transmission or storage.

Mbs Mega (million) bits per second. A unit of measure for expressing bit rates. MPEG-2 video encoding has a variable bit rate that averages 3.5Mbs.

MDF see "Medium Density Fiberboard"

Medium Density Fiberboard (MDF) Composite wood material from which most loudspeaker cabinets are made.

Meridian High Resolution (MHR) Proprietary high-resolution, multichannel, encrypted digital interface used between products made by Meridian Audio. Sanctioned by the DVD Forum for carrying high-resolution signals from a DVD-A player.

Meridian Lossless Packing (MLP) Data compression system used in DVD-Audio that reduces the bit rate with no loss in quality. The decompressed bitstream is bit-for-bit identical to the original bitstream. Contrasted with "lossy" compression systems that degrade fidelity.

metadata Non-audio information attached to a music file that contains artist, track, time, genre, and other information.

metallic An unpleasant harshness to a sound that is reminiscent of metal being struck.

MHR see "Meridian High Resolution"

midrange Audio frequencies in the middle of the audible spectrum, such as the human voice. Generally the range of frequencies from about 300Hz to 2kHz. Also: a driver in a loudspeaker that reproduces the range of frequencies in the middle of the audible spectrum.

millisecond One one-thousandth of a second.

minimonitor A small, stand-mounted loudspeaker.

MLP see "Meridian Lossless Packing"

modular A/V preamplifier An A/V preamplifier built with interchangeable modules for upgrading to future technologies.

monoblock A power amplifier with only one channel.

MOSFET (Metal-Oxide Semiconductor Field Effect Transistor) Type of transistor that is turned on (biased) by voltage rather than current. Contrast with "bipolar transistor."

most significant bit (MSB) In binary notation, the bit that has the highest value. In 16-bit digital audio, the first bit has the greatest effect on the numerical value of the 16-bit word. An example in decimal notation: in the decimal number 15,389, the digit in the ten-thousands column (1) has the most effect on the value of the number. Contrast with least significant bit (LSB).

moving-coil cartridge Transducer that converts stylus motion in a record groove to an electrical signal. Tiny coils attached to the cantilever are moved back and forth in a fixed magnetic field, inducing current flow through the coils.

moving-magnet cartridge Transducer that converts stylus motion in a record groove to an electrical signal. Tiny magnets attached to the cantilever are moved back and forth between fixed coils of wire, inducing current flow through the coils.

MP3 Low-bit-rate coding scheme used in portable audio players and Internet downloads.

MPX filter A switch on some tuners and cassette decks that invokes a 19kHz filter to remove the 19kHz pilot tone from the broadcast FM signal. An MPX filter is necessary when recording on a cassette deck because the pilot tone can interfere with correct Dolby noise-reduction tracking.

MSB see "Most Significant Bit"

multibit DAC Digital-to-analog converter employing a "ladder" of resistors, each resistor corresponding to one bit in the digital audio word. Each successive resistor value is twice that of the resistor below it on the rung, which allows half as much current flow with each step down the ladder. This resistance-doubling progression follows the binary progression 1, 2, 4, 8, 16, 32, 64, and so forth. Contrast with 1-bit DAC, in which the digital audio signal is represented by pulses just one bit high.

multichannel power amplifier A power amplifier with more than two channels, usually five or seven.

multichannel preamplifier A preamplifier with more than two channels, usually six.

multichannel sound Sound reproduction via more than two channels feeding more than two loudspeakers.

multipath In FM-radio transmission, two or more paths for the signal to travel between transmitter and receiver. Multipath is caused by mountains or buildings that reflect the radio signal; the receiving antenna picks up the directly broadcast signal along with the signal after it has been delayed by the reflections. Multipath introduces audible distortion. Multipath in television transmission is seen as "ghosting" in the picture.

multipath indicator Meter in better FM tuners that indicates the amount of multipath in the received signal.

multi-room A feature on some A/V products that lets you listen to two different sources in two different rooms.

NAS see "Network Attached Storage"

nasal Sonic description of midrange coloration (particularly loudspeakers) that makes vocalists sound congested.

Network Attached Storage (NAS) Type of storage device (usually hard-disc drive) that connects as a device attached to the network and can be addressed by other devices on the network. Contrast with Direct Attached Storage (DAS) in which the drive is addressable only by the host computer.

New Old Stock (NOS) Vacuum tubes that have never been used, but haven't been manufactured in a long time (perhaps decades).

noise shaping Digital audio technique in which quantization noise is shifted to the top octave of the audioband, or above the audioband, thus decreasing the audible noise. Used in 1-bit DACs and Direct Stream Digital.

NOS see "New Old Stock"

Nyquist Theorem Expresses relationship between a digital audio system's sampling frequency and the highest audio signal that can be preserved. Specifically, the sampling frequency must be at least twice as high as the highest audio frequency to be recorded. Named for telegraph engineer Harry Nyquist, who published his theorem in 1928.

octave The interval between two frequencies with a ratio of 2:1. The bottom octave in audio is 20Hz-40Hz; the top octave is 10kHz-20kHz.

off-axis response A loudspeaker's frequency response measured at the loudspeaker's sides. Contrast with on-axis, the loudspeaker's response directly in front of the baffle.

offset A bend in a pivoted tonearm that gives the arm an S or J shape and orients the cartridge at 25° relative to the arm. The orientation of the cartridge relative to the record groove.

ohm The unit of resistance to electrical current flow.

Ohm's law Expresses the relationship between voltage, current, and resistance in electrical circuits. If you know two of the values (current and resistance, for example), you can calculate the third value using Ohm's law. Formula is E=IXR, or voltage equal current flow multiplied by the resistance. E stands for electromotive force (voltage); I stands for current (for intensity of electrons); R stands for resistance.

omnipolar loudspeaker Trademarked name of Mirage Loudspeakers for its loudspeaker design technology that radiates sound in all directions.

on-axis response A loudspeaker's frequency response measured directly in front of the baffle.

one-note bass Sonic description of bass that seems to have just one pitch. Caused by excessive output over a narrow frequency band in loudspeakers. A "boom truck" is one-note bass taken to an extreme.

opaque Impression that the soundstage is cloudy, veiled, or thick. Contrast with "transparent."

output impedance Technically, the change in a component's output voltage in response to a change in load impedance. Practically, a component with a high output impedance can deliver less current to a load than a component with a low output impedance. Low output impedance is desirable.

output stage The last amplifier circuit in an audio component. The output stage in a CD player is an amplifier that drives the preamplifier. In power amplifiers, the output stage delivers current to the loudspeakers.

output transformer Transformer in tubed amplifiers that couples the output stage to the loudspeaker. Output transformers are required in tubed amplifiers to change the amplifier's high output impedance to a lower value that can better drive loudspeakers. The output transformer also blocks DC from appearing at the amplifier's output terminals.

overall perspective Describes a musical presentation's degree for forwardness. An up-front perspective is likened to sitting in row C of a concert hall; a laid-back perspective is likened to sitting in row W.

overdamped Technical term that describes a woofer's motion in reaction to a drive signal, specifically how long the driver continues to move after the drive signal has stopped. An overdamped loudspeaker has lean but articulate bass presentation. Contrast with "underdamped."

overhang In loudspeakers, woofer movement after the drive signal has ceased. In tonearms, the distance from the tonearm's pivot point to the cartridge.

oversampling digital filter Digital audio circuit that removes all energy above half the sampling frequency. An 8x-oversampling filter generates seven new samples for each incoming sample, thus increasing the sampling frequency by eight times. Oversampling shifts the spurious images to a higher frequency, where they are more easily filtered.

pace The sense of rhythmic drive or propulsion in music.

palpable Impression that the reproduced instrument or voice is tangible; literally, "touchable."

panning The movement of sounds and images from one location to another. Originally a camera term.

passband Range of frequencies allowed to pass by a filter.

passive level control Device that adjusts the volume of an audio system without using amplifiers or other active electronic components such as tubes or transistors. A passive level control is typically a box with a volume control.

passive radiator Diaphragm in some loudspeakers that isn't connected electrically, but is moved by air pressure inside the cabinet created by the woofer's motion. Also called an auxiliary bass radiator (ABR), the passive radiator covers what would have been the port in a reflex-loaded loudspeaker.

passive subwoofer A speaker for reproducing bass frequencies that must be powered by a separate power amplifier. Contrasted with "active" or "powered" subwoofers, which contain an integral amplifier.

PCM see "Pulse Code Modulation"

peak A short-term, high-level audio signal. Also: an excess of energy over a narrow frequency band (contrasted with "dip").

peaky Sonic description of a sound with excessive energy over a narrow frequency band.

pentode Vacuum tube with five elements: the cathode, plate, screen grid, control grid, and suppressor grid. More efficient than the simpler triode vacuum tube.

perceptual coding A method of reducing the number of bits needed to encode an audio or video signal by ignoring information unlikely to be heard or seen.

period The time it takes to complete one cycle of a sinewave.

phantom center-channel mode A setting on A/V receivers or A/V controllers invoked when no center-channel speaker is used.

phantom image The creation of an apparent sound source between two loudspeakers.

phase In a periodic wave, the fraction of a period that has elapsed. Describes the time relationship between two signals.

phase adjustment A control provided on some subwoofers that lets you delay the sound of the subwoofer slightly so that the subwoofer's output is in-phase (has the same time relationship) with the front speakers.

phase angle A measurement of how inductive or capacitive a loudspeaker's impedance is.

phase locked loop (PLL) A electronic circuit that compares the phase of an incoming signal with that of a reference frequency, thereby synchronizing the two signals.

pitch definition The ability to distinguish pitch in reproduced music, particularly in the bass. Some products (especially loudspeakers) obscure the individual pitches of notes.

pits and land Tiny depressions (pits) and flat areas (land) embedded in a CD or DVD that represent digital data.

pivoted tonearm A tonearm in which the cartridge and armtube traverse the record in an arc while maintaining a fixed pivot point. Contrasted with tangential-tracking tonearms.

planar loudspeaker Loudspeaker in which the driver or drivers are mounted in an open panel.

planar magnetic Type of driver in which conductors carrying the audio signal are bonded to a diaphragm. A subset of the ribbon driver, planar magnetic drivers are also called quasi-ribbons.

playback components Broadly speaking, the power amplifier and loudspeakers. Contrasted with source components (tuner, CD player, LP turntable, DVD player, etc.) and control components (preamplifier).

plinth The flat deck of a turntable beneath the platter.

PLL see "Phase Locked Loop"

pneumatic platform Device that isolates components from vibration with compressed air.

point-source loudspeaker Loudspeaker that emits sound from a point in space. Contrast with line-source loudspeaker.

port Opening in a loudspeaker cabinet that channels bass from inside the enclosure to outside the enclosure. Also called a "vent."

ported loudspeaker see "reflex-loaded loudspeaker"

port noise Noise generated by large air flows in a reflex-loaded loudspeaker's port. Also called "chuffing."

power amplifier An audio component that boosts a line-level signal to a powerful signal that can drive loudspeakers.

power bandwidth The range of frequencies over which an amplifier can deliver its rated power.

power handling A measure of how much amplifier power, in watts, a speaker can take before it is damaged.

power output A measure of a power amplifier's ability, in watts, to deliver electrical voltage and current to a speaker.

power supply Circuitry found in every audio component that converts 60Hz alternating current from the wall outlet into direct current that supplies the audio circuitry.

power transformer Device in a power supply that reduces the incoming voltage from 120V to a lower value.

preamplifier Component that receives signals from source components, selects a source for listening, controls the volume, and drives the power amplifier. Literally means "before the amplifier."

pre-preamplifier Component that boosts a tiny signal from a phono cartridge to a higher level for input to a phono preamplifier. Typically used to match a moving-coil cartridge output to a preamplifier designed for a moving-magnet input.

presence The sense that an instrument or voice is actually in the listening room.

presence region Band of frequencies in the midrange that contributes to presence.

psychoacoustics The study of human hearing.

Pulse Code Modulation (PCM) A method of representing an audio signal as a series of digital samples.

Pulse Width Modulation (PWM) A method of representing an audio signal digitally in which the signal's amplitude information is contained in the widths of a single-bit pulse.

punchy Sonic presentation having dynamic impact, particularly in the bass.

push-pull amplifier Amplifier in which pairs of transistors or vacuum tubes are arranged so that one transistor or tube amplifies the positive half of the audio waveform, and the other tube amplifies the negative half. Contrasted with single-ended amplifier.

PWM see "Pulse Width Modulation"

Q for Quality Factor. Numerical value describing woofer resonance in an enclosure. Technically, Q is the woofer's peak resonant frequency divided by the peak's bandwidth.

Quadraphonic Four audio channels. Two Quadraphonic formats were introduced in the early 1970s: the matrix SQ system and the discrete CD-4.

quantization Assigning a discrete numerical value to an analog function. In digital audio, the analog waveform's amplitude is converted to a number (quantized) each time a sample is taken.

quantization error Difference between the actual analog value and the number representing that analog value. Quantization error occurs when the analog value falls between two quantization steps; the quantizer assigns the closest number. Quantization error introduces noise and distortion in digital audio, often heard as a roughness at low signal levels, particularly during reverberation decay.

quantization word length The number of bits created by the A/D converter at each sample point. Compact disc records quantization words 16 bits in length.

quarter-wavelength bass trap Acoustical absorber in which an absorbent material is hung at a distance from a reflective surface. The peak absorption frequency occurs when the distance between the absorbent material and reflective surface is one-quarter the wavelength of sound, and at odd multiples of a quarter wavelength.

quasi-ribbon driver see "planar magnetic"

quieting sensitivity Tuner specification that states the voltage across the antenna required to produce an audio signal with a signal-to-noise ratio of 50dB. Contrast with the less stringent "usable sensitivity."

radiation pattern The way in which a speaker disperses sound.

Radio Data System System for broadcasting text or other data along with an FM radio broadcast. The receiving tuner must be equipped with RDS decoding to display the text or data.

Radio Frequency (RF) Most often refers to spurious noise with energy in the radio-frequency band.

rail The power-supply voltage fed to the audio circuits.

rail fuse Fuse sometimes found on the back of a power amplifier that is in series with the rail. The rail fuse will blow when excessive current is drawn by the amplifier's output stage.

rarefaction A component of a sound wave in which the air density is lower than normal atmospheric pressure. A woofer cone moving inward creates a rarefaction. One rarefaction and one compression make one complete cycle of a sound wave.

RCA jack A connector found on audio and video products. Signals transmitted via RCA jacks include line-level audio, composite video, and component video.

RDS see "Radio Data System"

rectifier A device that converts AC current into DC current. Found in all power supplies.

rectify To convert AC current into DC current.

Red Book Name for the compact disc specification, derived from the document's red cover. A "Red Book" disc is a conventional CD.

Red Book layer The information layer in a Super Audio CD that contains conventional 16-bit/44.1kHz digital audio.

re-equalization A Home THX technology that reduces the amount of treble on playback so that you hear a more natural-sounding reproduction when a film soundtrack is played back in the home.

reflex-loaded loudspeaker see "bass reflex"

resolution The quality of an audio component that reveals low-level musical information.

resonance Vibration of an object or air that is disproportionate in amplitude to the stimulus. A bell ringing at a certain frequency is an example of resonance; the pitch is the bell's resonant frequency.

reverberation Dense acoustical reflections in an acoustic space that become lower in amplitude and more closely spaced over time. The sound in a room after the sound source has stopped producing sound.

reverberation time The time it takes sound in a room to decrease in amplitude by 60dB. Symbol: RT_{60}.

RIAA Recording Industry Association of America

RIAA accuracy Flatness of the RIAA phono equalization circuit in a phono preamplifier.

RIAA equalization A treble boost and bass cut applied to the audio signal when a record is cut; a reciprocal treble cut and bass boost on playback restores flat response. RIAA equalization increases the playing time of an LP (because bass takes up the most room in the record groove) and decreases noise (because the treble cut on playback also reduces record surface noise).

ribbon loudspeaker Loudspeaker in which the diaphragm is electrically conductive and carries the audio signal. Usually made from a long, thin strip of aluminum.

rip To transfer data from a storage medium (usually CD) to a mass storage device (usually a hard-disc drive).

ripping NAS A Network Attached Storage device that contains an optical disc drive and can transfer the data from the optical disc to the NAS.

rolled off Sonic description of reduced energy at the frequency extremes (bass or treble). A loudspeaker whose treble is rolled off sounds dull.

rolloff Reduction in energy at the frequency extremes, or the effect of a filter; e.g. the crossover produces a 12dB per octave rolloff above 2kHz.

room correction see "DSP room correction"

room gain Increase in bass level when a loudspeaker plays in a room compared with the loudspeaker's bass level in an anechoic chamber. The room's walls increase the amount of bass heard; the closer to the walls the loudspeaker is placed, the greater the room gain.

room resonance modes Excessive acoustical energy at certain frequencies when the air in a room is excited by the sound of a loudspeaker.

rumble Low-frequency noise associated with LP playback.

SACD see "Super Audio CD"

sampling The process of converting an analog audio signal into digital form by taking periodic "snapshots" of the audio signal at some regular interval. Each snapshot (sample) is assigned a number that represents the analog signal's amplitude at the moment the sample was taken.

sampling frequency The rate at which samples are taken when converting analog audio to digital audio. Expressed in samples per second, or, more commonly, in Hertz; i.e., the CD format's sampling frequency is 44.1kHz.

sampling-rate converter Circuit that changes the sampling frequency of a digital audio signal.

satellite speaker A small speaker with limited bass output designed to be used with a subwoofer.

screechy Unpleasant character of upper midrange or treble that makes instruments or voices have some of the sound of noisy car brakes. Most often applied to the sound of violins.

secondary windings Coils of wire in a transformer that aren't physically connected to the AC line, but that supply voltage and current to the power supply through electromagnetic induction. Power transformers often have multiple secondary windings to increase the isolation between a component's multiple power supplies.

selectable IF bandwidth A tuner feature that adjusts the bandwidth in the tuner's intermediate frequency (IF) stage for best sound quality and minimum interference from adjacent stations. Found on better tuners.

selectivity Tuner specification describing the tuner's ability to reject unwanted stations. Good selectivity is important to those who live in cities, where stations are closely spaced on the broadcast spectrum.

sensitivity 1) A measure of how much sound a speaker produces for a given amount of input power. Speaker sensitivity is measured by driving a speaker with 1W of power and measuring the sound-pressure level from a distance of 1 meter. 2) A measure of an FM tuner's ability to pull in weak stations. Tuner sensitivity is important if you live a long distance from FM transmitters.

servo An electrical or mechanical system that controls the speed or position of a moving device by forcing it to conform to a desired speed or position. In a servo-controlled turntable, the platter's rotational speed is compared with a reference frequency, and a correction signal is applied to the drive motor to keep the two frequencies identical.

servo-driven woofer A woofer in which an accelerometer (a device that converts motion to an electrical signal) attached to the voice coil sends to an amplifier information about its position and motion. The amplifier then applies a correction signal so that the woofer's motion matches the characteristics of the audio signal.

shielded loudspeaker A loudspeaker lined with metal to contain magnetic energy within the speaker. Shielded loudspeakers are used in home theater because the speakers' magnetic energy can distort a video monitor's picture.

sibilance *s* and *sh* sounds in spoken word or singing.

signal-to-noise ratio Numerical value expressing in decibels the difference in level between an audio component's noise floor and some reference signal level.

silky Sonic description of midrange or treble sounds lacking hardness or glare.

single-ended amplifier Amplifier in which both half-cycles of the audio waveform are amplified by the output tube or transistors. Contrasted with "push-pull amplifier."

single-presentation listening Evaluating audio components by listening to just that component rather than in comparison with other components.

six-channel input A preamplifier input comprising six discrete jacks that will accept the six discrete analog outputs from a multichannel disc player.

skating Force generated in LP playback that pulls the tonearm toward the record center.

skin effect Phenomenon in cables in which high frequencies travel along the conductor's surface and low frequencies travel through the conductor's center.

slam Sonic descriptSion of large dynamic impact in the bass that the listener feels in his body.

Smart Radio see "Radio Data System" (RDS)

smooth Sonic description of a presentation lacking peaks and dips in the frequency response.

solid-state drive Storage medium in which data are stored in memory rather than on a moving mechanism such as a hard-disc drive.

soundfield see "soundstage"

sound-pressure level (SPL) A measure of loudness expressed in decibels (dB).

soundstage The impression of soundspace existing in three dimensions in front of or around the listener.

source components Components that provide audio or video signals to the rest of the system. CD players, turntables, music servers, and FM tuners are audio source components.

source switching Function performed by preamplifier or integrated amplifier that selects which source component's signals are fed to the speakers.

spade lug A speaker termination with a flat area that fits around a binding post.

S/PDIF interface Standardized method of transmitting digital audio from one component to another. Stands for Sony/Philips Digital Interface Format.

speaker see "loudspeaker"

SPL see "Sound-Pressure Level"

SPL meter A device for measuring the Sound-Pressure Level created by an audio source.

sprung turntable A turntable in which the platter and armboard are mounted on a sub-chassis that floats within the base on springs. Contrasted with unsprung turntables.

SSD see "solid-state drive"

standing wave Stationary area of high and low acoustical pressure in a room caused by interaction of the sound with the room's boundaries.

stator The element in an electrostatic loudspeaker driver that remains stationary.

step-up transformer Transformer with a higher output voltage than the input voltage. Sometimes used between a moving-coil cartridge and a moving-magnet phono input.

triode The simplest vacuum tube, employing just three elements: the cathode, plate, and control grid.

tube see "vacuum tube"

tube damper Device that fits over a vacuum tube to reduce vibration. Some tube dampers also act as heatsinks.

tweeter A speaker driver designed to reproduce treble signals.

two-way speaker A loudspeaker that splits the frequency spectrum into two parts, bass and treble, for reproduction by two or more drivers.

UHJ In Ambisonics, the mono and stereo compatible combination of the W (mono pressure signal), X (sides), and Y (front to back) signals.

ultra-linear A method of operating a pentode vacuum tube that combines the output power of a pentode with the distortion characteristics of a triode. Invented by David Hafler and Herbert Keroes in 1951.

unbalanced connection Connection method in which the audio signal is carried on two conductors, called signal and ground. Contrasted with balanced connection, in which the audio signal is carried on three conductors.

underdamped In loudspeakers, woofer movement after the drive signal has stopped.

unipivot Type of tonearm bearing in which the arm rests on, and pivots around, a single point. Contrast with gimbal bearing.

uni-polar A loudspeaker that directs its acoustic output in one direction. Contrast with dipolar.

universal disc player Source component that plays CD, DVD-Video, DVD-Audio, and SACD.

Universal Serial Bus (USB) Computer interface format that has been adapted for digital audio data transmission. See also "Adaptive USB" and "Asynchronous USB".

unsprung turntable Turntable in which the platter and tonearm are connected directly to the turntable base rather than suspended on a sub-chassis. Contrast with "sprung turntable."

upsampling Technique of increasing the sampling frequency of a digital audio signal so that it can be converted to analog from the higher frequency.

uptilted A sound having too much treble and not enough bass.

usable sensitivity Tuner specification that states the voltage across the antenna required to produce an audio signal with a signal-to-noise ratio of 30dB. Contrast with the more stringent "quieting sensitivity."

USB See "Universal Serial Bus"

user interface The controls and displays on a product and their logic and ease of use.

vacuum hold-down Mechanism for achieving tight coupling between an LP and platter by creating a vacuum just above the platter.

vacuum tube Device for amplifying audio in which the active elements are enclosed in a glass envelope devoid of air.

value product Audio component that seeks to achieve maximum sonic performance for the lowest price by cost-effective design and by eschewing expensive chassis.

veiled Impression of a haze or veil between you and the musical presentation. Contrast with "transparent."

vented loudspeaker see "reflex loudspeaker"

vertical tracking angle (VTA) The angle at which the stylus sits in a record's groove. Adjusted by moving the tonearm pivot point up or down.

vertical tracking force (VTF) Pressure applied by gravity to the stylus in a record groove.

vertical venetian-blind effect A change in perceived tonal balance when the listener moves sideways in the listening position when listening to electrostatic loudspeakers.

video display A device that converts a video signal into a visual image.

video monitor A direct-view television set with video input and output jacks.

vivid Musical presentation in which every sound is clearly audible. An overly vivid presentation is analytical and unmusical.

voice coil Coil of wire inside a loudspeaker driver through which current from the power amplifier flows.

volt Unit of electromotive force. The difference in potential required to make one Ampere of current flow through one ohm of resistance. See also "voltage."

voltage Analogous to electrical pressure. Voltage exists between two points when one point has an excess of electrons in relation to the other point. A battery is a good example: the negative terminal has an excess of electrons in relation to the positive terminal. If you connect a piece of wire between a battery's positive and negative terminals, voltage pushes current through the wire. One volt across 1 ohm of resistance produces a current of 1 Ampere.

voltage source Technical description of a power amplifier that can maintain its output voltage regardless of the load impedance.

VTA see "Vertical Tracking Angle"

VTF see "Vertical Tracking Force"

wall-stud resonance Acoustic resonance of a listening room's walls when struck by sound. Occurs only in drywall-on-stud construction.

watermarking Technique of embedding digital codes disguised as noise in the audio waveform to control how content is copied, and to trace the source of music piracy. Also describes a visible imprint on the surface of a legitimate SACD that disappears on a copy.

watt The unit of electrical power, defined as the power dissipated by 1 Ampere of current flowing through 1 ohm of resistance.

WAV Digital audio file protocol in which the data are in raw PCM format.

wavelength The distance between successive cycles of a sinewave or other periodic motion.

weighting curve A filter applied to noise or sound-pressure level measurements that approximates the perceived loudness or noise level.

weighty A full and robust bass presentation.

Windows Media Audio Lossless File format that provides perfect bit-for-bit accuracy to the source data with an approximately 40% reduction in file size.

woofer Driver in a loudspeaker system that reproduces bass.

XLR jack and plug Three-pin connector that usually carries a balanced audio signal.

zero-crossing distortion Discontinuity in a musical waveform in push-pull amp.

Index

A

B

C

D

E

F

G

H

M

N

O

P

Q

R

T

U

V

W

Z